双语教材

活页·微课版

多轴数控编程与加工典型案例教程

Typical Case Book of Multi-axis NC Programming and Machining

主　编　季业益　石皋莲　孙书娟

副主编　朱　伟　丁倩倩　殷　铭

参　编　顾　涛　丁云鹏　李春雷　周　挺

主　审　耿　哲

苏州大学出版社
Soochow University Press

图书在版编目(CIP)数据

多轴数控编程与加工典型案例教程 = Typical Case Book of Multi - axis NC Programming and Machining : 活页·微课版 / 季业益, 石皋莲, 孙书娟主编. -- 苏州: 苏州大学出版社, 2023.7
ISBN 978-7-5672-4455-9

Ⅰ. ①多… Ⅱ. ①季… ②石… ③孙… Ⅲ. ①数控机床 - 程序设计 - 教材②数控机床 - 加工 - 教材 Ⅳ. ①TG659

中国国家版本馆 CIP 数据核字(2023)第 117544 号

书　　名: Duozhou Shukong Biancheng Yu Jiagong Dianxing Anli Jiaocheng
多轴数控编程与加工典型案例教程
Typical Case Book of Multi-axis NC Programming and Machining

主　　编: 季业益　石皋莲　孙书娟
责任编辑: 征　慧　沈　琴

出版发行: 苏州大学出版社(Soochow University Press)
社　　址: 苏州市十梓街1号　邮编: 215006
印　　刷: 苏州市古得堡数码印刷有限公司
邮购热线: 0512-67480030
销售热线: 0512-67481020

开　　本: 787 mm×1 092 mm　1/16　印张: 24　字数: 458 千
版　　次: 2023 年 7 月第 1 版
印　　次: 2023 年 7 月第 1 次印刷
书　　号: ISBN 978-7-5672-4455-9
定　　价: 69.80 元

若有印装错误,本社负责调换
苏州大学出版社营销部　电话: 0512-67481020
苏州大学出版社网址　http://www.sudapress.com
苏州大学出版社邮箱　sdcbs@suda.edu.cn

前　言

Foreword

本书深刻践行习近平总书记在党的二十大报告中提出的“加快建设国家战略人才力量，努力培养造就更多大师、战略科学家、一流科技领军人才和创新团队、青年科技人才、卓越工程师、大国工匠、高技能人才”“加强人才国际交流”“打造具有国际竞争力的数字产业集群”等精神，立足于“智改数转”产业变革对多轴数控编程与加工技术岗位人才素质提出的更高要求，实施价值引领推进课程改革，着力培养“德技双馨”的技能型人才。

本书以NX软件为平台，与企业合作开发课程内容，实践案例来自企业真实产品，具有很强的专业性和实用性；基于项目任务，在展开项目中教学、在解决问题中学习、在完成任务中提高，有利于激发读者探究性学习的兴趣，培养团队合作和创新精神。

全书共分3个模块6个学习项目，模块1主要以导板、玩具相机凸模2个三轴铣削加工项目为例，重点介绍了NX CAM三轴数控加工的基础及操作流程，引导读者入门；模块2则通过滚轴、定位夹具2个四轴铣削加工项目，重点介绍了NX四轴数控加工的设置管理与刀具轨迹生成和验证等知识与方法；模块3以壳体、叶轮2个五轴铣削加工项目为例，详细讲解了五轴铣削加工编程、后置处理定制和NX仿真加工等。

The book deeply implements the document spirit of "Speed (ing) up efforts to build a contingent of personnel with expertise of strategic importance and cultivate greater numbers of master scholars, science strategists, first-class scientists and innovation teams, young scientists, outstanding engineers, master craftsmen, and highly-skilled workers" " increasing international personnel exchanges" "building an internationally competitive digital industrial cluster" proposed in the report to the 20th National Congress of the Communist Party of China. Based on the industrial transformation of "intelligent and digital development" , higher requirements are put forward for the quality of talents in multi-axis CNC programming and processing technology positions, and the implementation of value-leading curriculum reform is promoted to cultivate technical talents with "virtue and technology".

This book takes NX software as the platform and develops course content in cooperation with enterprises. Practice cases come from real products of enterprises, which are highly professional and practical. Based on the project task, teaching in the process of developing tasks, learning in the process of solving problems and improving in the process of completing tasks are conducive to stimulating readers' interest in inquiry-based learning and cultivating the spirit of team cooperation and innovation.

The book is divided into three modules which consist of six learning projects. The first module mainly introduces the foundation and operation of NX CAM 3-axis CNC machining by taking guide plate and toy camera punch as examples. The second module mainly introduces the knowledge and methods of setting management and tool path generation and verification of NX 4-axis NC machining through two 4-axis milling items, namely roller and positioning fixture. Module 3 explains the 5-axis milling programming, post-processing customization and NX simulation machining in detail by using shell and impeller as two 5-axis milling processing items.

每个项目都阐述了学习目标、项目导读、任务描述、项目实施等，并在项目实施中重点讲解加工工艺制定、加工程序编制及仿真加工等，以便读者进行有针对性的操作，从而掌握学习重点和难点。每个项目后面都有专家点拨、课后训练，提示和辅助读者理解操作要领、使用技巧和注意事项。本书以中英文对照的形式呈现，旨在帮助读者在掌握专业技能的同时，提高国际交流水平，中英文部分因软件版本略有差异，故表述略有不同。本书最后提供了中英文微课视频，读者可扫描二维码获得相关知识点和操作视频。

本书由校企人员联合创作编写。中文部分模块 1 由季业益编写，模块 2 由殷铭编写，模块 3 由石皋莲编写；英文部分模块 1 由孙书娟编写，模块 2 由朱伟编写，模块 3 由丁倩倩编写；顾涛、丁云鹏、李春雷、周挺等教师和企业工程技术人员参与了本书项目的加工工艺制定、加工程序编制及学习资源建设等工作；季业益、石皋莲、孙书娟担任本书主编，编制教材体例、编写大纲并进行统稿工作；耿哲担任主审。

本书在编写过程中参考了大量的文献资料，在此向相关作者致以诚挚的谢意！

衷心感谢西门子工业软件（上海）有限公司、英格索兰苏州吴江有限公司、苏州微创关节医疗科技有限公司等企业无私提供实践案例。

由于编者水平有限，谬误之处在所难免，恳请相关专家和广大读者批评指正。

Each project elaborates learning objectives, project guidance, task description, and project implementation, and focuses on the development of processing technology, processing program and simulation machining, so that readers can carry out targeted operations and conquer the learning difficulties. Each project is followed by expert reviews and practice to help readers deepen their understanding of operation essentials and precautions. The book provides video learning resources, and relevant knowledge points and operation videos can be obtained by scanning QR code.

This book was written by college and enterprise staff. The Chinese Module 1 was written by Ji Yeyi, Module 2 by Yin Ming and Module 3 by Shi Gaolian. The English Module 1 was written by Sun Shujuan, Module 2 by Zhu Wei and Module 3 by Ding Qianqian. Gu Tao, Ding Yunpeng, Li Chunlei, Zhou Ting and other teachers and enterprise engineers and technicians participated in the process formulation, programming and learning resource construction of this book. Ji Yeyi, Shi Gaolian and Sun Shujuan served as the chief editors of the book, compiling the book structure and the outline and coordinating the draft. Geng Zhe served as the chief reviewer.

We have consulted a large number of references during the writing of this book, and we would like to express our sincere thanks to the authors of these references.

We sincerely thank Siemens Industrial Software (Shanghai) Co. LTD, Ingersoll Rand Suzhou Co., LTD, Suzhou MicroPort Co., LTD and other enterprises for selflessly providing practical cases.

Due to the limitation of our knowledge, errors are inevitable, we sincerely invite experts and readers to criticize and help improve the book.

目 录

Contents

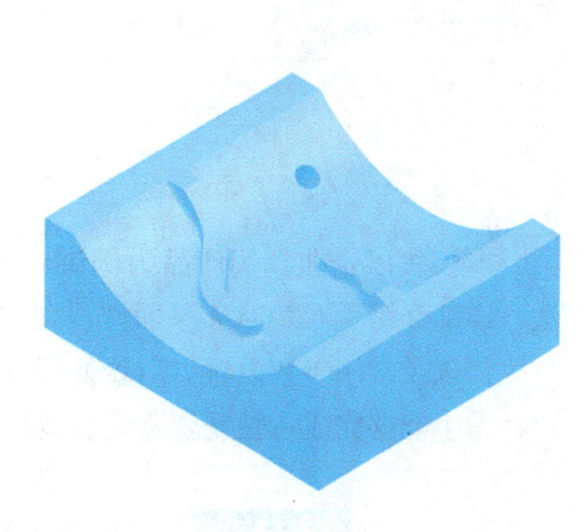

模块 1 三轴铣削加工

本模块以企业真实产品为例讲述 NX CAM 三轴铣削数控编程与仿真加工方法，详细介绍了 NX CAM 平面铣、型腔铣、固定轴曲面轮廓铣和点位加工等加工方式，以及常用参数设置、后置处理方法、编程操作技巧等。通过对本模块的学习，学生能够完成三轴铣削零件的数控编程与仿真加工。

项目 1 导板的数控编程与仿真加工

学习目标

能力目标：能运用 NX 软件完成导板的数控编程与仿真加工。

知识目标：掌握面铣、平面铣、精加工壁、底壁加工等几何体的设置；
掌握加工边界的创建方法；
掌握切削参数的设置方法；
掌握非切削运动的设置方法；
掌握啄钻、铣孔等点位加工的参数设置。

素质目标：激发学生的学习兴趣，培养团队合作精神和创新精神。

项目导读

导板是机械结构中出现频率较高的一类零件，这类零件的特点是结构比较简

Module 1 3-axis Milling

This project takes real products of the enterprises as examples to describe the NC programming, simulation and processing methods of NX CAM 3-axis milling. It introduces in detail the processing methods such as planar milling, cavity milling, fixed axis surface contour milling and point processing. It also introduces common parameter setting, post-processing methods, programming operation skills, etc. Through the study of this module, students can complete the NC programming and simulation of 3-axis milling parts.

Project 1 Programming and Simulation Machining of Guide Plate

Learning Objectives

Capacity Objective: Complete model programming and simulation machining with NX software.

Knowledge Objective: Master the geometry settings such as face milling, plane milling, finishing wall and bottom machining;
Master the method of creating boundary;
Master the setting method of cutting parameters;
Master the setting method of non-cutting motion;
Master the parameter setting of point processing such as pecking drilling and hole milling.

Quality Objective: Stimulate students' interest in learning and cultivate the spirit of teamwork as well as innovation.

Project Guidance

Guide plate is a kind of parts with high frequency in mechanical structure. The

单，零件整体外形成块状，零件上一般会有导向槽、腔体、连接孔、减轻孔等特征。在编程与加工过程中要特别注意导向槽的加工精度和表面粗糙度。

任务描述

学生以企业制造部门MC数控程序员的身份进入NX CAM功能模块，根据导板的特征，制定合理的工艺路线，创建表面铣、平面铣、点位加工操作，设置必要的加工参数，生成刀具轨迹，检验刀具轨迹是否正确合理，并对操作过程中存在的问题进行研讨和交流，通过相应的后处理生成数控加工程序，并运用机床加工零件。

项目实施

按照零件加工要求，制定导板加工工艺；编制导板加工程序；完成导板的仿真加工后，处理得到数控加工程序，完成零件加工。

一、制定加工工艺

1. 导板结构分析

该连接块结构比较简单，主要有导向槽、开口腔体、连接孔、减轻孔等特征组成，主要加工内容为外形、槽、腔体和孔。

2. 毛坯选用

零件材料是由厚度为26 mm的45#钢板切割而成，尺寸为140 mm×120 mm×55 mm。零件四周单边最小余量为4 mm，零件厚度方向为了保证零件的装夹，余量为6 mm。

3. 加工工序卡制定

零件选用立式三轴联动机床加工，平口钳夹持，遵循先粗后精、先面后孔的加工原则。加工工序卡如表1–1所示。

characteristics of this kind of parts are that the structure is relatively simple, the overall shape of the parts is block, and the parts generally have the characteristics of the guide groove, the cavity, the connecting hole, the lightening hole and so on. In the process of programming and machining, special attention should be paid to the machining accuracy and surface roughness of the guide groove.

Task Description

Students operate NX CAM functional modules as MC programmers in the enterprise manufacturing department. According to the characteristics of model parts, establish a reasonable processing route, create face milling, plane milling, point processing operations, set necessary processing parameters, generate tool path and check the generated tool path. Besides, students should discuss the problems occurred in the operation. By selecting the corresponding post-processor to generate NC machining programs, students import them into machine tools to complete the parts processing.

Project Implementation

Firstly, formulate processing technic under processing requirements. Secondly, program for machining model. Finally, complete simulation machining, and then import the NC machining program obtained by post processor to the machine tool to complete machining.

I. Formulate Processing Technic

1. Structural analysis of guide plate

The structure of the connecting block is relatively simple. It is mainly composed of the guide groove, the open cavity, the connecting hole, the lightening hole and other features. The main processing contents are the shape, the groove, the cavity and the hole.

2. Blank selection

The part material is cut from a 45# steel plate with a thickness of 26 mm, and the size is 140 mm × 120 mm × 55 mm. The minimum allowance on one side around the part is 4 mm, and the allowance in the thickness direction of the part is 6 mm to ensure the clamping of the part.

3. Formulation of processing procedure card

The parts are processed by vertical 3-axis linkage machine tools, clamped by flat pliers, and follow the processing principle of rough before fine, face before hole. The processing procedure is shown in Table 1-1.

表 1-1　加工工序卡

零件号：263863		工序名称：导板铣削加工		工艺流程卡_工序单	
材料：45#		页码：1	工序号：01		版本号：0
夹具：平口钳		工位：MC	数控程序号：		
刀具及参数设置					
刀具号	刀具规格	加工内容	主轴转速/rpm	进给速度/mmpm	
T01	D20R2	外轮廓粗加工	1800	1200	
T01	D20R2	开口腔粗加工	1800	1200	
T02	D6R0-ROUGH	导轨槽粗加工	2200	1000	
T03	D16R0	顶面和腔底面精加工	2200	1000	
T03	D16R0	外轮廓精加工	2200	1000	
T04	D6R0-FINISH	导轨槽精加工	3600	800	
T06	ZXZ10	钻中心孔	1200	100	
T05	D10DRILL	钻D10通孔	800	100	
T07	D10R0	铣D24通孔	1800	800	
02					
01					
更改号	更改内容		批准	日期	
拟制：	日期：	审核：	日期：	批准：	日期：

二、加工准备

（1）启动 NX，单击【文件】→【打开】，选择“导板 .prt”文件，如图 1-1 所示，单击【OK】，打开零件模型。

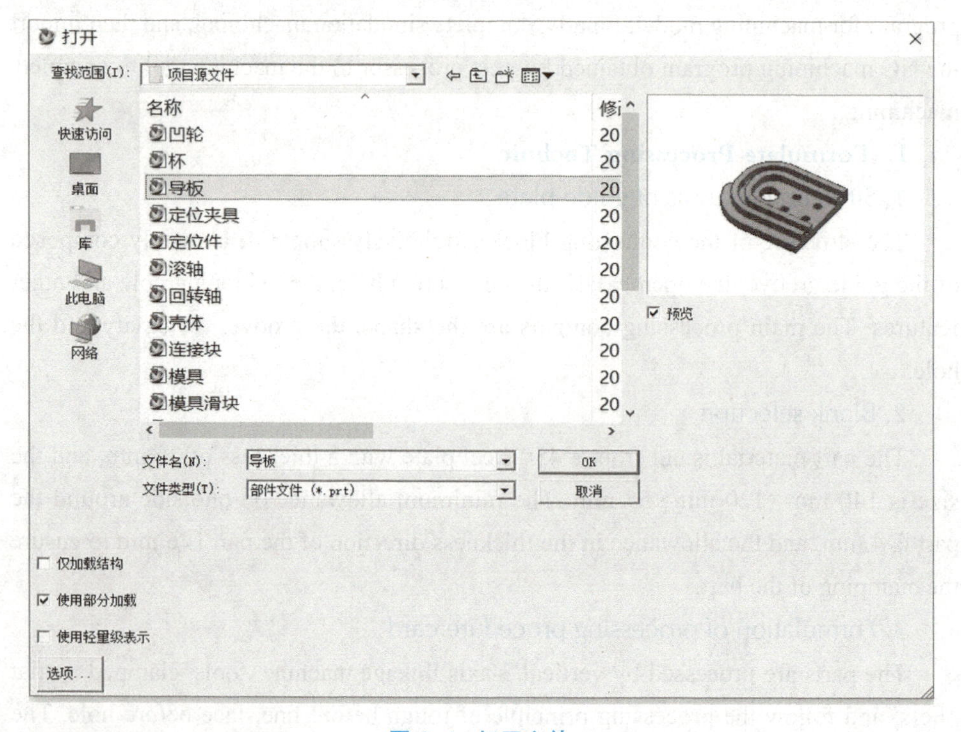

图 1-1　打开文件

Table 1-1　Processing Procedure Card

Part number: 263863			Name of process: Milling of guide plate		Process card-Process sheet
Material: 45#		Page number: 1		Procedure number: 01	Version number: 0
Fixture: Parallel-jaw vice		Work station: MC		CNC program number:	
Tool and parameter setting					
Tool number	Tool specification	Processing content	Spindle speed /rpm	Feed speed /mmpm	
T01	D20R2	Contour rough machining	1800	1200	
T01	D20R2	Part opening rough machining	1800	1200	
T02	D6R0-ROUGH	Rough machining of rail groove	2200	1000	
T03	D16R0	Finish machining of part top surface and cavity bottom surface	2200	1000	
T03	D16R0	Part contour finishing	2200	1000	
T04	D6R0-FINISH	Finishing of rail groove	3600	800	
T06	ZXZ10	Drill center hole	1200	100	
T05	D10DRILL	Drill D10 thru Hole	800	100	
T07	D10R0	Mill D24 thru Hole	1800	800	
02					
01					
Change number	Change content	Approve	Date		
Draws:	Date: / review:	Date: / Approve:	Date:		

Ⅱ. Preparation for Processing

(1) As shown in Fig. 1-1, start NX, click "File" "Open", select model "Guide Plate" (.prt file), and click "OK" . Then, the part model can be seen in WCS.

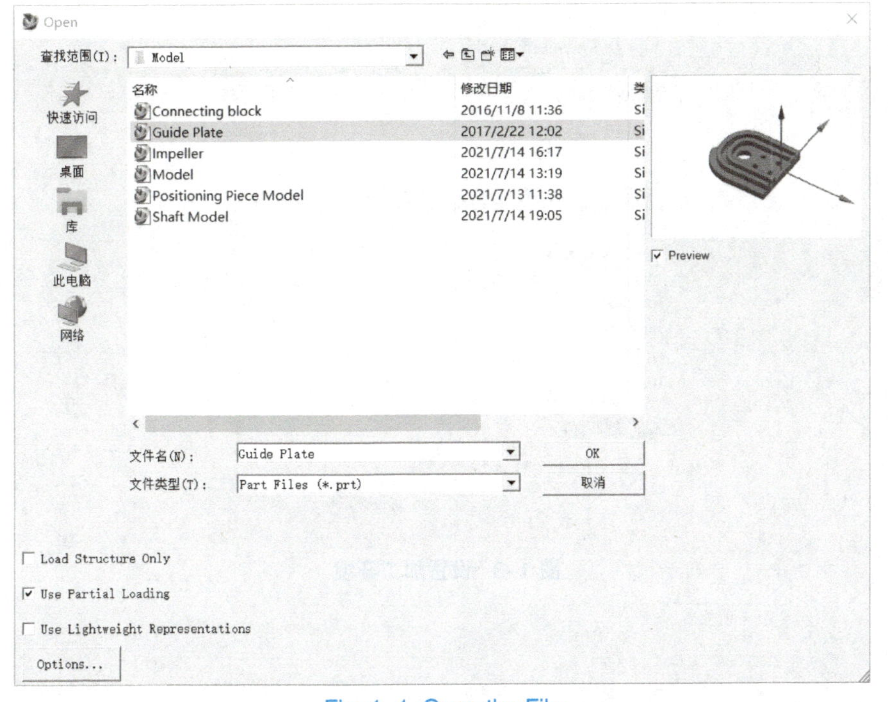

Fig. 1-1　Open the File

（2）选择【应用模块】选项卡，单击【加工】，进入加工环境，如图 1-2 所示。

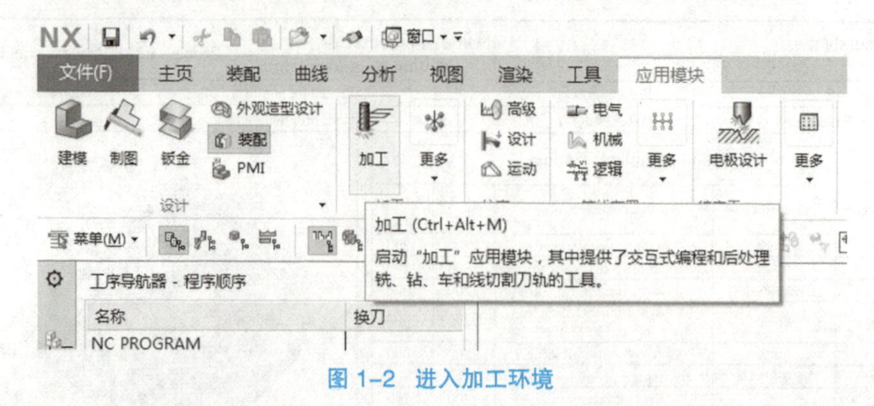

图 1-2 进入加工环境

（3）在弹出的【加工环境】对话框中，在【CAM 会话配置】选项中选择【cam_general】，在【要创建的 CAM 设置】选项中选择【mill_planar】，如图 1-3 所示，单击【确定】。

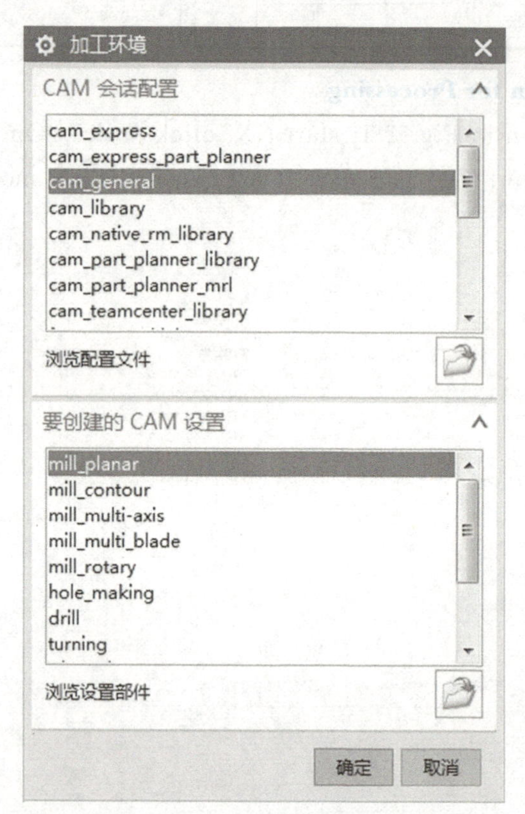

图 1-3 设置加工环境

(2) Select "Application", and click "Manufacturing". The manufacturing menu can be seen as shown in Fig. 1-2.

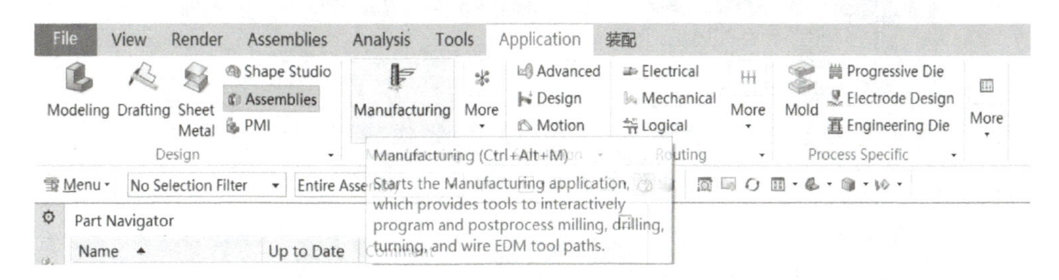

Fig. 1-2 Enter Manufacturing Environment

(3) As shown in Fig. 1-3, in "Machining Environment" window, select "cam_general" in "CAM Session Configuration" tab, and then, select "milll_planar" in "CAM Setup to Create" tab. Finally, click "OK".

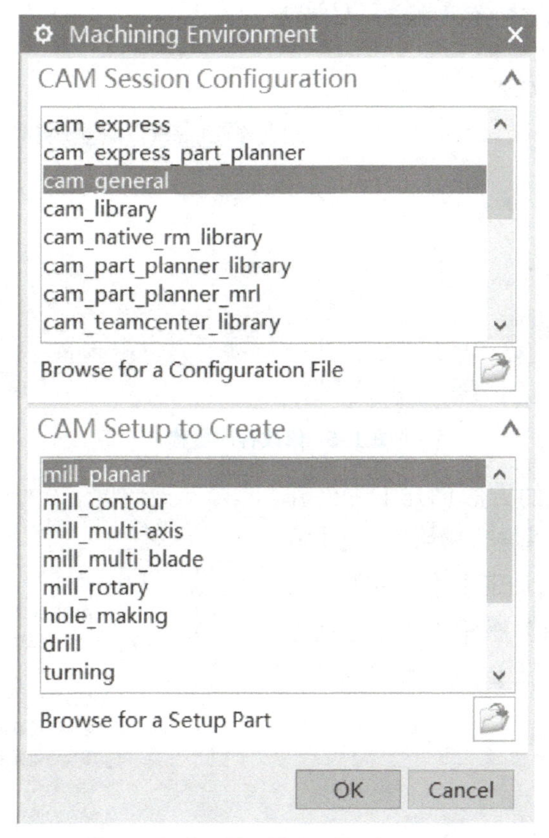

Fig. 1-3 Set Machining Environment

（4）单击【创建程序】，如图 1–4 所示。

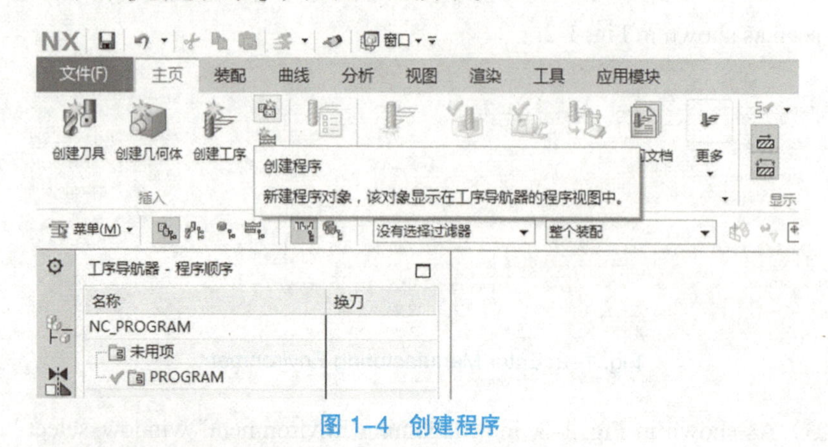

图 1–4　创建程序

（5）修改【名称】为【零件外轮廓粗加工】，单击【应用】，弹出【程序】对话框，如图 1–5 所示，单击【确定】。

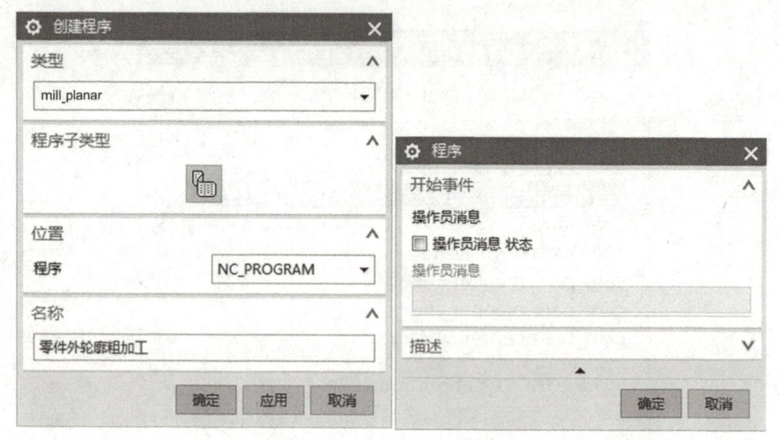

图 1–5　修改程序名称

（6）用同样的方法创建【零件开口腔粗加工】【零件导轨槽粗加工】【零件顶面和腔底面精加工】【零件外轮廓精加工】【零件导轨槽精加工】【钻中心孔】【钻 D10 通孔】【铣 D24 通孔】，如图 1–6 所示。

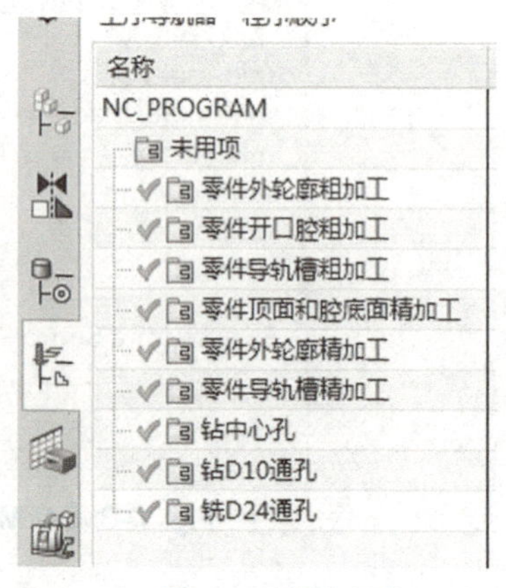

图 1–6　创建程序

(4) As shown in Fig. 1-4, click "Create Program".

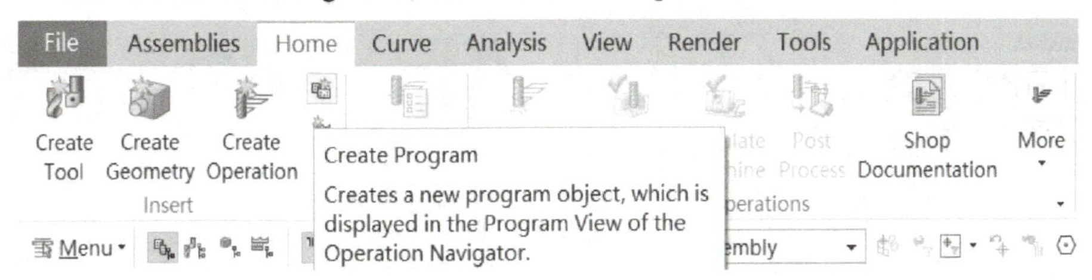

Fig. 1-4 Create Program

(5) As shown in Fig. 1-5, modify name as "Contour Rough Machining" in the dialog of "Name". Click "Apply". Click "OK".

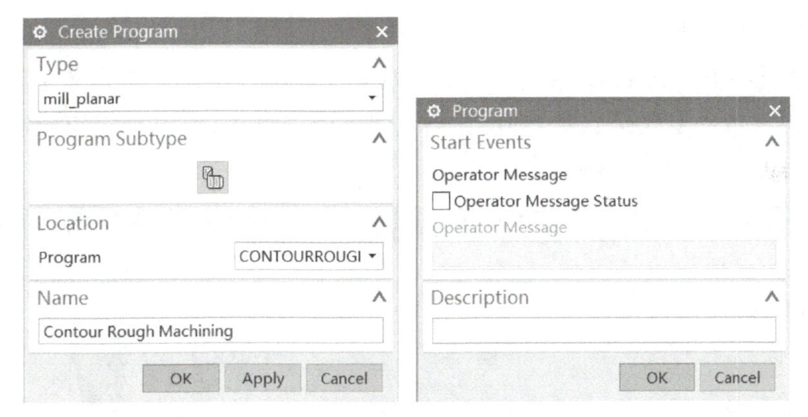

Fig. 1-5 Modify Program Name

(6) Similarly, create programs and modify the names as "Part Opening Rough Machining" "Rough Machining of Rail Groove" "Finish Machining of Part Top Surface and Cavity Bottom Surface" "Part Contour Finishing" "Finishing of Rail Groove" "Drill Center Hole" "Drill D10 Thru Hole" "Mill D24 Thru Hole" respectively. The program groups are shown in Fig. 1-6.

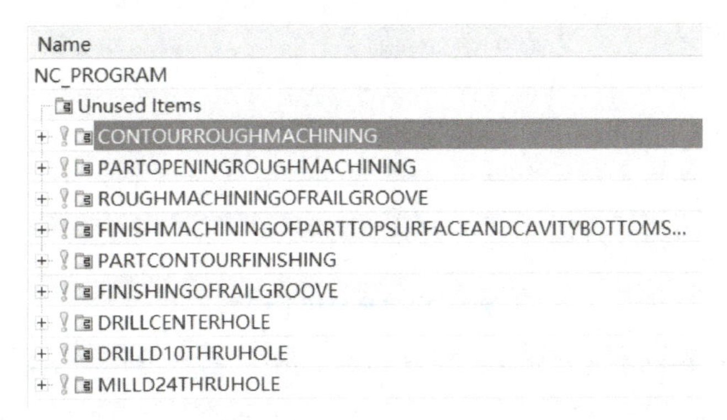

Fig. 1-6 Create Programs

（7）单击【机床视图】，单击【创建刀具】，如图 1-7 所示。

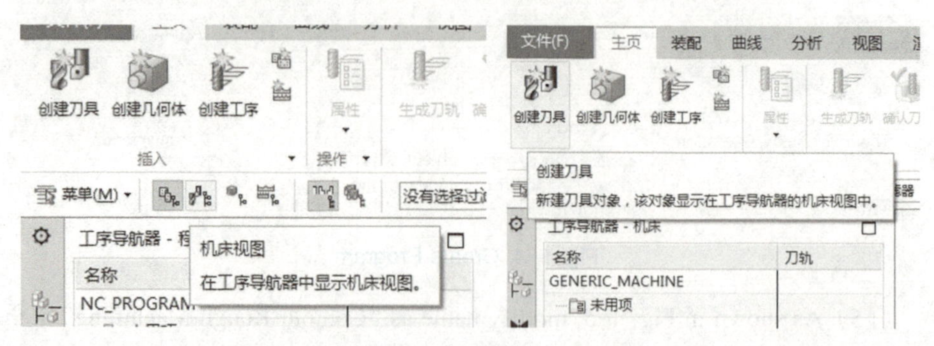

图 1-7 创建刀具

（8）在打开的【创建刀具】对话框中，【类型】选择【mill_planar】，修改刀具【名称】为【D20R2】，单击【应用】。弹出【铣刀 -5 参数】对话框，设置刀具参数：【直径】为【20】，【下半径】为【2】，其他参数为默认值，如图 1-8 所示，单击【确定】。

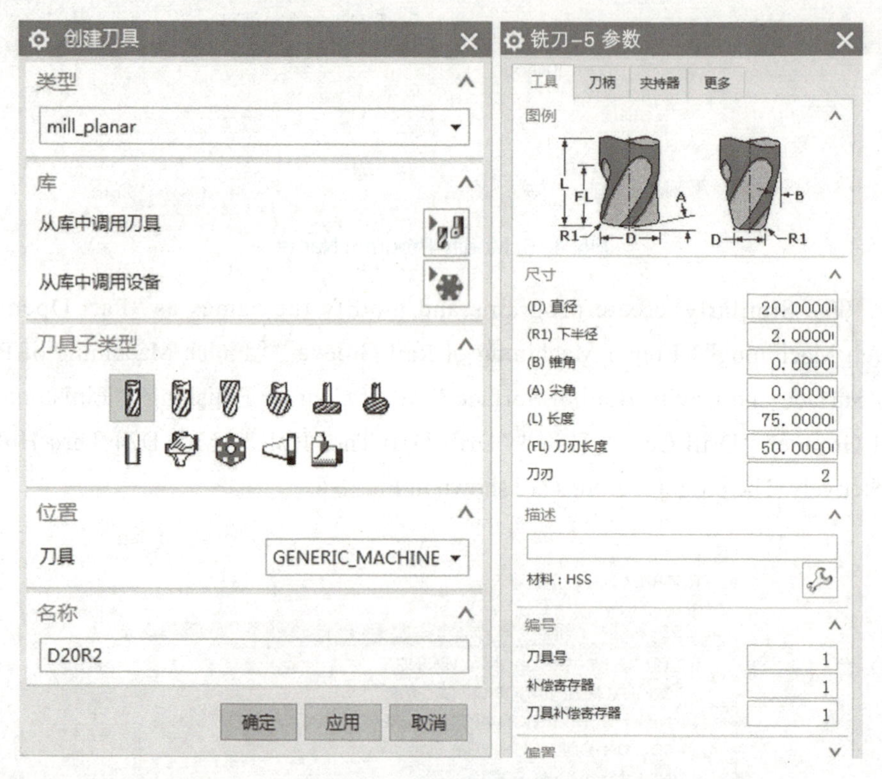

图 1-8 刀具参数（1）

（9）用同样的方法创建【D6】的端铣刀用于开粗，创建【D16】的端铣刀，创建【D6】的端铣刀用于精加工。创建【D10】的端铣刀用来精加工 D24 的孔，单击【确定】。

(7) Click "Machine Tool View", and then click "Create Tool" as shown in Fig.1-7.

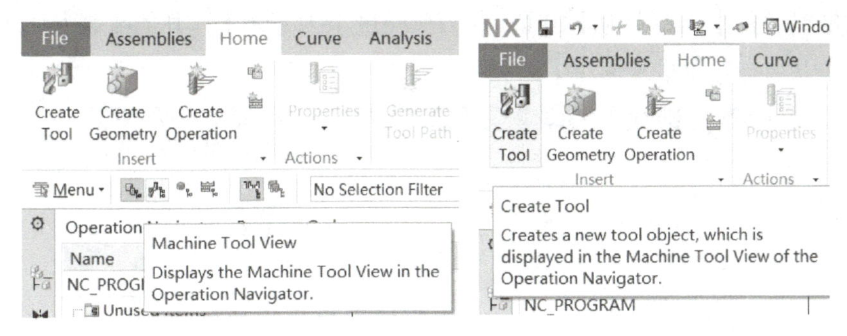

Fig. 1-7 Create Tool

(8) As shown in Fig. 1-8, in the "Create Tool" window, select "mill_planar" in "Type" and modify "Name" as "D20R2", then click "OK". In "Milling Tool-5 Parameters" window, "Diameter" value is 20 and "Lower Radius" value is 2. Other parameters default, click "OK" to close the window.

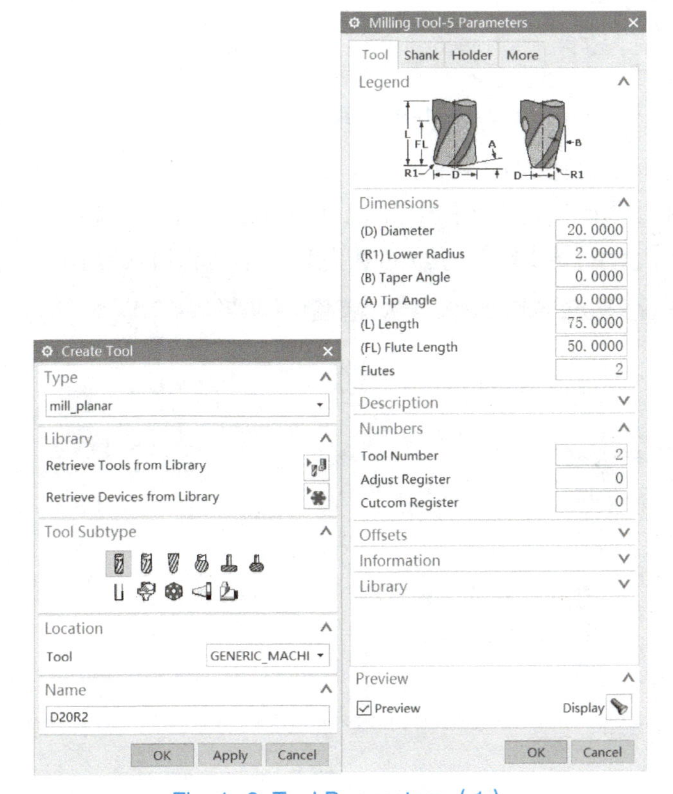

Fig. 1-8 Tool Parameters (1)

(9) Similarly, create tool and modify name as "D6" "D16" "D10" and modify diameters respectively.

（10）回到【创建刀具】对话框，【类型】选择【drill】，【刀具子类型】选择【中心钻】，修改刀具【名称】为【ZXZ10】，单击【应用】。弹出【钻刀】对话框，设置【直径】为【10】，【刀尖角度】为【90】，其他参数为默认值，如图 1-9 所示，单击【确定】。

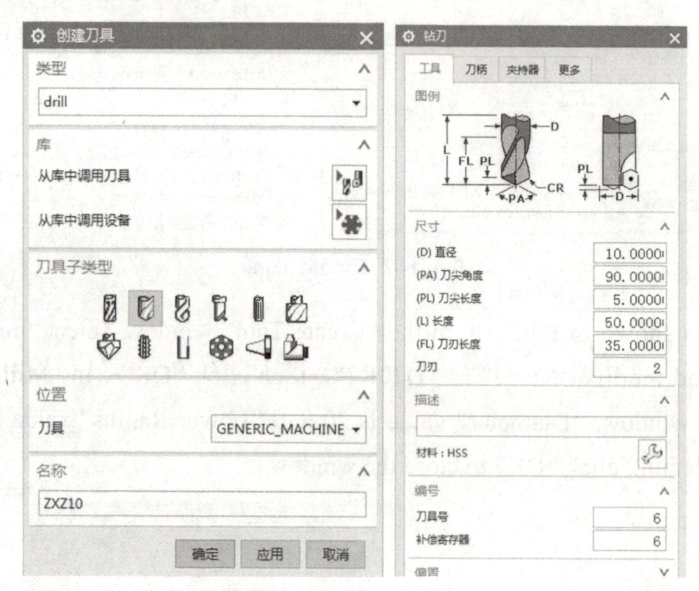

图 1-9 刀具参数（2）

（11）返回【创建刀具】对话框，【类型】选择【drill】，【刀具子类型】选择【麻花钻】，修改刀具【名称】为【D10DRILL】，单击【应用】。弹出【钻刀】对话框，设置【直径】为【10】，其他参数为默认值，如图 1-10 所示，单击【确定】。

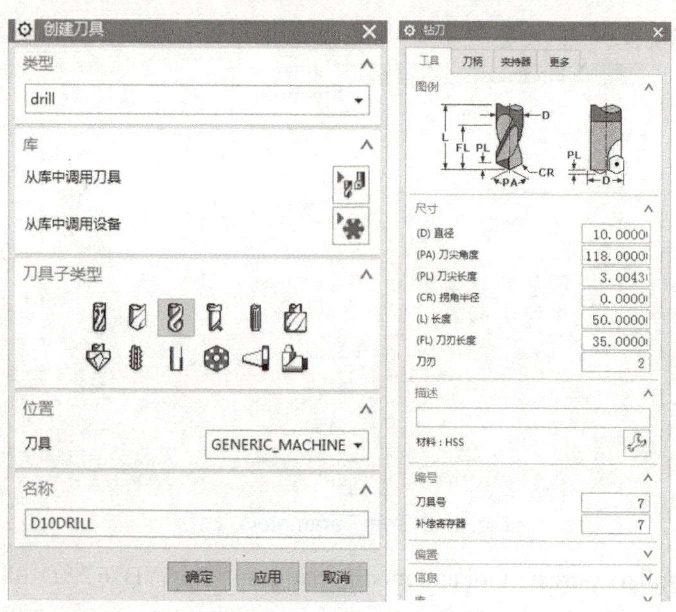

图 1-10 刀具参数（3）

(10) Click "Create Tool" button and select "drill" in "Type" and select "spotdrilling_tool" in "Tool Subtype". Further, modify tool name as "ZXZ10". Click "Apply". As shown in Fig. 1-9, in "Drilling Tool" window, "Diameter" value is 10 and "Point Angle" value is 90. Other parameters default, click "OK" to close the window.

Fig. 1-9 Tool Parameters (2)

(11) Click "Create Tool" button and select "drill" in "Type" and select "drilling_tool" in "Tool Subtype". Further, modify tool name as "D10Drill". Click "Apply". As shown in Fig. 1-10, in "Drilling Tool" window, "Diameter" value is 10. Other parameters default, click "OK" to close the window.

Fig. 1-10 Tool Parameters (3)

（12）进入【几何视图】，双击【MCS_MILL】，弹出【MCS 铣削】对话框，选择【指定 MCS】下的绝对坐标系，【安全距离】设为【50】，如图 1-11 所示，单击【确定】。

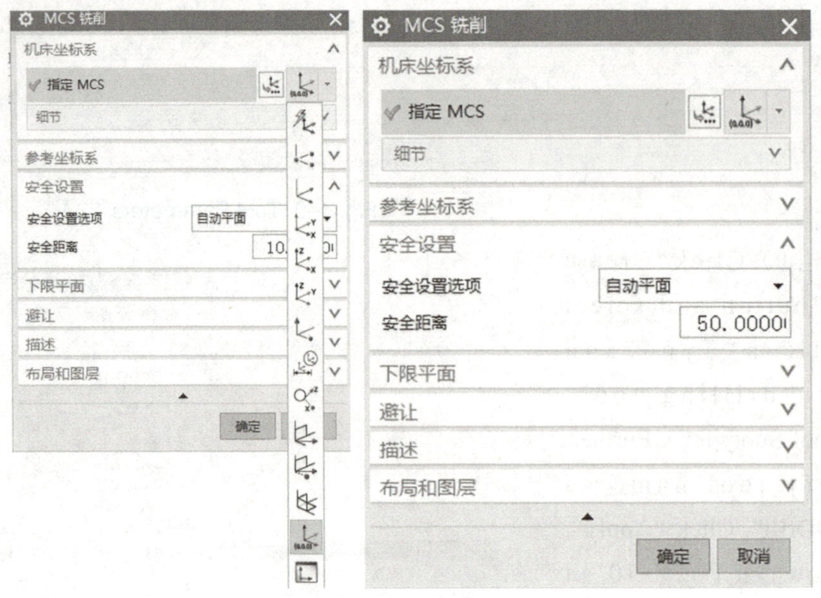

图 1-11　设置坐标

（13）双击【WORKPIECE】选项，弹出【工件】对话框，单击【指定部件】中的【选择或编辑部件几何体】，弹出【部件几何体】对话框，选择零件实体，如图 1-12 所示，单击【确定】。

（12）Click “Geometry View” and then double click “MCS_MILL”. In “MCS Mill” window, click “CSYS Dialog”, select “MCS-MILL”, and change “Safe Clearance Distance ” to 50 as shown in Fig. 1-11 , and then click “OK” to close the window.

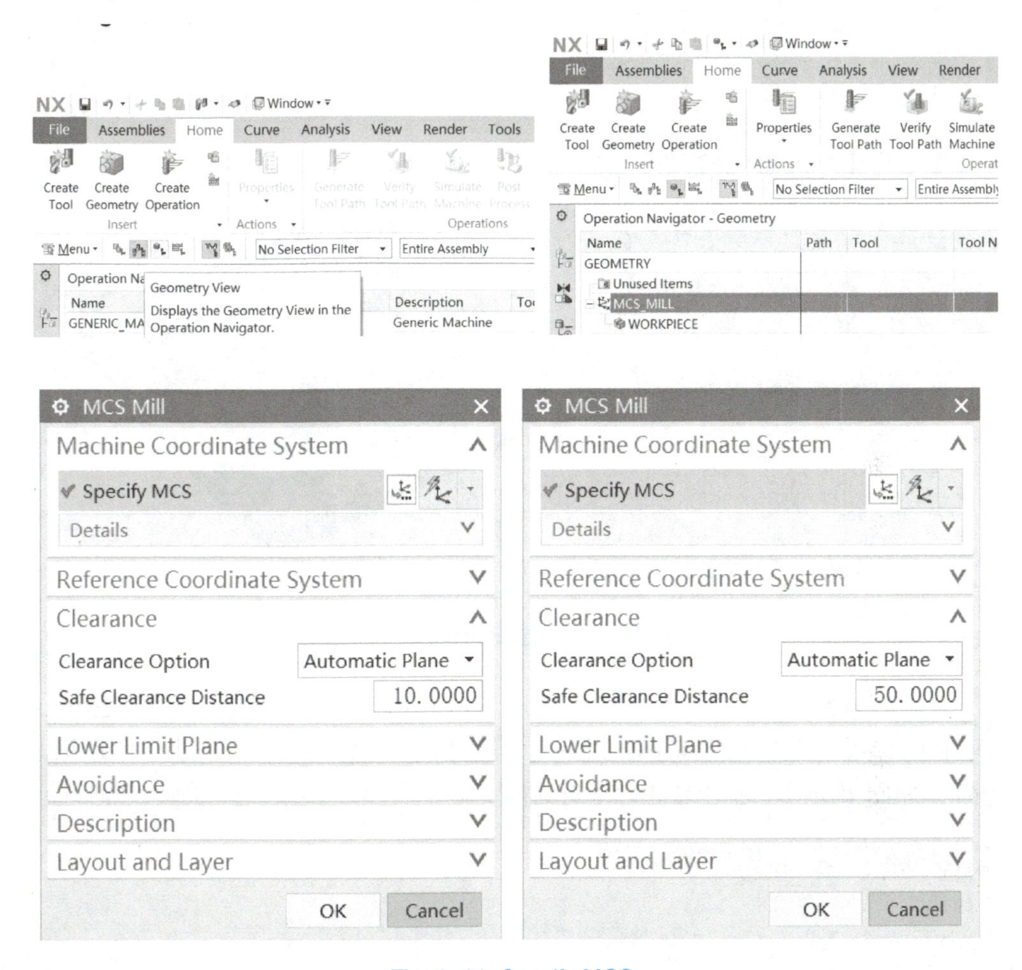

Fig. 1–11　Specify MCS

（13）Double click “WORKPIECE” and click “Select or Edit the Part Geometry” in “Workpiece” window. In “Part Geometry” window, select the solid part as shown in Fig. 1-12. Click “OK” to close the window.

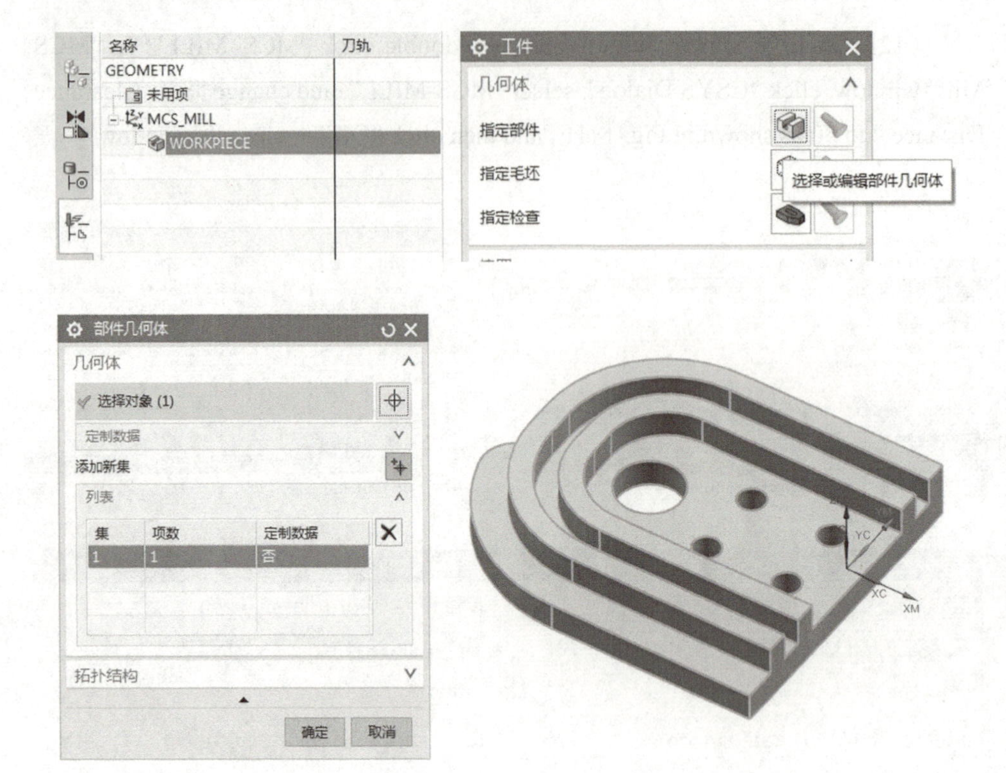

图 1–12 选择零件实体

（14）返回到【工件】对话框，单击【指定毛坯】中的【选择或编辑毛坯几何体】，弹出【毛坯几何体】对话框，【类型】选择【包容块】，如图 1–13 所示，依次单击【确定】，退出【工件】对话框。

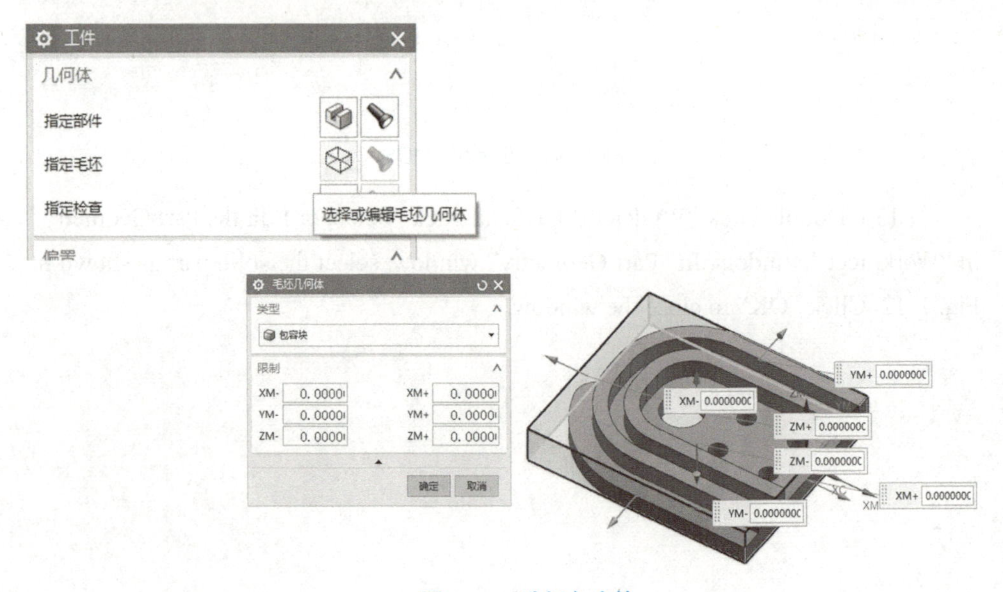

图 1–13 选择包容块

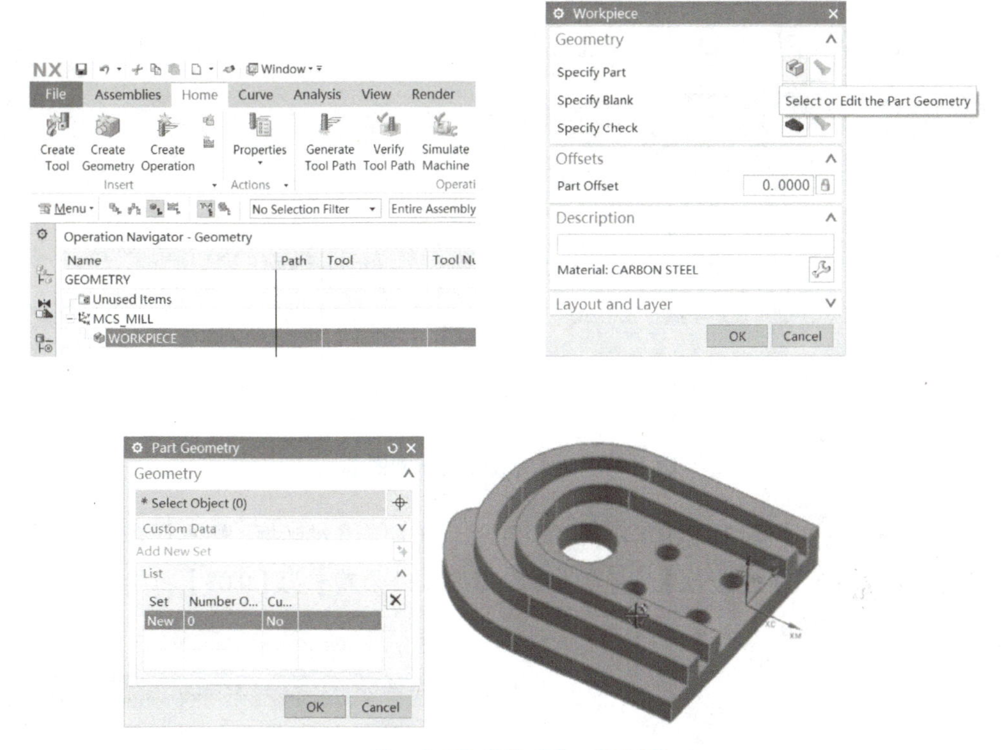

Fig. 1–12 Select the Solid Part

（14） Click "Workpiece" and click "Select or Edit the Blank Geometry" in "Workpiece" window. In "Blank Geometry" window, select " Bounding Block" as type as shown in Fig. 1-13. Click "OK".

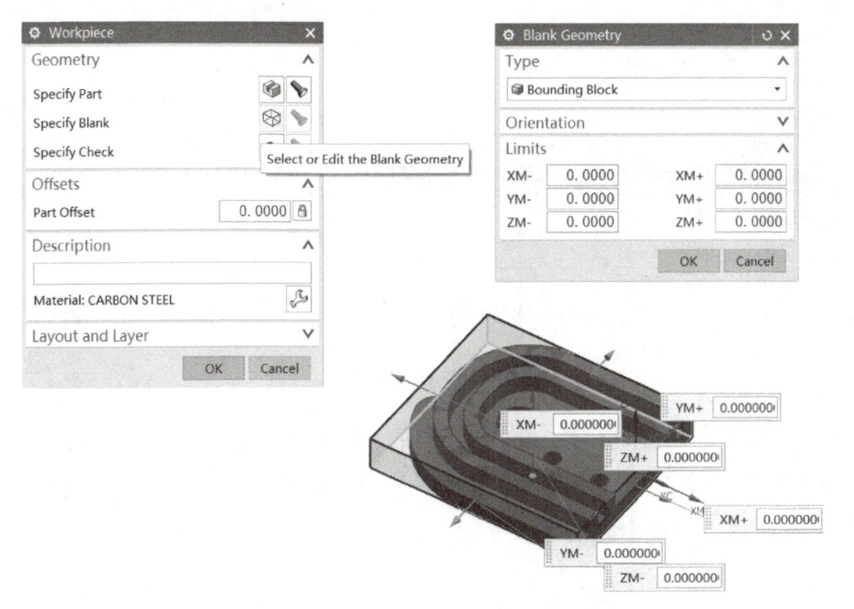

Fig. 1–13 Select Bounding Block

（15）单击【加工方法】，进入【加工方法视图】，如图 1-14 所示。

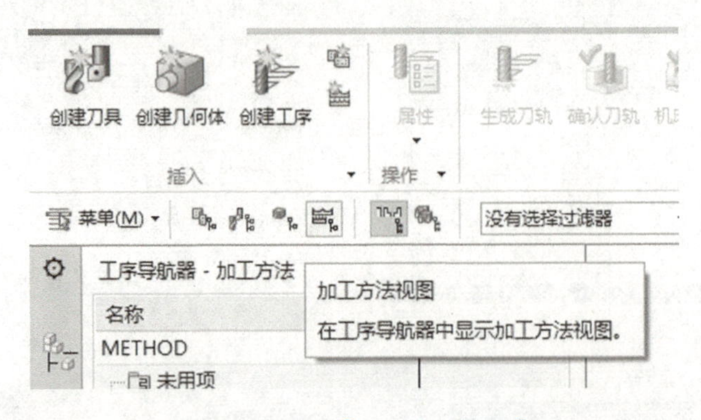

图 1-14　进入【加工方法视图】

（16）双击【MILL_ROUGH】，弹出【铣削粗加工】对话框，修改【部件余量】为【0.5】，修改【内公差】为【0.03】，修改【外公差】为【0.03】，如图 1-15 所示，单击【确定】。

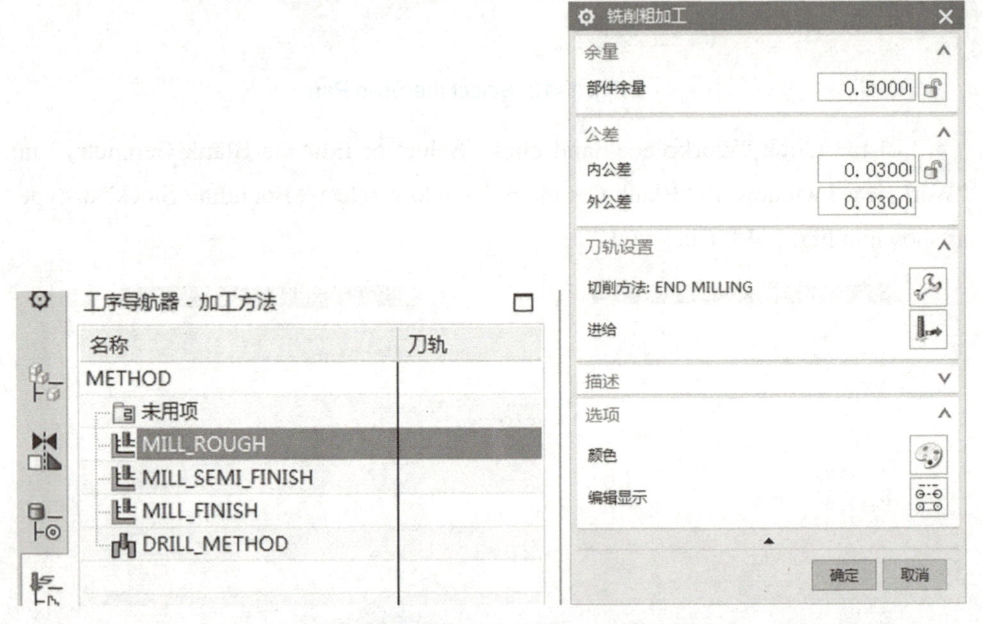

图 1-15　铣削粗加工

（17）双击【MILL_SEMI_FINISH】，弹出【铣削半精加工】对话框，修改【部件余量】为【0.1】，修改【内公差】为【0.01】，修改【外公差】为【0.01】，如图 1-16 所示，单击【确定】。

(15) Click "Machining Method View" button as shown in Fig. 1-14.

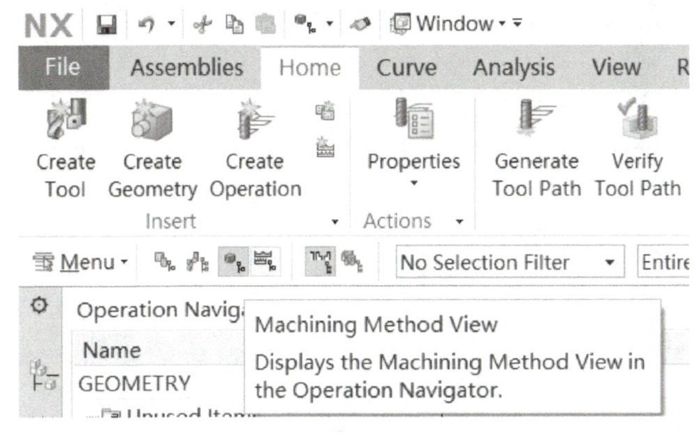

Fig. 1-14 Enter Machining Method View

(16) Double click "MILL_ROUGH" to modify "Part Stock" value as 0.5, both "Intol Tolerance" and "Outtol Tolerance" values as 0.03 as shown in Fig. 1-15. Click "OK".

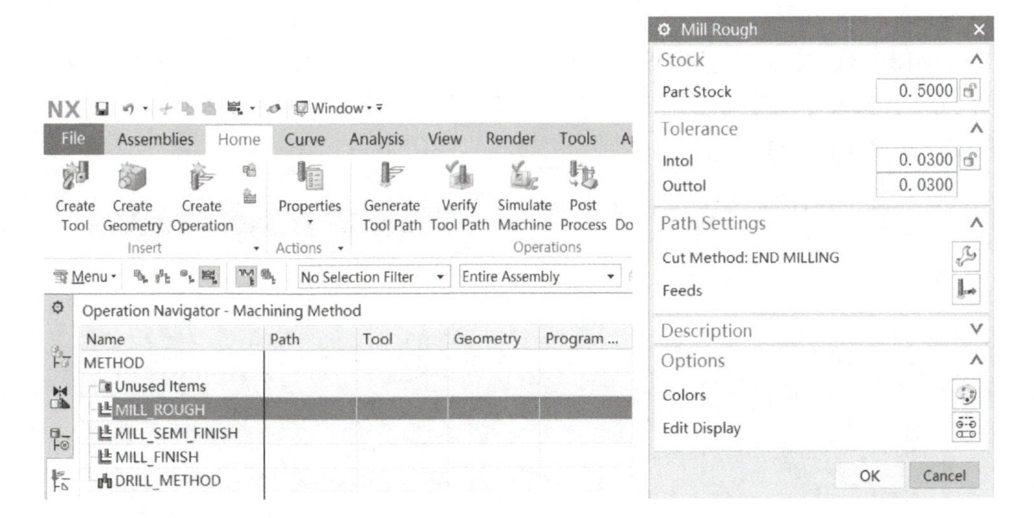

Fig. 1-15 Mill Rough

(17) Double click " MILL_SEMI_FINISH " to modify "Part Stock" value as 0.1, both "Intol Tolerance" and "Outtol Tolerance" values as 0.01 as shown in Fig. 1-16. Click "OK" to close the window.

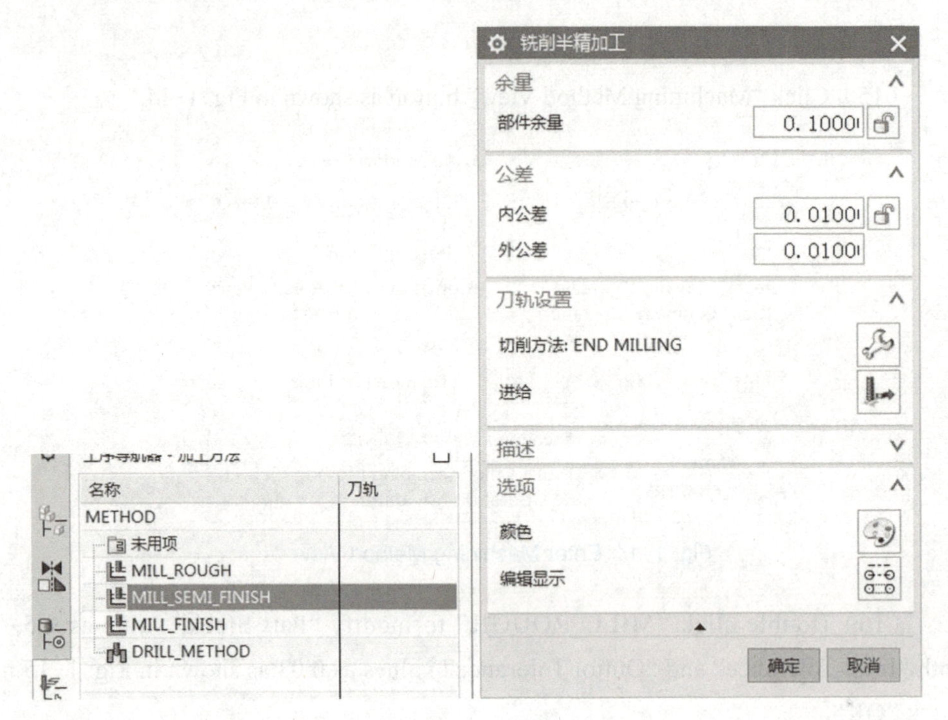

图 1-16 铣削半精加工

（18）双击【MILL _FINISH】，弹出【铣削精加工】对话框，修改【部件余量】为【0】，修改【内公差】为【0.003】，修改【外公差】为【0.003】，如图 1-17 所示，单击【确定】。

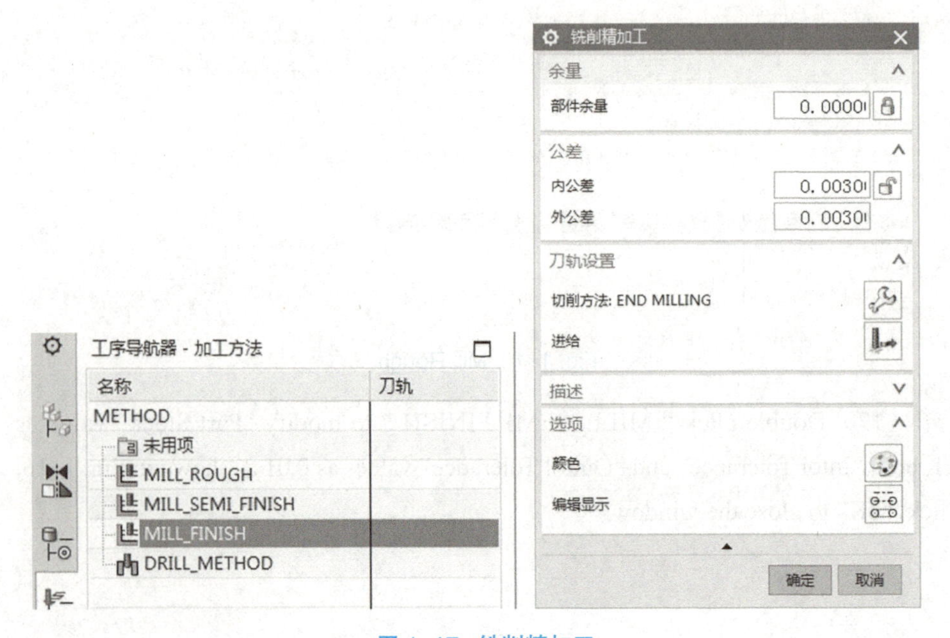

图 1-17 铣削精加工

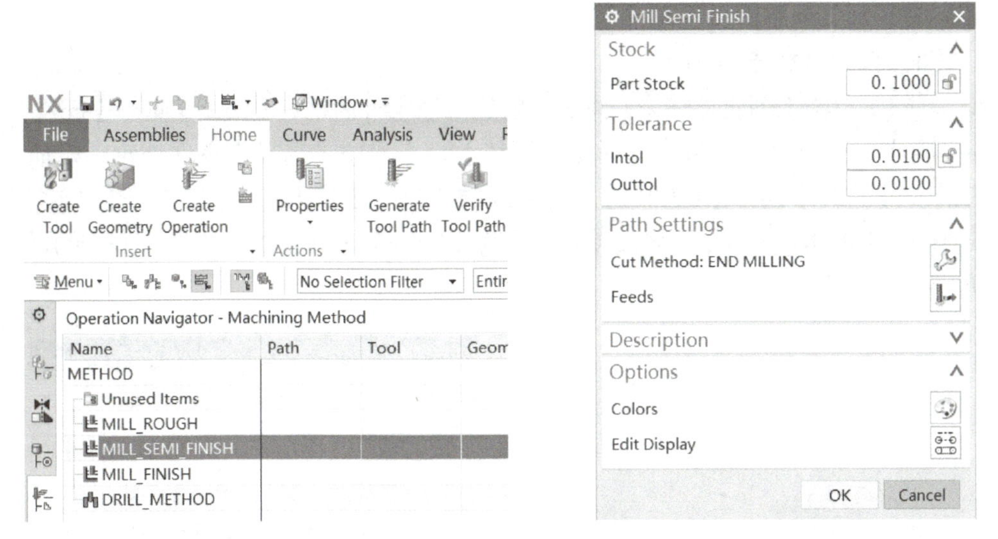

Fig. 1-16　Mill Semi Finish

（18） Double click " MILL_FINISH " to modify both "Intol Tolerance" and "Outtol Tolerance" values as 0.003 as shown in Fig. 1-17. Click "OK".

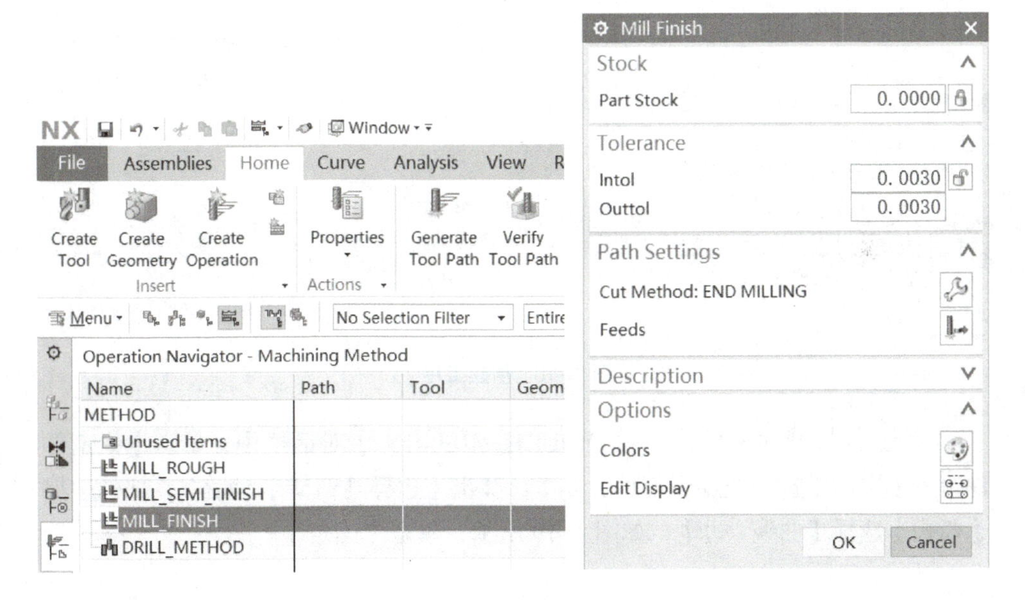

Fig. 1-17　Mill Finish

三、加工程序编制

1. 零件外轮廓粗加工

（1）单击【创建工序】，弹出【创建工序】对话框，【类型】选择【mill_planar】，【工序子类型】选择【平面铣】，【程序】选择【零件外轮廓粗加工】，【刀具】选择【D20R2】，【几何体】选择【WORKPIECE】，【方法】选择【MILL_ROUGH】，修改【名称】为【Rough_mill-1】，如图 1-18 所示，单击【确定】。

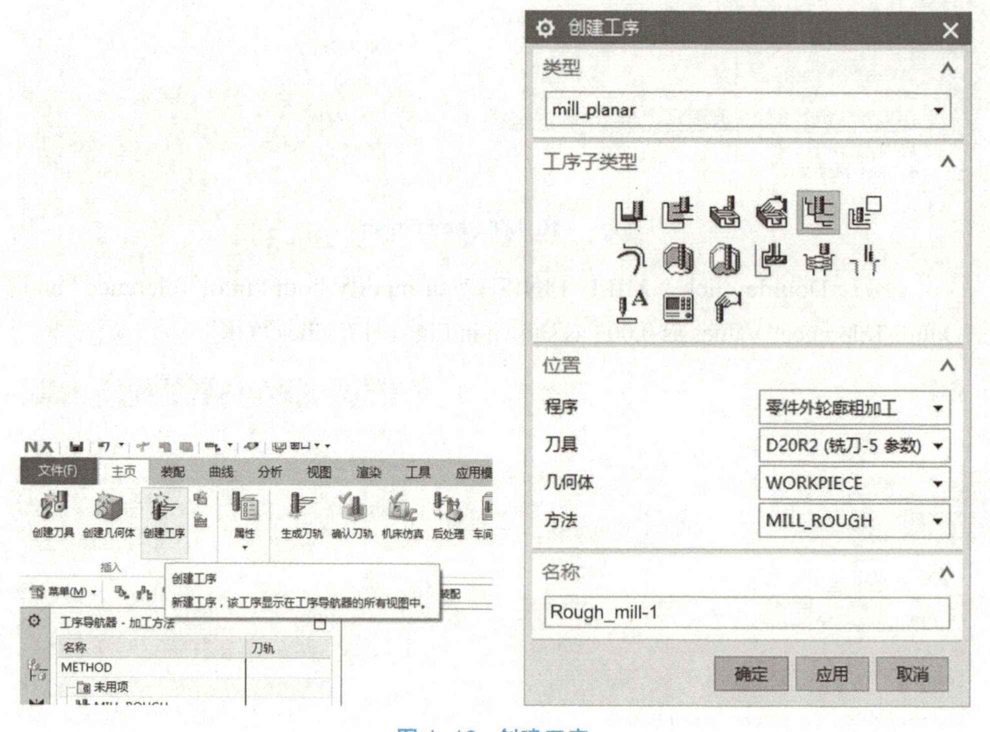

图 1-18　创建工序

（2）在打开的【平面铣 -［ROUGH_MILL-1］】对话框中，单击【指定部件边界】中的【选择或编辑部件边界】，弹出【边界几何体】对话框，部件边界【模式】选择【曲线 / 边】，如图 1-19 所示，单击【确定】。

Ⅲ. Programming

1.Contour Rough Machining

(1) As shown in Fig. 1-18, click "Create Operation", and then select "mill_planar" in pull-down box of "Type". Select operation subtype as "Planar Mill", program as "CONTOUR ROUGH MACHINING", tool as "D20R2", geometry as "WORKPIECE" and method as "MILL_ROUGH". Modify name as "Rough_mill-1", and then click "OK" .

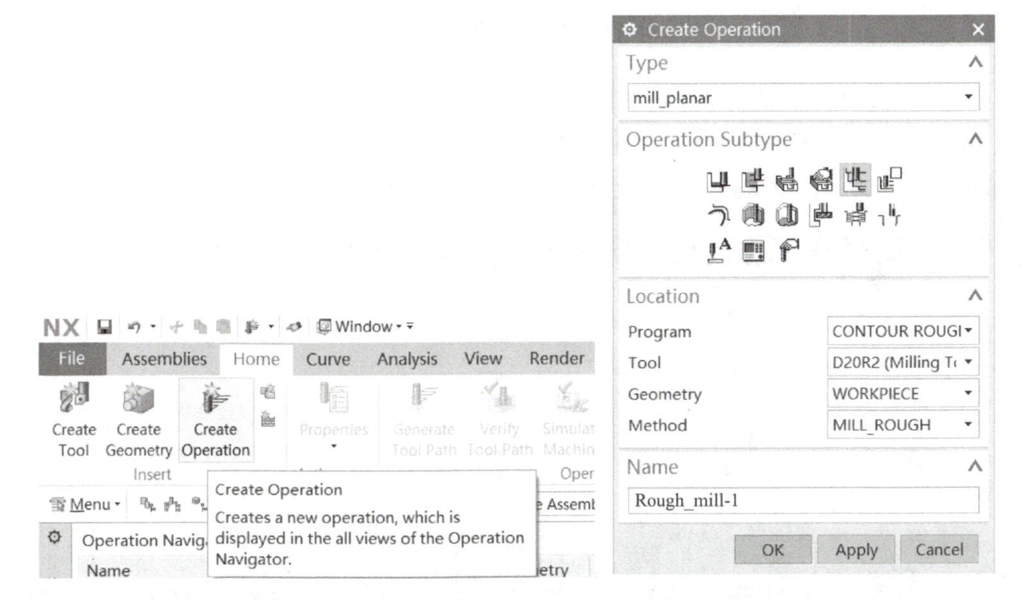

Fig. 1-18 Create Operation

(2) As shown in Fig. 1-19, click edit button "Specify Part Boundaries". Then, in "Boundary Geometry" window, select the mode as "Curves/Edges", and then the "Create Boundary" window pops up.

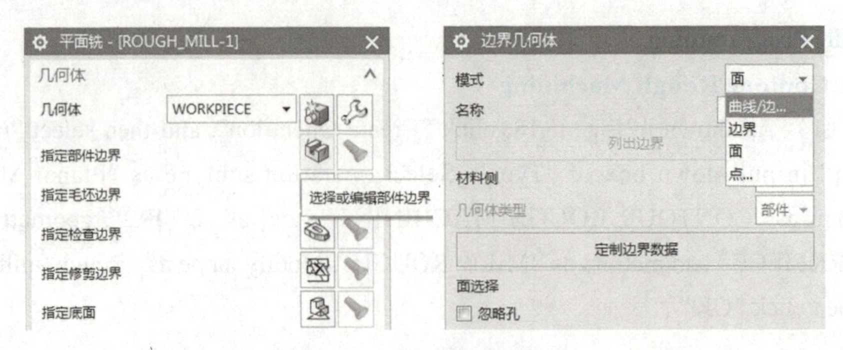

图 1-19　指定部件边界

（3）打开【创建边界】对话框，【刨】选择【用户定义】，弹出【刨】对话框，【类型】选择【XC-YC 平面】，【距离】为【1 mm】，单击【确定】。选择边界，如图 1-20 所示，依次单击【确定】。

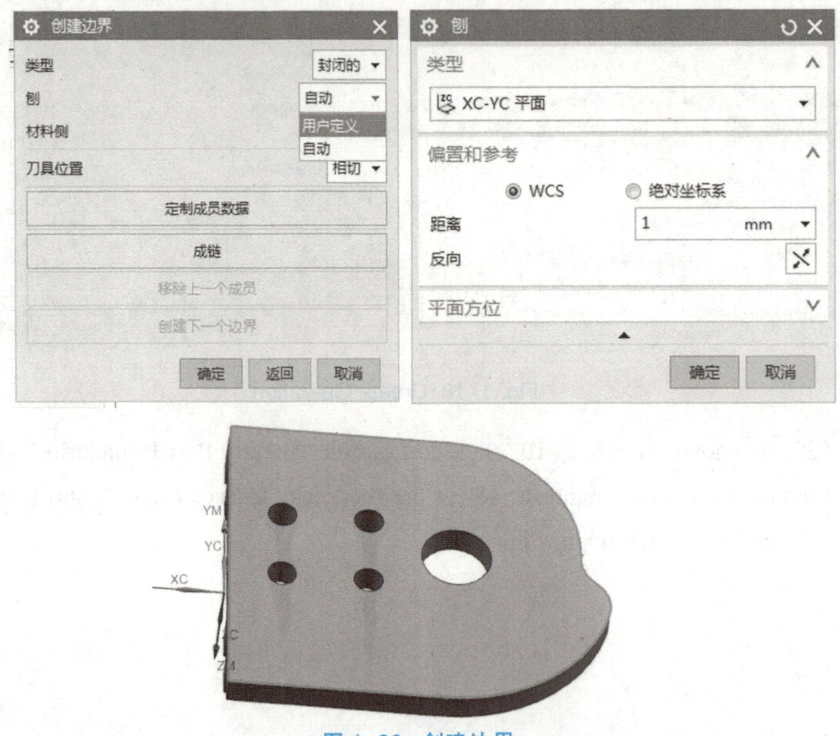

图 1-20　创建边界

（4）单击【指定部件边界】，单击【附加】，选择台阶面，如图 1-21 所示，依次单击【确定】。

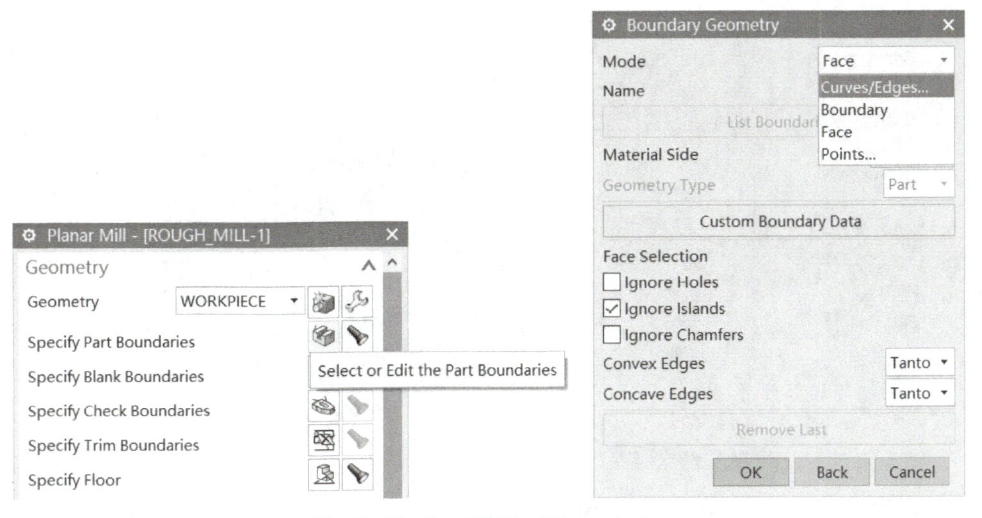

Fig. 1-19 Specify Part Boundaries

(3) Click "Create Boundary". Select "User-Defined" in the option of "Plane". The "Plane" window pops up. Then select "XC-YC Plane" as type and modify distance as 1 mm. Click "OK" and select the boundary as shown in Fig. 1-20.

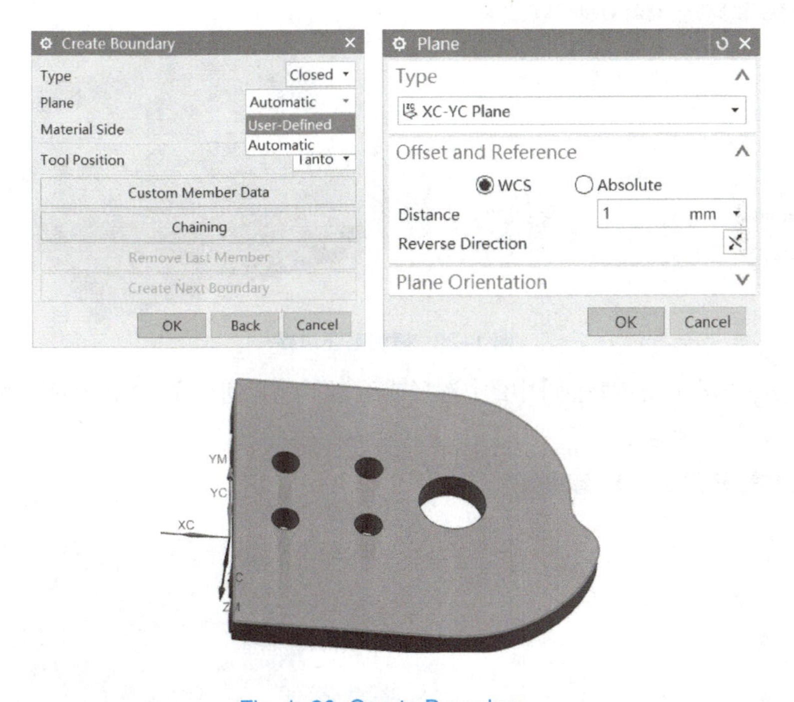

Fig. 1-20 Create Boundary

(4) Click "Edit Boundary" and click button "Append", and select surface of the step as shown in Fig. 1-21. Click "OK".

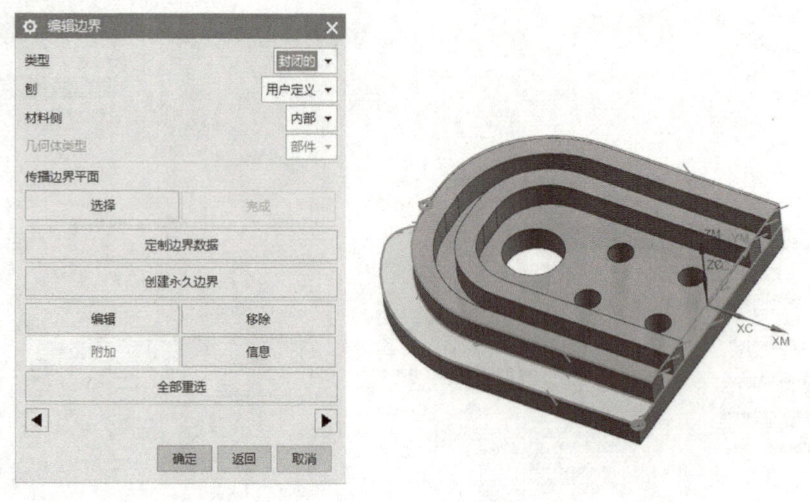

图 1-21　附加边界

（5）单击【指定毛坯边界】中的【选择或编辑毛坯边界】，选择图中毛坯上表面（毛坯可在建模环境下创建），如图 1-22 所示，单击【确定】。

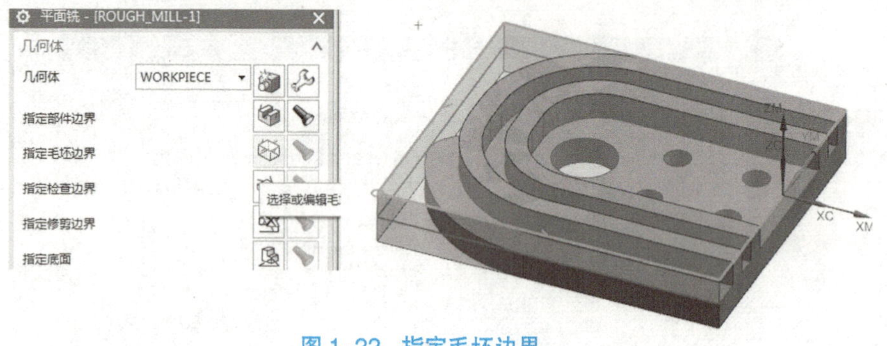

图 1-22　指定毛坯边界

（6）单击【指定底面】中的【选择或编辑平面几何体】，选择底面，如图 1-23 所示，单击【确定】。

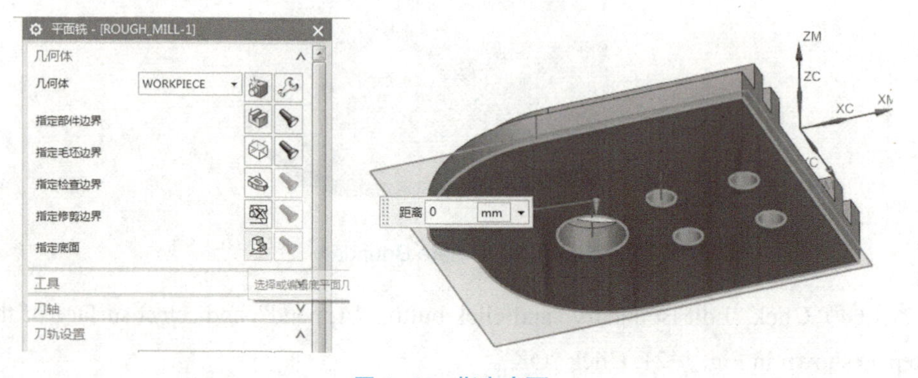

图 1-23　指定底面

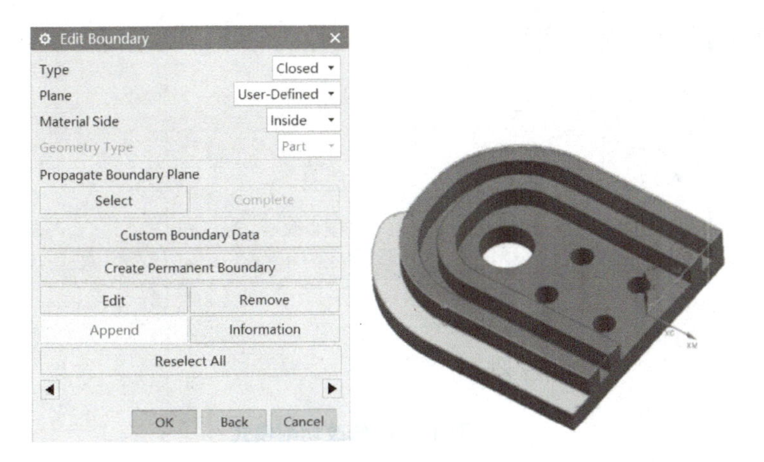

Fig. 1-21　Append Boundary

(5) Then click edit button "Specify Blank Boundaries", and select blank boundary as shown in Fig. 1-22. Click "OK".

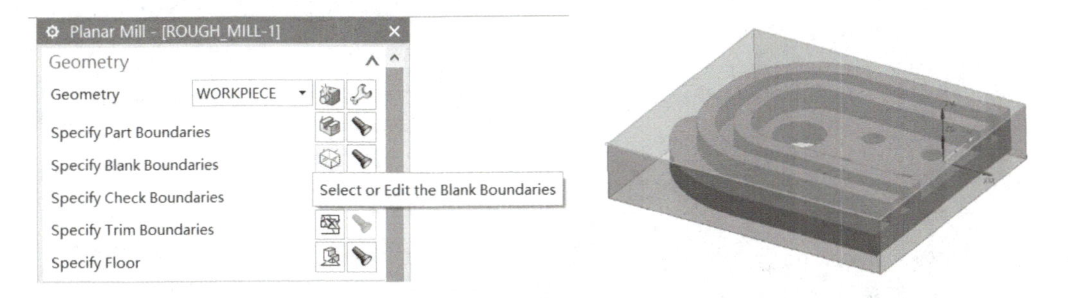

Fig. 1-22　Specify Blank Boundaries

(6) As shown in Fig. 1-23, click the button "Specify Floor" and select the bottom of the part. Click "OK".

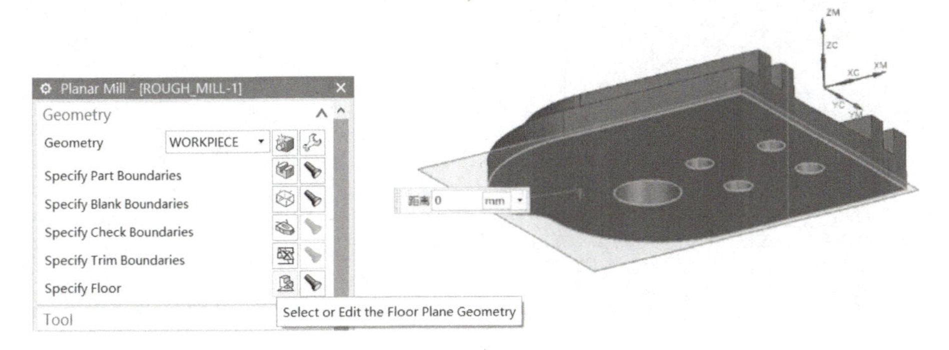

Fig. 1-23　Specify Floor

（7）设置【切削模式】为【跟随部件】，如图 1-24 所示。

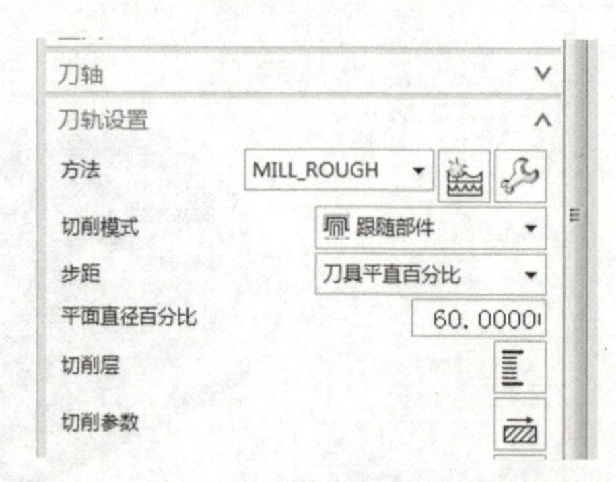

图 1-24 修改切削模式

（8）单击【切削层】，打开【切削层】对话框，【类型】选择【恒定】，设置【每刀切削深度】中的【公共】为【1.5】，如图 1-25 所示，单击【确定】。

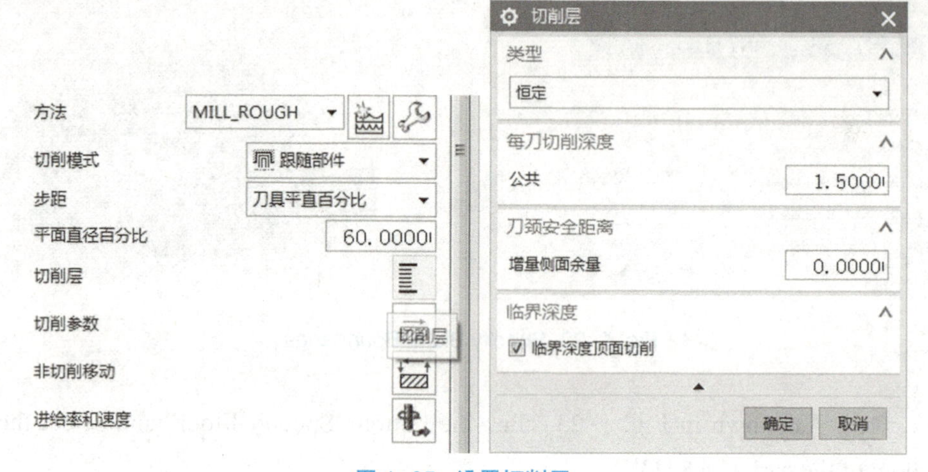

图 1-25 设置切削层

(7) As shown in Fig. 1-24, modify "Cut Pattern" as "Follow Part".

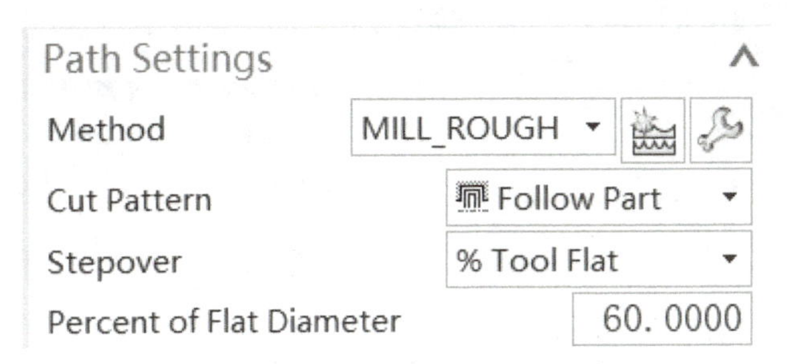

Fig. 1-24 Modify Cut Pattern

(8) As shown in Fig. 1-25, click "Cut Levels", then select "Constant" as the type, and modify "Common Depth Per Cut" as 1.5. Click "OK".

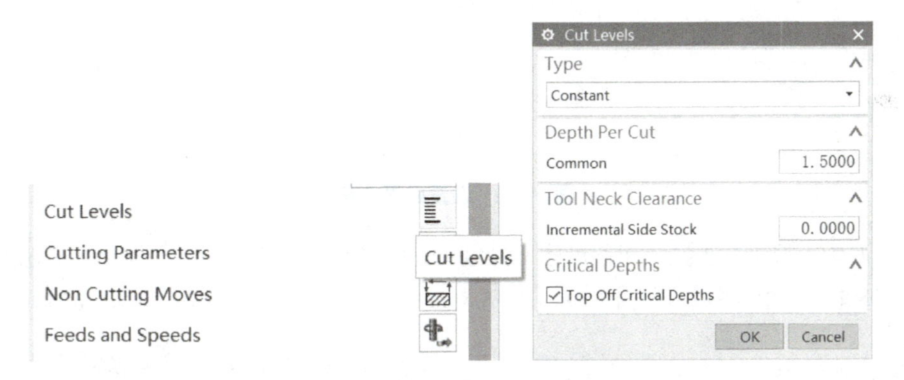

Fig. 1-25 Modify Cut Levels

（9）单击【切削参数】，打开【切削参数】对话框，设置【最终底面余量】为【0.2】，如图 1-26 所示，单击【确定】。

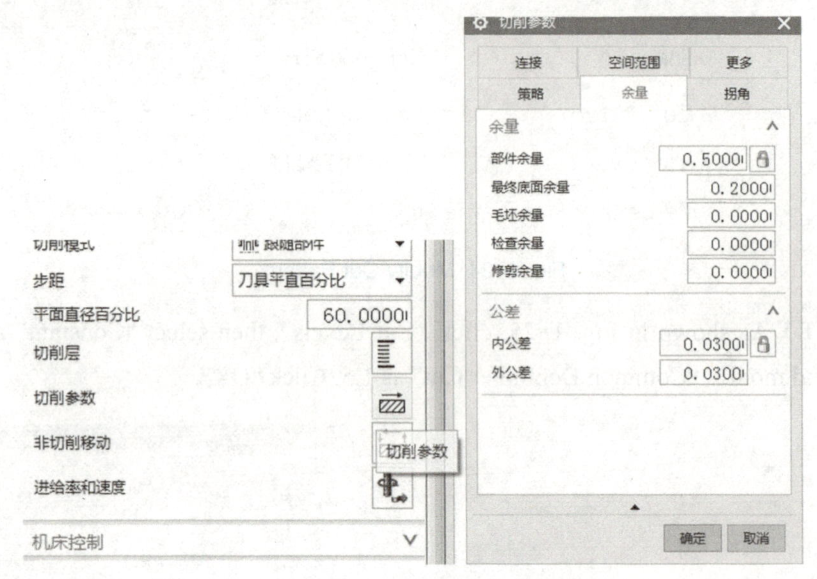

图 1-26 设置底面余量

（10）单击【进给率和速度】，打开【进给率和速度】对话框，设置【主轴速度】为【1800】，进给率【切削】设为【1200 mmpm】，如图 1-27 所示，单击【确定】。

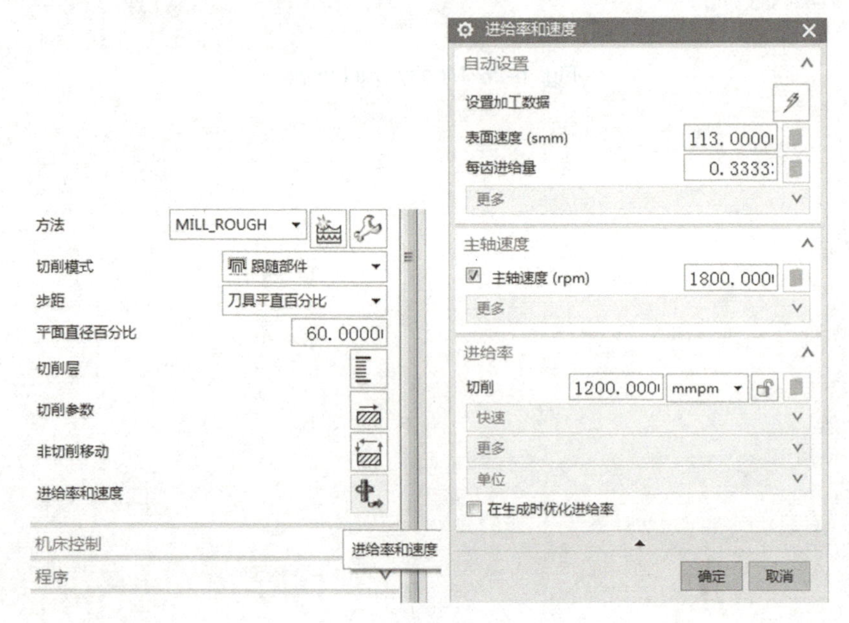

图 1-27 设置进给率和速度

(9) As shown in Fig. 1-26, click "Cutting Parameters" button, and in "Stock" tab, modify "Final Floor Stock" as 0.2. Click "OK".

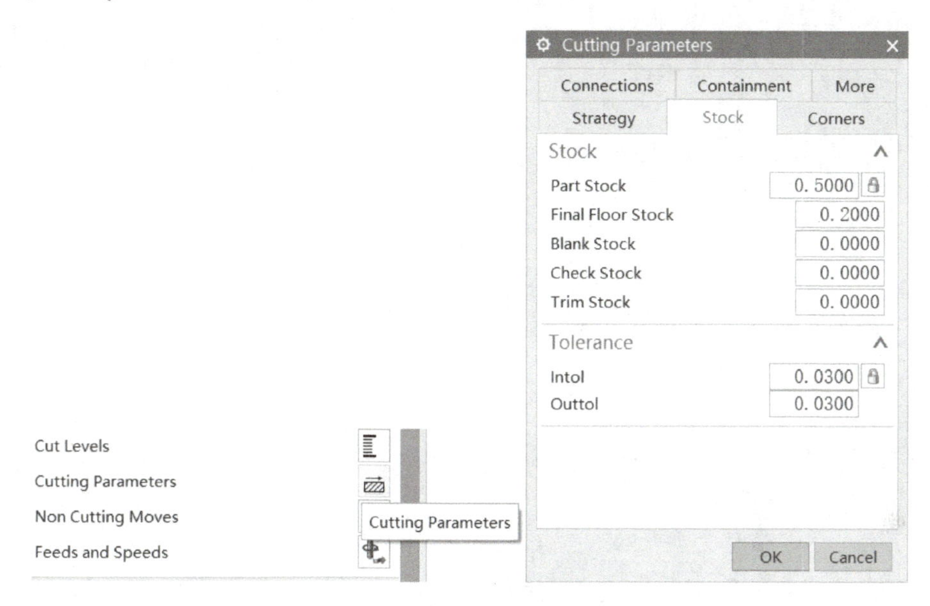

Fig. 1-26 Modify Floor Stock

(10) Click "Feeds and Speeds", modify "Spindle Speed" as 1, 800 and "Feed Rates" as 1, 200 mmpm as shown in Fig. 1-27. Click "OK".

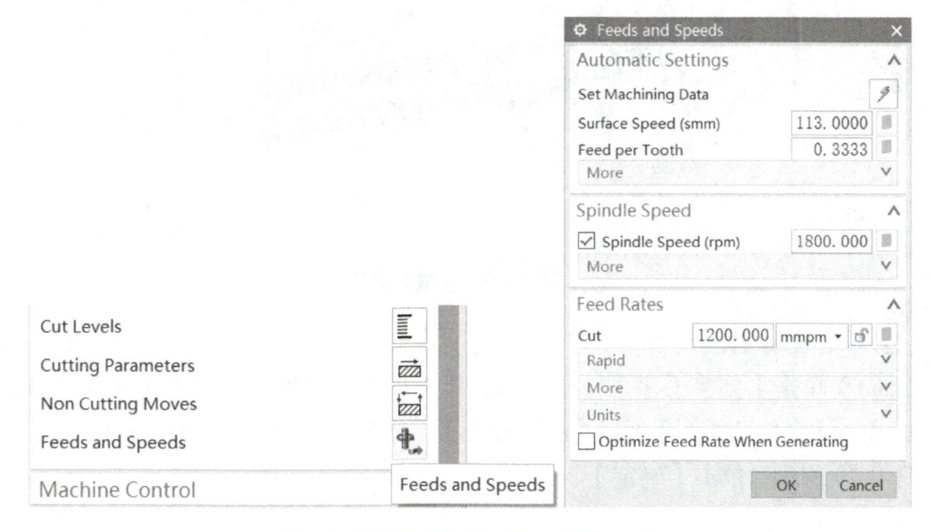

Fig. 1-27 Modify Feeds and Speeds

（11）单击【生成】，查看生成的刀具轨迹，如图 1-28 所示，单击【确定】。

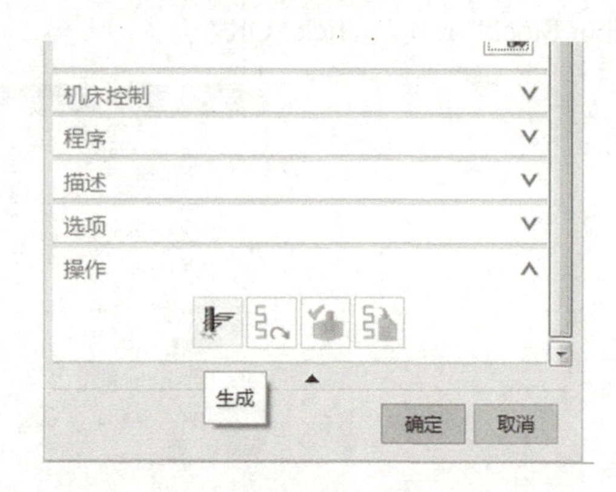

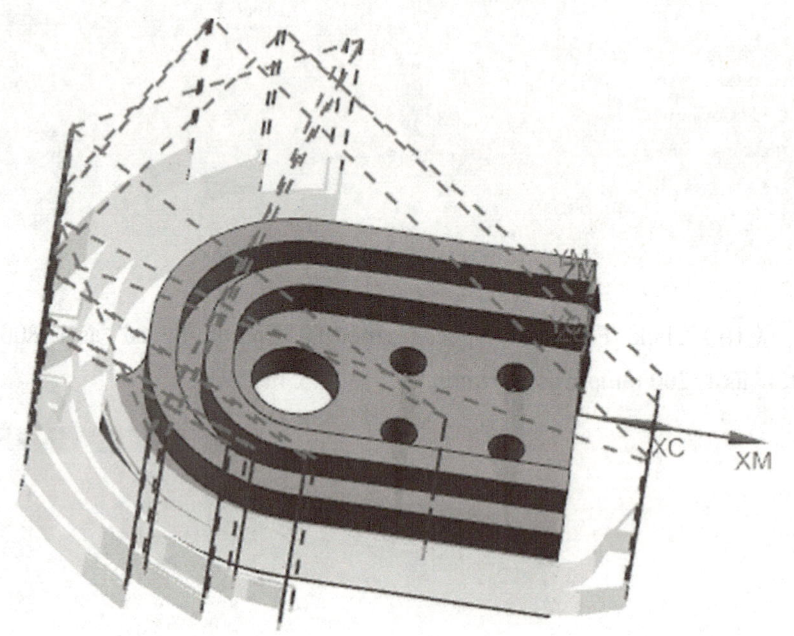

图 1-28　生成刀具轨迹

2. 零件开口腔粗加工

（1）单击【创建工序】，打开【创建工序】对话框，【类型】选择【mill_planar】，【工序子类型】选择【底壁加工】，修改【名称】为【Rough_mill-2】，如图 1-29 所示，单击【确定】。

(11) Click "Generate" to view the tool path as shown in Fig. 1-28 and then click "OK".

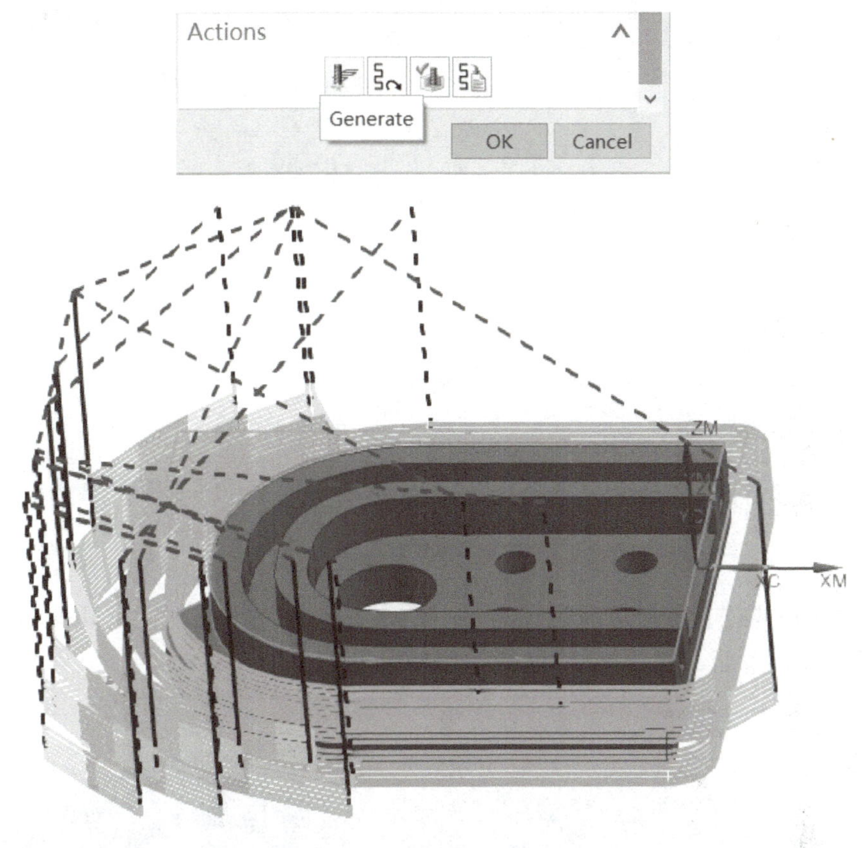

Fig. 1-28 Generate Tool Path

2. Part Opening Rough Machining

(1) As shown in Fig. 1-29, click "Create Operation", and then select "mill_planar" in pull-down box of "Type". Select operation subtype as "Floor and Wall", program as "PART OPENING ROUGH MACHINING", tool as "D20R2", geometry as " WORKPIECE" and method as "MILL_ROUGH". Modify name as "Rough_mill-2", and then click "OK".

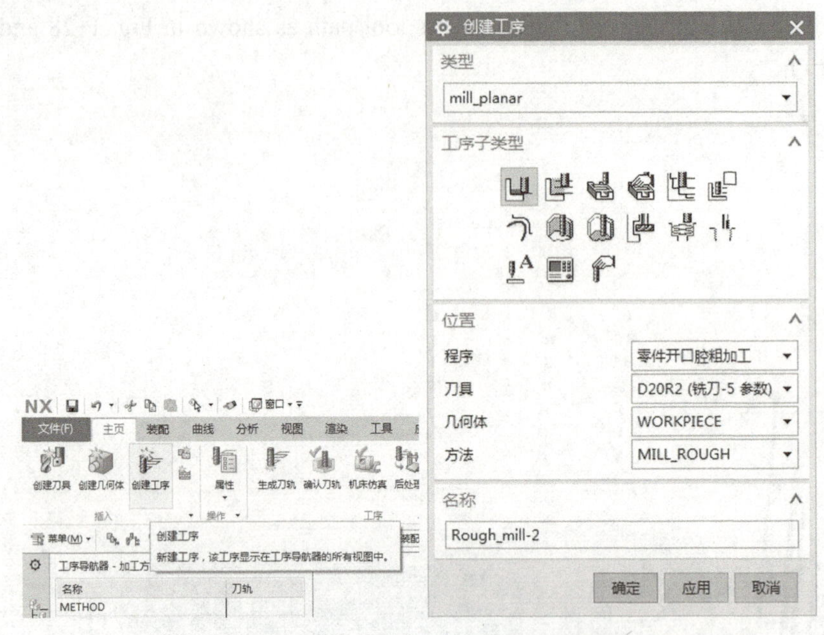

图 1-29 创建工序

（2）单击【指定切削区底面】中的【选择或编辑切削区域几何体】，选择工件台阶面，如图 1-30 所示，单击【确定】。

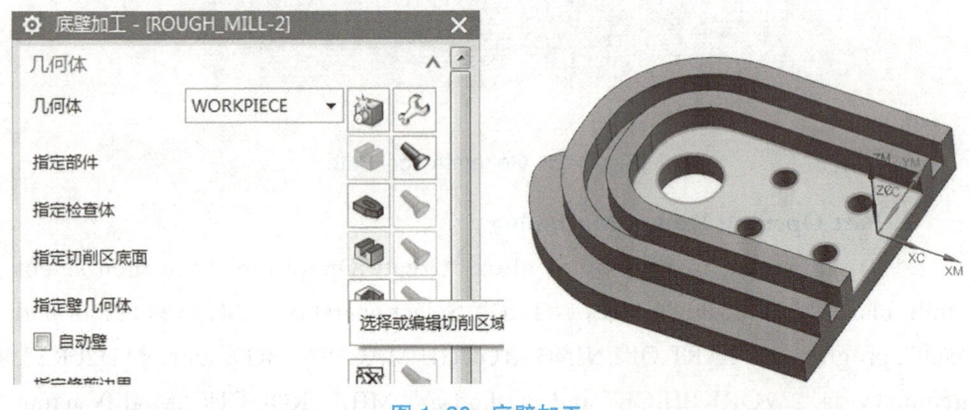

图 1-30 底壁加工

（3）勾选【自动壁】。【切削模式】选择【跟随周边】，【步距】为【刀具平直百分比】，【平面直径百分比】为【60】，【底面毛坯厚度】为【10】，【每刀切削深度】为【1.5】，如图 1-31 所示。

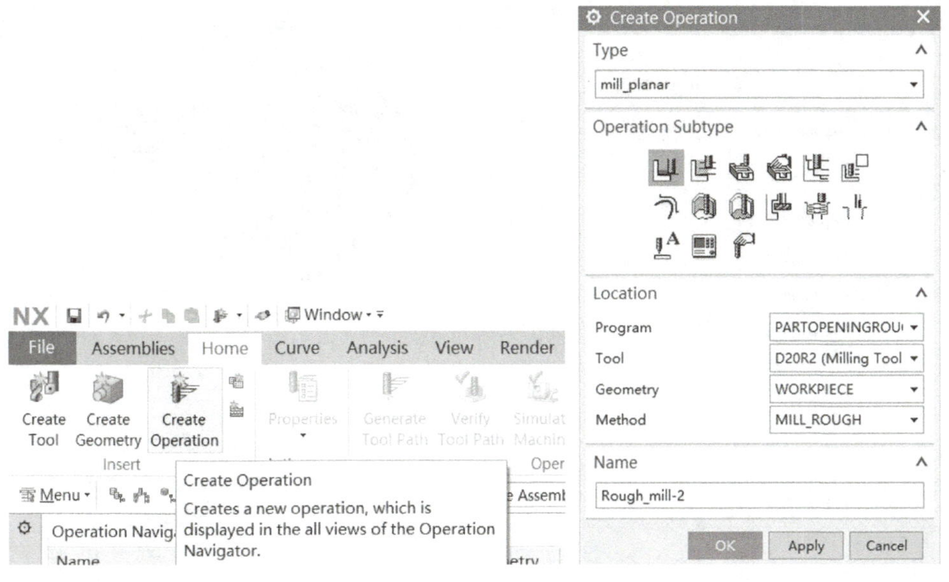

Fig. 1-29　Create Operation

（2）As shown in Fig. 1-30, click the button “Specify Floor” and select surface of the step. Click “OK”.

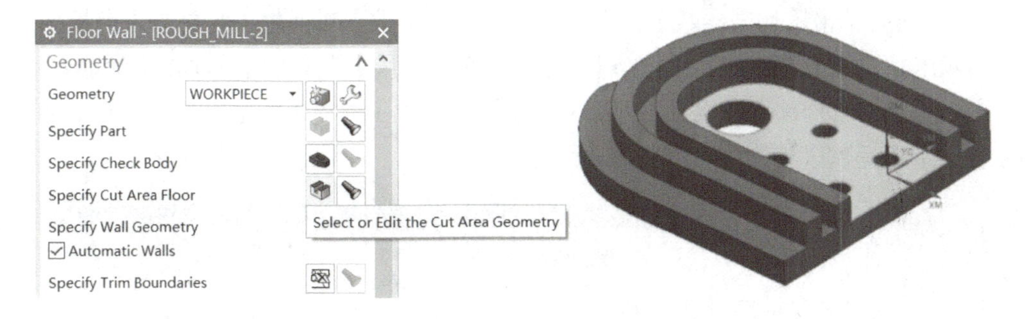

Fig. 1-30　Specify Floor

（3）As shown in Fig. 1-31, tick the option “Automatic Walls”. Modify “Cut Pattern” as “Follow Periphery”, modify “Percent of Flat Diameter” as 60. “Floor Blank Thickness” as 10 and “Depth Per Cut” as 1.5.

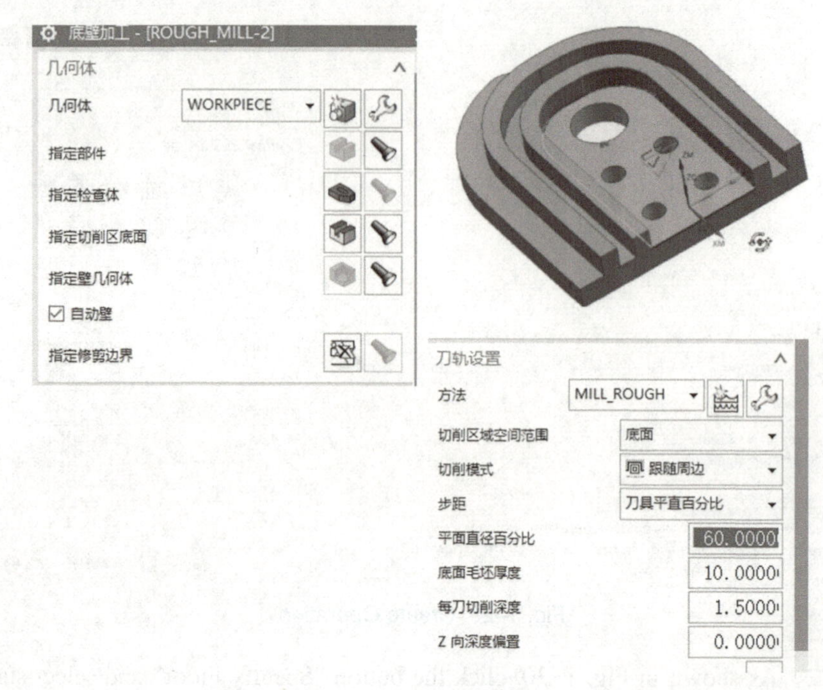

图 1–31 选择切削模式

（4）单击【切削参数】，在打开的【切削参数】对话框中，单击【余量】选项卡，设置【壁余量】为【0.5】，【最终底面余量】为【0.2】，单击【确定】。单击【策略】选项卡，设置【刀路方向】为【向内】，单击【确定】。【非切削移动】选择默认参数，如图 1–32 所示。

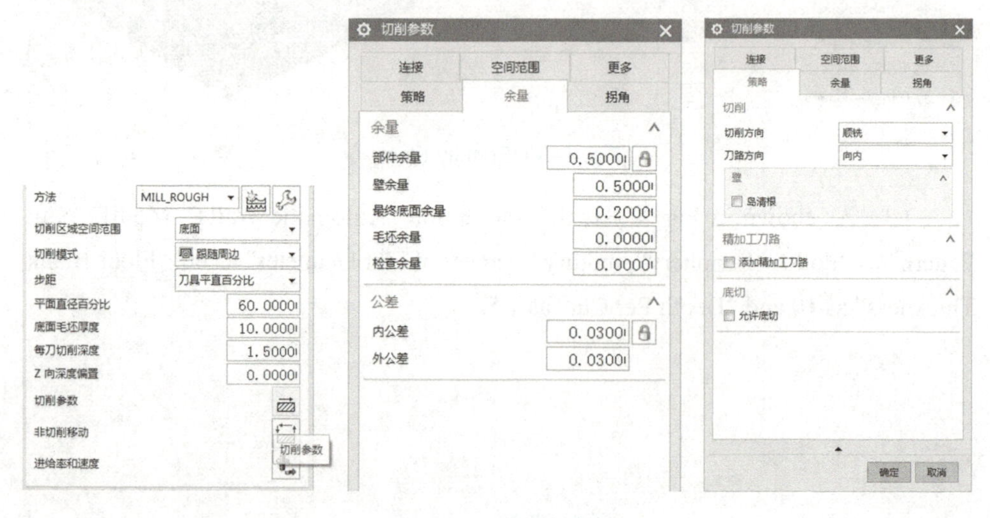

图 1–32 设置切削参数

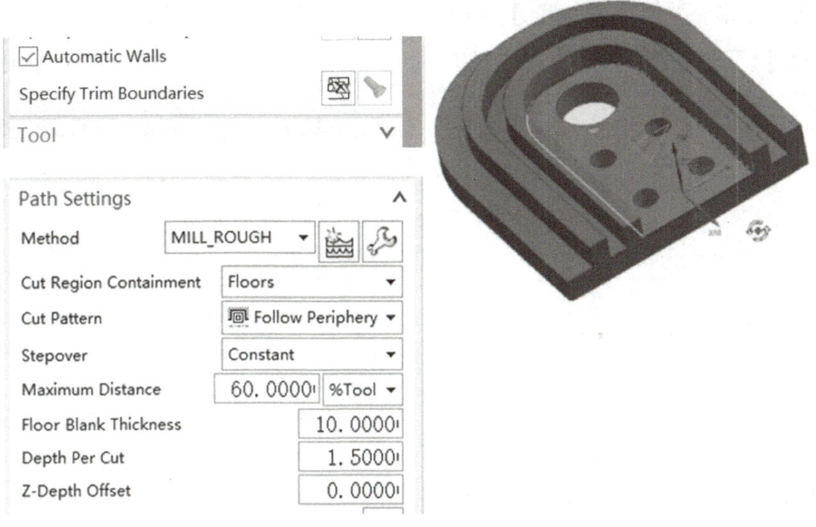

Fig. 1–31 Modify Cut Pattern

(4) As shown in Fig. 1-32, click "Cutting Parameters" button, and in "Stock" tab, modify "Wall Stock" as 0.5 and "Final Floor Stock" as 0.2. In "Strategy" tab, select "Inward" as "Pattern Direction". Click "OK".

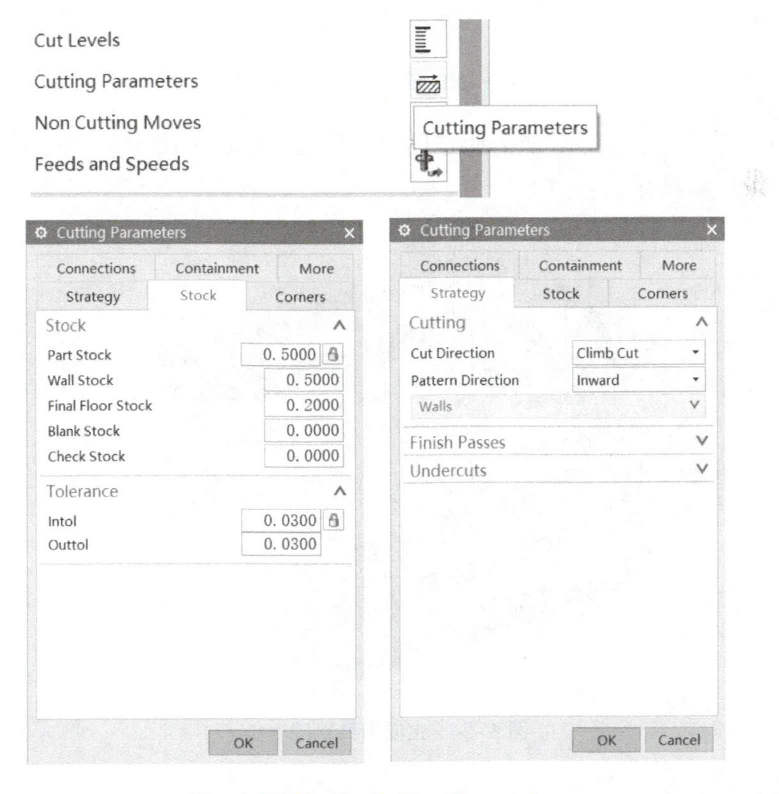

Fig. 1–32 Modify Cutting Parameters

（5）单击【进给率和速度】，打开【进给率和速度】对话框，设置【主轴速度】为【1800】，进给率【切削】为【1200 mmpm】，如图 1-33 所示，单击【确定】。

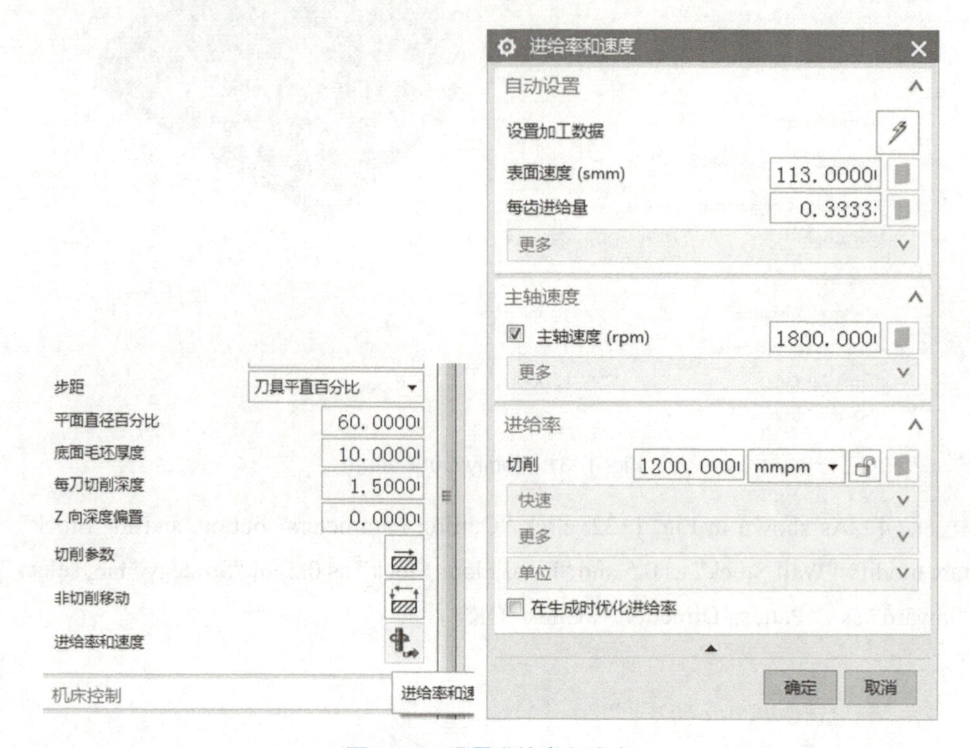

图 1-33 设置进给率和速度

（6）单击【生成】，查看生成的刀具轨迹，单击【确定】，完成零件的粗加工如图 1-34 所示。

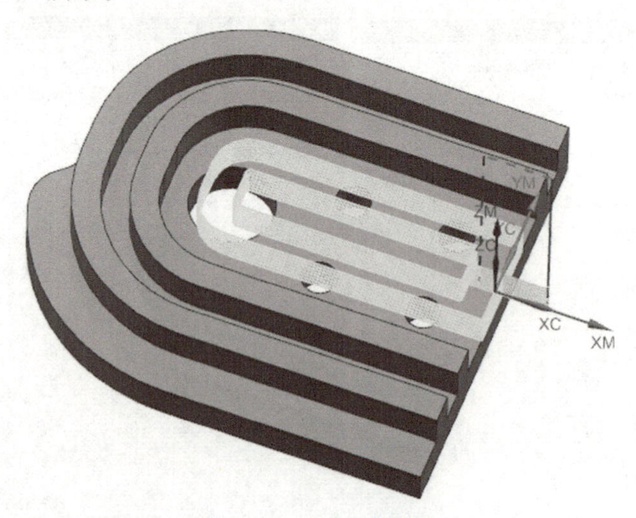

图 1-34 生成刀具轨迹

（5）Click “Feeds and Speeds”, modify “Spindle Speed” as 1,800 and “Feed Rates” as 1,200 mmpm as shown in Fig. 1-33. Click “OK”.

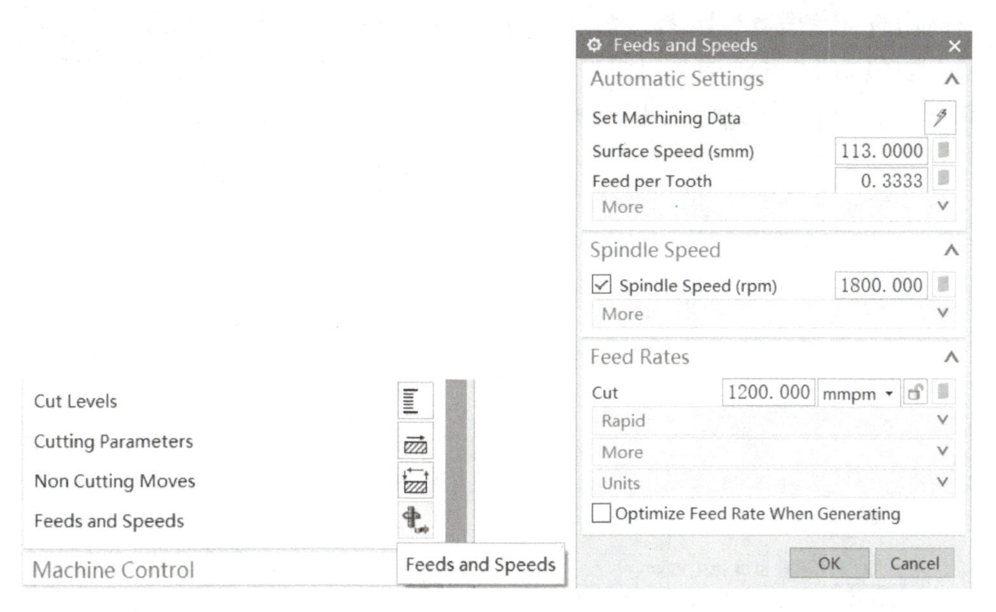

Fig. 1-33 Modify Feeds and Speeds

（6）Click “Generate” to view the tool path as shown in Fig. 1-34, and then click “OK”.

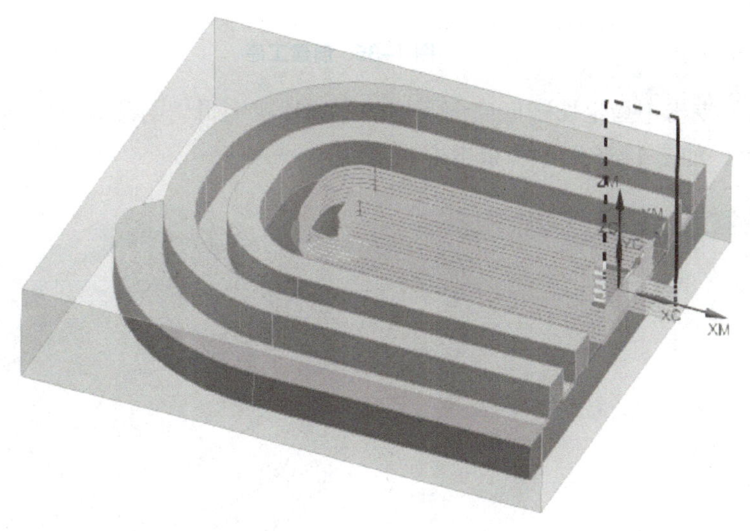

Fig. 1-34 Generate Tool Path

3. 零件导轨槽粗加工

（1）单击【创建工序】，打开【创建工序】对话框，【类型】选择【mill_planar】,【工序子类型】选择【底壁加工】,【程序】选择【零件导轨槽粗加工】,【刀具】选择【D6 R0-ROUGH】，【几何体】选择【WORKPIECE】，【方法】选择【MILL_ROUGH】，修改【名称】为【Rough_mill-3】，如图 1-35 所示，单击【确定】。

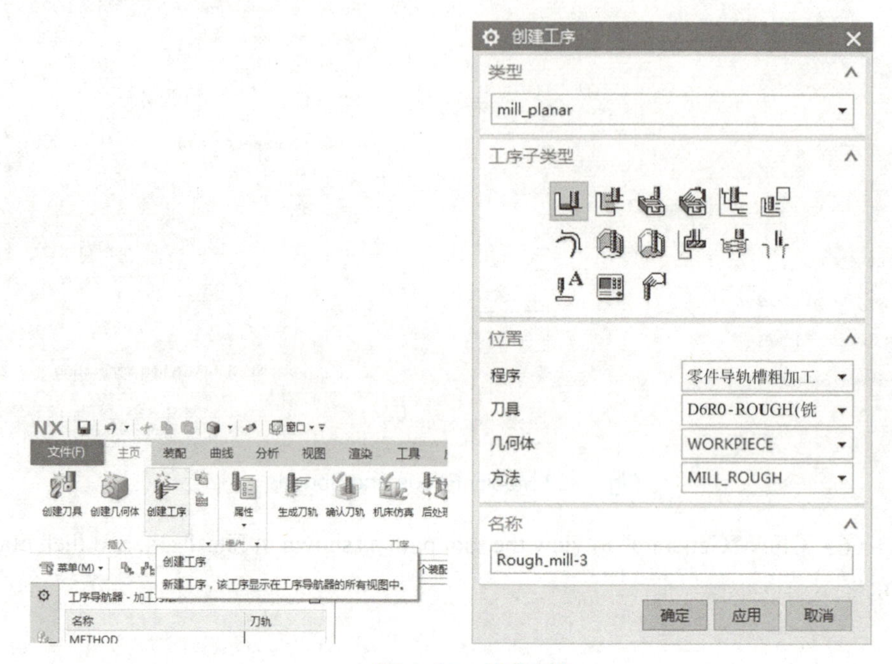

图 1-35 创建工序

（2）单击【指定切削区底面】中的【选择或编辑切削区域几何体】，选择槽的底面，如图 1-36 所示，单击【确定】。

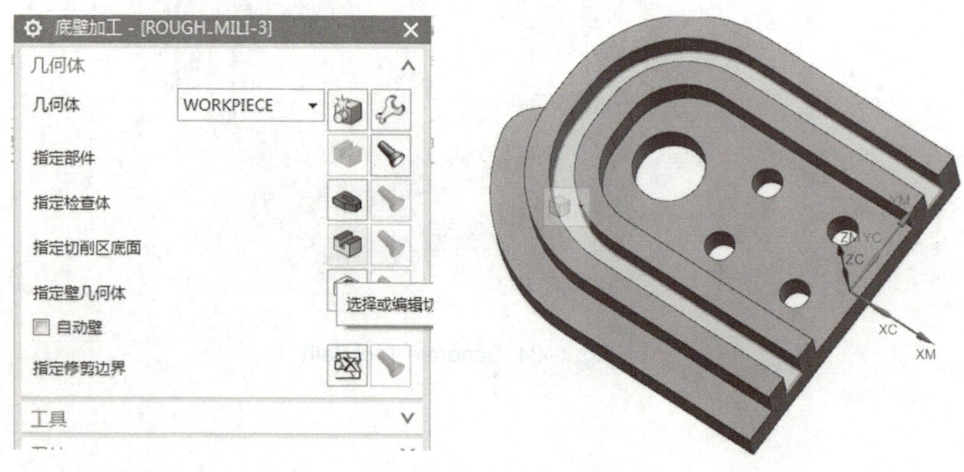

图 1-36 指定切削区底面

3. Rough Machining of Rail Groove

(1) As shown in Fig. 1-35, click “Create Operation”, and then select “mill_planar” in pull-down box of “Type”. Select operation subtype as “Floor and Wall”, program as “ROUGH MACHINING OF RAIL GROOVE”, tool as “D6R0-ROUGH”, geometry as “WORKPIECE” and method as “MILL_ROUGH”. Modify name as “Rough_mill-3”, and then click OK”.

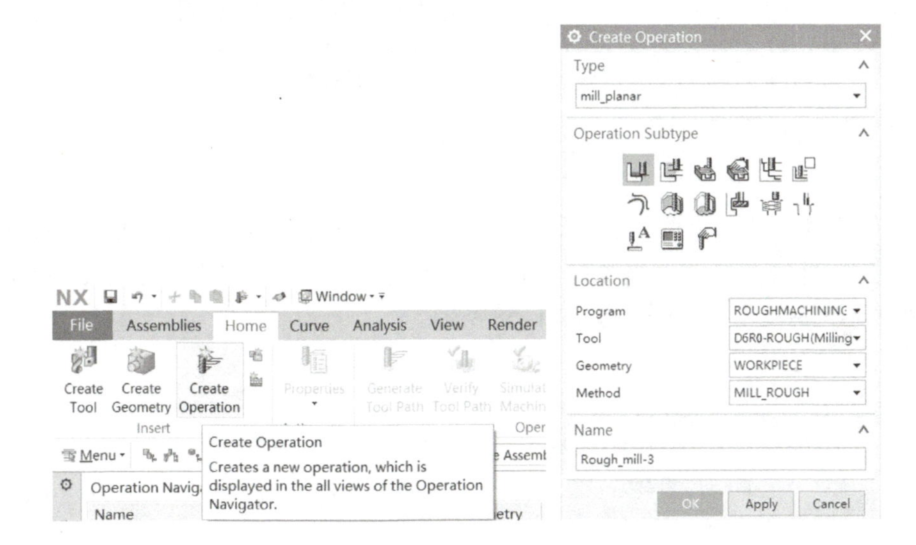

Fig. 1-35 Create Operation

(2) As shown in Fig. 1-36, click the button “Specify Cut Area Floor” and select the bottom of the groove. Click “OK”.

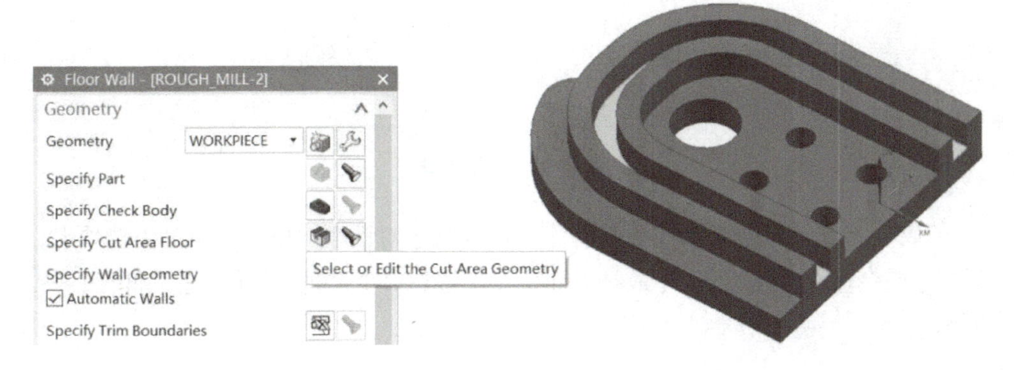

Fig. 1-36 Specify Cut Area Floor

（3）勾选【指定壁几何体】下的【自动壁】，如图 1–37 所示。

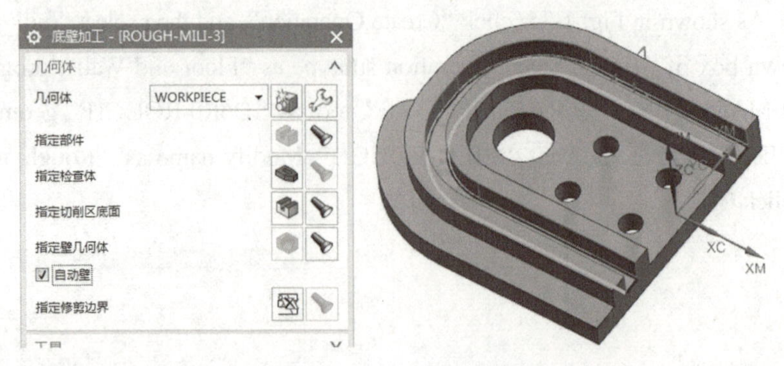

图 1–37 指定壁几何体

（4）单击【刀轨设置】，设置【切削模式】为【轮廓】，【步距】为【刀具平直百分比】，【平面直径百分比】为【60】，【底面毛坯厚度】为【10】，【每刀切削深度】为【1】，如图 1–38 所示。

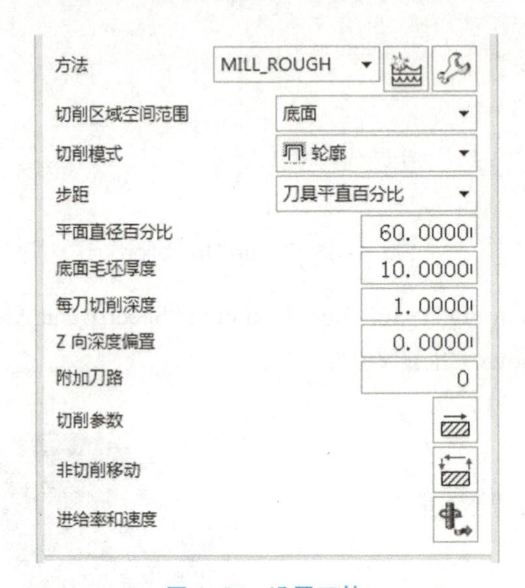

图 1–38 设置刀轨

（5）单击【切削参数】，打开【切削参数】对话框，单击【余量】选项卡，设置【壁余量】为【0.3】，【最终底面余量】为【0.1】，单击【确定】，如图 1–39 所示。

（3）Tick the option “Automatic Walls” as shown in Fig. 1-37.

Fig. 1-37 Tick Automatic Walls

（4）Click “Path Settings”. Modify “Cut Pattern” as “Profile”, select “% Tool Flat” as “Stepover”, modify “Percent of Flat Diameter” as 60, “Floor Blank Thickness” as 10 and “Depth Per Cut” as 1 as shown in Fig. 1-38.

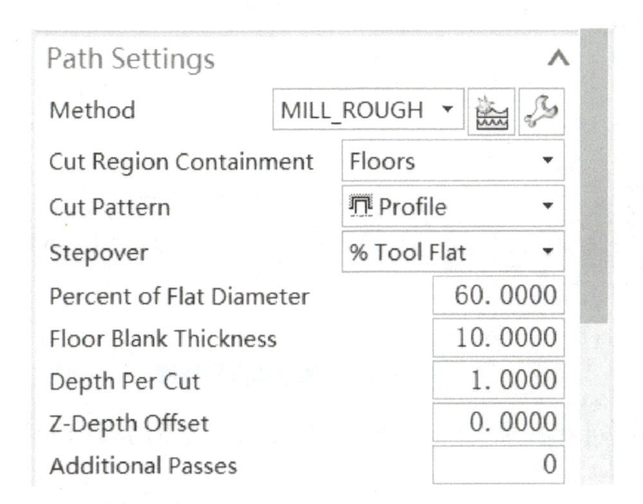

Fig. 1-38 Modify Path Settings

（5）As shown in Fig. 1-39, click “Cutting Parameters” button, and in “Stock” tab, modify “Wall Stock” as 0.3 and “Final Floor Stock” as 0.1. Click “OK”.

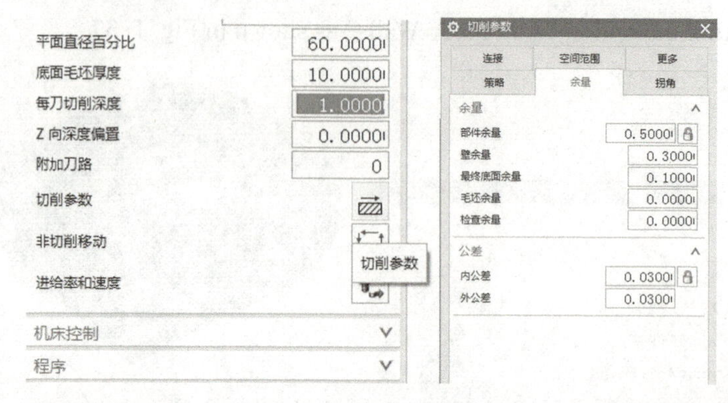

图 1–39　设置切削参数

（6）单击【进给率和速度】，打开【进给率和速度】对话框，设置【主轴速度】为【2800】，进给率【切削】为【1000 mmpm】，单击【确定】，如图 1–40 所示。

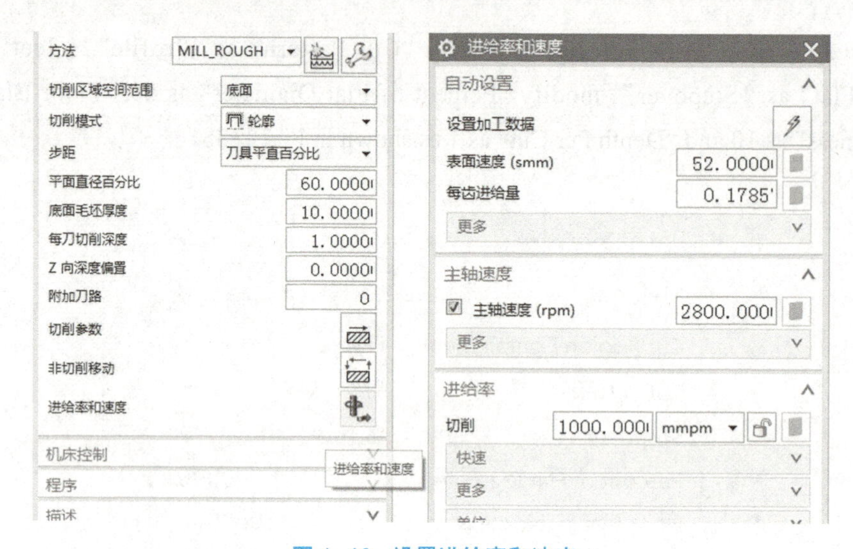

图 1–40　设置进给率和速度

（7）单击【生成】，查看生成的刀具轨迹，如图 1–41 所示，单击【确定】。

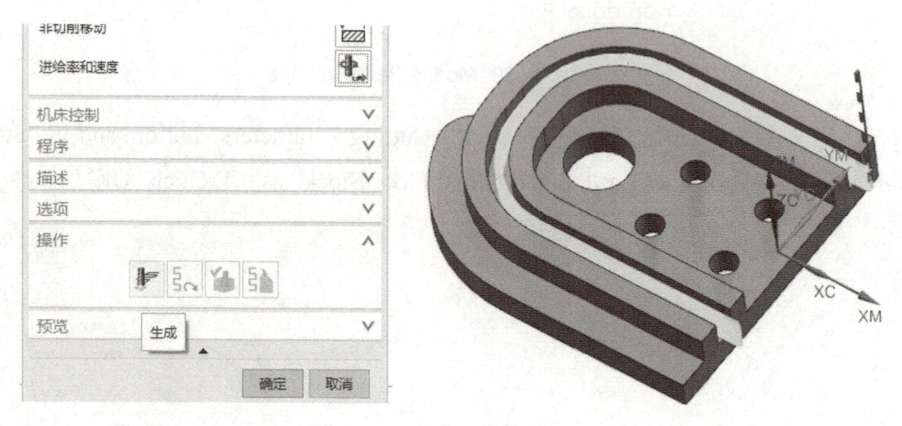

图 1–41　生成刀具轨迹

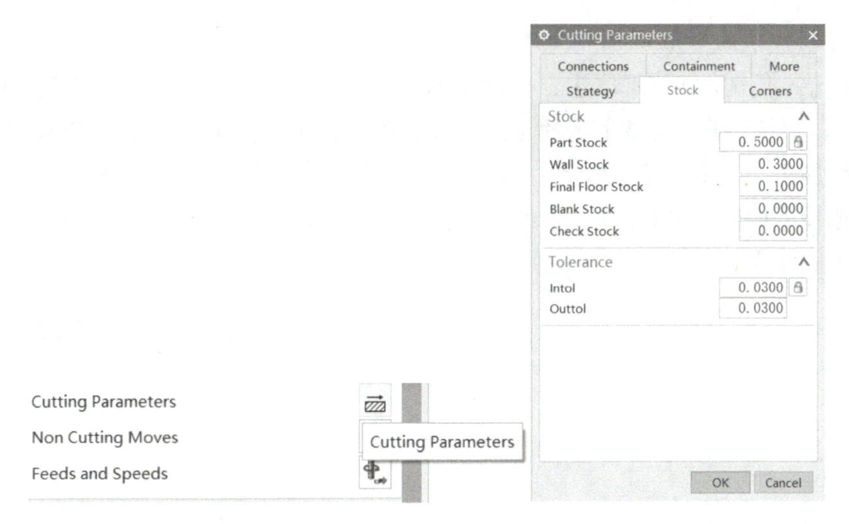

Fig. 1-39 Modify Cutting Parameters

(6) Click "Feeds and Speeds", modify "Spindle Speed" as 2,800 and "Feed Rates" as 1,000 mmpm as shown in Fig. 1-40. Click "OK".

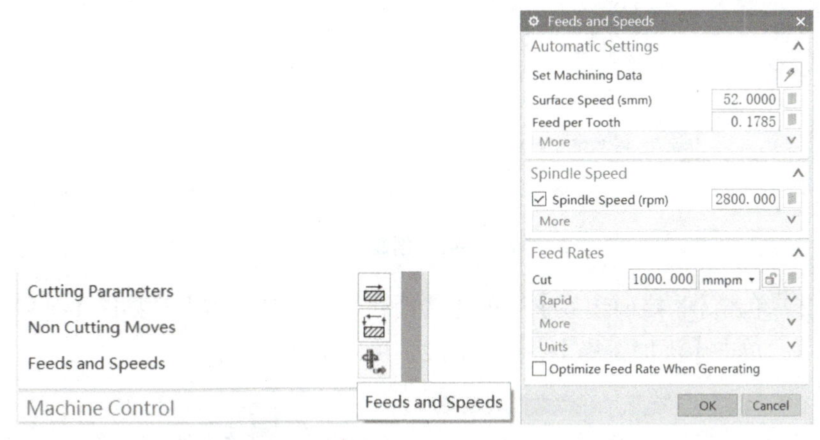

Fig. 1-40 Modify Feeds and Speeds

(7) Click "Generate" to view the tool path as shown in Fig. 1-41, and then click "OK".

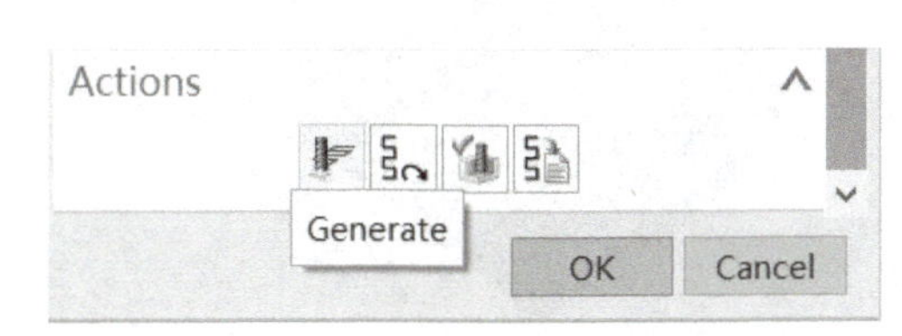

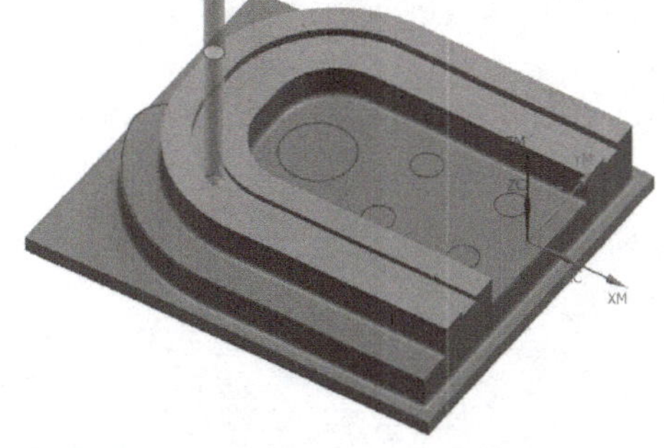

Fig. 1-41 Generate Tool Path

4. 零件顶面和腔底面精加工

（1）单击【创建工序】，打开【创建工序】对话框，【类型】选择【mill_planar】，【工序子类型】选择【边界面铣削】，【程序】选择【零件顶面和腔底面精加工】，【刀具】选择【D16】，【几何体】选择【WORKPIECE】，【方法】选择【MILL_FINISH】，修改【名称】为【Finish_mill-1】，如图 1-42 所示，单击【确定】。

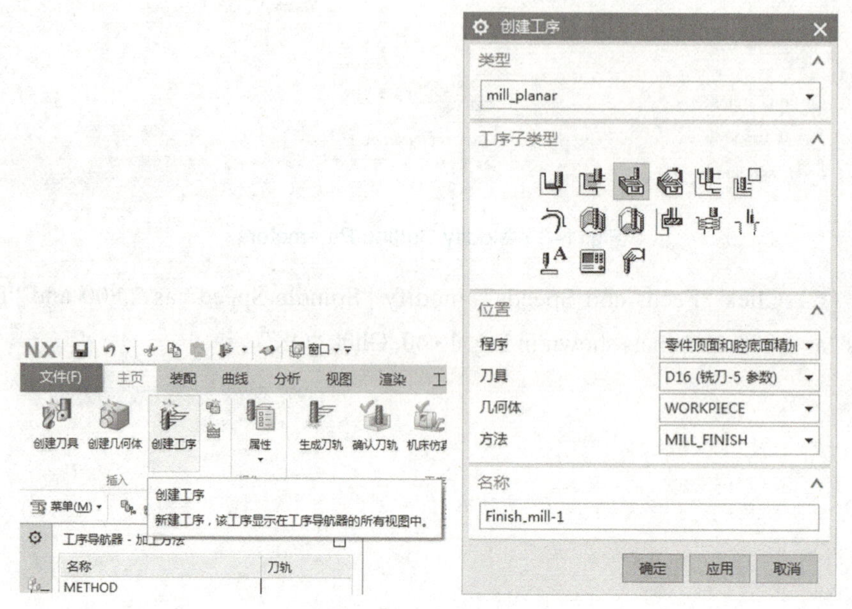

图 1-42 创建工序

（2）单击【指定面边界】中的【选择或编辑面几何体】，如图 1-43 所示，选择面 1，单击【添加新集】，选择面 2，单击【确定】。

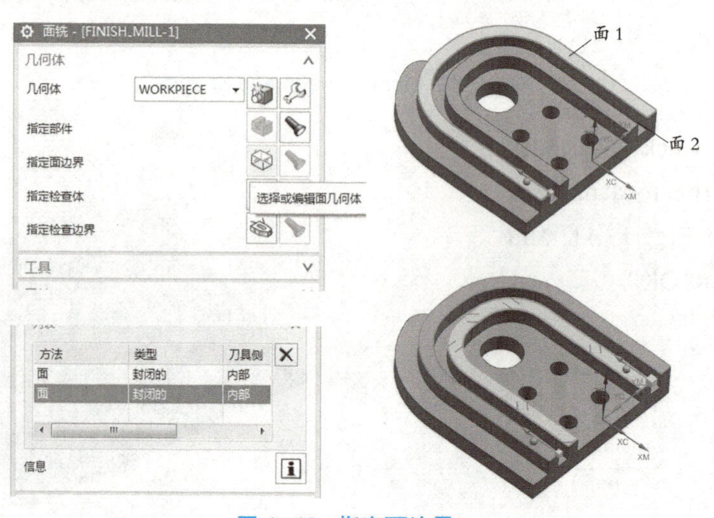

图 1-43 指定面边界

4. Finish Machining of Part Top Surface and Cavity Bottom Surface

(1) As shown in Fig. 1-42, click "Create Operation", and then select "mill_planar" in pull-down box of "Type". Select operation subtype as "Face Milling with Boundaries", program as "FINISH MACHINING OF PART TOP SURFACE AND CAVITY BOTTOM SURFACE ", tool as "D16", geometry as " WORKPIECE" and method as "MILL_FINISH". Modify name as "Finish_mill-1", and then click "OK".

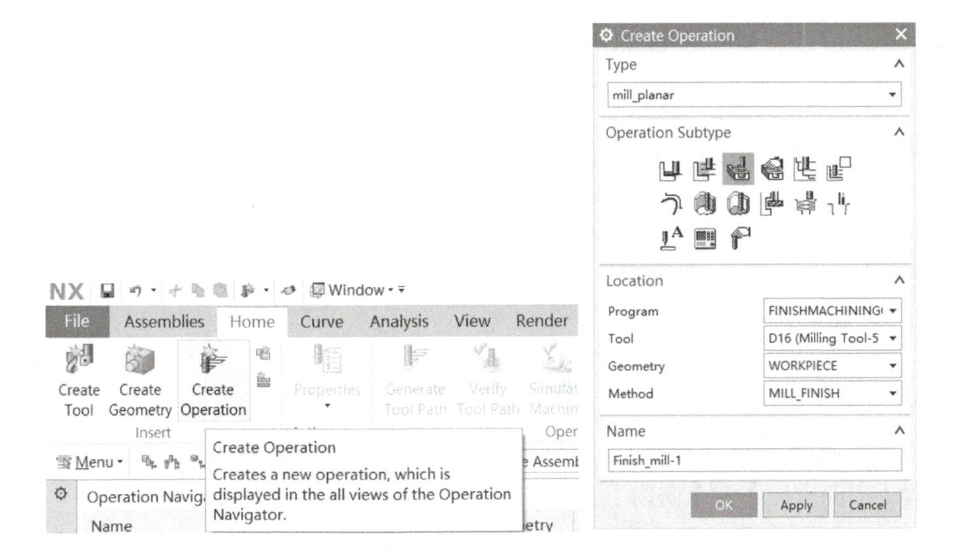

Fig. 1-42 Create Operation

(2) As shown in Fig. 1-43, click edit button "Specify Face Boundaries", select the surface as shown below, click "Add New Set" and then select the other surface. Click "OK".

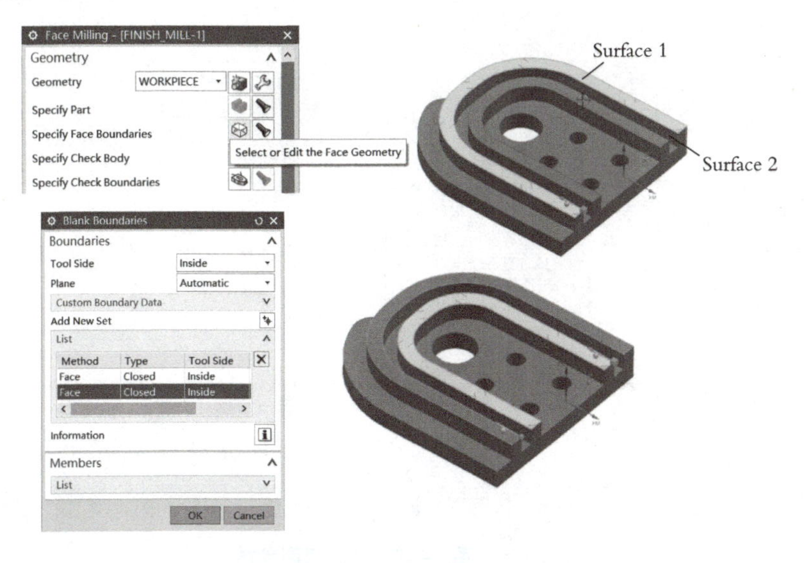

Fig. 1-43 Specify Face Boundaries

（3）单击【刀轨设置】，打开【刀轨设置】对话框，设置【切削模式】为【跟随周边】，【步距】为【刀具平直百分比】，【平面直径百分比】为【60】，【毛坯距离】为【1】，【每刀切削深度】为【0.5】，【最终底面余量】为【0.1】，如图1-44所示。

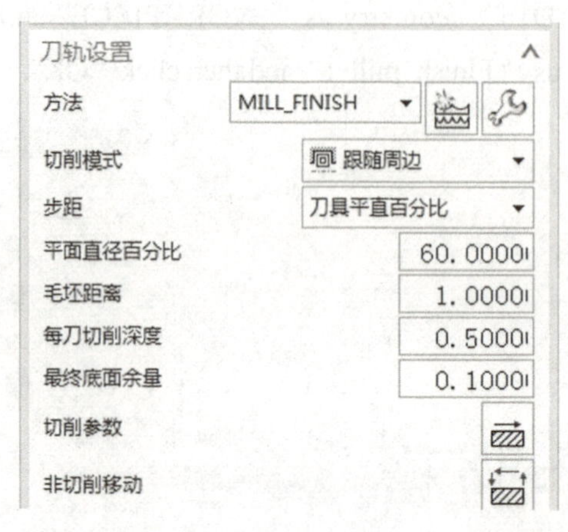

图1-44 设置刀轨

（4）单击【切削参数】，打开【切削参数】对话框，单击【策略】选项卡，设置【刀路方向】为【向内】，单击【确定】，如图1-45所示。

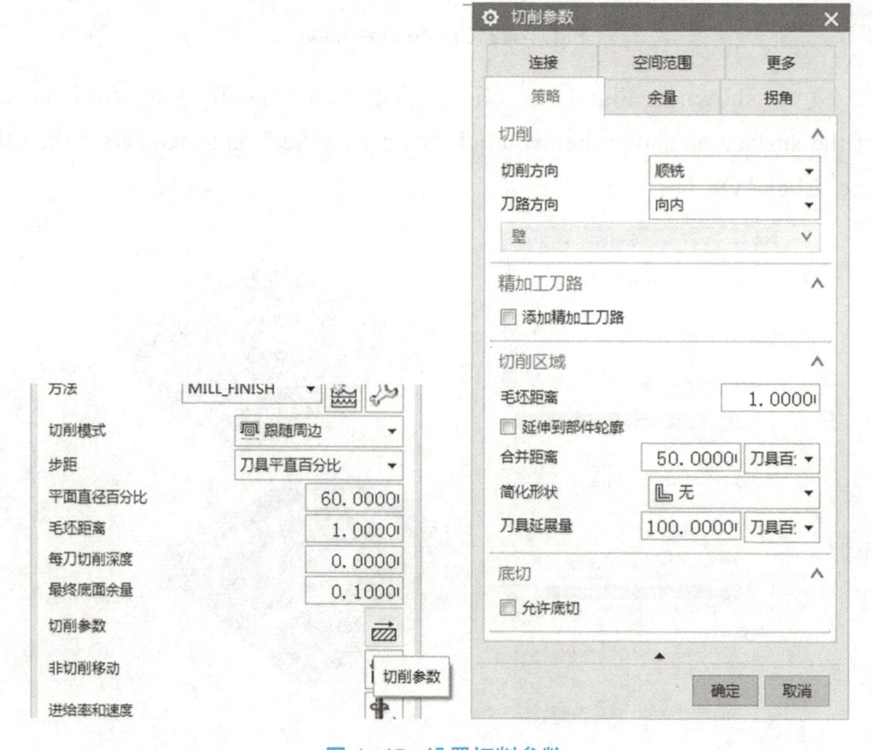

图1-45 设置切削参数

(3) As shown in Fig. 1-44, click "Path Settings". Modify "Cut Pattern" as "Follow Periphery", select "% Tool Flat" as "Stepover", and modify "Percent of Flat Diameter" as 60, "Blank Distance" as 1 and "Depth Per Cut" as 0.5 and "Final Floor Stock" as 0.1.

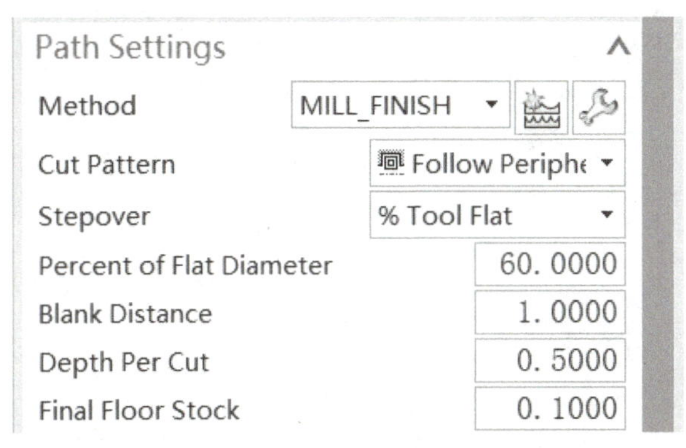

Fig. 1-44 Modify Path Settings

(4) As shown in Fig. 1-45, click "Cutting Parameters" button, in "Strategy" tab, select "Inward" as "Pattern Direction". Click "OK".

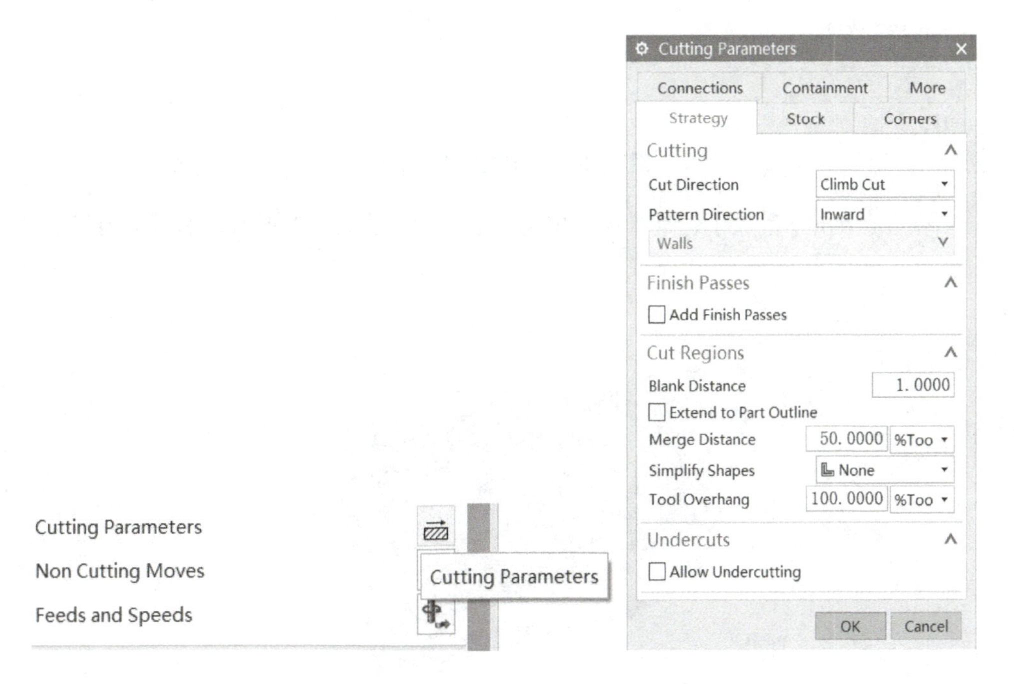

Fig. 1-45 Specify Cutting Parameters

（5）单击【进给率和速度】，打开【进给率和速度】对话框，设置【主轴速度】为【2200】，进给率【切削】为【1000 mmpm】，单击【确定】，如图 1–46 所示。

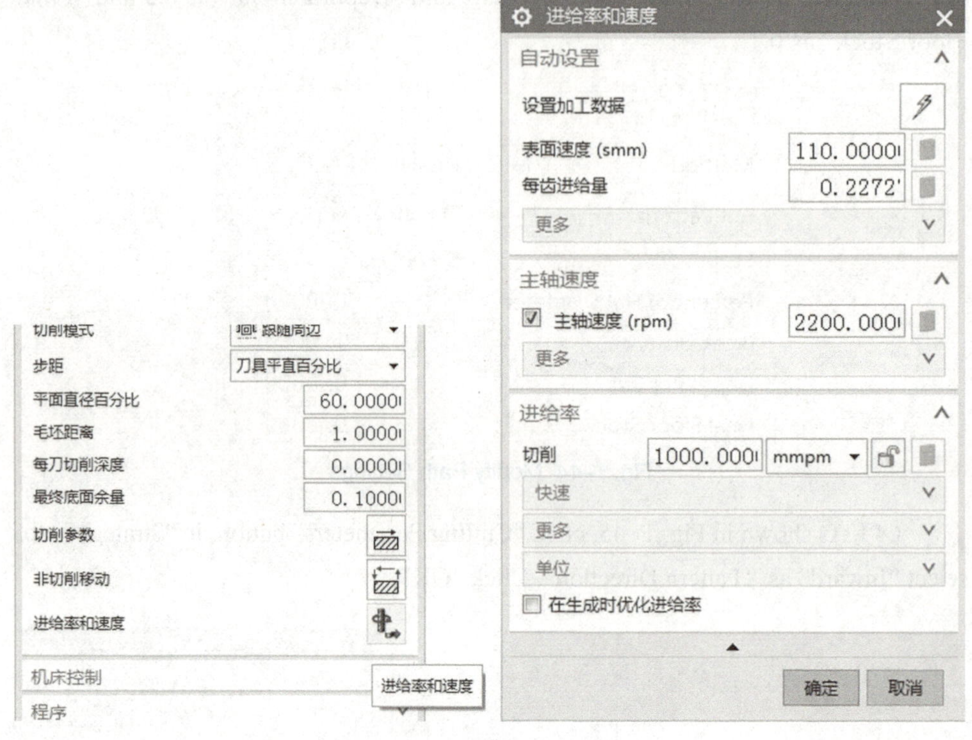

图 1–46　设置进给率和速度

（6）单击【生成】，查看生成的刀具轨迹，如图 1–47 所示，单击【确定】。

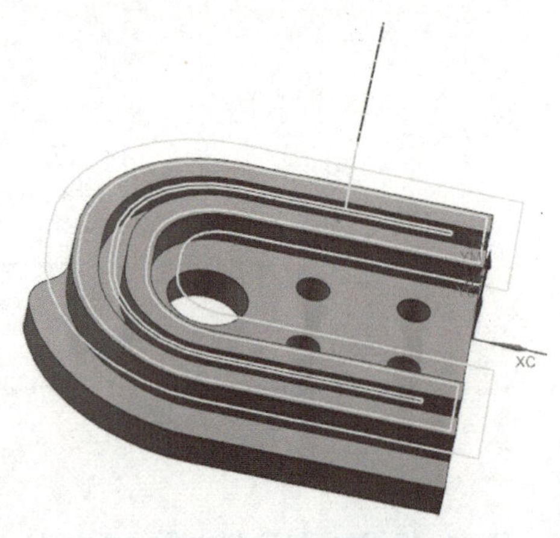

图 1–47　生成刀具轨迹

（7）复制工序名【FINISH_MILL-1】，粘贴，更改工序名为【FINISH_MILL-2】。

(5) Click "Feeds and Speeds", modify "Spindle Speed" as 2,200 and "Feed Rates" as 1,000 mmpm as shown in Fig. 1-46. Click "OK".

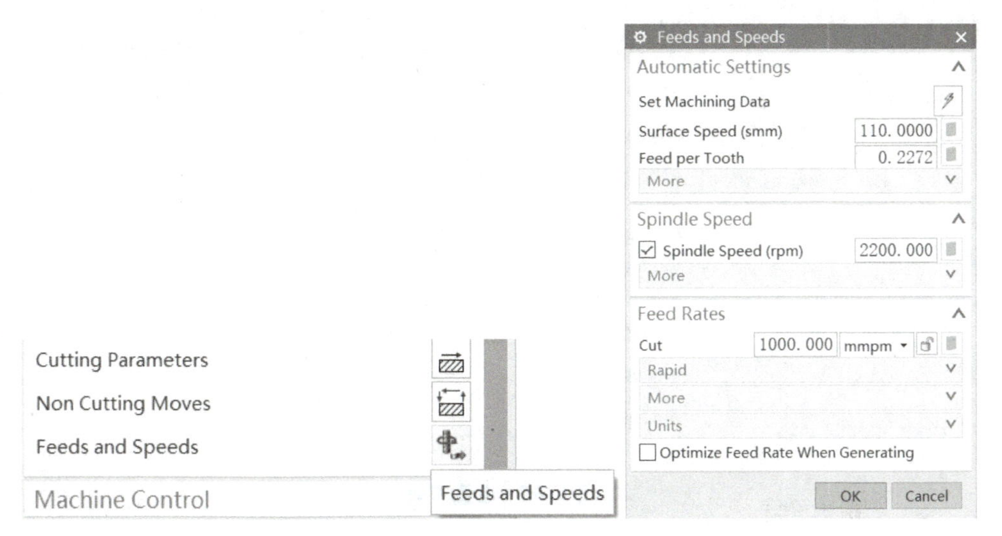

Fig. 1-46 Modify Feeds and Speeds

(6) Click "Generate" to view the tool path as shown in Fig. 1-47, and then click "OK".

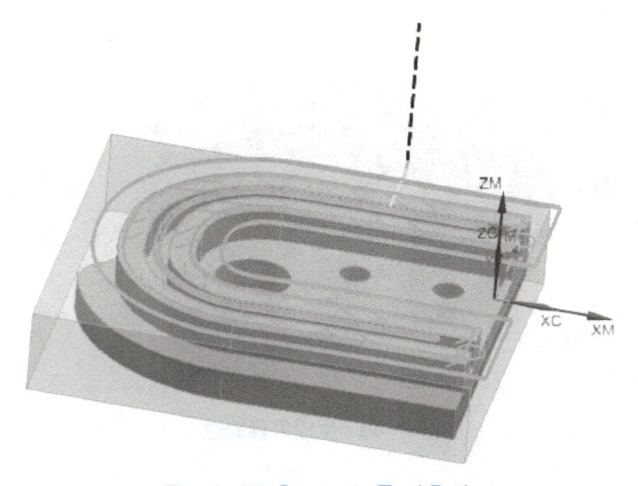

Fig. 1-47 Generate Tool Path

(7) Copy program " FINISH_MILL-1" program and paste it. Modify the pasted file name as " FINISH_MILL-2".

（8）双击工序名，设置【毛坯距离】为【0.1】，【每刀切削深度】为【0.1】，【最终底面余量】为 0，如图 1-48 所示。

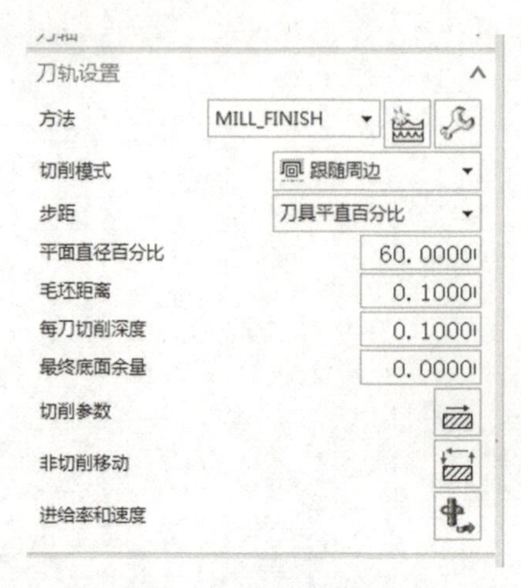

图 1-48 设置刀轨

（9）单击【生成】，查看生成的刀具轨迹，如图 1-49 所示，单击【确定】。

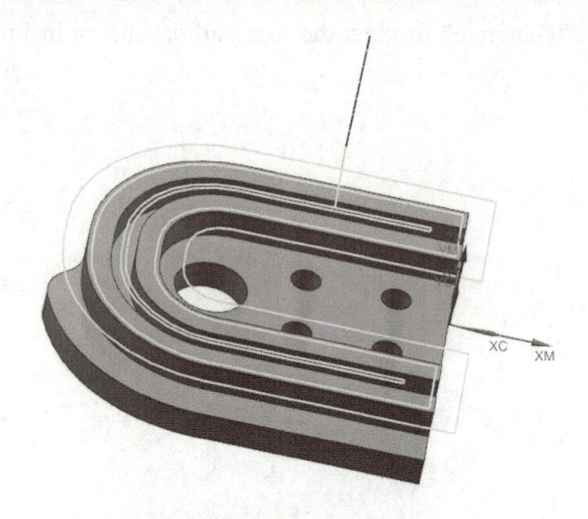

图 1-49 生成刀具轨迹

（10）单击【创建工序】，打开【创建工序】对话框，【类型】选择【mill_planar】，【工序子类型】选择【底壁加工】，【刀具】选择【D16R0】，【几何体】选择【WORKPIECE】，【方法】选择【MILL_FINISH】，修改【名称】为【Finish_mill-3】，如图 1-50 所示，单击【确定】。

(8) As shown in Fig. 1-48, double click the program and modify "Blank Distance" as 0.1, "Depth Per Cut" as 0.1 and "Final Floor Stock" as 0. Click "OK".

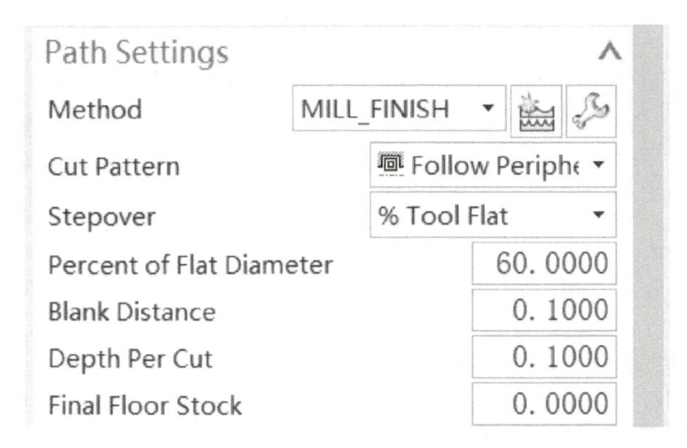

Fig. 1-48 Modify Path Settings

(9) Click "Generate" to view the tool path as shown in Fig. 1-49, and then click "OK".

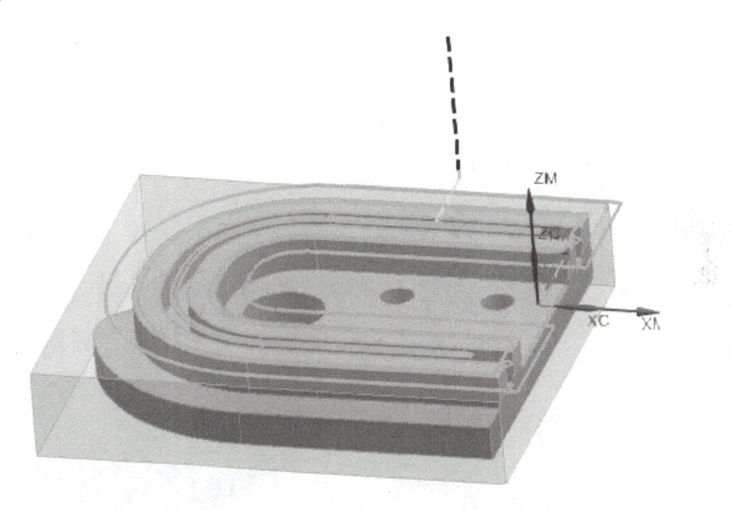

Fig. 1-49 Generate Tool Path

(10) As shown in Fig. 1-50, click "Create Operation", and then select "mill_planar" in pull-down box of "Type". Select operation subtype as "Floor and Wall", tool as "D16R0", geometry as " WORKPIECE" and method as "MILL_FINISH". Modify name as "Finish_mill-3", and then click "OK".

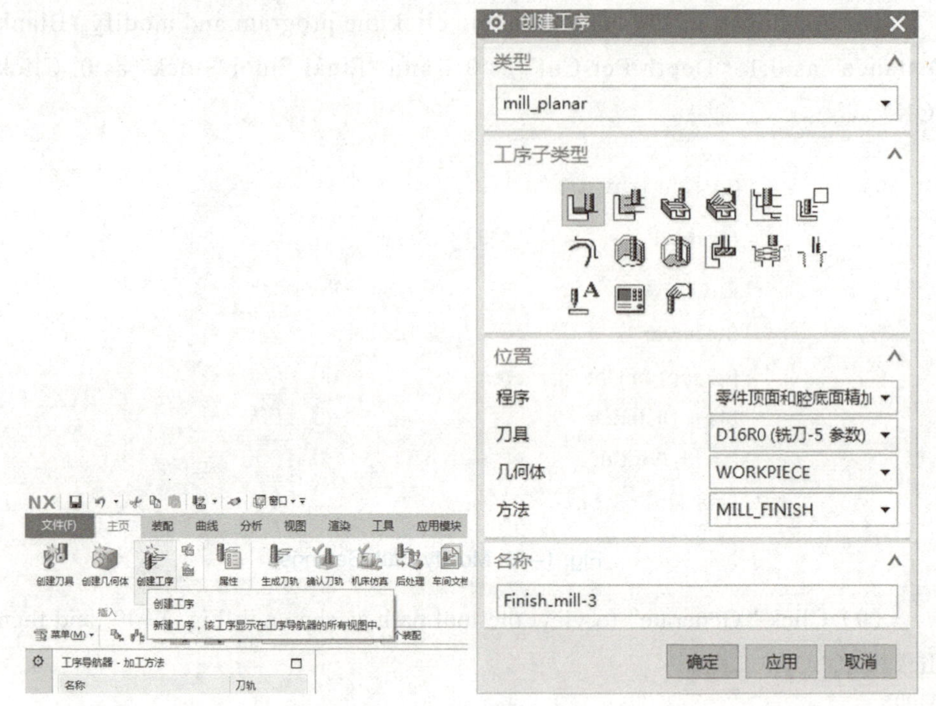

图 1-50 创建工序

（11）单击【指定切削区底面】中的【选择或编辑切削区底面】，选择底面，如图 1-51 所示，单击【确定】。

图 1-51 指定切削区底面

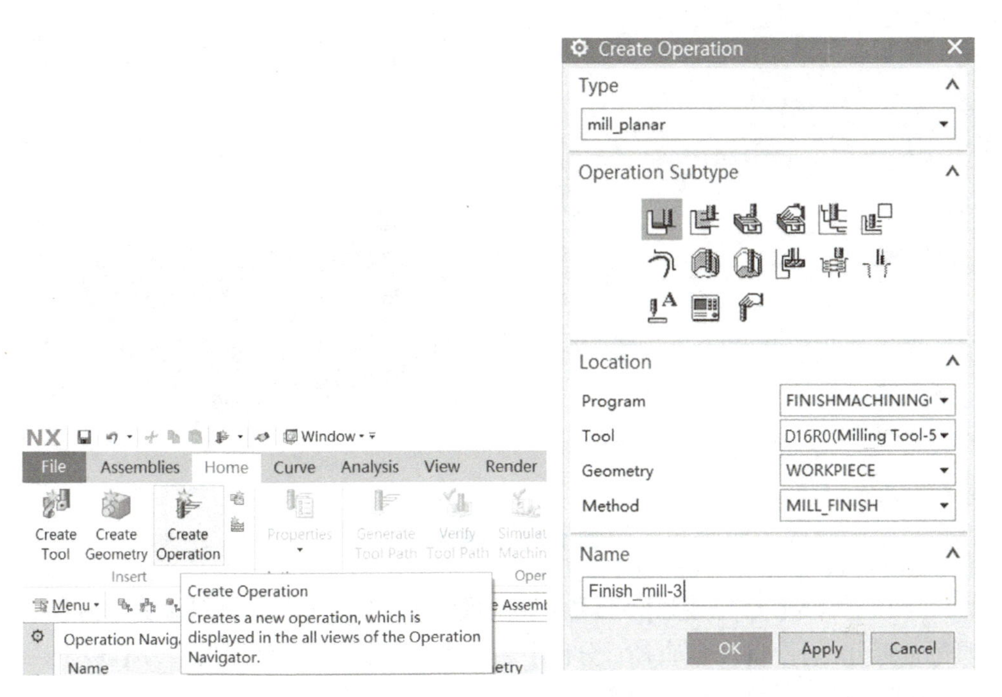

Fig. 1-50 Create Operation

(11) Click "Specify Cut Area Floor" and select the bottom as shown in Fig. 1-51. Click "OK".

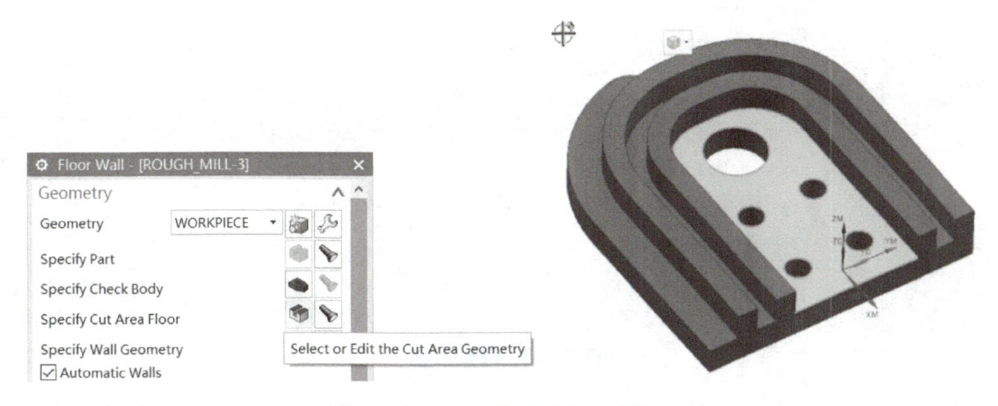

Fig. 1-51 Specify Cut Area Floor

（12）勾选【指定壁几何体】下的【自动壁】，如图 1–52 所示。

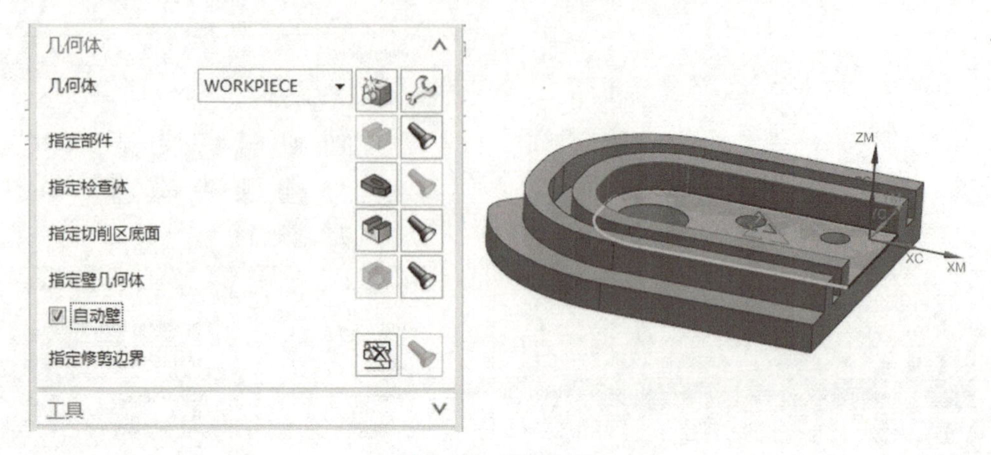

图 1–52 底壁加工

（13）单击【刀轨设置】，设置【切削模式】为【跟随周边】，【步距】为【刀具平直百分比】，【平面直径百分比】为【60】，【底面毛坯厚度】为【0.2】，【每刀切削深度】为【0.1】，如图 1–53 所示，单击【确定】。

图 1–53 设置刀轨

（14）单击【切削参数】，打开【切削参数】对话框，单击【策略】选项卡，设置【刀路方向】为【向内】，单击【余量】选项卡，设置【壁余量】为【1】，如图 1–54 所示，单击【确定】。

(12) As shown in Fig. 1-52, tick the option "Automatic Walls".

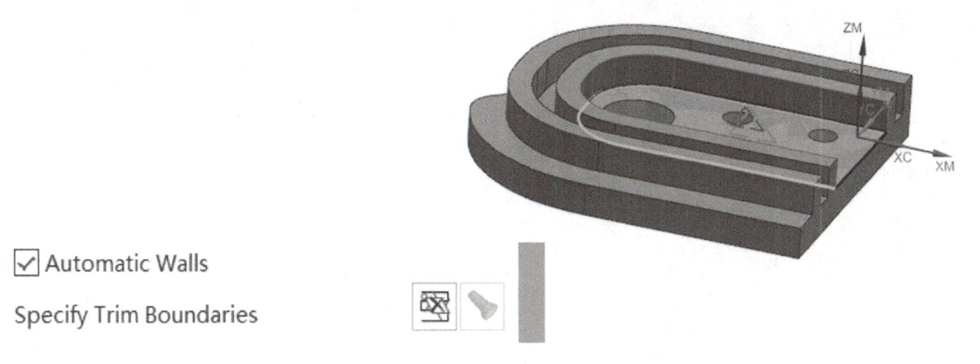

Fig. 1-52　Tick Automatic Walls

(13) Click "Path Settings". Modify "Cut Pattern" as "Follow Periphery", select "% Tool Flat" as "Stepover", and modify "Percent of Flat Diameter" as 60, "Floor Blank Thickness" as 0.2 and "Depth Per Cut" as 0.1 as shown in Fig. 1-53.

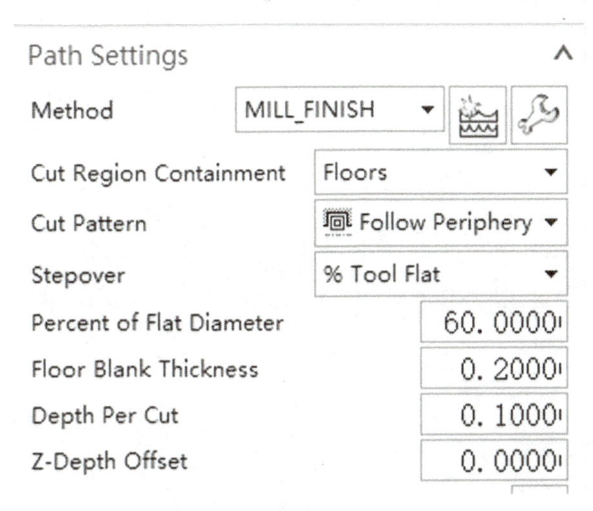

Fig. 1-53　Modify Path Settings

(14) As shown in Fig. 1-54, click "Cutting Parameters" button, in "Strategy" tab, select "Inward" as "Cut Direction". Click "Stock" and modify "Wall Stock" as 1. Click "OK".

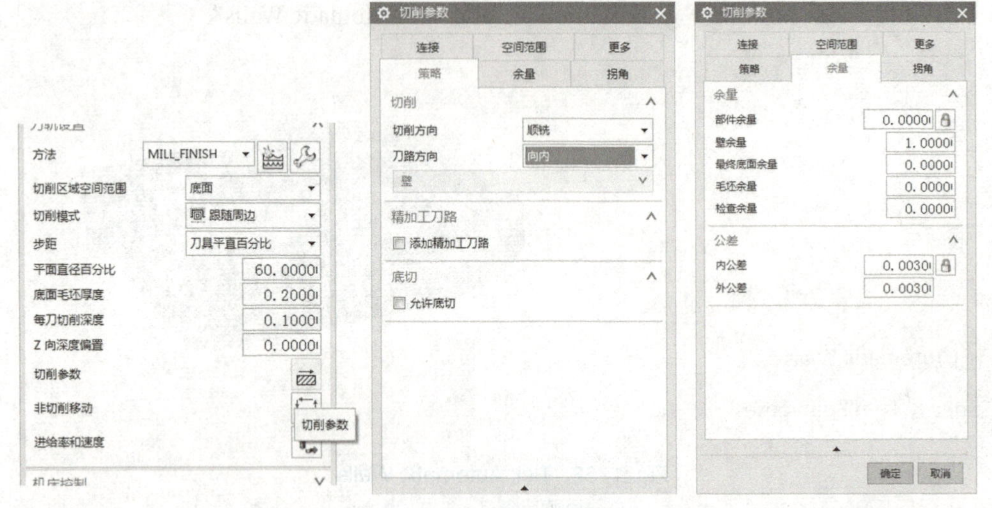

图 1-54 设置切削参数

（15）选择【进给率和速度】，打开【进给率和速度】对话框，设置【主轴速度】为【2200】，进给率【切削】为【1000 mmpm】，单击【确定】，如图 1-55 所示。

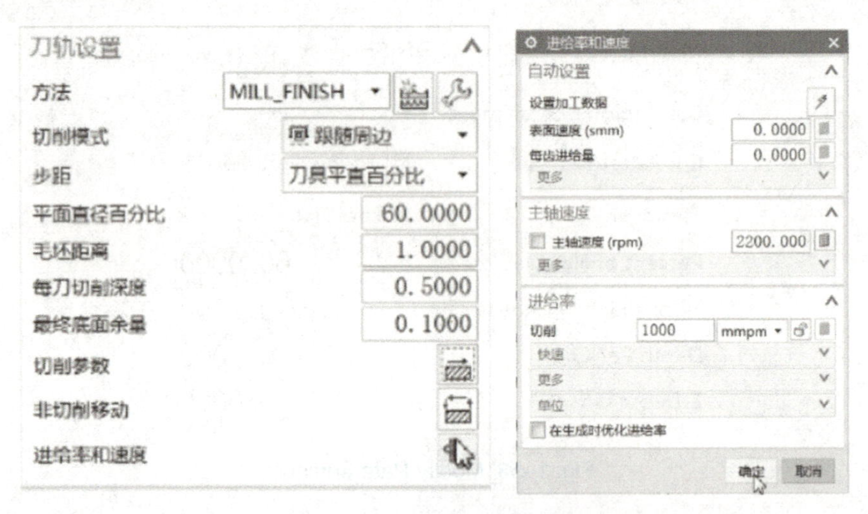

图 1-55 设置进给率和速度

（16）单击【生成】，查看生成的刀具轨迹，如图 1-56 所示，单击【确定】。

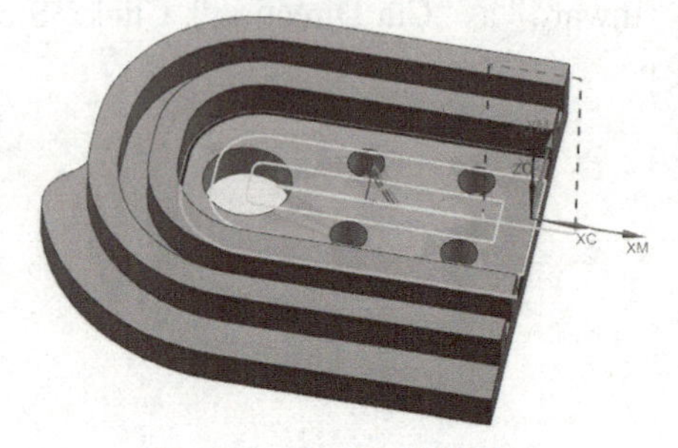

图 1-56 生成刀具轨迹

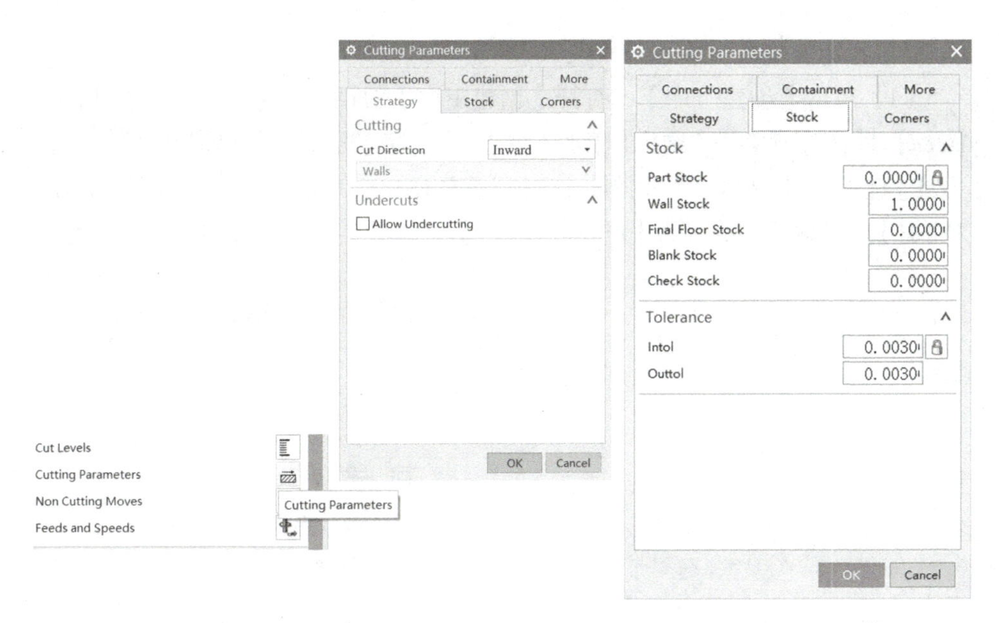

Fig. 1-54 Specify Cutting Parameters

(15) As shown in Fig. 1-55, click "Feeds and Speeds", modify "Spindle Speed" as 2,200 and "Feed Rates" as 1,000 mmpm. Click "OK".

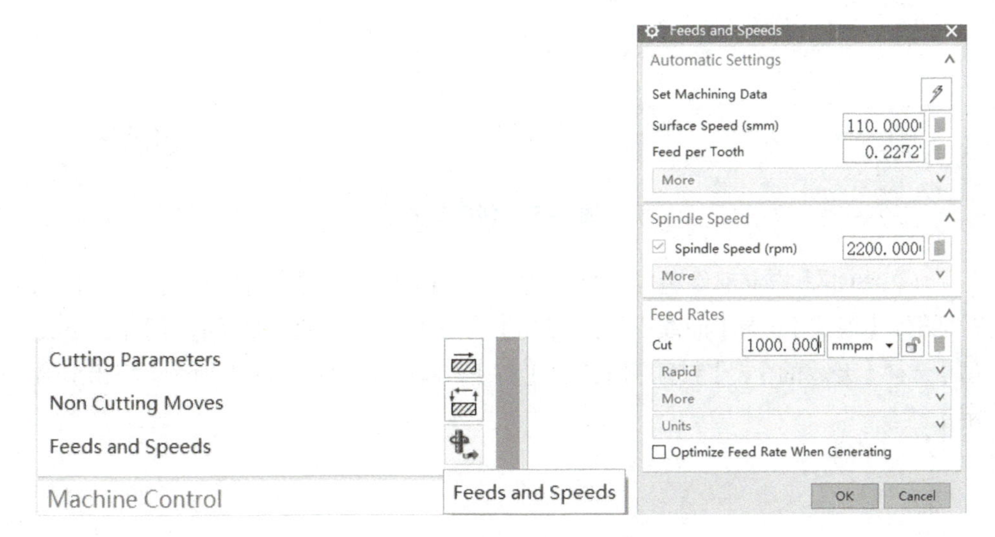

Fig. 1-55 Modify Feeds and Speeds

(16) Click "Generate" to view the tool path as shown in Fig. 1-56, and then click "OK".

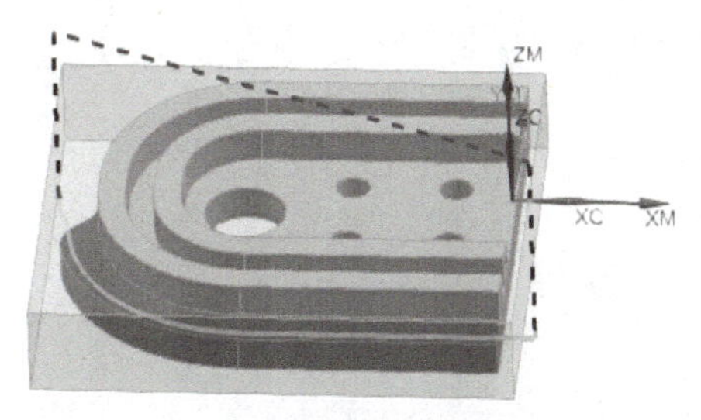

Fig. 1-56 Generate Tool Path

5. 零件外轮廓精加工

（1）单击【创建工序】，打开【创建工序】对话框，【类型】选择【mill_planar】，【工序子类型】选择【精加工壁】，【程序】选择【零件外轮廓精加工】，【刀具】选择【D16R0】，【几何体】选择【WORKPIECE】，【方法】选择【MILL_FINISH】，修改【名称】为【Finish_mill-5】，如图 1-57 所示，单击【确定】。

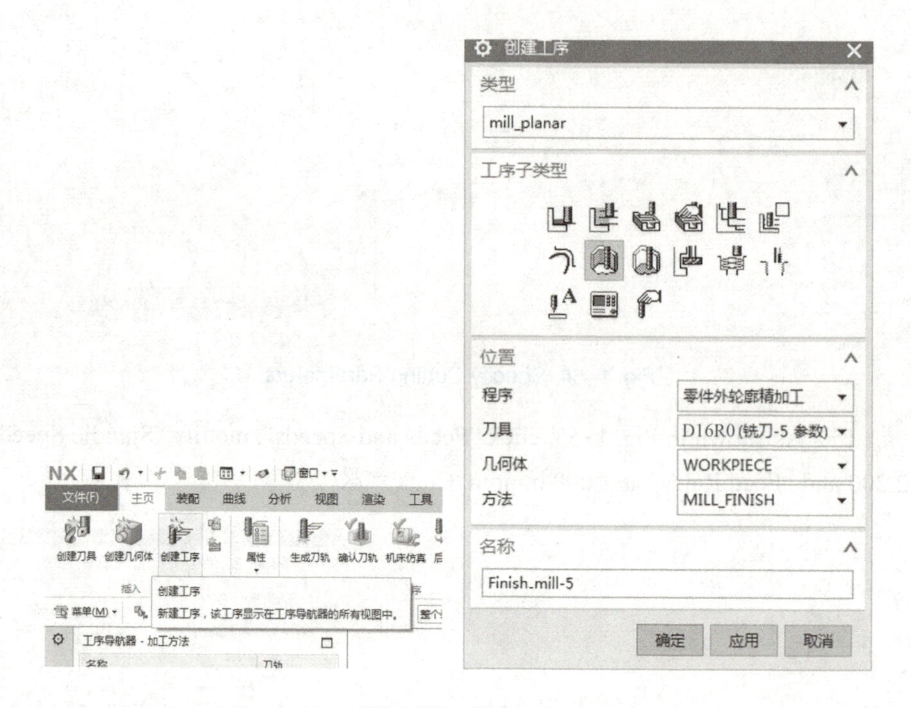

图 1-57　创建工序

（2）单击【指定部件边界】中的【选择或编辑部件边界】，打开【边界几何体】对话框，【模式】选择【曲线/边】，单击【确定】。打开【创建边界】对话框，【类型】选择【开放的】，【材料侧】选择【右】，选择边，如图 1-58 所示，单击【确定】。

5. Part Contour Finishing

（1）As shown in Fig. 1-57, click “Create Operation”, and then select “ mill_planar” in pull-down box of “Type”. Select operation subtype as “Finish Walls”, program as “Part contour finishing”, tool as “D16R0”, geometry as “ WORKPIECE ” and method as “MILL_FINISH”. Modify name as “ Finish_mill-5”, and then click “OK”.

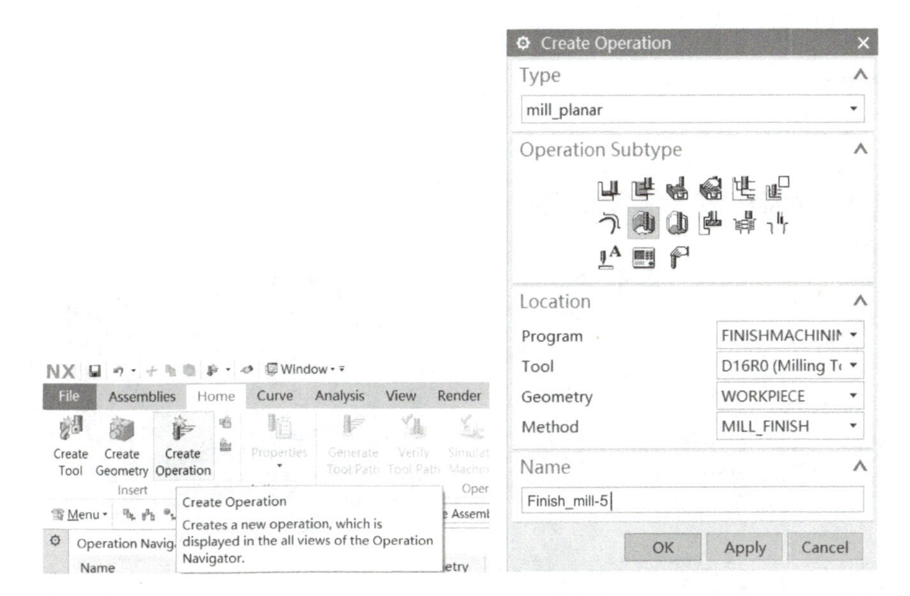

Fig. 1-57 Create Operation

（2）As shown in Fig. 1-58, click edit button “Specify Part Boundaries”. Then, in “Boundary Geometry” window, select the mode as “Curves/Edges”, then the “Create Boundary” window pops up, select “Open” as “Type” and “Right” as “Material Side”. Click “OK”.

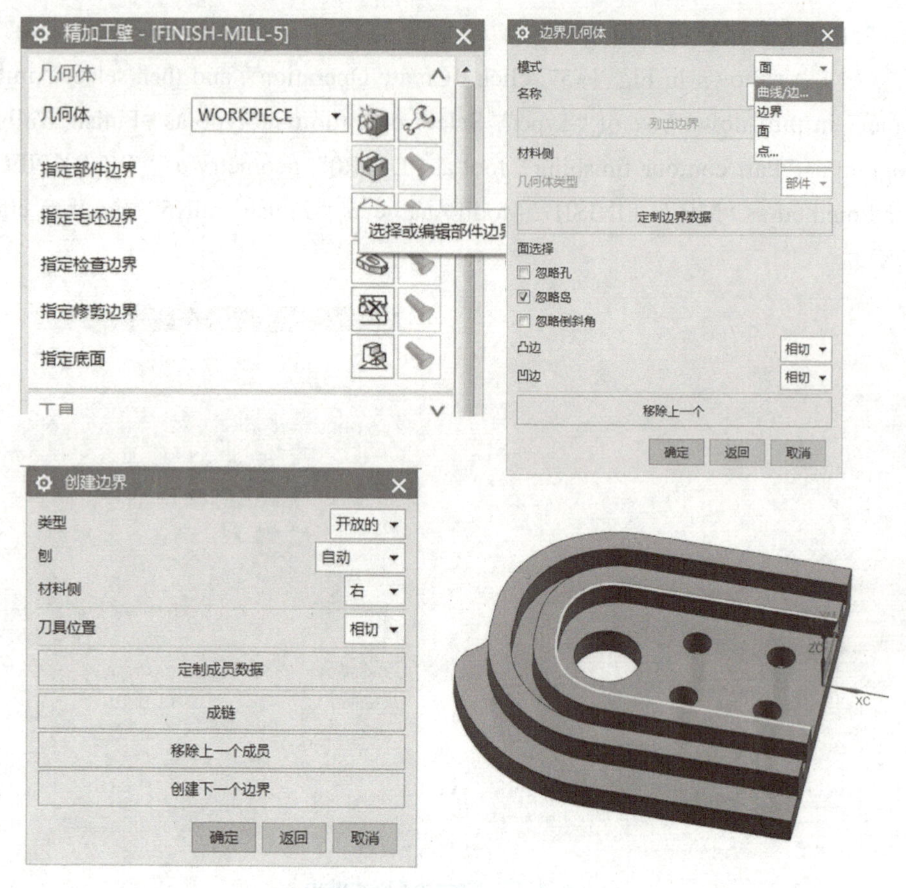

图 1–58 指定部件边界

（3）单击【指定部件边界】中的【选择或编辑部件边界】，打开【编辑边界】对话框，单击【编辑】，打开【编辑成员】对话框，选择第一条刀轨，单击【起点】，打开【修改边界起点】对话框，选中【延伸】和【距离】，并设置【距离】为【5】，如图 1–59 所示，单击【确定】。

图 1–59 编辑边界

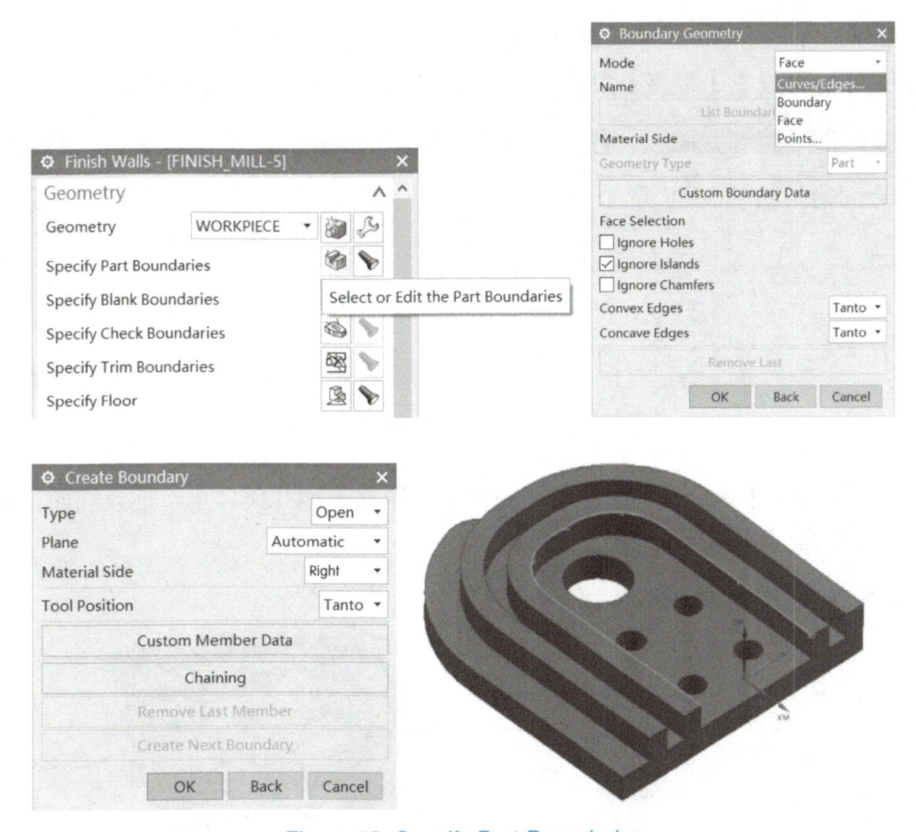

Fig. 1–58　Specify Part Boundaries

(3) As shown in Fig. 1-59, click button "Specify Part Boundaries" again and click the button "Edit" in "Edit Boundary" window. Click "Start Point" and select "Extend", then modify the distance as 5. Click "OK".

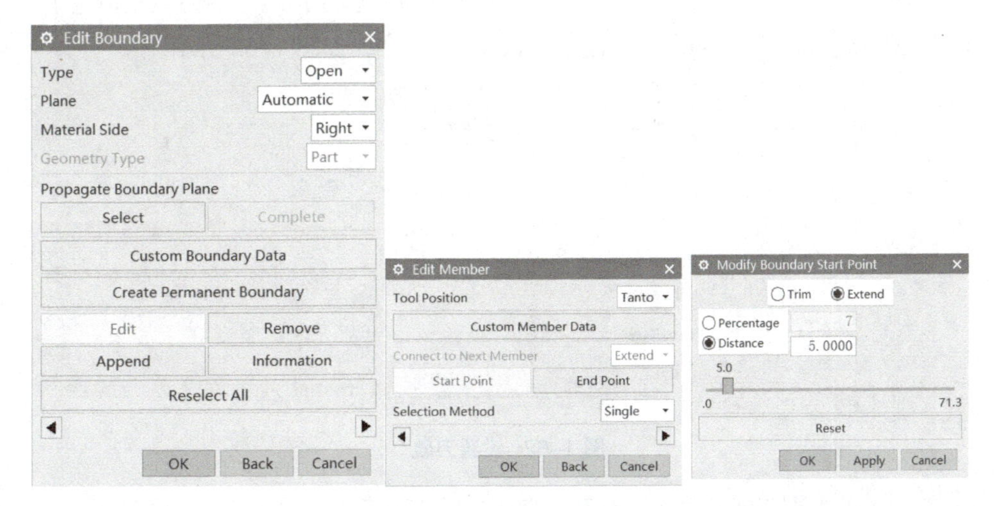

Fig. 1–59　Edit Boundary

（4）单击【下一步】，选择最后一条刀轨，单击【终点】，选中【延伸】和【距离】，并设置【距离】为【80】，如图 1–60 所示，单击【确定】。

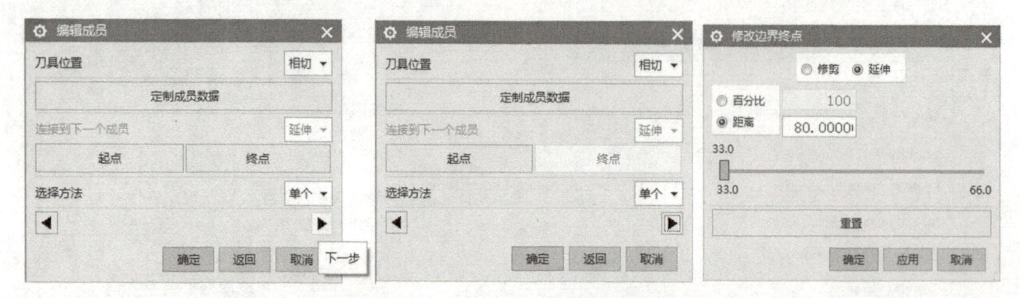

图 1–60 设置刀具位置

（5）单击【指定底面】中的【选择或编辑底平面几何体】，选择底面，如图 1–61 所示，单击【确定】。

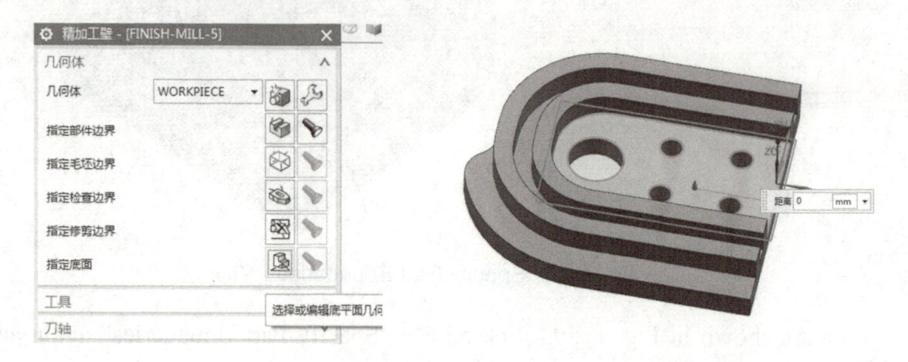

图 1–61 指定底面

（6）单击【刀轨设置】，设置【切削模式】为【轮廓】，【步距】为【恒定】，【最大距离】为【0.1 mm】，【附加刀路】为【1】，如图 1–62 所示。

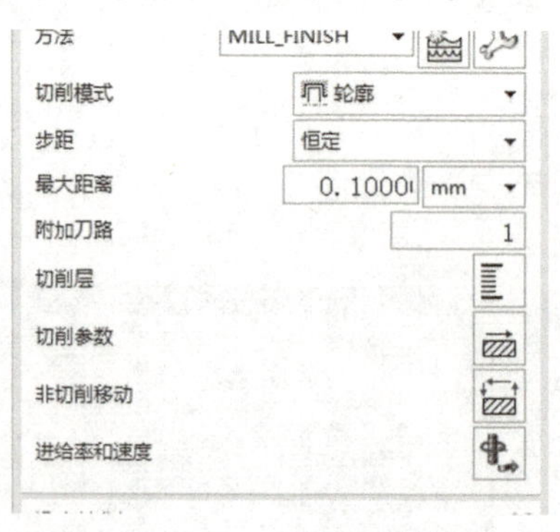

图 1–62 设置刀轨

（7）单击【切削参数】，打开【切削参数】对话框，单击【余量】选项卡，设置【最终底面余量】为【0】，如图 1–63 所示，单击【确定】。

(4) Then, click "next" to the find the final tool path. Click "End Point" and select "Extend", then modify the distance as 80 as shown in Fig. 1-60. Click "OK".

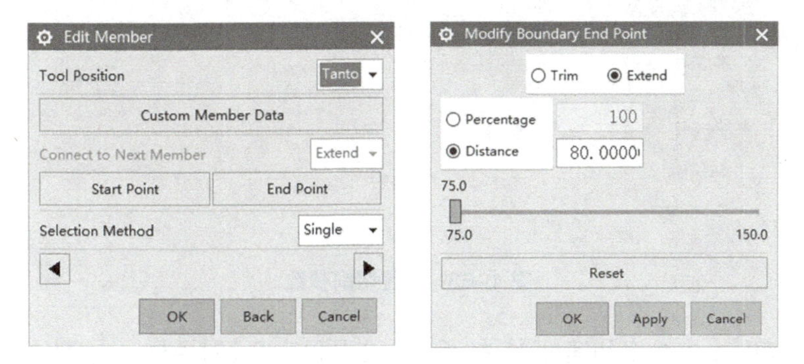

Fig. 1-60 Modify Tool Position

(5) As shown in Fig. 1-61, click the button "Specify Floor" and select the surface. Click "OK".

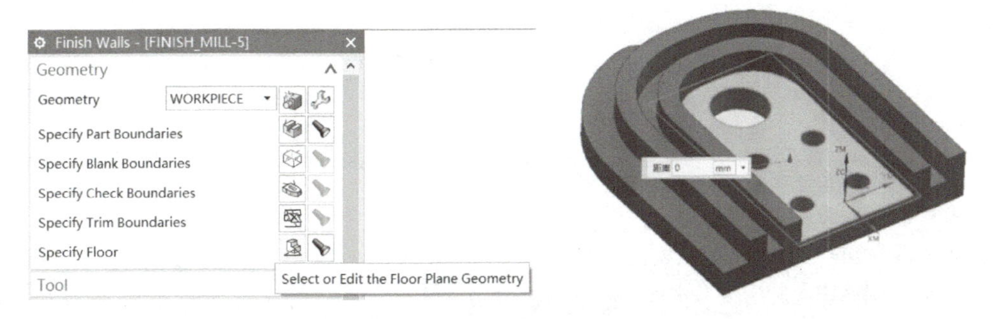

Fig. 1-61 Specify Floor

(6) As shown in Fig. 1-62, modify "Cut Pattern" as "Profile". Modify the type of "Stepover" as "Constant" and modify "Maximum Distance" as 0.1 mm and "Additional Passes" as 1.

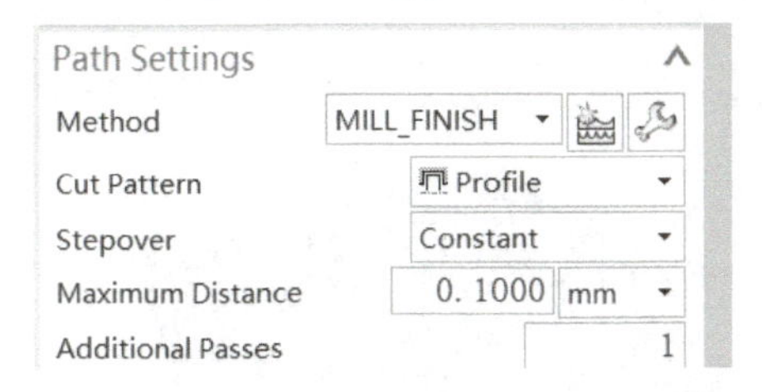

Fig. 1-62 Modify Path Settings

(7) As shown in Fig. 1-63, click "Cutting Parameters" button, and in "Stock" tab, modify "Final Floor Stock" as 0. Click "OK". Keep parameters of "Cutting Parameters" default.

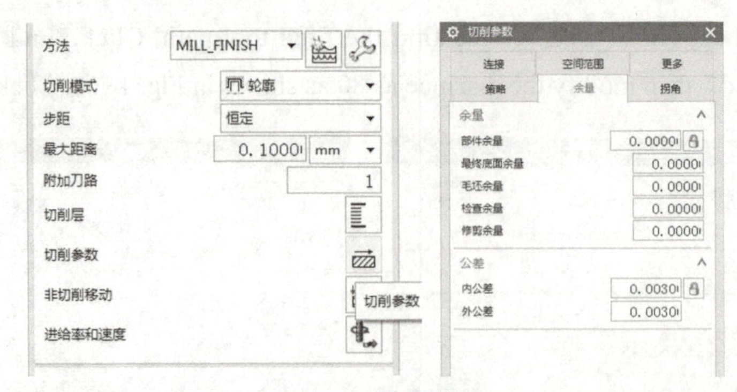

图 1-63　设置切削参数

（8）单击【进给率和速度】，打开【进给率和速度】对话框，设置【主轴速度】为【2200】，进给率【切削】为【1000 mmpm】，单击【确定】，如图 1-64 所示。

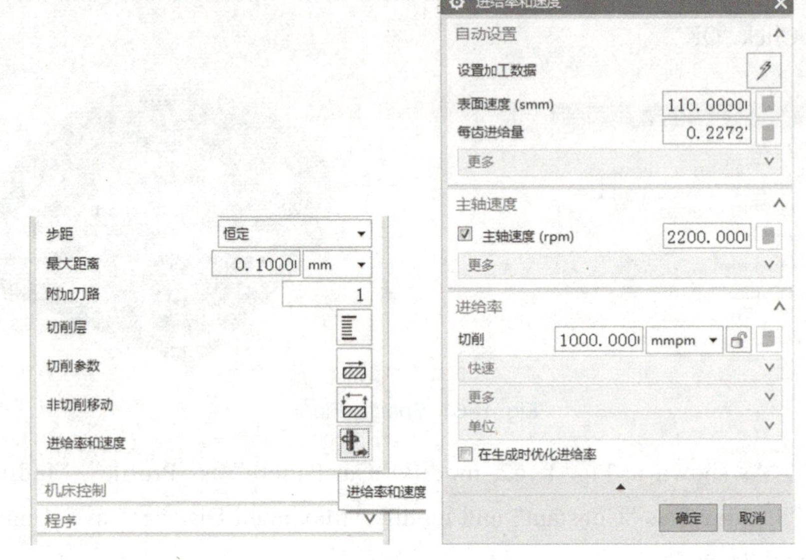

图 1-64　设置进给率和速度

（9）单击【生成】，查看生成的刀具轨迹，如图 1-65 所示，单击【确定】。

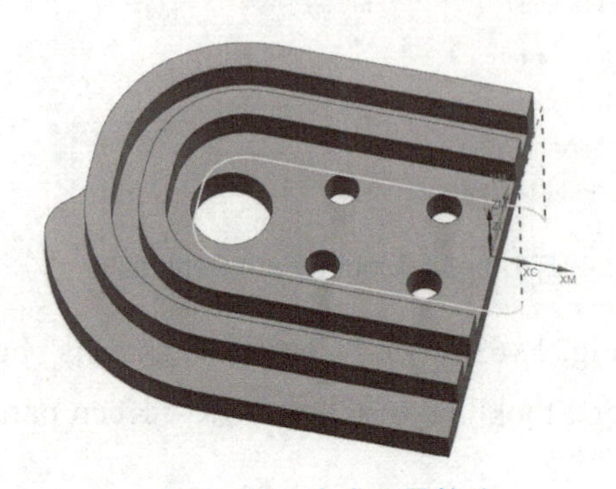

图 1-65　生成刀具轨迹

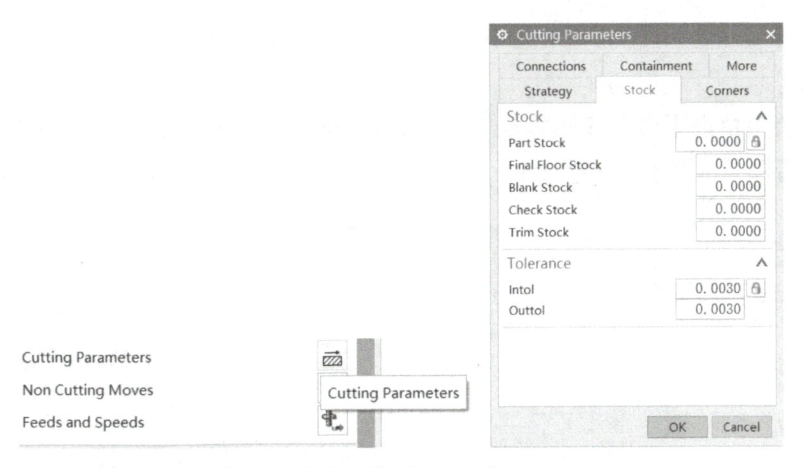

Fig. 1-63 Modify Cutting Parameters

(8) Click "Feeds and Speeds", modify "Spindle Speed" as 2,200 and "Feed Rates" as 1,000 mmpm as shown in Fig. 1-64. Click "OK".

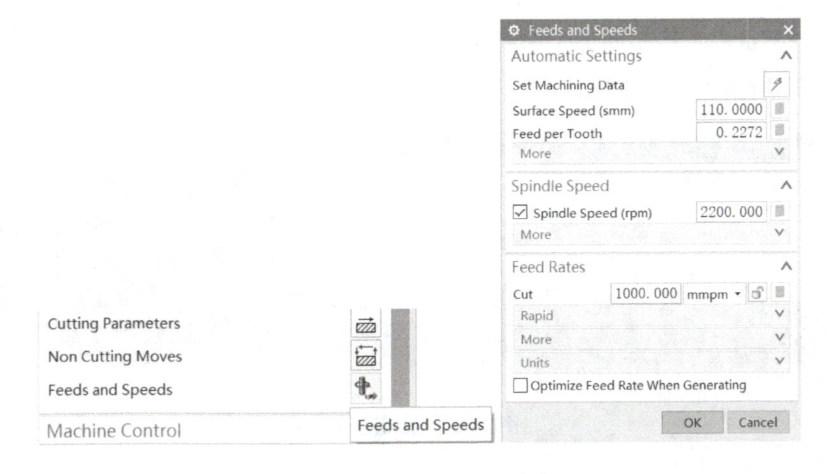

Fig. 1-64 Modify Feeds and Speeds

(9) Click "Generate" to view the tool path as shown in Fig. 1-65, and then click "OK".

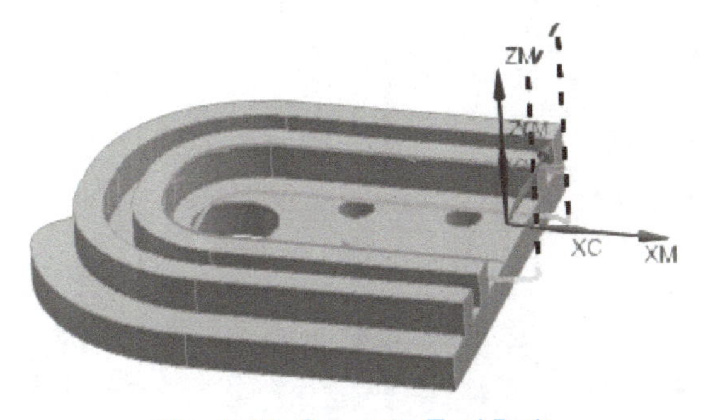

Fig. 1-65 Generate Tool Path

（10）复制工序名【FINISH_MILL-5】，粘贴，更改工序名为【FINISH_MILL-6】。

（11）单击【指定部件边界】中的【选择或编辑部件边界】，打开【编辑边界】对话框，单击【全部重选】，选择曲线，依次单击【确定】，如图1-66所示。

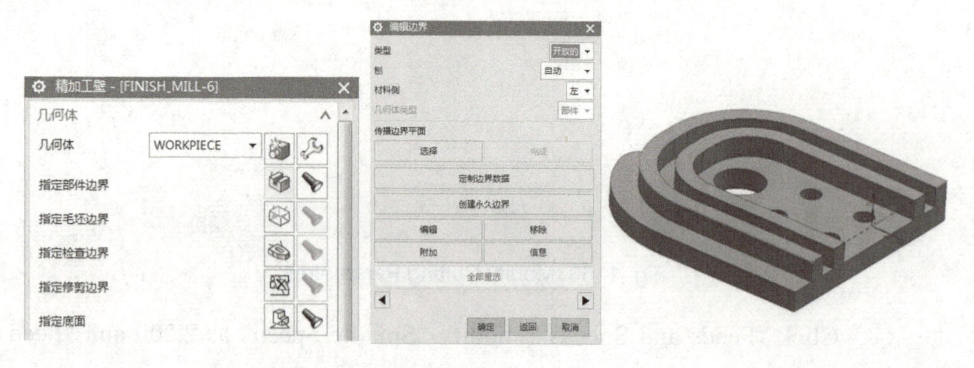

图1-66 指定部件边界

（12）单击【指定底面】中的【选择或编辑底平面几何体】，选择台阶面，如图1-67所示，单击【确定】。

图1-67 指定底面

（13）单击【非切削移动】，打开【非切削移动】对话框，单击【进刀】选项卡，取消选中【修剪至最小安全距离】，刀路会有更多的延伸，如图1-68所示，单击【确定】。

(10) Copy program " FINISH_MILL-5" program and paste it. Modify the pasted file name as " FINISH_MILL-6".

(11) Click edit button "Specify Part Boundaries". Click "Reselect All" and select the curves as shown in Fig. 1-66. Click "OK".

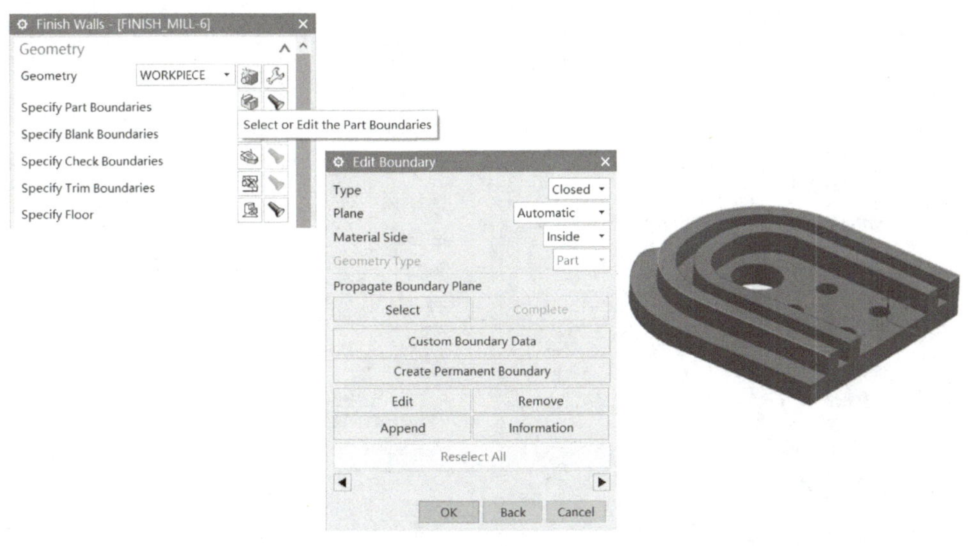

Fig. 1-66　Specify Part Boundaries

(12) As shown in Fig. 1-67, click the button "Specify Floor" and select surface of the step. Click "OK".

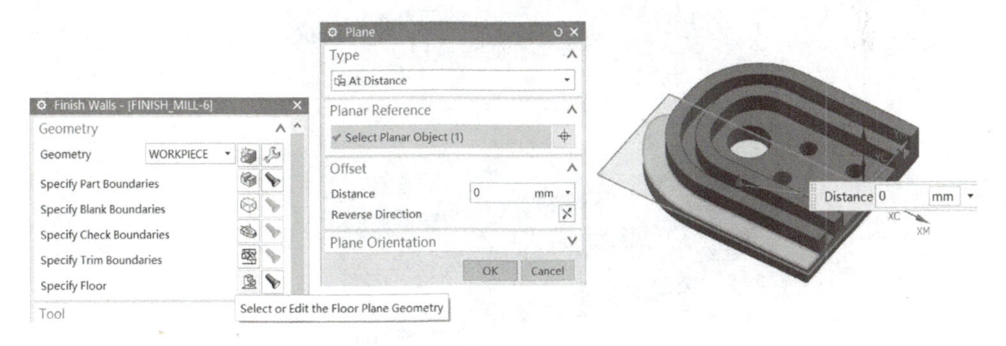

Fig. 1-67　Specify Floor

(13) As shown in Fig. 1-68, click "Non Cutting Moves" button. In the tab of "Engage", uncheck the option of "Trim to Minimun Clearance" in the area of "Open Area". Thus, the tool path will extend more.

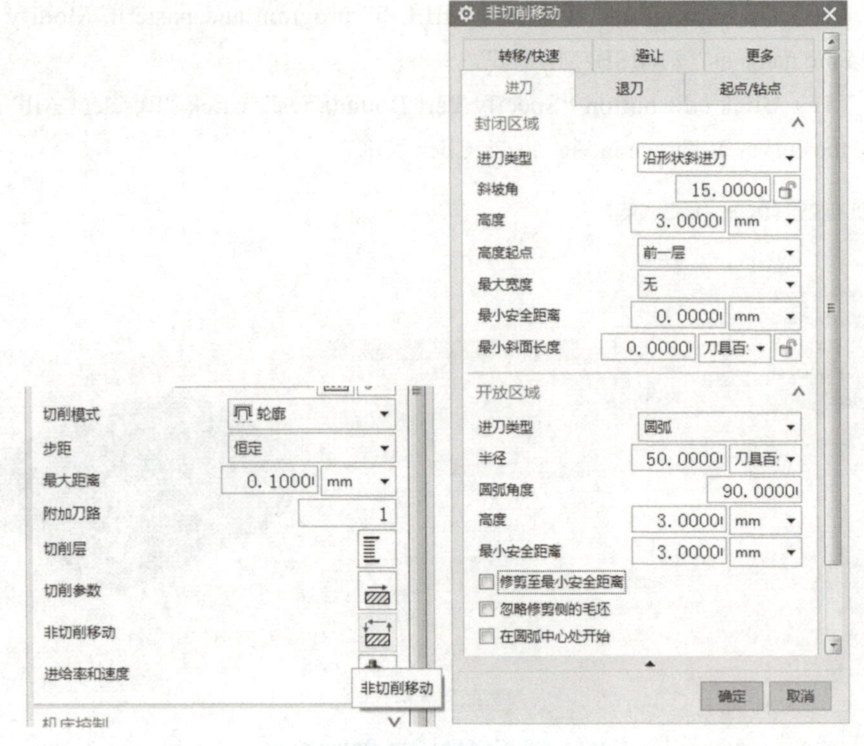

图 1-68　设置非切削移动

（14）单击【生成】，查看生成的刀具轨迹，如图 1-69 所示，单击【确定】。

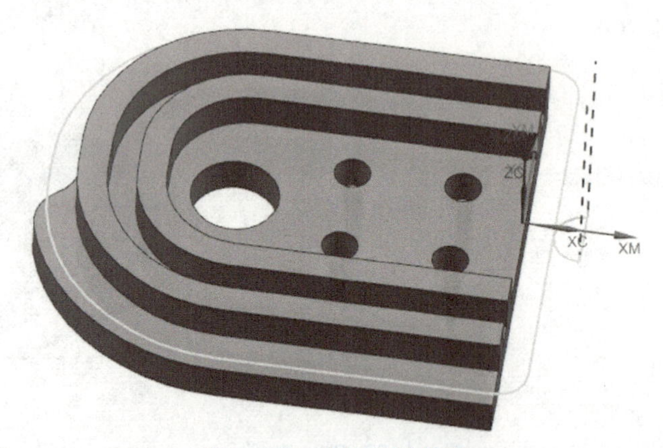

图 1-69　生成刀具轨迹

（15）复制工序名【FINISH_MILL-6】，粘贴，更改工序名称为【FINISH_MILL-7】。

（16）单击【指定部件边界】中的【选择或编辑部件边界】，打开【编辑边界】对话框，单击【全部重选】，如图 1-70 所示。

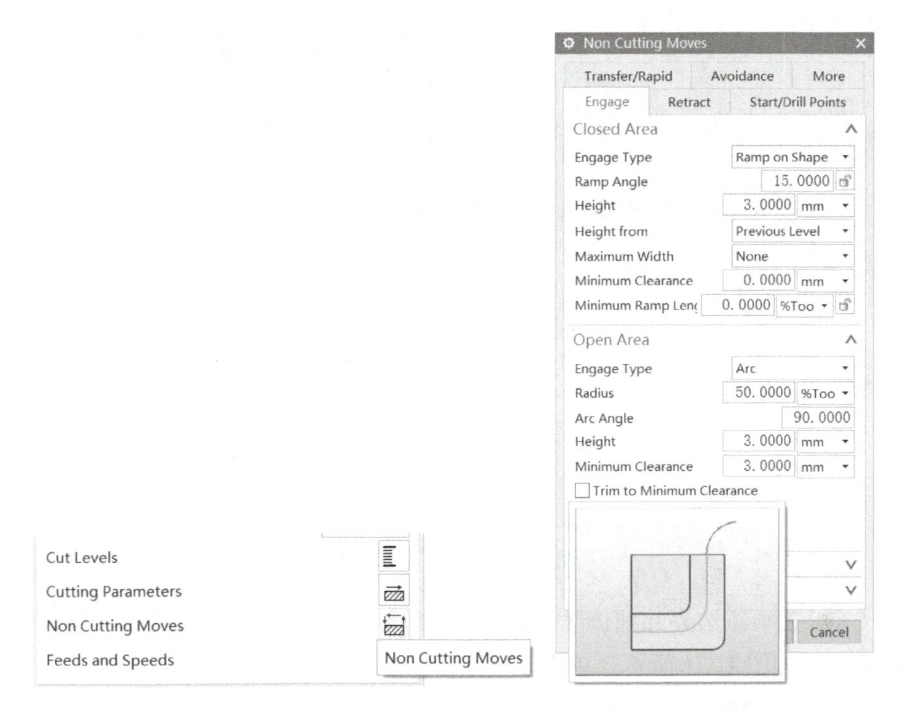

Fig. 1–68 Specify Non Cutting Moves

(14) Click "Generate" to view the tool path as shown in Fig. 1-69, and then click "OK".

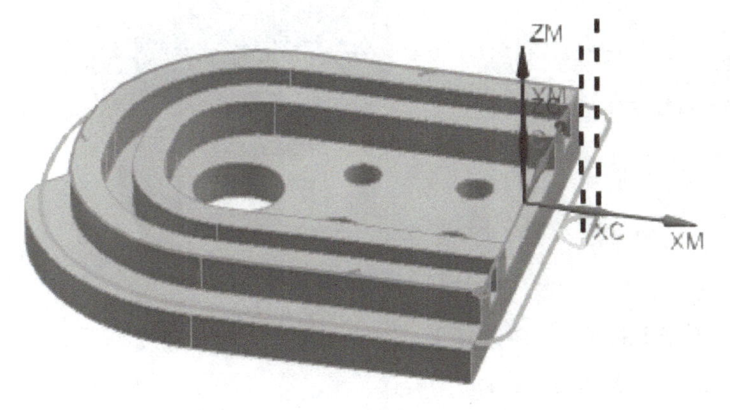

Fig. 1–69 Generate Tool Path

(15) Copy program " FINISH_MILL-6" program and paste it. Modify the pasted file name as " FINISH_MILL-7".

(16) As shown in Fig. 1-70, click edit button "Specify Part Boundaries". Click "Reselect All" and then click "OK".

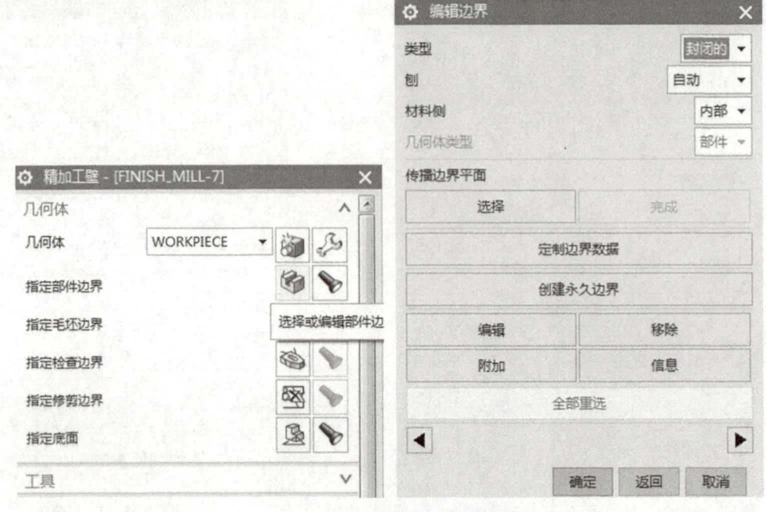

图 1–70　编辑部件边界

（17）打开【边界几何体】对话框，勾选【忽略孔】，设置【模式】为【面】，选择底面，如图 1–71 所示，单击【确定】。

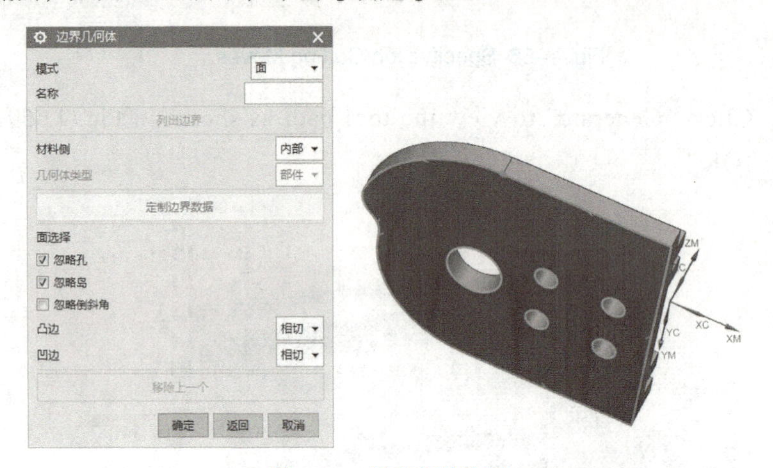

图 1–71　指定部件边界

（18）【刨】选择【用户定义】，选择台阶面，如图 1–72 所示，单击【确定】。

图 1–72　编辑边界

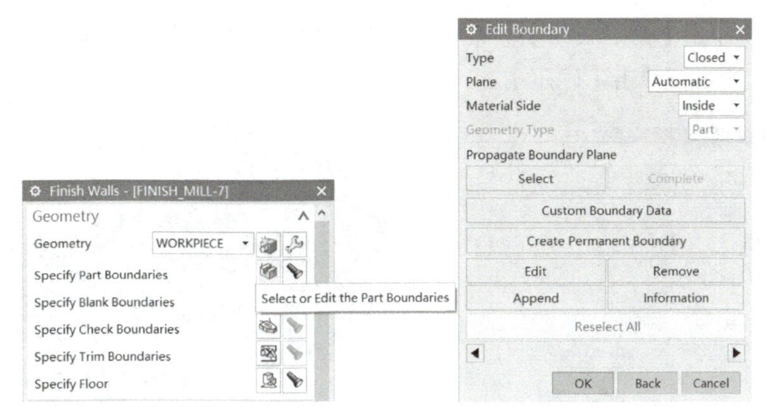

Fig. 1–70 Specify Part Boundaries

(17) As shown in Fig. 1-71, tick the option "Ignore Holes". And then select the bottom of the part.

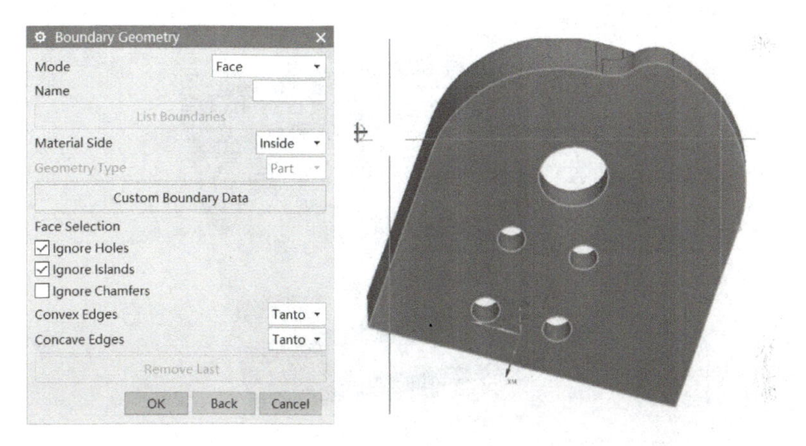

Fig. 1–71 Specify Boundary Geometry

(18) Select " User-Defined " in the option of "Plane" and then select surface of the step as shown in Fig. 1-72. Click "OK".

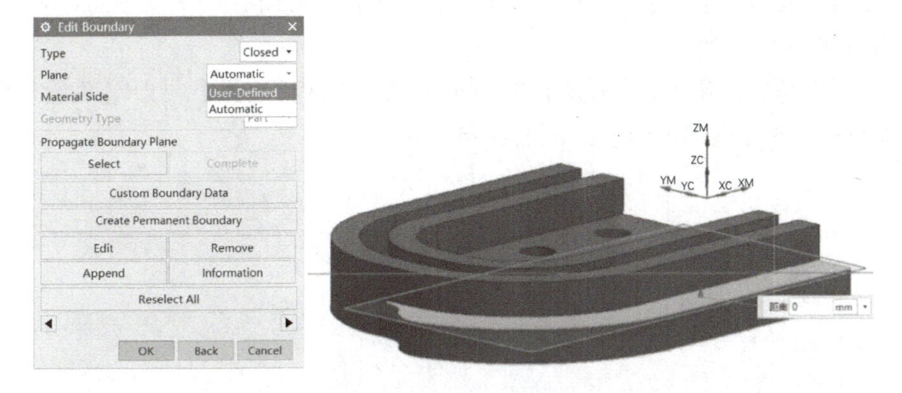

Fig. 1–72 Edit Boundary

（19）单击【指定底面】中的【选择或编辑底平面几何体】，选择零件底面，如图 1–73 所示，单击【确定】。

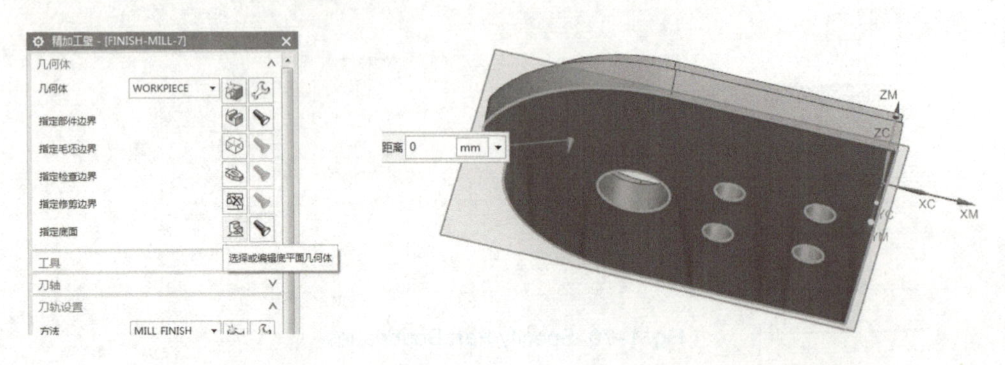

图 1–73　指定底面

（20）单击【生成】，查看生成的刀具轨迹如图 1–74 所示，单击【确定】。

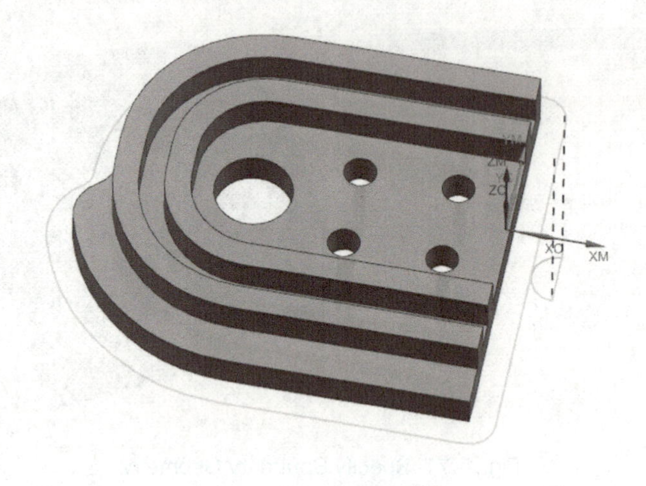

图 1–74　生成刀具轨迹

6. 零件导轨槽精加工

（1）复制工序名【FINISH_MILL–5】，选择【内部粘贴】，更改工序名称为【FINISH_MILL–8】。

（2）单击【指定部件边界】中的【选择或编辑部件边界】，打开【编辑边界】对话框，单击【全部重选】，如图 1–75 所示。

(19) As shown in Fig. 1-73, click the button "Specify Floor" and select the bottom of the part. Click "OK".

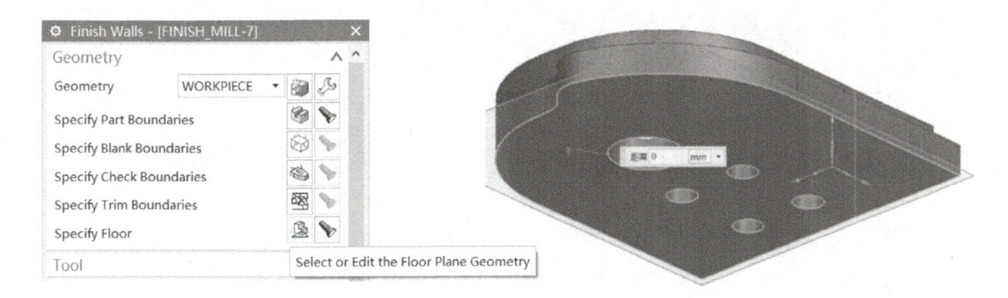

Fig. 1-73 Specify Floor

(20) Click "Generate" to view the tool path as shown in Fig. 1-74, and then click "OK".

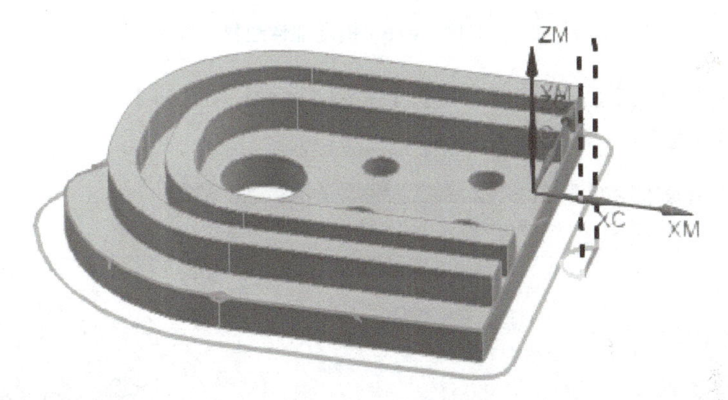

Fig. 1-74 Generate Tool Path

6. Finishing of Rail Groove

(1) Copy program " FINISH_MILL-5" program and paste it. Modify the pasted file name as " FINISH_MILL-8".

(2) As shown in Fig.1-75, click edit button "Specify Part Boundaries". Click "Reselect All" and then click "OK".

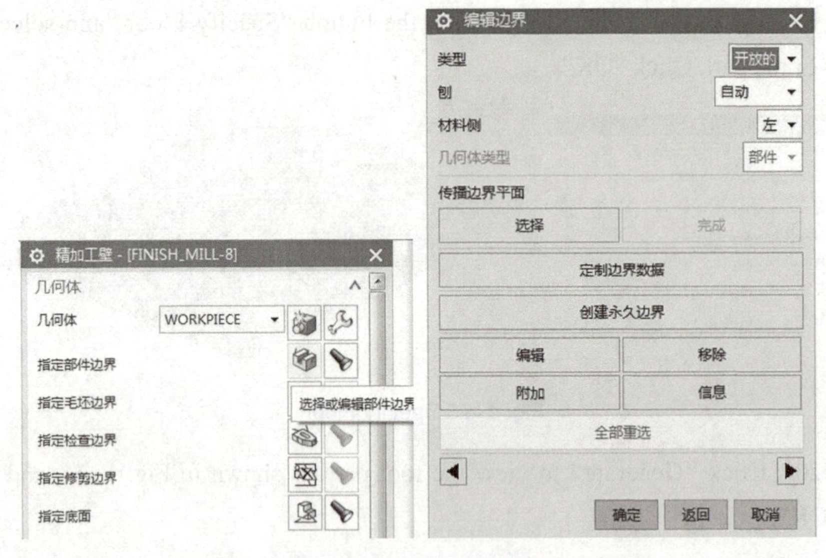

图 1–75　指定部件边界

（3）【模式】选择【曲线/边】,【类型】选择【开放的】,【材料侧】选择【右】，选择边界，如图 1–76 所示。

图 1–76　创建边界

（4）单击【创建下一个边界】，选择边界，如图 1–77 所示，依次单击【确定】。

图 1–77　创建下一个边界

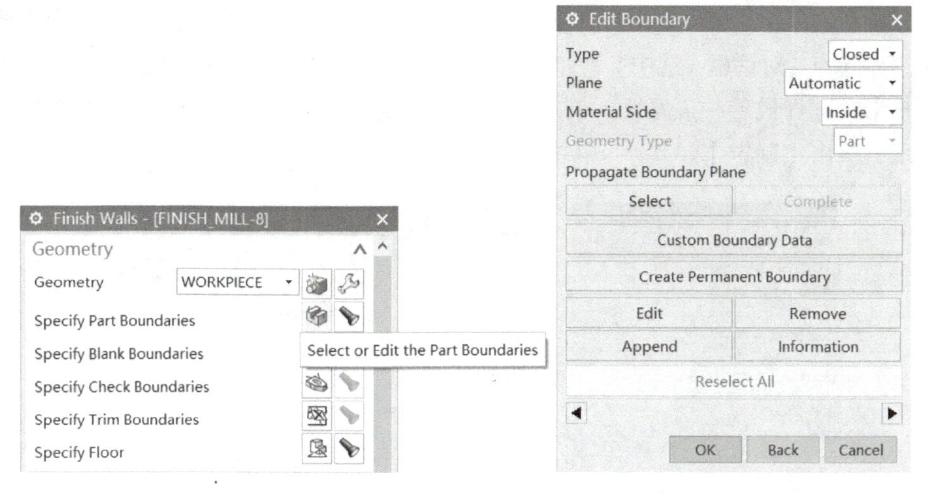

Fig. 1–75 Specify Part Boundaries

(3) As shown in Fig. 1-76, select the mode as "Curves/Edges", then the "Create Boundary" window pops up. Select "Open" as "Type" and "Right" as "Material Side". Click "OK".

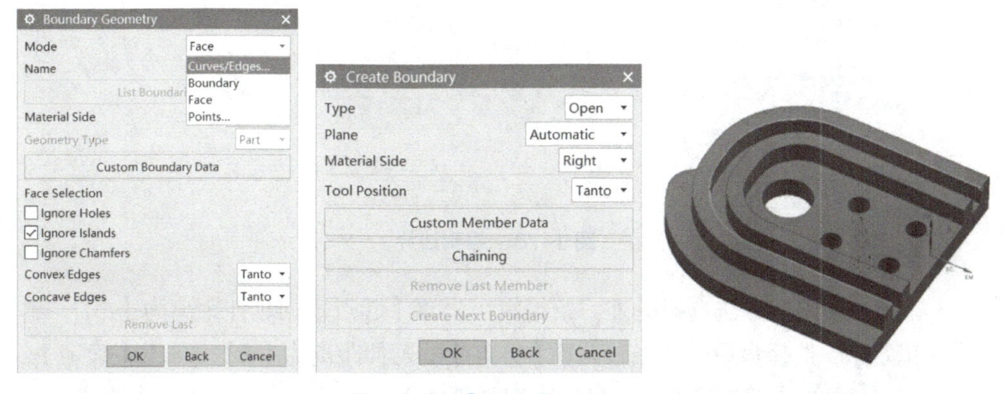

Fig. 1–76 Create Boundary

(4) Click "Create Next Boundary" and then select the boundary as shown in Fig. 1-77. Click "OK".

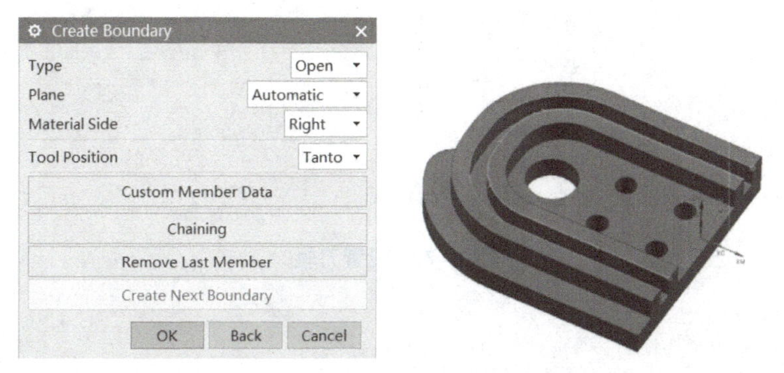

Fig. 1–77 Create Next Boundary

（5）单击【编辑】，打开【编辑成员】对话框，单击【起点】，打开【修改边界起点】对话框，选中【延伸】和【距离】，并设置【距离】为【5】。单击【下一步】，选择最后一条刀轨，单击【终点】，设置【距离】为【80】，如图1-78所示，单击【确定】。

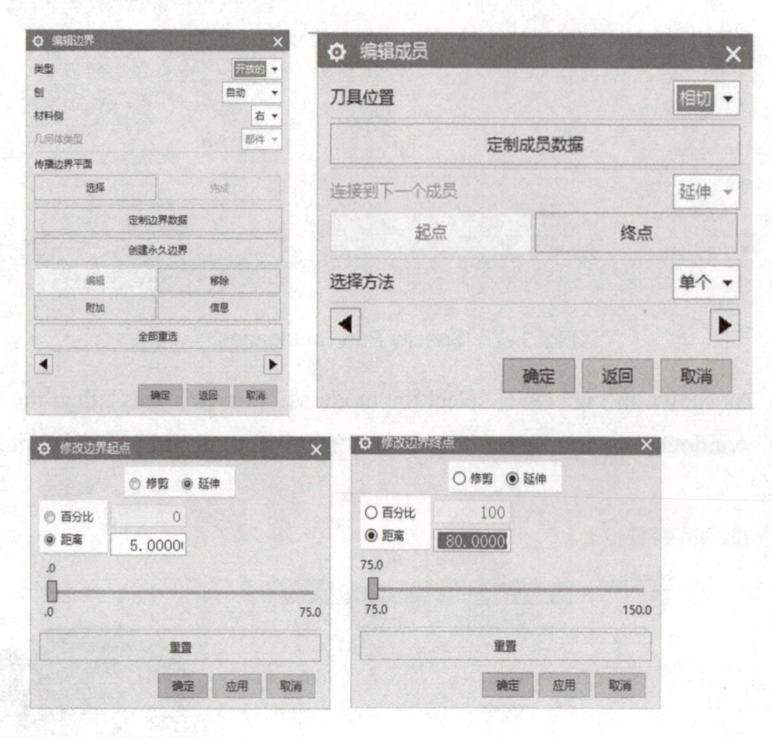

图1-78　编辑边界

（6）用同样方法选择另外一条编辑，设置【起点】的延伸距离为【5】，单击【确定】，选择最后一条线，设置【终点】的延伸距离为【5】，单击【确定】。

（7）单击【刀轨设置】，设置【切削模式】为【轮廓】，【步距】为【刀具平直百分比】，【平面直径百分比】为【50】，如图1-79所示。

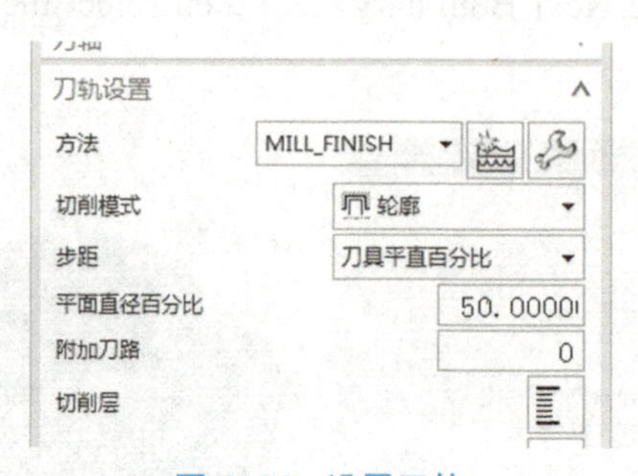

图1-79　设置刀轨

(5) Click the button "Edit", click "Start Point" and select "Extend". Then modify the distance as 5. Click "OK". Then, Click "Next" to find the final tool path. Click "End Point" and select "Extend", and then modify the distance as 80 as shown in Fig 1-78. Click "OK".

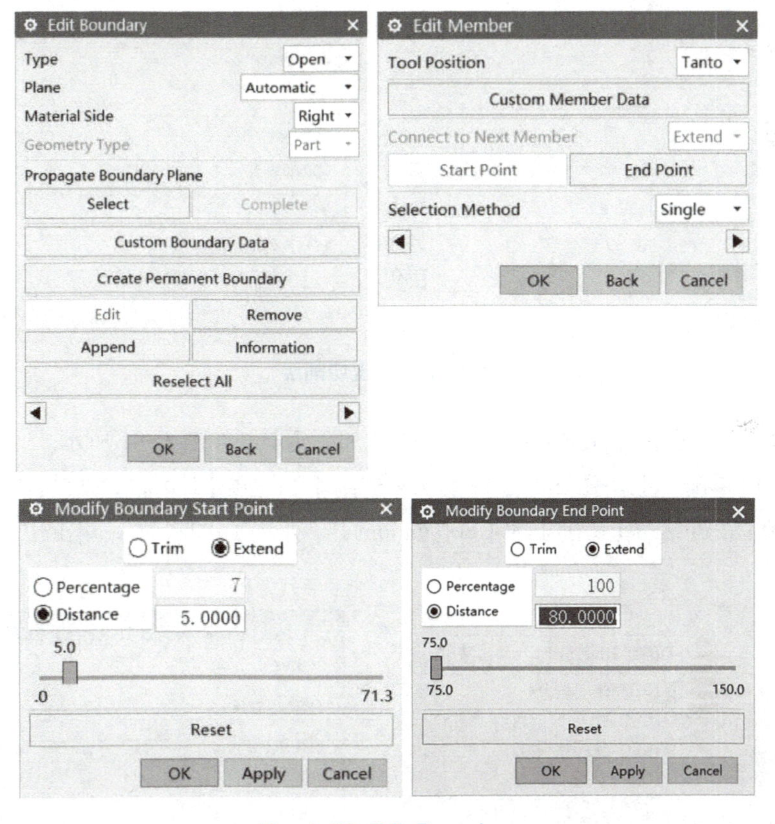

Fig. 1-78 Edit Boundary

(6) Similarly, edit another boundary, click "Start Point" and select "Extend", then modify the distance as 5. Click "OK". Then, click "Next" to find the final tool path. Click "End Point" and select "Extend". Then modify the distance as 5.

(7) Modify "Cut Pattern" as "Profile", select "% Tool Flat" as "Stepover", modify "Percent of Flat Diameter" as 50 as shown in Fig. 1-79.

Fig. 1-79 Modify Path Settings

（8）单击【切削层】，打开【切削层】对话框，设置类型为【恒定】，每刀切削深度【公共】为【1】，如图 1-80 所示，单击【确定】。

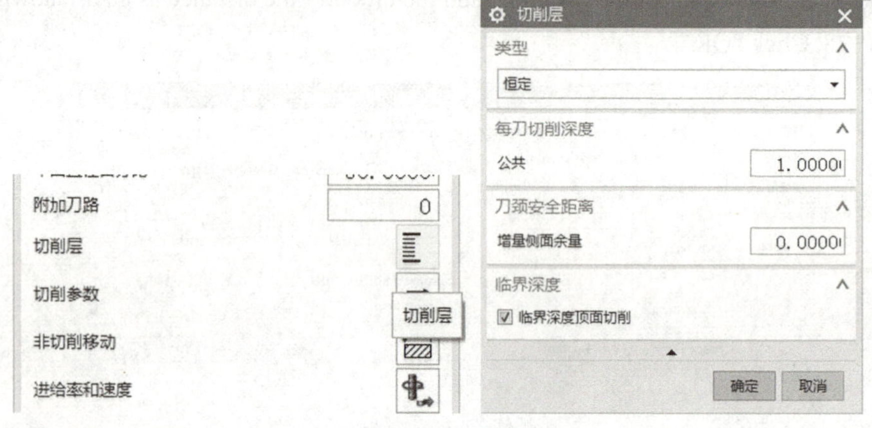

图 1-80 设置切削层

（9）刀具选择【D6R0-FINISH】的精加工刀具，如图 1-81 所示。

（10）单击【进给率和速度】，打开【进给率和速度】对话框，设置【主轴速度】为【3600】，进给率【切削】为【800 mmpm】，如图 1-82 所示，单击【确定】。

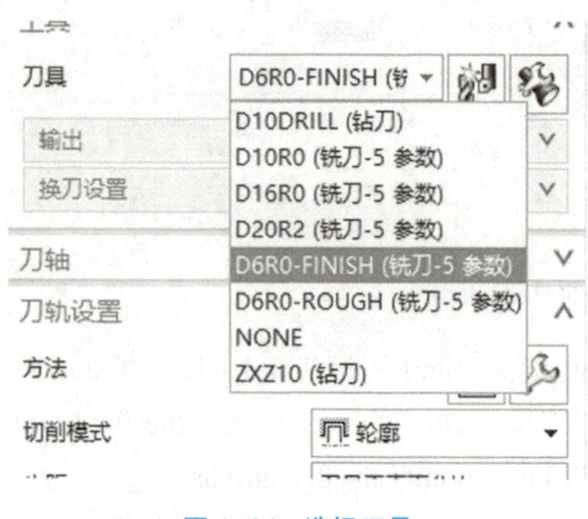

图 1-81 选择刀具

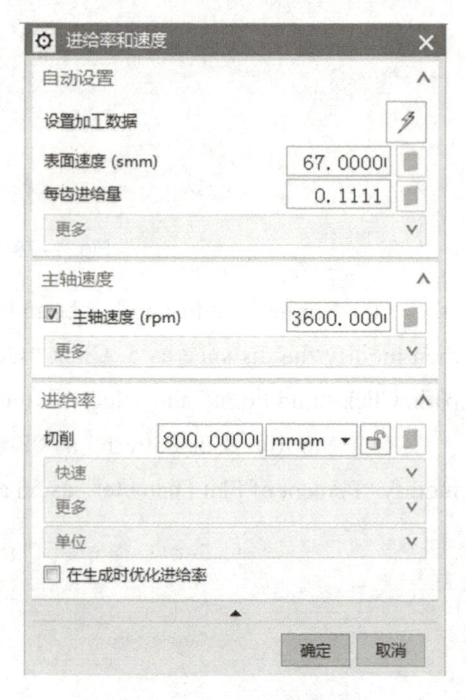

图 1-82 设置进给率和速度

(8) As shown in Fig. 1-80, click "Cut Levels", select "Constant" as the type, and modify "Common Depth Per Cut" as 1. Click "OK".

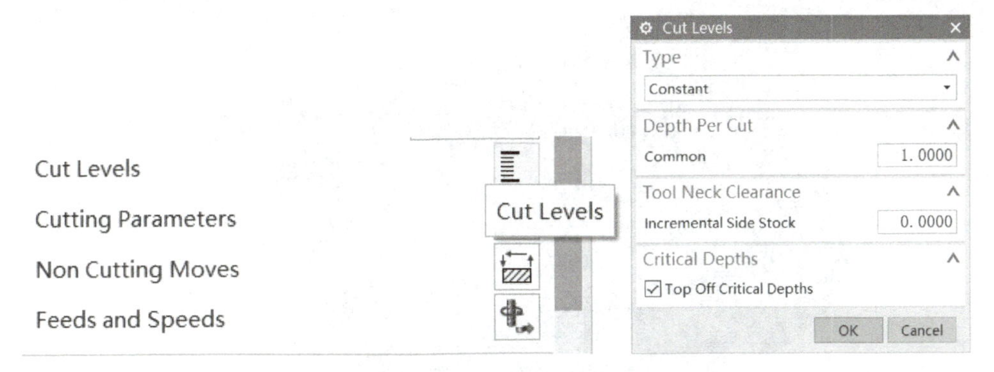

Fig. 1-80 Modify Cut Levels

(9) As shown in Fig. 1-81, modify "Tool" as "D6R0-FINISH".

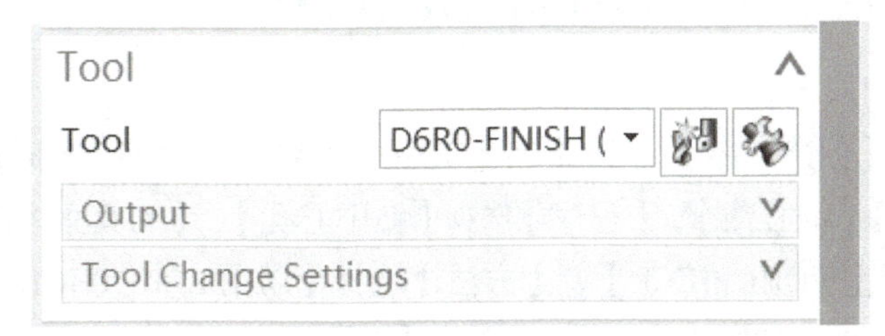

Fig. 1-81 Modify Tool

(10) Click "Feeds and Speeds", modify "Spindle Speed" as 3,600, and "Feed Rates" as 800 mmpm as shown in Fig. 1-82. Click "OK".

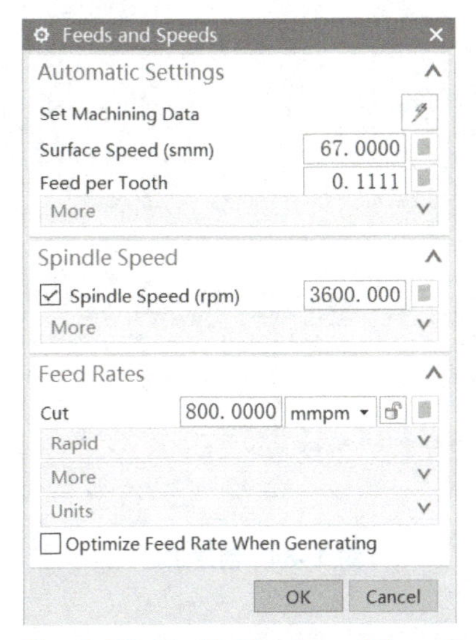

Fig. 1-82 Modify Feeds and Speeds

（11）单击【生成】，查看生成的刀具轨迹，如图 1-83 所示，单击【确定】。

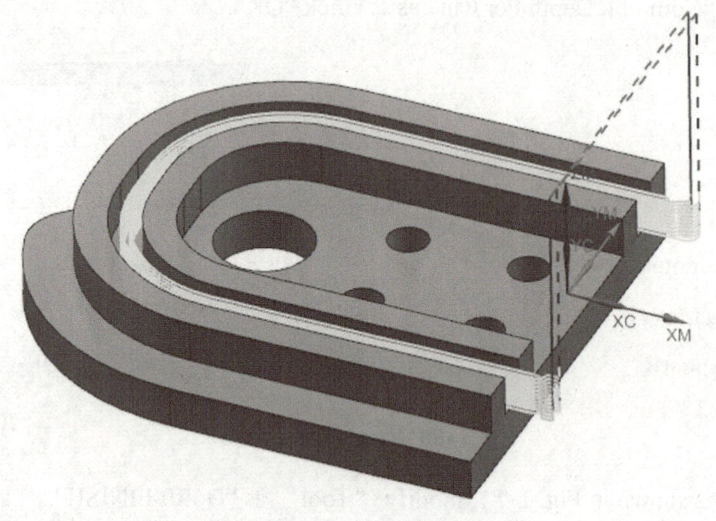

图 1-83 生成刀具轨迹

7. 钻中心孔

（1）单击【创建工序】，打开【创建工序】对话框，【类型】选择【drill】，【工序子类型】选择【定心钻】，【程序】选择【钻中心孔】，【刀具】选择【ZXZ10】，【几何体】选择【WORKPIECE】，【方法】选择【DRILL_METHOD】，修改【名称】为【Drill-1】，如图 1-84 所示，单击【确定】。

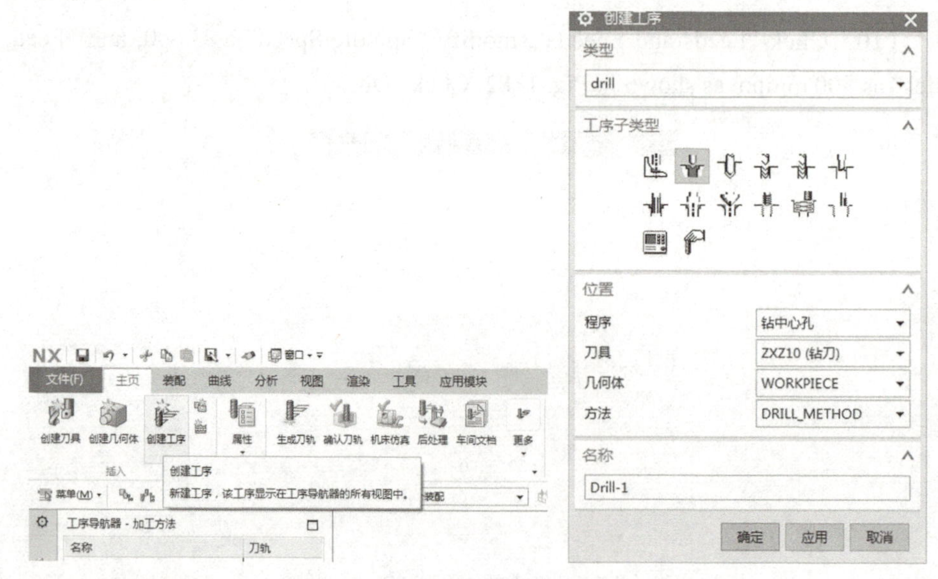

图 1-84 创建工序

(11) Click "Generate" to view the tool path as shown in Fig. 1-83, and then click "OK".

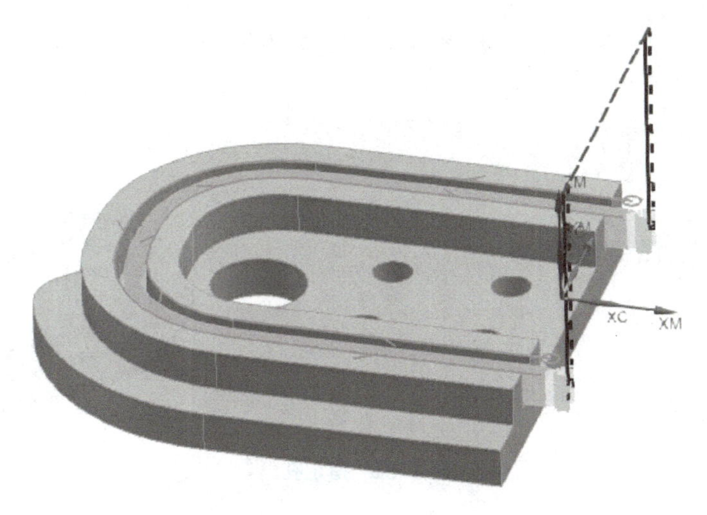

Fig. 1-83 Generate Tool Path

7. Drill Center Hole

(1) As shown in Fig. 1-84, click "Create Operation", and then select "drill" in pull-down box of "Type". Select operation subtype as "Spot Drilling", program as "DRILL CENTER HOLE", tool as "ZXZ10", geometry as "WORKPIECE" and method as " DRILL_METHOD ". Modify name as " DRILL-1", and then click "OK".

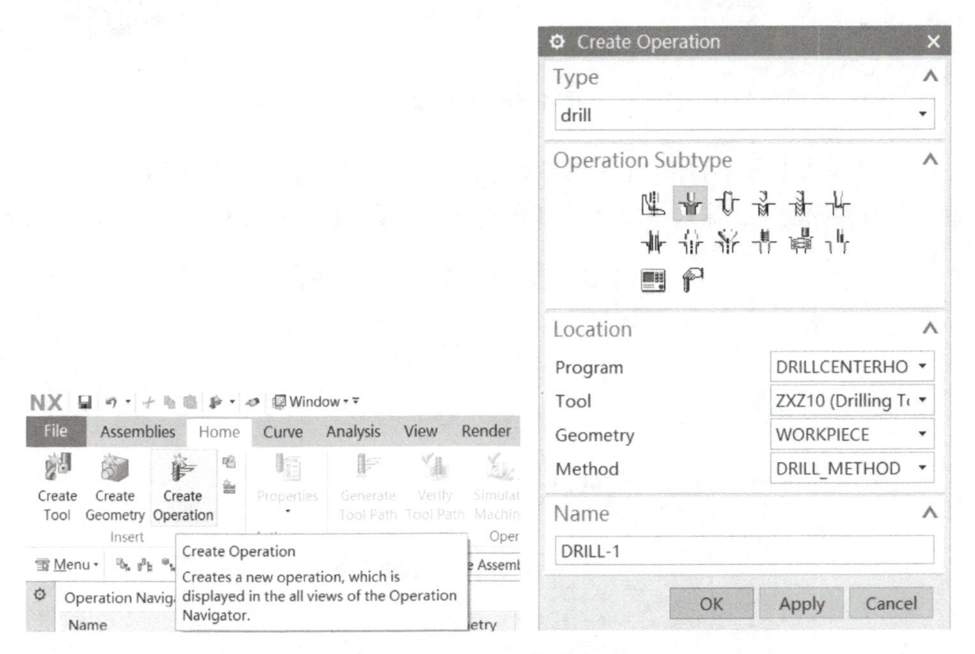

Fig. 1-84　Create Operation

（2）单击【指定孔】，单击【选择】，选择 5 个孔，如图 1–85 所示，依次单击【确定】。

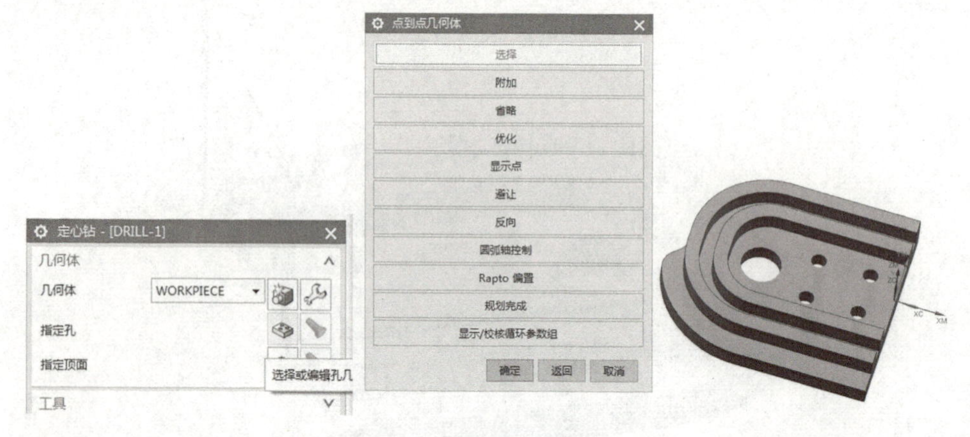

图 1–85 指定孔

（3）单击【指定顶面】中的【选择或编辑部件表面几何体】，打开【顶面】对话框，【顶面选项】选择【面】，选择面，如图 1–86 所示，单击【确定】。

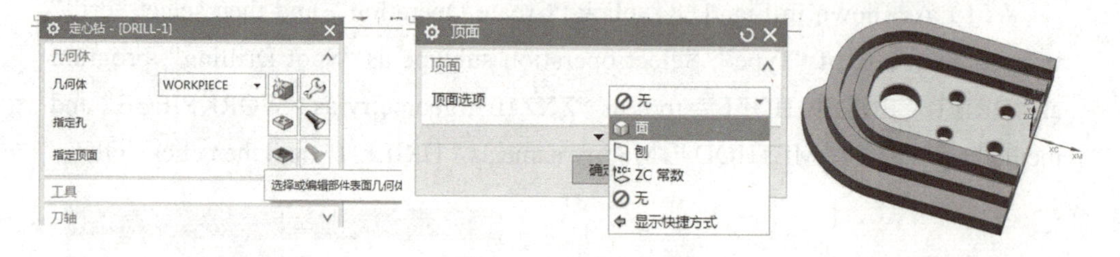

图 1–86 指定顶面

（4）单击【循环】中的【编辑参数】，单击【确定】，单击【Depth（Tip）】，单击【刀尖深度】，输入【1.5】，如图 1–87 所示，单击【确定】。

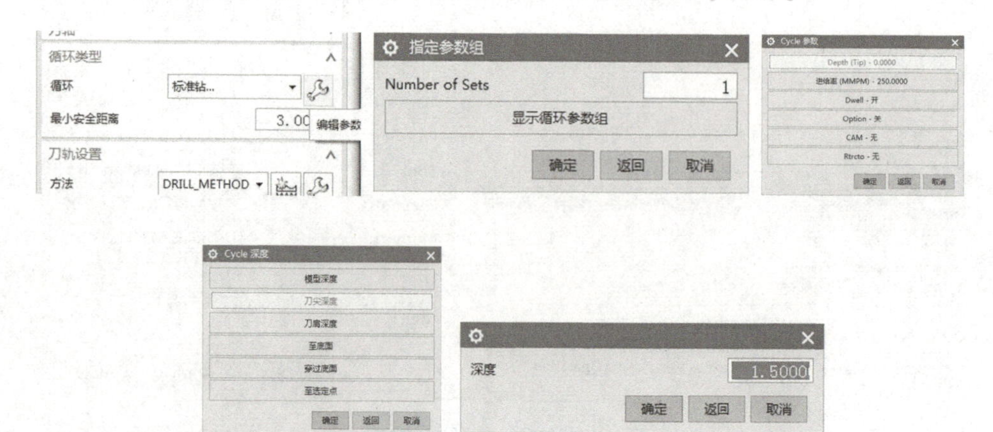

图 1–87 刀尖深度设置

(2) Click "Specify Holes" and then click "Select". Select boundary of the hole as shown in Fig. 1-85. Click "OK".

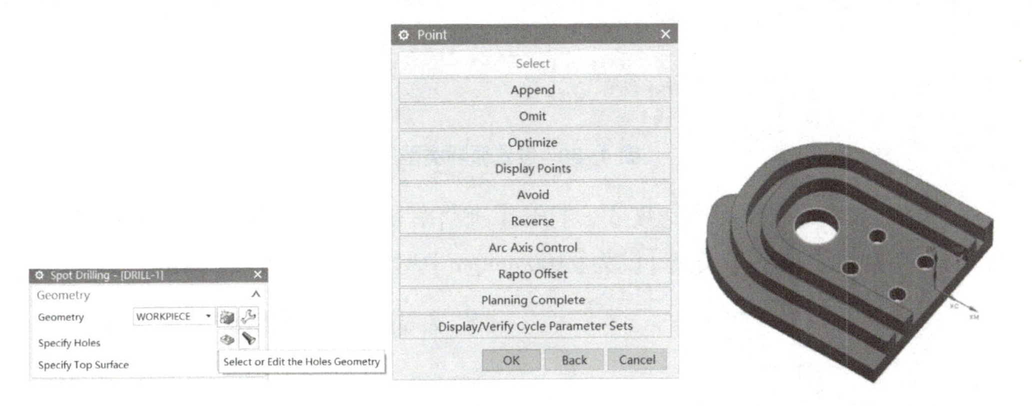

Fig. 1-85 Specify Holes

(3) As shown in Fig. 1-86, click button "Specify Top Surface", and select "Face" as the top surface option and select the surface as shown in Fig. 1-86. Click "OK".

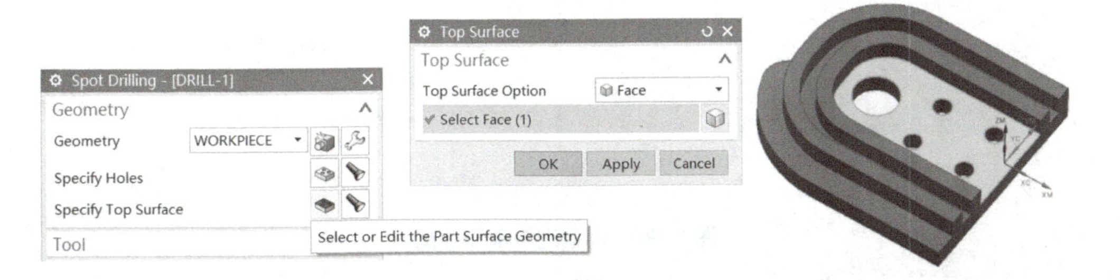

Fig. 1-86 Specify Top Surface

(4) As shown in Fig. 1-87, click "Edit parameters" button in "Cycle Type" tab. Then, click "Depth (Tip)" to choose "Model Depth". Modify "Tool Tip Depth" as 1.5. Click "OK".

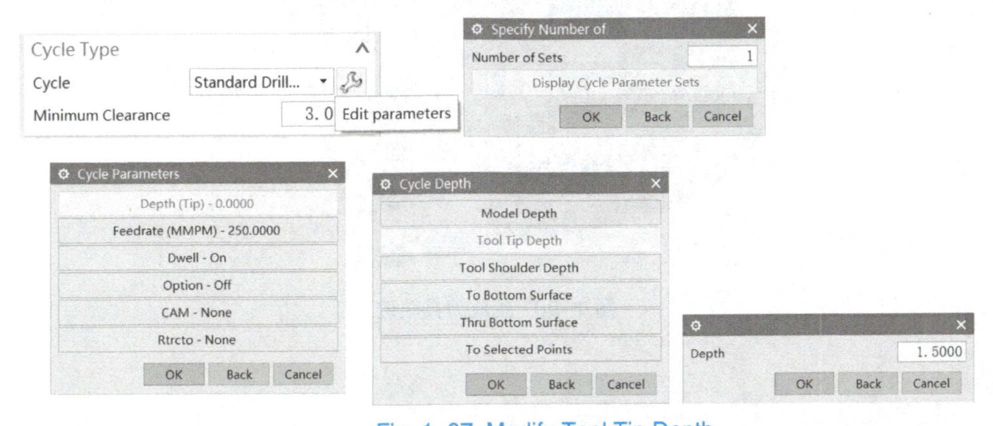

Fig. 1-87 Modify Tool Tip Depth

（5）设置【最小安全距离】为【15】，如图 1-88 所示。

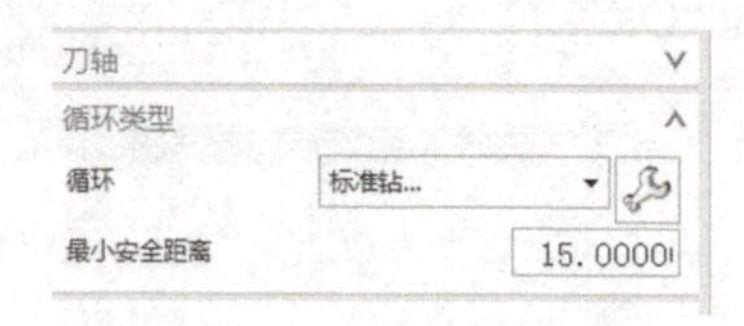

图 1-88 设置循环类型

（6）单击【进给率和速度】，打开【进给率和速度】对话框，【主轴速度】设为【1200】，进给率【切削】设为【100 mmpm】，如图 1-89 所示，单击【确定】。

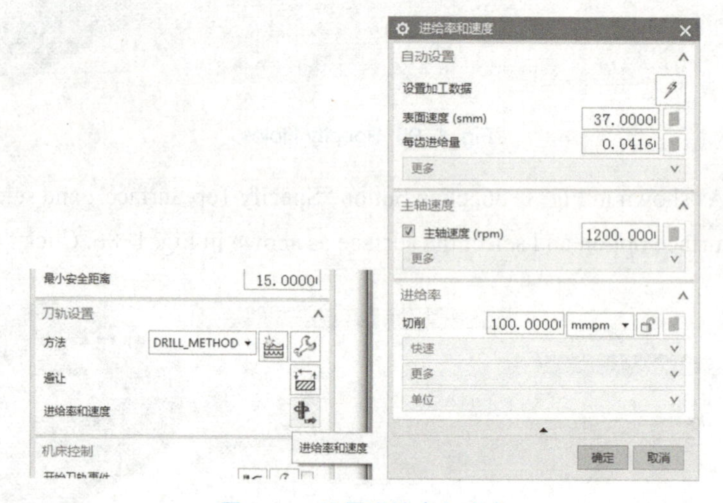

图 1-89 设置进给率和速度

（7）单击【生成】，查看生成的刀具轨迹，如图 1-90 所示，单击【确定】。

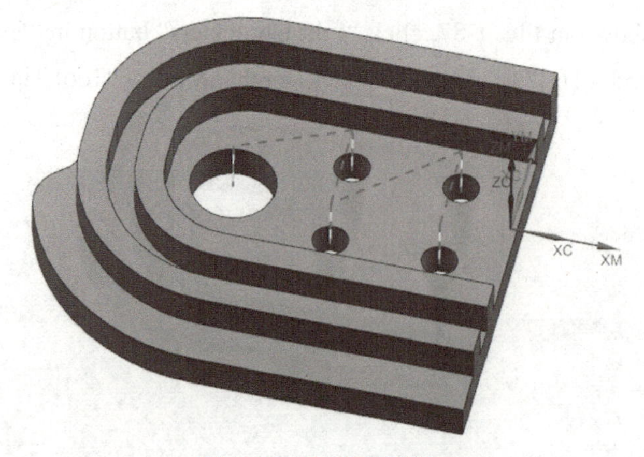

图 1-90 生成刀具轨迹

(5) Modify "Minimum Clearance" as 15 as shown in Fig. 1-88.

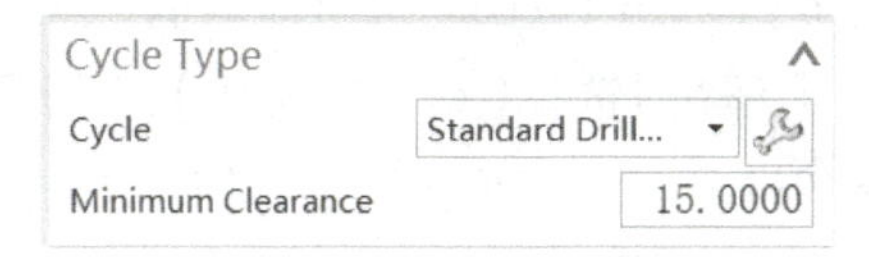

Fig. 1-88 Modify Cycle Type

(6) Click "Feeds and Speeds" button. Then, set "Spindle Speed" value as 1,200 and "Cut" value as 100 mmpm. Then, click "OK". The parameter settings are shown in Fig. 1-89.

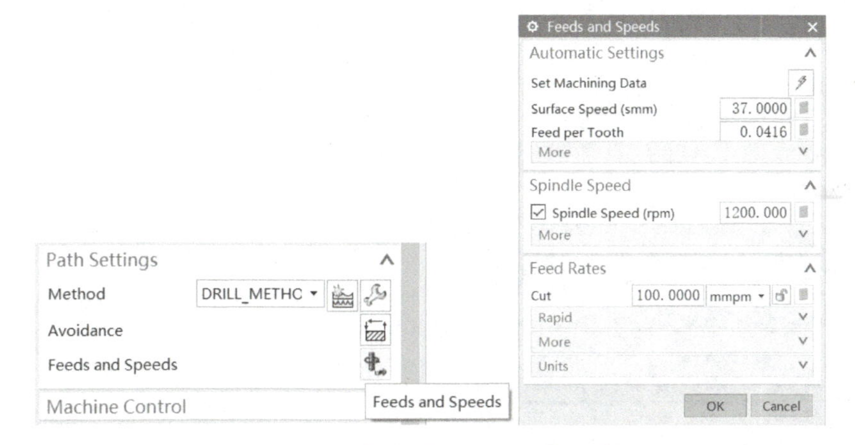

Fig. 1-89 Modify Feeds and Speeds

(7) Click "Generate" to view the tool path as shown in Fig. 1-90, and then click "OK".

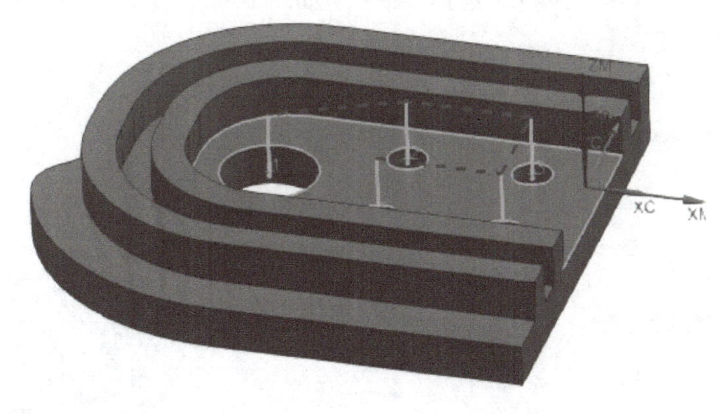

Fig. 1-90 Generate Tool Path

8. 钻 D10 通孔

（1）单击【创建工序】，打开【创建工序】对话框，【类型】选择【drill】，【工序子类型】选择【啄钻】，【程序】选择【钻 D10 通孔】，【刀具】选择【D10DRILL】，【几何体】选择【WORKPIECE】，【方法】选择【DRILL_METHOD】，修改【名称】为【Drill-2】，如图 1-91 所示，单击【确定】。

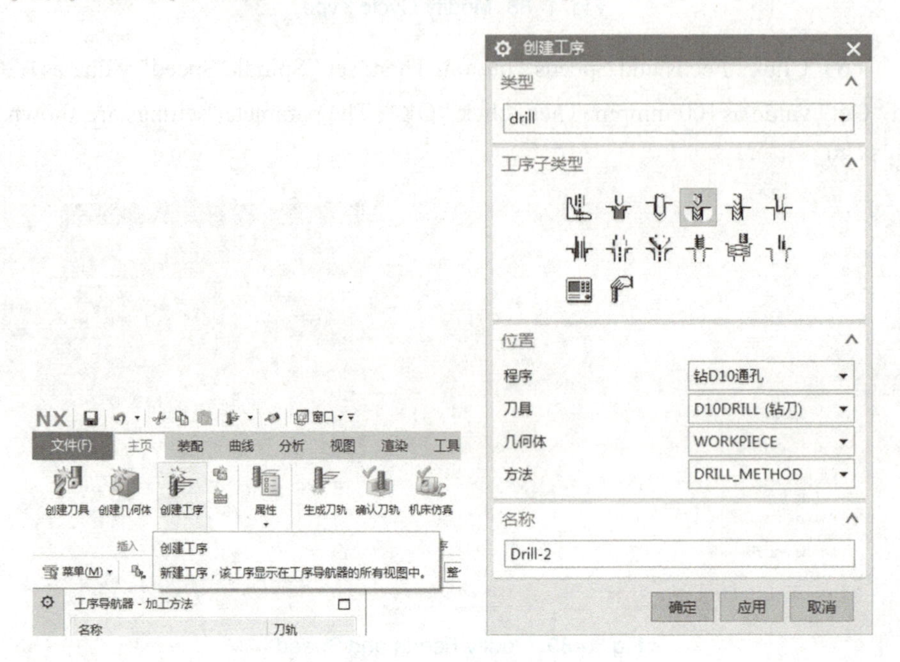

图 1-91　创建工序

（2）单击【指定孔】中的【选择或编辑孔几何体】，在打开的【点到点几何体】对话框中，单击【选择】，选择 5 个孔，如图 1-92 所示，依次单击【确定】。

图 1-92　指定孔

8. Drill D10 Thru Hole

(1) As shown in Fig. 1-91, click "Create Operation", and then select "drill" in pull-down box of "Type". Select operation subtype as "Peck Drilling", program as "DRILL D10 THRU HOLE", tool as "D10DRILL", geometry as "WORKPIECE" and method as "DRILL_METHOD". Modify name as "DRILL-2", and then click "OK".

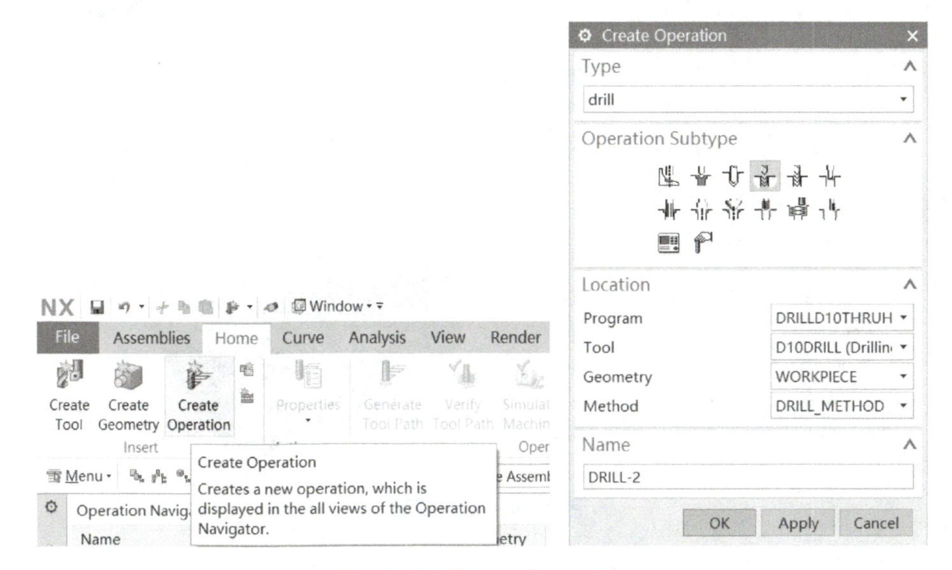

Fig. 1-91　Create Operation

(2) Click "Specify Holes" and then click "Select". Select boundary of the hole as shown in Fig. 1-92. Click "OK".

Fig. 1-92　Specify Holes

（3）单击【指定顶面】中的【选择或编辑部件表面几何体】，打开【顶面】对话框，【顶面选项】选择【面】，选择面，如图 1–93 所示，单击【确定】。

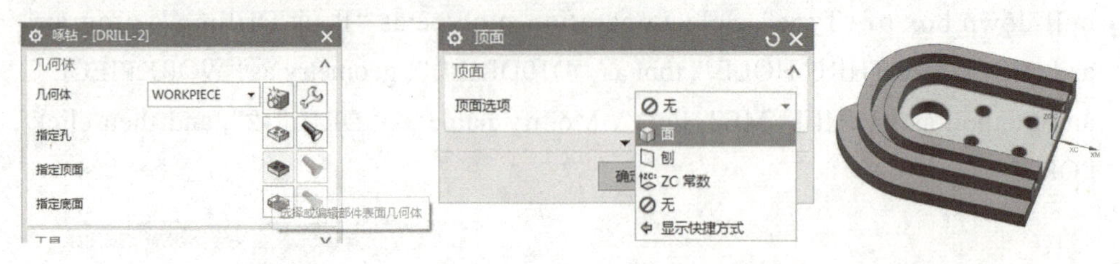

图 1–93　指定顶面

（4）单击【指定底面】中的【选择或编辑底面几何体】，选择零件底面，如图 1–94 所示，单击【确定】。

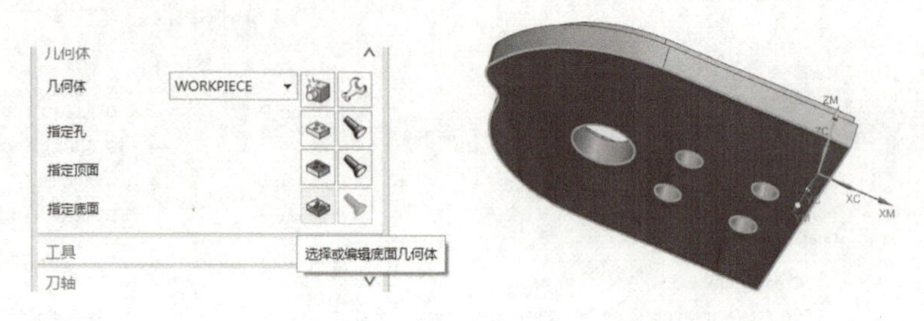

图 1–94　指定底面

（5）单击【循环】中的【编辑参数】，在打开的【指定参数组】对话框中，单击【确定】，在打开的【Cycle 参数】对话框中，单击【Depth- 模型深度】，在打开的【Cycle 深度】对话框中，单击【穿过底面】，如图 1–95 所示，依次单击【确定】。

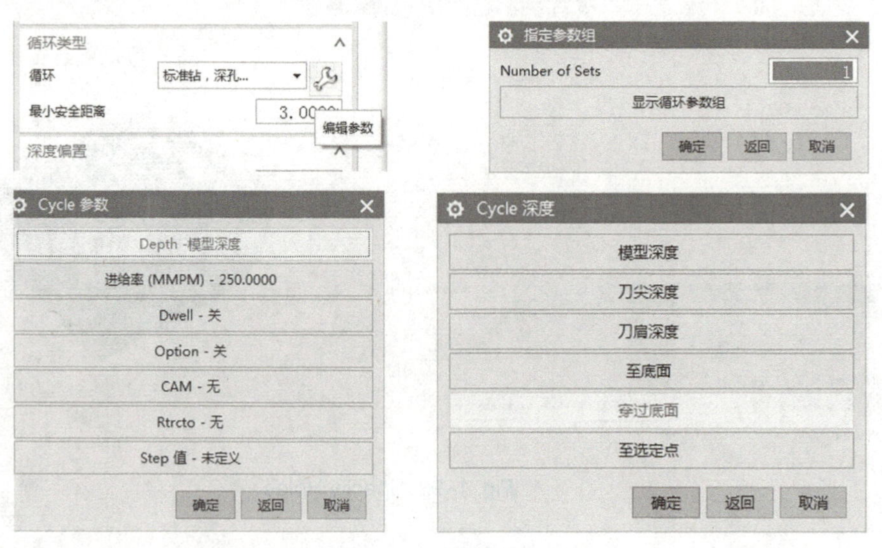

图 1–95　指定循环

(3) As shown in Fig. 1-93, click button “Specify Top Surface”, and select “Face” as the top surface option and select the surface. Click “OK”.

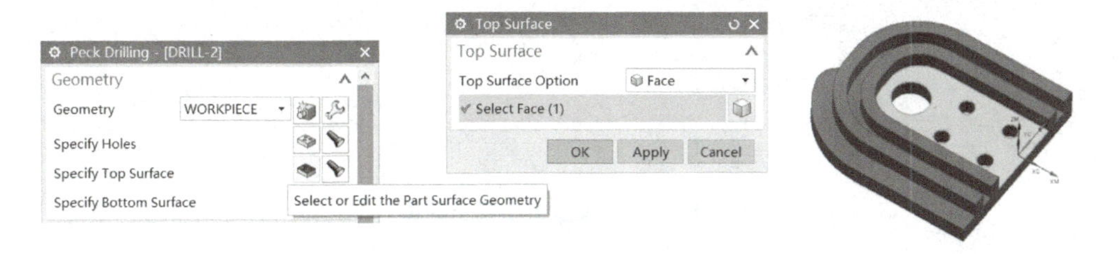

Fig. 1-93 Specify Top Surface

(4) Click button “Specify Bottom Surface”, and select “Face” as the bottom surface option and select the surface as shown in Fig. 1-94. Click “OK”.

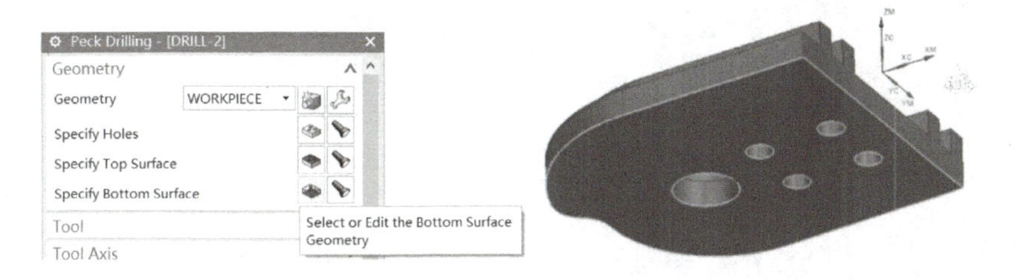

Fig. 1-94 Specify Bottom Surface

(5) Click “Cycle Type” tab. Click “Edit Parameters” button. Then, click “Depth (Tip)” to choose “Thru Bottom Surface”. Click “OK”. The steps are shown in Fig. 1-95.

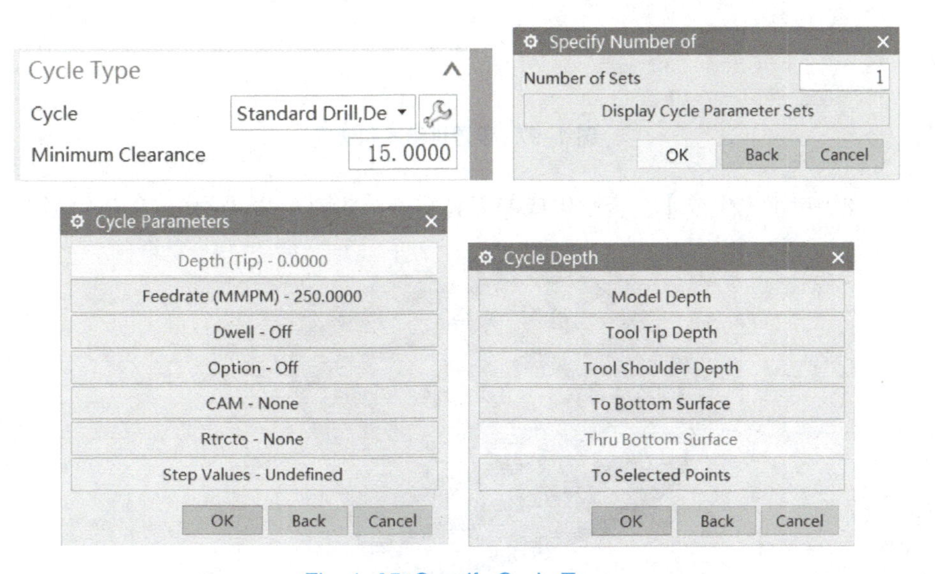

Fig. 1-95 Specify Cycle Type

（6）【最小安全距离】设置为【15】，如图 1-96 所示。

图 1-96 设置最小安全距离

（7）单击【进给率和速度】，打开【进给率和速度】对话框，【主轴速度】设为【800】，进给率【切削】设为【100 mmpm】，如图 1-97 所示，单击【确定】。

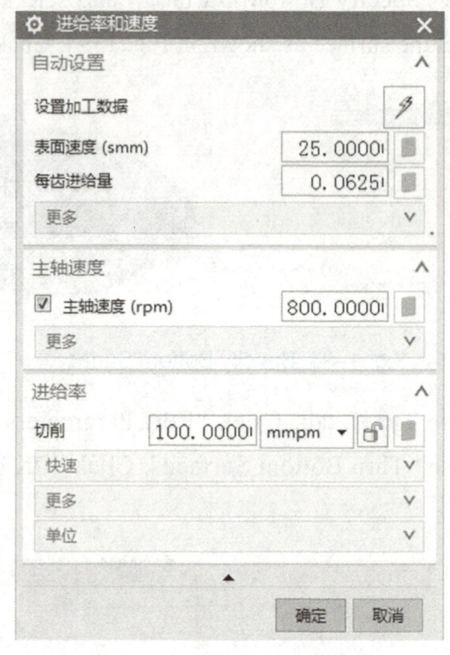

图 1-97 设置进给率和速度

（8）单击【生成】，查看生成的刀具轨迹，如图 1-98 所示，单击【确定】。

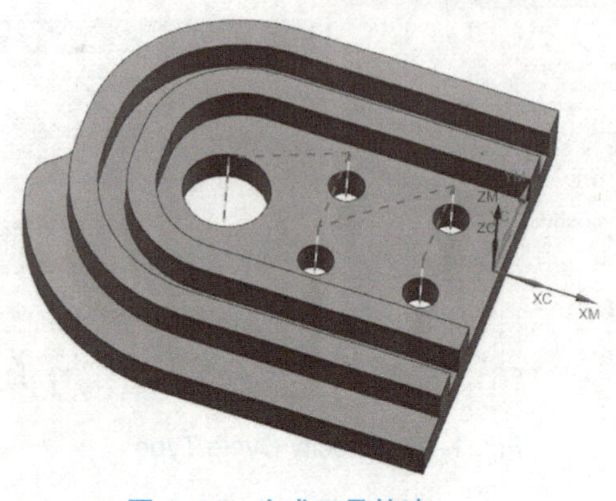

图 1-98 生成刀具轨迹

（6）Modify "Minimum Clearance" as 15 as shown in Fig. 1-96.

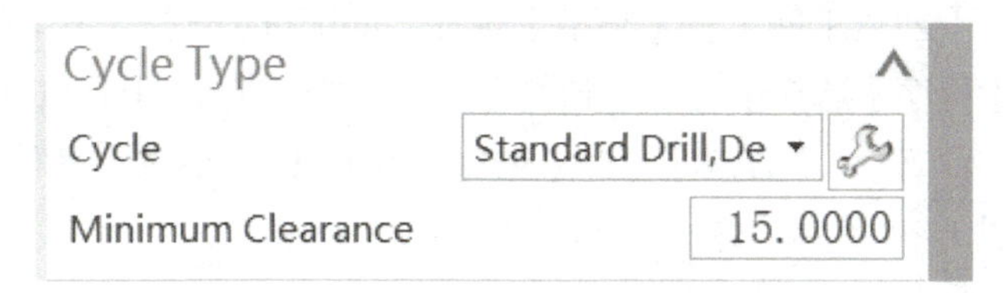

Fig. 1-96　Modify Minimum Clearance

（7）Click "Feeds and Speeds", modify "Spindle Speed" as 800 and "Feed Rates" as 100 mmpm as shown in Fig. 1-97. Click "OK".

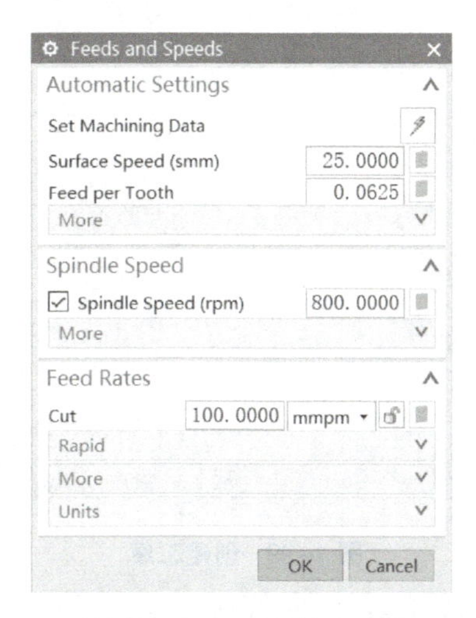

Fig. 1-97　Modify Feeds and Speeds

（8）Click "Generate" to view the tool path as shown in Fig. 1-98, and then click "OK".

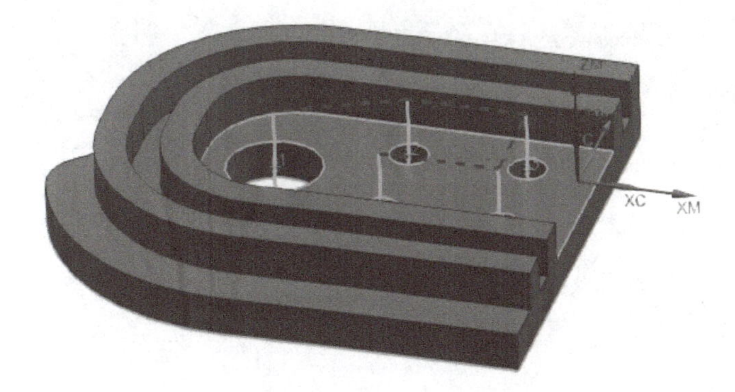

Fig. 1-98　Generate Tool Path

9. 铣 D24 通孔

（1）单击【创建工序】，【类型】选择【mill_planar】，【工序子类型】选择【孔铣】，【程序】选择【铣 D24 通孔】，【刀具】选择【D10R0】，【几何体】选择【WORKPIECE】，【方法】选择【MILL_FINISH】，修改【名称】为【Finish_mill-9】，如图 1-99 所示，单击【确定】。

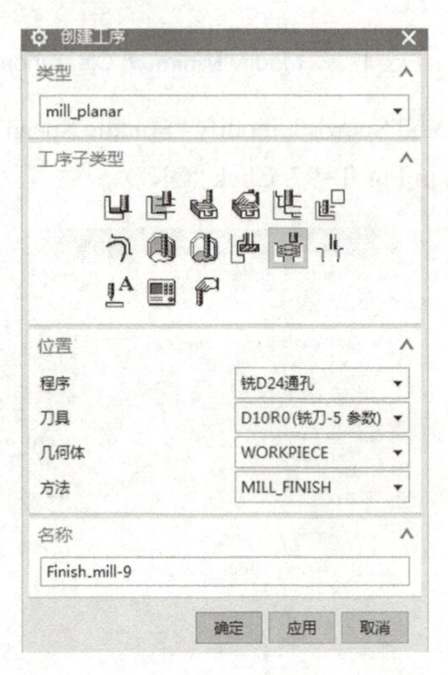

图 1-99 创建工序

（2）单击【指定特征几何体】中的【选择或编辑特征几何体】，选择特征面，修改【深度】为【10.5】，如图 1-100 所示，单击【确定】。

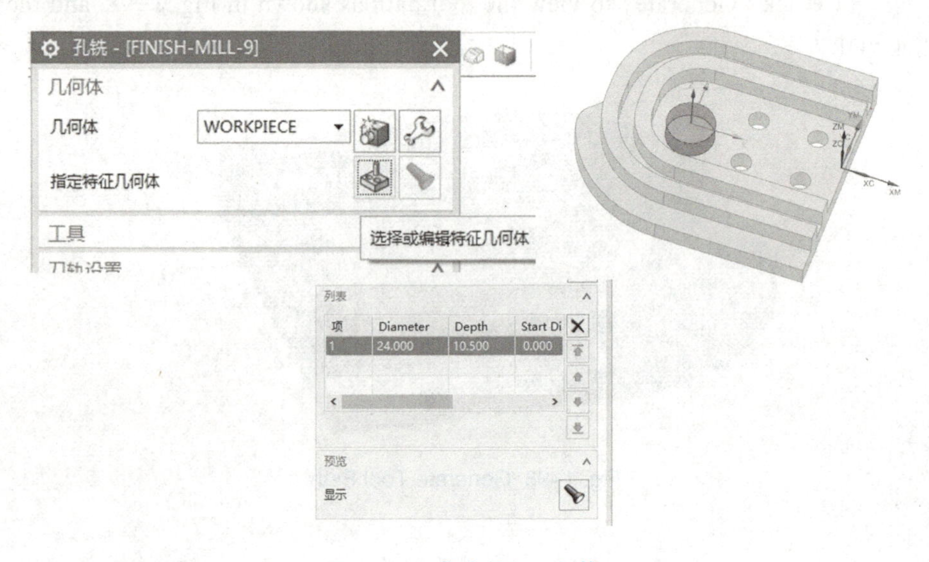

图 1-100 指定特征几何体

9. Mill D24 Thru Hole

(1) As shown in Fig. 1-99, click "Create Operation", and then select "mill_planar" in pull-down box of "Type". Select operation subtype as "Hole Milling", program as "MILL D24 THRUHOLE", tool as "D10R0", geometry as "WORKPIECE" and method as "Mill_Finish". Modify name as "Finish_mill-9", and then click "OK".

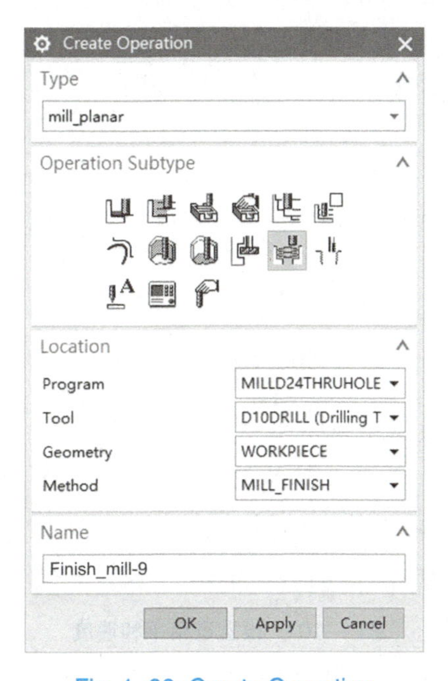

Fig. 1-99 Create Operation

(2) Click "Specify Feature Geometry", select the side wall of the hole as shown in Fig. 1-100. Modify "Depth" as 10.5. Click "OK".

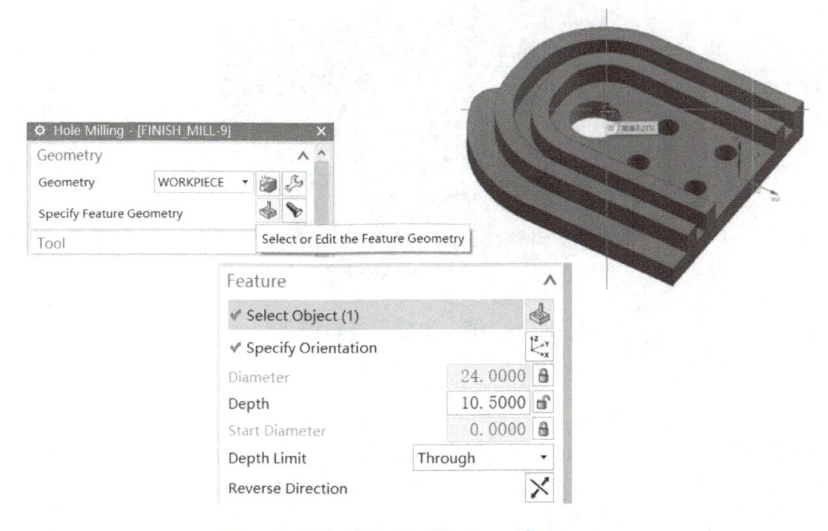

Fig. 1-100 Specify Feature Geometry

（3）单击【进给率和速度】，打开【进给率和速度】对话框，修改的【主轴速度】为【1800】，进给率【切削】为【800 mmpm】，如图 1-101 所示。

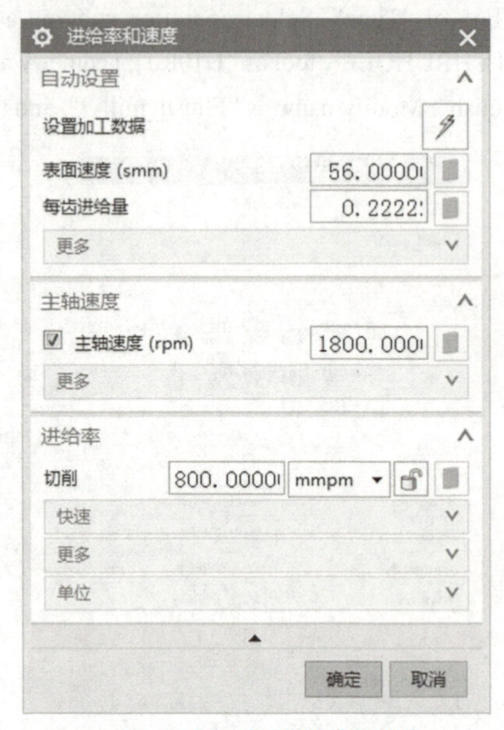

图 1-101　设置进给率和速度

（4）单击【生成】，查看生成的刀具轨迹，如图 1-102 所示，单击【确定】。

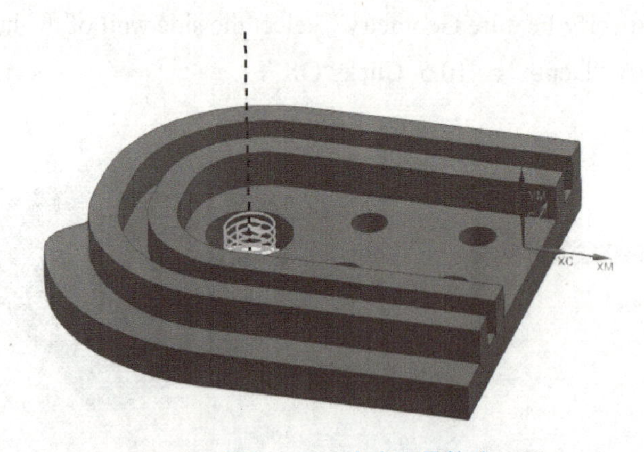

图 1-102　生成刀具轨迹

(3) Click “Feeds and Speeds”, modify “Spindle Speed” as 1,800 and “Feed Rates” as 800 mmpm as shown in Fig. 1-101. Click “OK”.

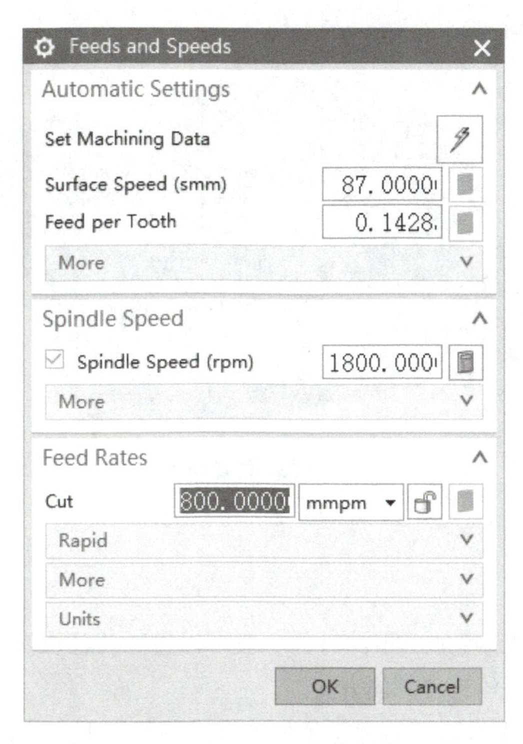

Fig. 1-101　Modify Feeds and Speeds

(4) Click “Generate” to view the tool path as shown in Fig. 1-102, and then click “OK”.

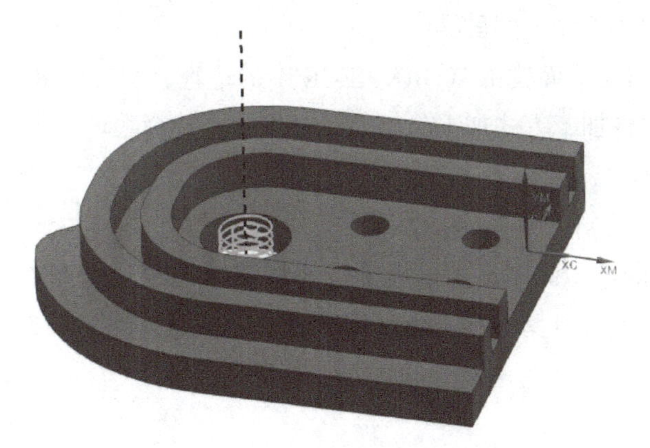

Fig. 1-102　Generate Tool Path

四、仿真加工

单击【NC_PROGRAM】，单击【确认刀轨】，选择【3D 动态】，单击【播放】，如图 1-103 所示。

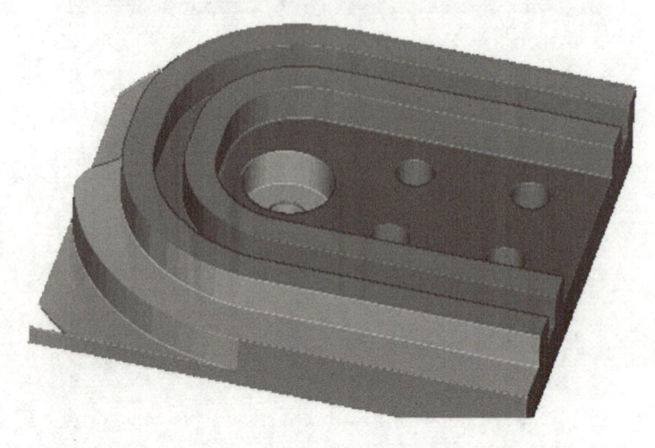

图 1-103 仿真结果

专家点拨

（1）使用自动编程加工零件时，一般可以遵循“轻拉快跑”的原则，也就是“小切削量、大进给速度”的方式。

（2）NX CAM 中材料侧的意思是加工过后需要留下来的材料，所以加工孔、腔体内部轮廓时，材料侧应该设为外，而加工岛、凸台等外部轮廓时，材料侧应该设为内。

（3）NX CAM 平面铣中，边界的平面是用来定义边界的开始加工高度，而底面是用来设置边界加工的最终深度。

（4）NX CAM 平面铣的 WORKPIECE 中的毛坯，只在仿真时起作用，当操作中使用跟随部件加工方式加工凸台时，必须重新选择毛坯边界。

课后训练

（1）根据图 1-104 所示的盖板零件的特征，制定合理的工艺路线，设置必要的加工参数，生成刀具轨迹、通过相应的后处理生成数控加工程序，并运用机床加工零件。

Ⅳ. Simulation Machining

Click "NC_PROGRAM", and then "Verify Tool Path" button. Select "3D Dynamic" and click "Play" button to view the simulation result. The simulation result is shown as in Fig. 1-103.

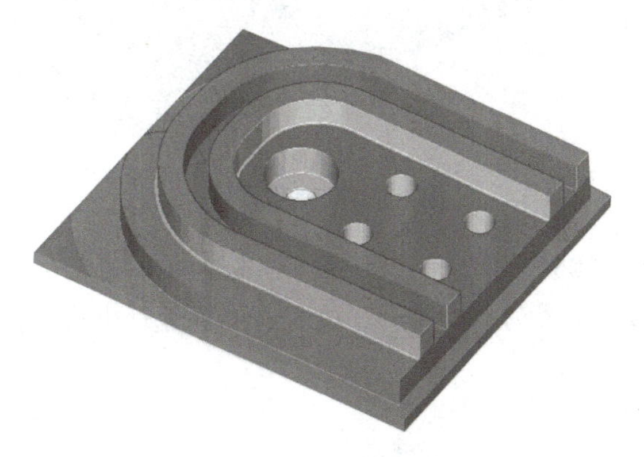

Fig. 1-103 Simulation Result

(1) When using automatic programming to process parts, you can generally follow the principle of "light pulling and fast running", that is, the way of decreasing cutting amount and increasing feed speed.

(2) In NX CAM, the material side means the material that needs to be left after processing. Therefore, when processing holes or internal cavities, the material side should be set as the outside, while when processing islands or boss, the material side should be set as the inside.

(3) In NX CAM, the plane of the boundary is used to define the starting machining height of the boundary, and the bottom is used to set the final machining depth of the boundary.

(4) In NX CAM , blank in workpiece of plane milling only works in simulation. If the following part processing method is used in operation, the blank boundary must be reselected.

Practice

(1) Make a reasonable processing technic, set necessary parameters, generate tool path, generate NC program through corresponding post-processor, and use machine tools to machine parts according to the characteristics of disc parts as shown in Fig. 1-104 .

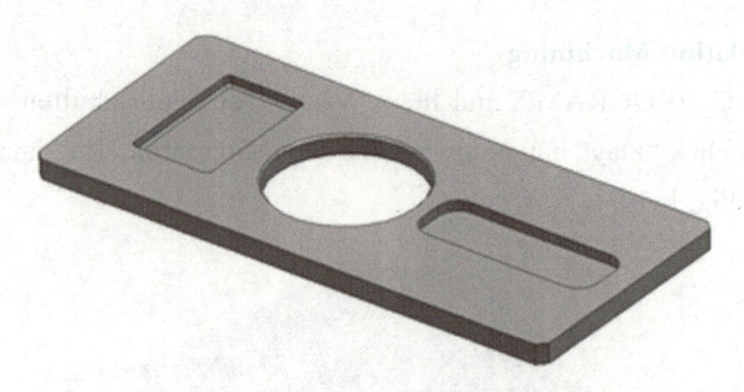

图 1-104 盖板零件

（2）根据图 1-105 所示的齿形压板零件的特征，制定合理的工艺路线，设置必要的加工参数，生成刀具轨迹、通过相应的后处理生成数控加工程序，并运用机床加工零件。

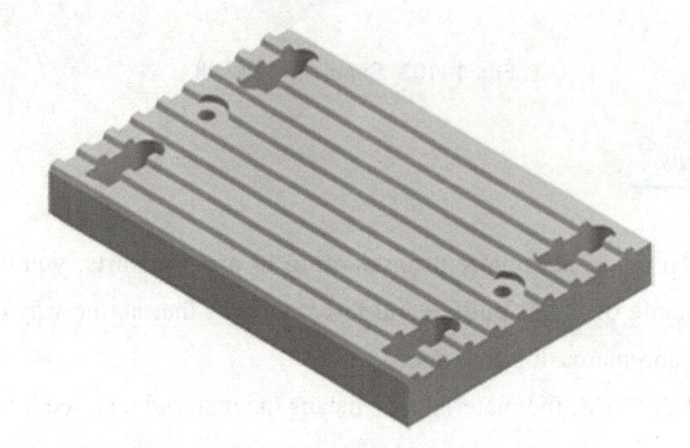

图 1-105 齿形压板零件

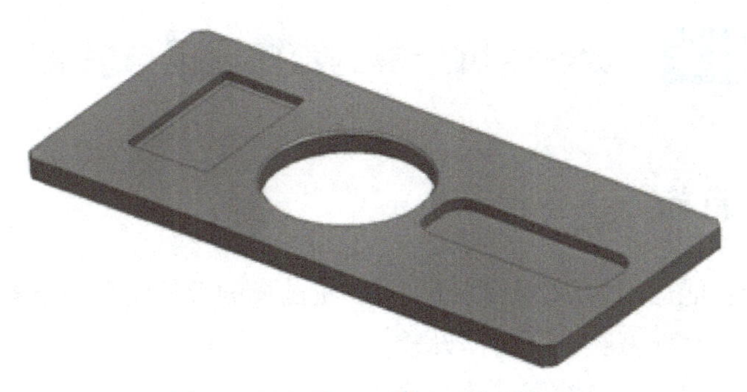

Fig. 1–104　Cover Plate Disc Parts

（2）Make a reasonable processing technic, set necessary parameters, generate tool path, generate NC program through corresponding post-processor, and use machine tools to machine parts according to the characteristics of disc parts as shown in Fig. 1-105.

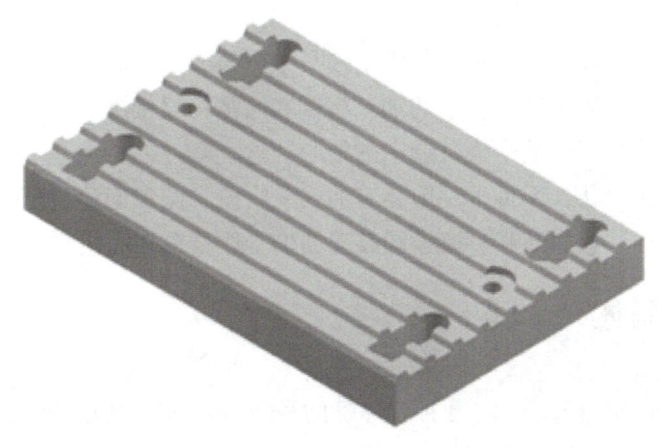

Fig. 1–105　Jugged Press Plate Disc Parts

项目 2 玩具相机凸模的数控编程与仿真加工

学习目标

能力目标：能运用 NX 软件完成玩具相机凸模的数控编程与仿真加工。

知识目标：掌握型腔铣几何体设置；

掌握固定轴区域轮廓铣几何体设置；

掌握清根加工几何体设置；

掌握非切削运动设置方法。

素质目标：激发学生的学习兴趣，培养团队合作和创新精神。

项目导读

注塑模具中的成型件凸模，是整个模具中的核心零件之一，也是数控加工中出现频率较高的一类零件。凸模的加工精度和质量将直接影响塑料件的精度和质量，因此，凸模加工是整个模具制造的核心。这类零件的特点是形状比较复杂，零件整体外形成块状，零件上一般会有曲面、台阶、圆弧面等特征。在编程与加工过程中要特别注意曲面和过渡小圆弧面的加工质量和表面粗糙度。

任务描述

学生以企业制造部门 MC 数控程序员的身份进入 NX CAM 功能模块，根据玩具相机凸模的特征，制定合理的工艺路线，创建型芯铣、等轴轮廓铣、参考刀具清根铣等加工操作，设置必要的加工参数，生成刀具轨迹，检验刀具轨迹是否正确合理，以及对操作过程中存在的问题进行研讨和交流。通过相应的后处理生成数控加工程序，并运用机床加工零件。

Project 2 NC Programming and Simulation Machining of Terrace Die for Toy Camera

Learning Objectives

Capacity Objective: Complete model programming and simulation machining of terrace die for toy camera with NX software.

Knowledge Objective: Master cavity milling geometry settings;

Master the fixed axis area contour milling geometry settings;

Master the geometry setting of flowcut;

Master the setting method of non-cutting move.

Quality Objective: Stimulate students' interest in learning and cultivate the spirit of teamwork as well as innovation.

Project Guidance

The terrace die of molded parts in injection mold is one of the core parts in the whole mold, and it is also a kind of parts with high frequency in NC machining. The machining accuracy and quality of terrace die will directly affect the accuracy and quality of plastic parts. Therefore, terrace die machining is the core of the whole mold manufacturing. The characteristics of this kind of parts are that the shape is complex, the overall shape of the parts is block, and the parts generally have the characteristics of curved surface, step, arc surface and so on. In the process of programming and machining, special attention should be paid to the machining quality and surface roughness of curved surface and transition small arc surface.

Task Description

Students operate NX CAM functional modules as MC programmers in enterprise manufacturing department. According to the characteristics of model parts, establish a reasonable processing route, create machining operations such as core milling, contour milling, reference tool flowcut, and set necessary machining parameters, generate tool path and check the generated tool path. Further, students should discuss the problems occurred in the operation. By selecting the corresponding post-processor to generate NC machining programs, students import them into machine tools to complete the parts processing.

项目实施

按照零件加工要求，制定玩具相机凸模加工工艺；编制玩具相机凸模加工工程序；完成玩具相机凸模的仿真加工，后处理得到数控加工程序，完成零件加工。

一、制定加工工艺

1. 玩具相机凸模零件分析

该零件形状比较复杂，主要由曲面、台阶、圆弧面、凹腔、分型面等组成，主要加工内容为凹腔、曲面、过渡面、分型面。经过对零件的分析，可知零件上最小的内凹圆弧半径为 2 mm，所以清根加工时刀具半径不能大于 2 mm。

2. 毛坯选用

零件材料为模具钢，尺寸为 160 mm × 110 mm × 43 mm。零件长、宽尺寸已经精加工到位，无须再次加工，零件厚度方向有 0.5 mm 余量，底面已经进行过精加工，无须再次加工。

3. 加工工序卡制定

零件选用立式三轴联动机床加工，平口钳夹持，遵循先粗后精加工原则。制定的加工工序卡如表 2–1 所示。

表 2–1　加工工序卡

零件号：26578932		工序名称：玩具相机凸模铣削加工			工艺流程卡_工序单
材料：模具钢		页码：1		工序号：01	版本号：0
夹具：平口钳		工位：MC		数控程序序号：	
刀具及参数设置					
加工内容	刀具号	刀具规格	主轴转速/rpm	进给速度/mmpm	
型面粗加工-1	T01	D25R5	1800	1200	
型面粗加工-2	T02	D10R1-ROUGH	2600	1400	
型面粗加工-3	T03	D6R0.5	3200	1000	
型面半精加工	T04	D6R3-SEMI-FINISH	3500	1600	
精铣平面	T07	D10R1-FINISH	2600	1400	
型面精加工-1	T05	D6R3-FINISH	3500	1600	
型面精加工-2	T05	D6R3-FINISH	3500	1600	
清根	T06	D3R1.5	4500	1000	
更改号	更改内容	批准	日期		
拟制： 日期：	审核： 日期：	批准：	日期：		

Project Implementation

Firstly, formulate processing technic under processing requirements. Secondly, program for machining model. Finally, complete simulation machining , and then import the NC machining program obtained by post processor to the machine tool to complete machining.

I. Formulate Processing Technic

1. Analysis of punch parts of toy camera

The shape of the part is complex. It is mainly composed of curved surface, step, arc surface, cavity, parting surface and other features. The main processing contents are cavity, curved surface, transition surface and parting surface. Through the analysis of the part, it can be seen that the minimum concave arc radius on the part is 2 mm, so the tool radius cannot be greater than 2 mm during flowcut.

2. Blank selection

The part material is die steel, and the size is 160 mm × 110 mm × 43 mm. The length and width of the part have been finished. There is a 0.5 mm allowance in the thickness direction of the part. The bottom surface has been finished and no machining is required.

3. Formulation of processing procedure card

The parts are processed by vertical 3-axis linkage machine tool, clamped by flat pliers, and follow the processing principle of rough before fine, face before hole. The processing procedure card is shown in Table 2-1.

Table 2-1　Processing Procedure Card

Part number: 26578932		Name of process: Milling of terrace die for toy camera		Process card - Process sheet	
Material: Die steel		Page number: 1		Procedure number: 01	Version number: 0
Fixture: Parallel-jaw vice		Work station: MC		CNC program number:	

Tool and parameter setting

Tool number	Tool specification	Processing content	Spindle speed /rpm	Feed speed /mmpm
T01	D25R5	Rough machining-1	1800	1200
T02	D10R1-ROUGH	Rough machining-2	2600	1400
T03	D6R0. 5	Rough machining-3	3200	1000
T04	D6R3-SEMI-FINISH	Semi-finishing	3500	1600
T07	D10R1-FINISH	Finishing face milling	2600	1400
T05	D6R3-FINISH	Finishing-1	3500	1600
T05	D6R3-FINISH	Finishing-2	3500	1600
T06	D3R1. 5	Flowcut	4500	1000

Change number	Change content		Approve	Date
02				
01				
Draws:	Date: review:	Date:	Approve:	Date:

二、加工准备

（1）启动NX，单击【打开】，选择【玩具相机凸模.prt】文件，如图2-1所示，单击【OK】，打开零件模型。

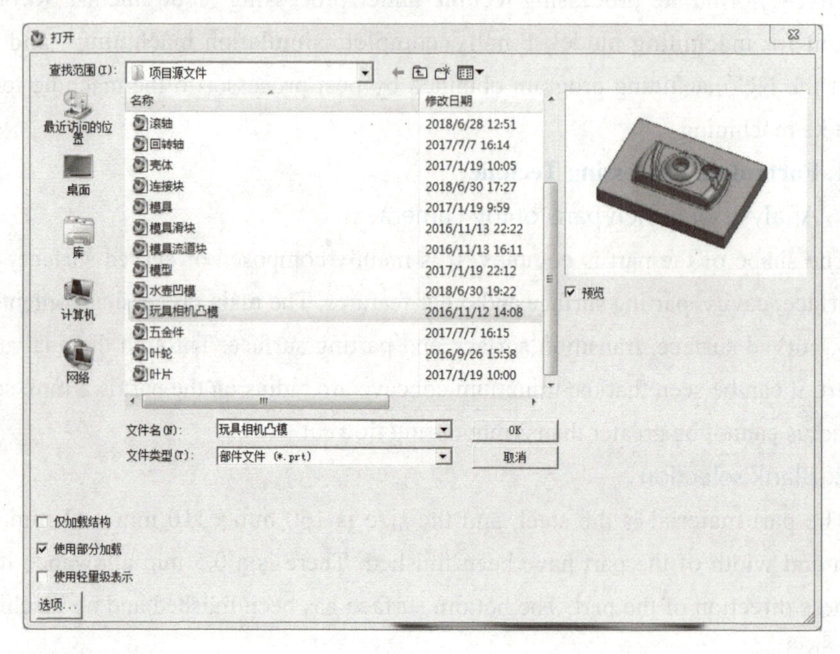

图 2-1　打开文件

（2）选择【应用模块】选项卡，单击【加工】，进入加工环境，如图2-2所示。

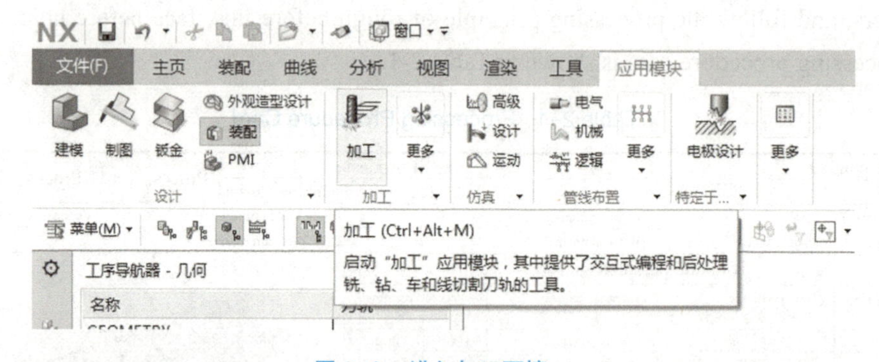

图 2-2　进入加工环境

（3）在弹出的【加工环境】对话框中，在【CAM会话配置】选项中选择【cam_general】，在【要创建的CAM设置】选项中选择【mill_planar】，如图2-3所示，单击【确定】。

Ⅱ. Preparation for Processing

(1) As shown in Fig. 2-1, start NX, click "Open", select model "Terrace Die for Toy Camera" (.prt file), and click "OK". Then, the part model can be seen in WCS.

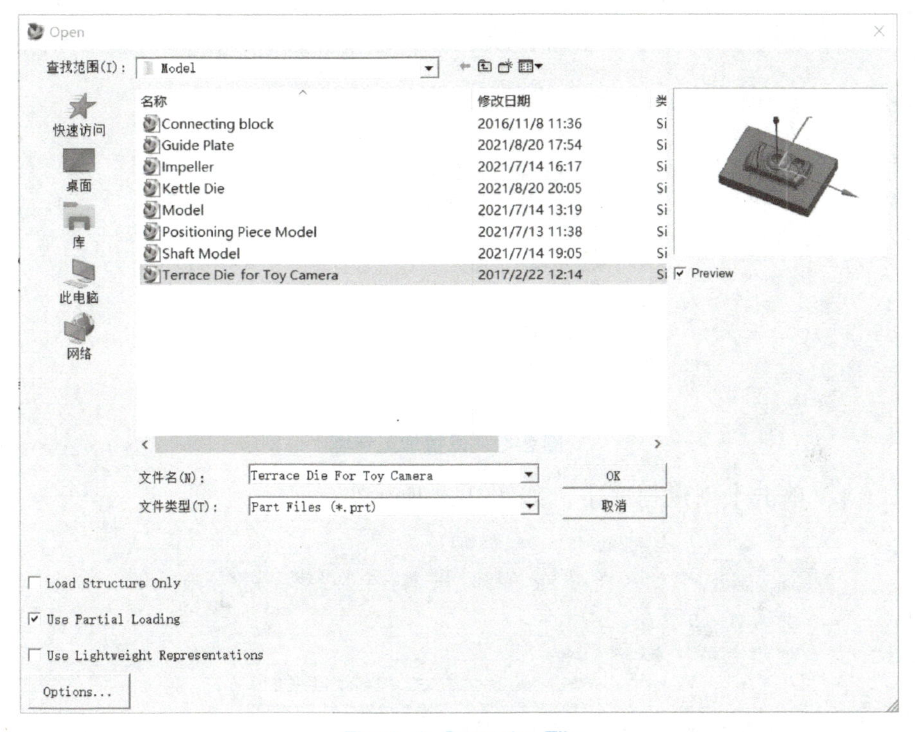

Fig. 2-1　Open the File

(2) Select "Application", and click "Manufacturing". The manufacturing menu can be seen as shown in Fig. 2-2.

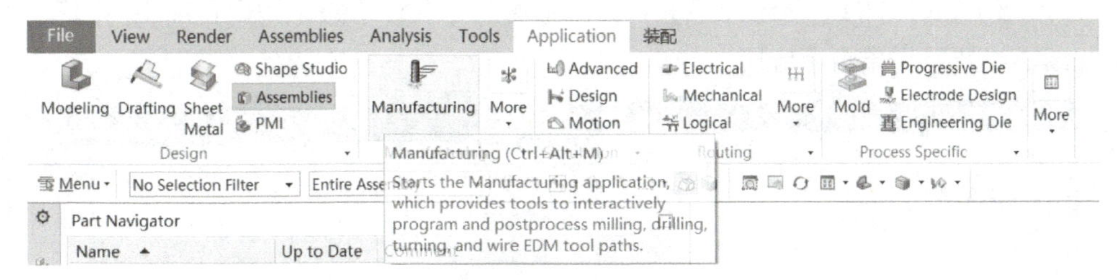

Fig. 2-2　Enter Manufacturing Environment

(3) As shown in Fig. 2-3, in "Machining Environment" window, select "cam_general" in "CAM Session Configuration" tab, and then, select "mill_planar" in "CAM Setup to Create" tab. Finally, click "OK".

图 2-3　设置加工环境

（4）单击【创建程序】，如图 2-4 所示。

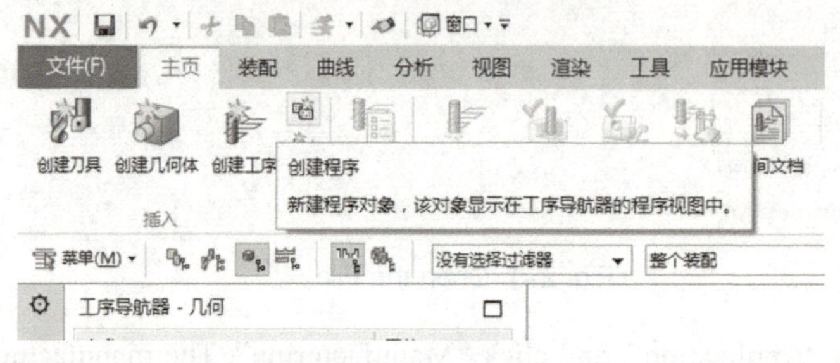

图 2-4　创建程序

（5）在打开的【创建程序】对话框中修改【名称】为【型面粗加工 -1】，如图 2-5 所示，单击【应用】。

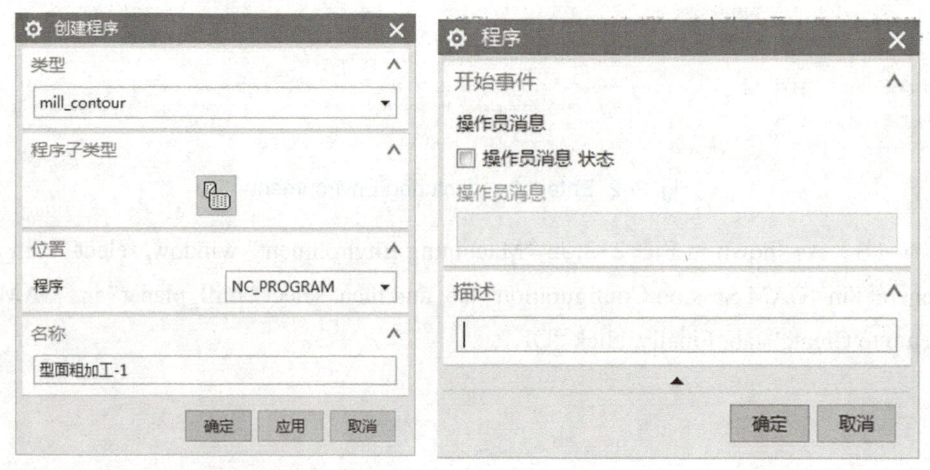

图 2-5　粗加工

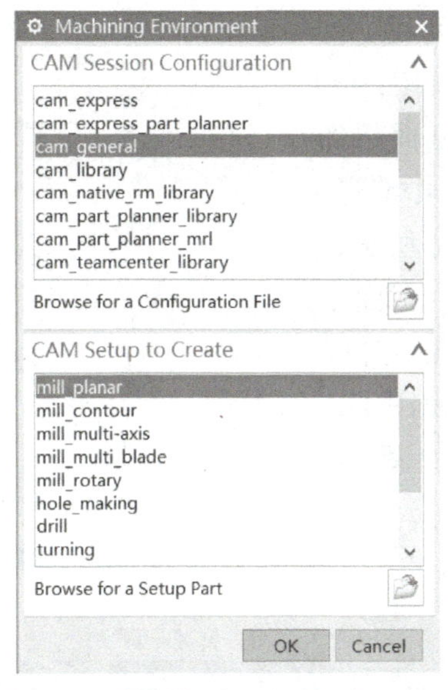

Fig. 2-3 Set Machining Environment

(4) As shown in Fig. 2-4, click "Create Program".

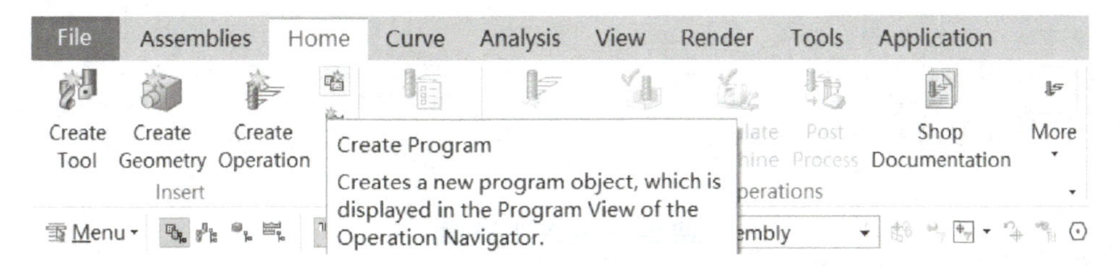

Fig. 2-4 Create Program

(5) As shown in Fig. 2-5, modify name as "ROUGH_MILL-1" in the dialog of "Name". Click "Apply".

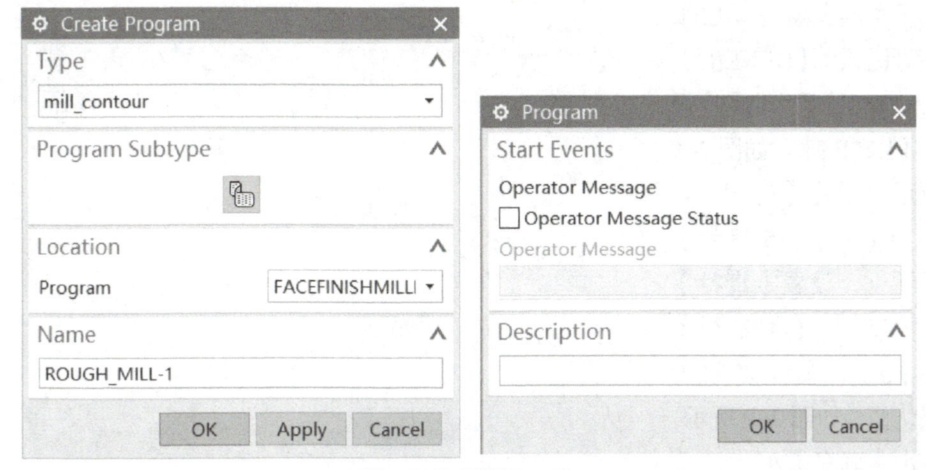

Fig. 2-5 Mill Rough

（6）同样的方法创建【型面粗加工 –2】【型面粗加工 –3】【型面半精加工】【型面精加工 –1】【型面精加工 –2】【清根】【精铣平面】，如图 2–6 所示，将【精铣平面】插到【型面半精加工】后面，单击【确定】。

（7）单击【机床视图】，单击【创建刀具】，如图 2–7 所示。

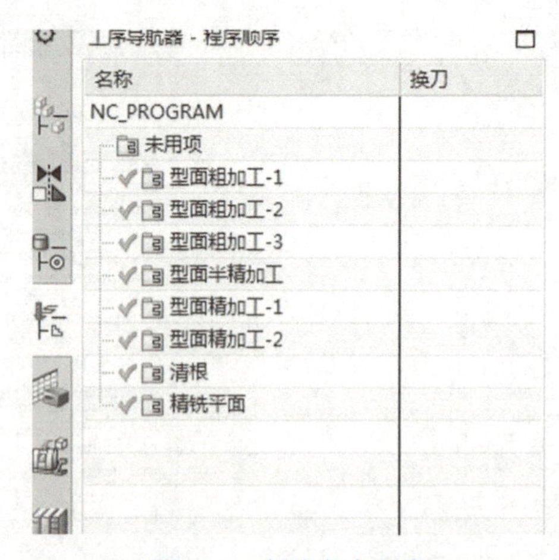

图 2–6　创建多个程序

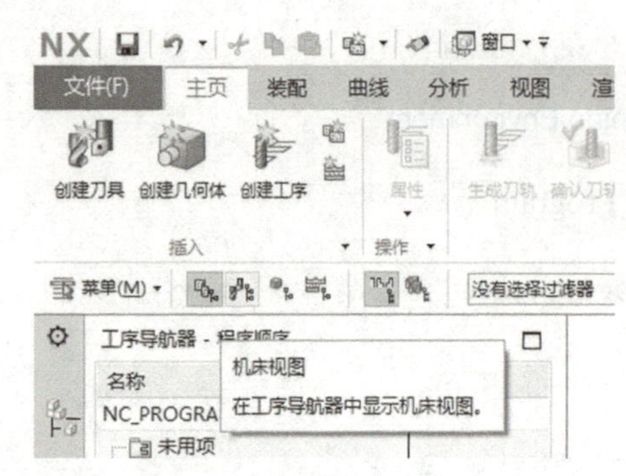

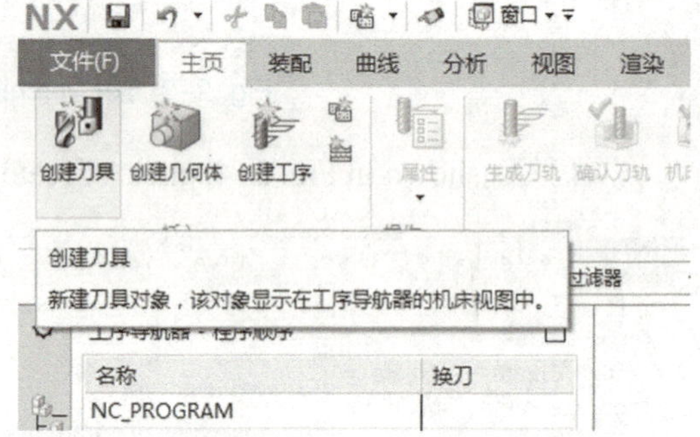

图 2–7　进入机床视图

（8）在【创建刀具】对话框中，【类型】选择【mill_planar】，【刀具子类型】选择【MILL】，【刀具】设为【GENERIC_MACHINE】，【名称】设为【D25R5】，如图 2–8 所示，单击【应用】。弹出【铣刀 –5 参数】对话框，设置【直径】为【25】，【下半径】为【5】，其他参数为默认值，如图 2–9 所示，单击【确定】。

图 2–8　创建刀具

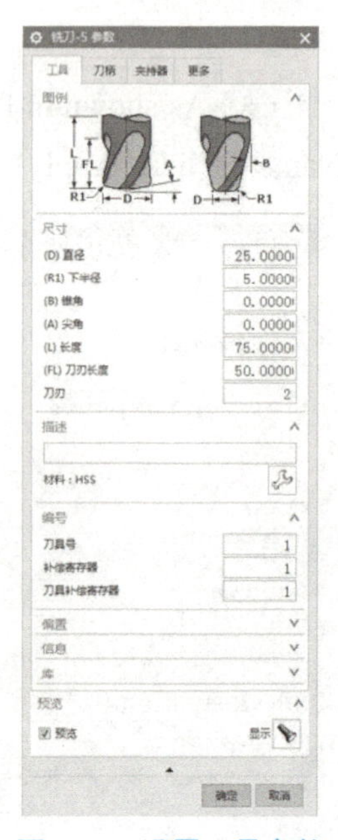

图 2–9　设置刀具参数

(6) Similarly, create programs and modify the names as "ROUGH MILL-2" "ROUGH MILL-3" "SEMI-ROUGH MILL" "FINISHING-1" "FINISHING-2" "FLOWCUT" "FACE FINISH MILLING". The program groups are shown in Fig. 2-6.

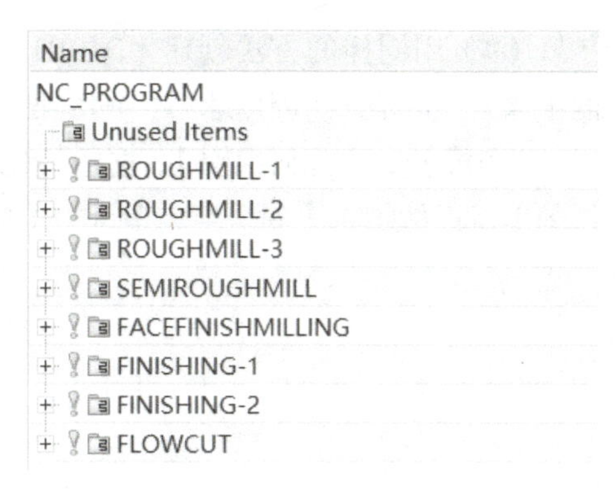

Fig. 2-6 Create Programs

(7) Click "Machine Tool View", and then click "Create Tool", as shown in Fig. 2-7.

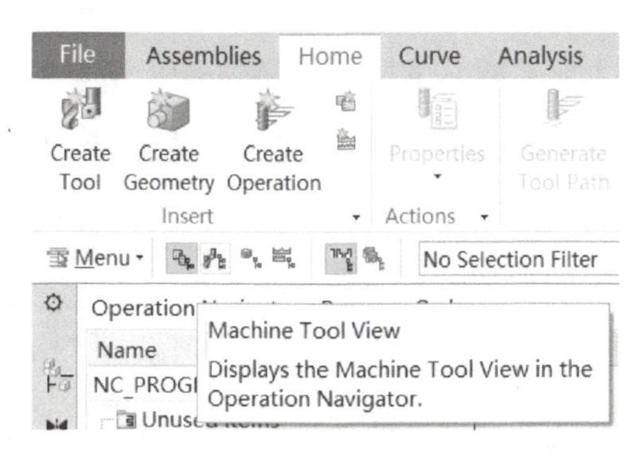

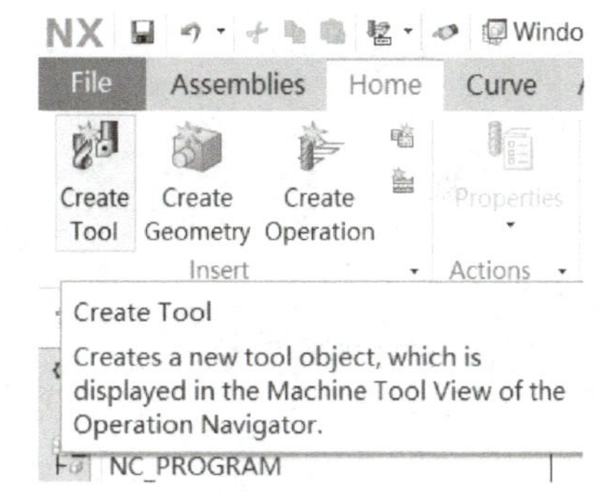

Fig. 2-7 Enter Machine Tool View

(8) As shown in Fig. 2-8, in the "Create Tool" window, select "MILL" in "Tool Subtype", "Tool Location" as "GENERIC_MACHINE", and modify "Name" as "D25R5", then click "OK". In "Milling Tool-5 Parameters" window as shown in Fig.2-9, "Diameter" value is 25 and "Lower Radius" value is 5. Other parameters default, click "OK" to close the window.

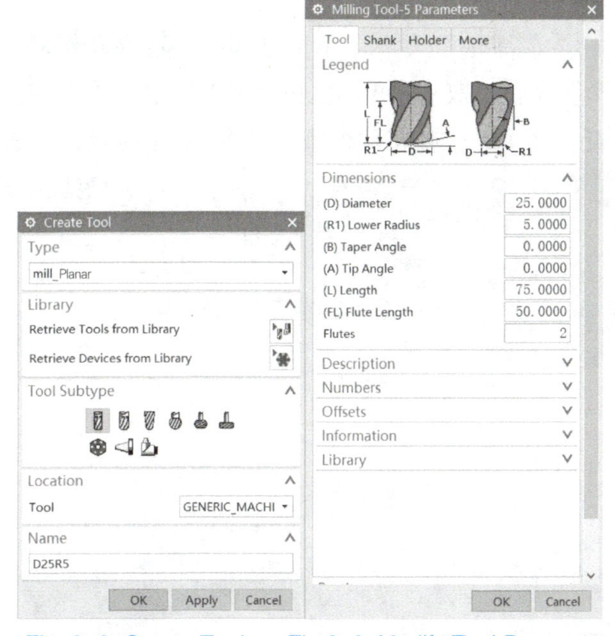

Fig. 2-8 Create Tool   Fig.2-9 Modify Tool Parameters

（9）用同样的方法创建【D10R1-ROUGH】【D6R0.5】【D10R1-FINISH】的牛鼻刀。同样方法创建球刀【D6R3-SIMI-FINISH】【D6R3-FINISH】【D3R1.5】。

（10）进入【几何视图】，选择【坐标】，选择【绝对坐标】，在打开的【MCS铣削】对话框中将【安全距离】设为【50】，如图2-10所示，单击【确定】。

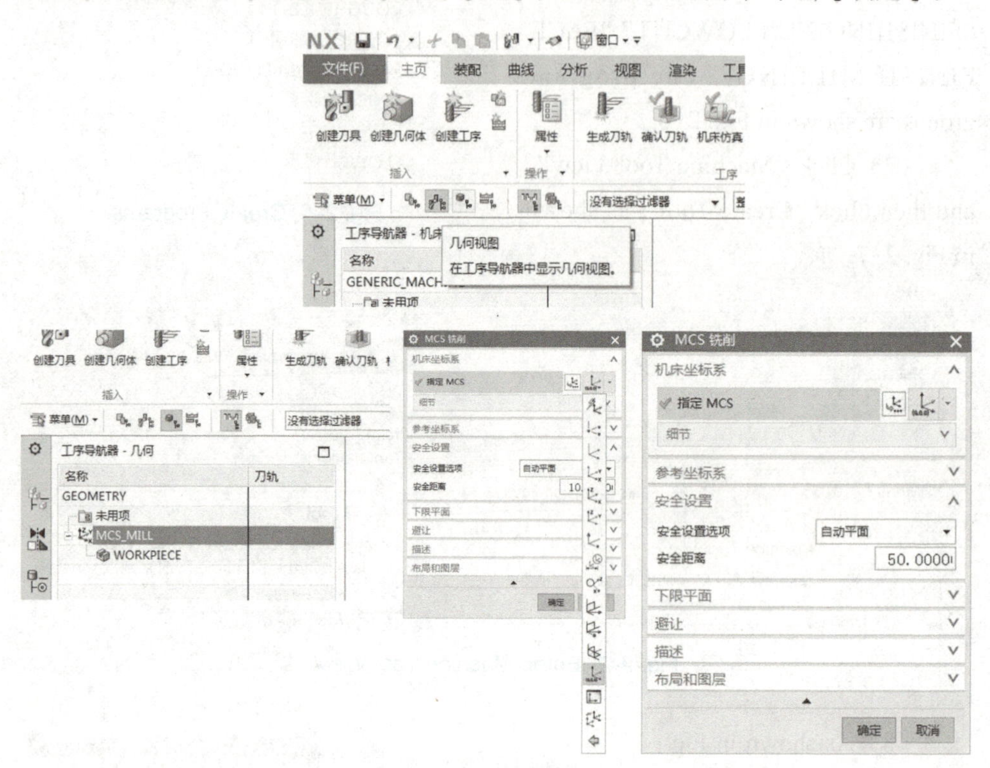

图2-10 建立坐标系

（11）双击【WORKPIECE】，单击【指定部件】中的【选择或编辑部件几何体】，选择工件模型，如图2-11所示，单击【确定】。

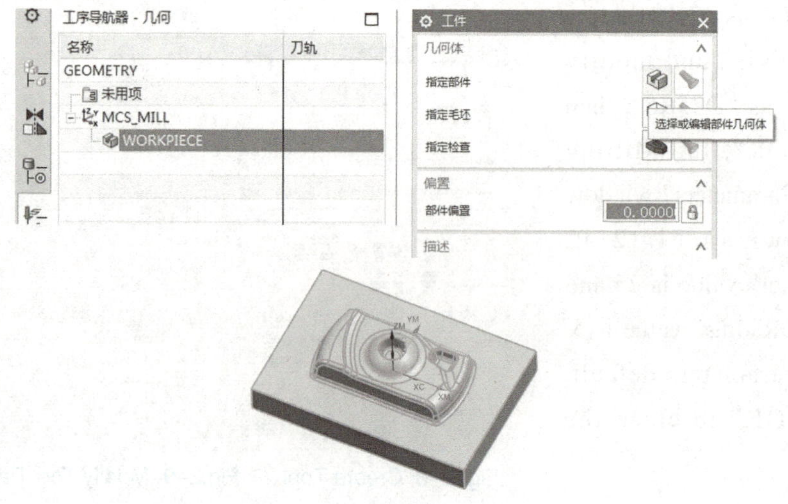

图2-11 指定部件几何体

（9）Create new tools named “D10R1-ROUGH” “D6R0.5” “D10R1-FINISH” in the same way, and create balltool “D6R3-SIMI-FINISH”“D6R3-FINISH”“D3R1.5”.

（10）Click “Geometry View” and then double click “MCS_MILL”, in “MCS Mill” window, click “CSYS Dialog”, select “Absolute-Displayed Part”, and change “Safe Clearance Distance ” to 50 as shown in Fig. 2-10, and then click “OK” to close the window.

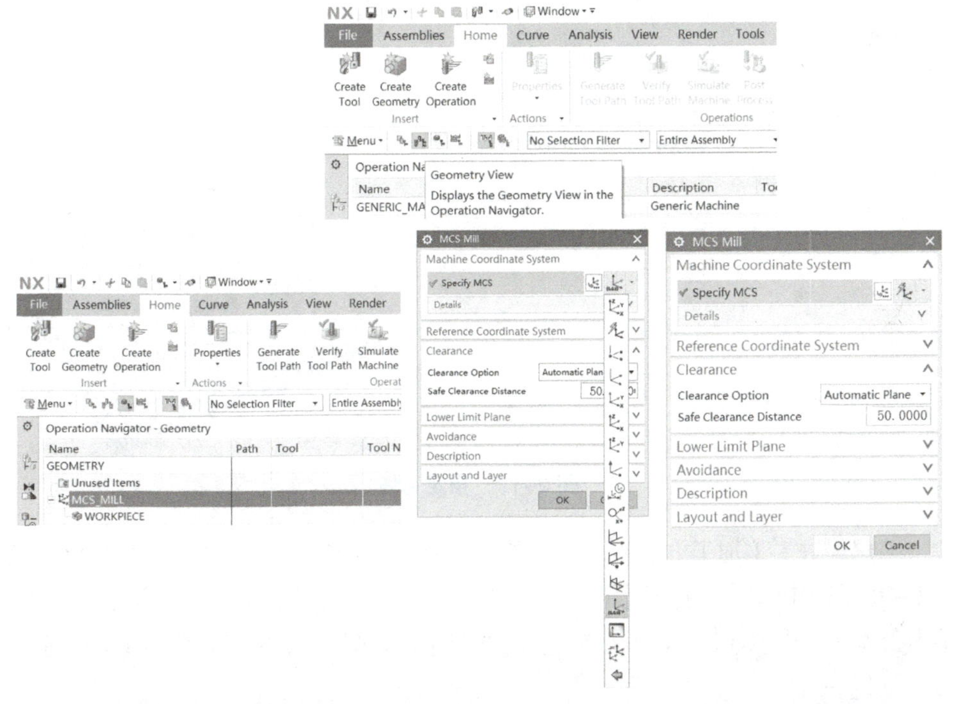

Fig. 2-10　Set MCS

（11）Double click “WORKPIECE” and click “Select or Edit the Part Geometry” in “Workpiece” window. In “Part Geometry” window, select the solid part as Fig. 2-11. Click “OK” to close the window.

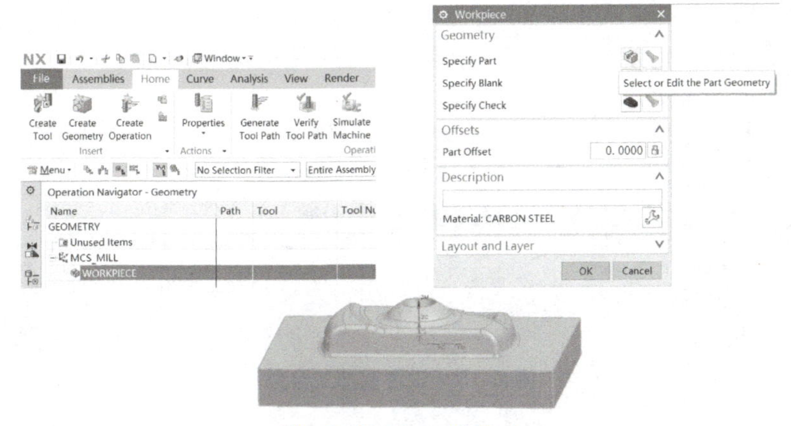

Fig. 2-11　Specify Part

（12）单击【指定毛坯】中的【选择或编辑毛坯几何体】，打开【毛坯几何体】对话框，【类型】选择【包容块】，如图 2-12 所示，单击【确定】。

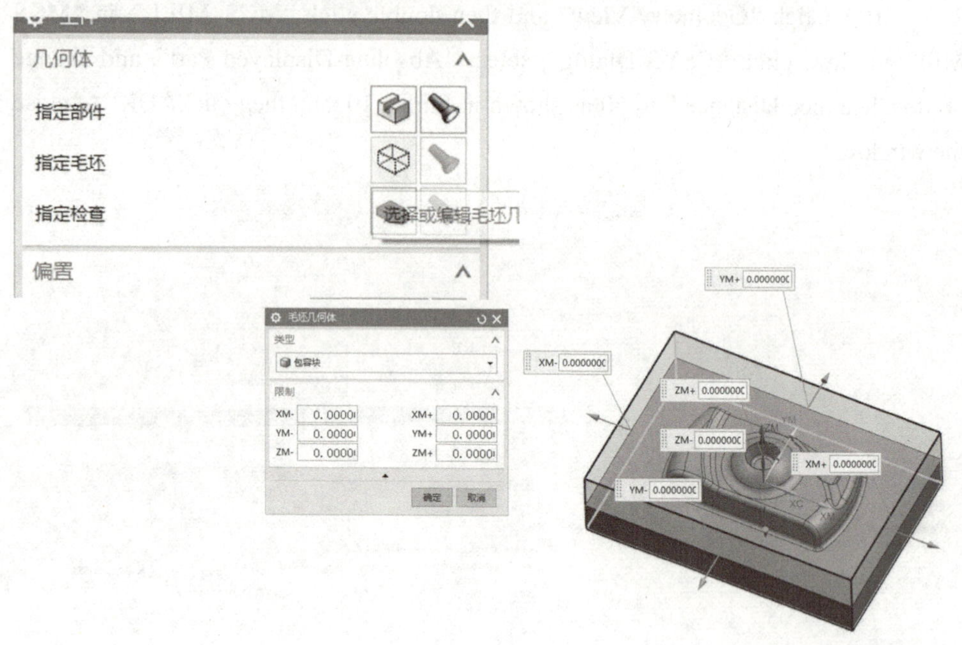

图 2-12　指定毛坯

（13）进入【加工方法视图】，修改粗加工、半精加工、精加工的参数。双击【MILL_ROUGH】，打开【铣削粗加工】对话框，设置【部件余量】为【0.5】，【内公差】为【0.03】，【外公差】为【0.03】，如图 2-13 所示，单击【确定】。

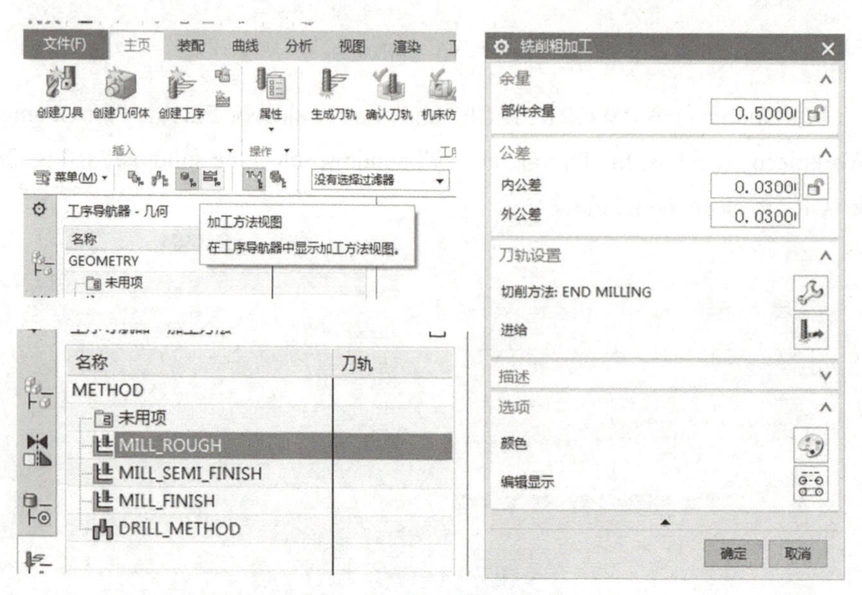

图 2-13　设置粗加工参数

(12) Click "WORKPIECE" and click "Select or Edit the Blank Geometry" in "Workpiece" window. In "Blank Geometry" window, select "Bounding Block" as type as shown in Fig. 2-12. Click "OK".

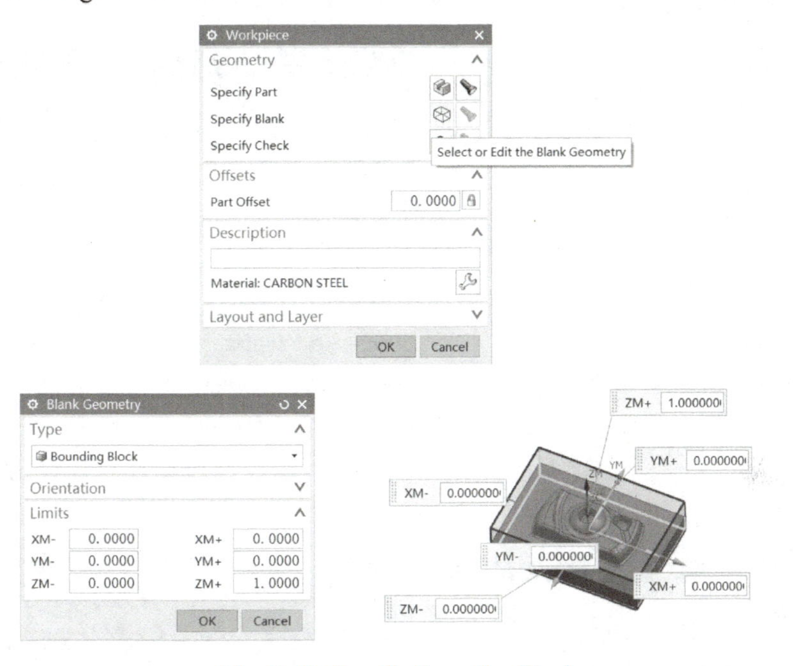

Fig. 2-12 Specify Bounding Block

(13) Click "Machining Method View" button, Double click "MILL_ROUGH" to modify "Part Stock" value as 0.5, both "Intol Tolerance" and "Outtol Tolerance" values as 0.03 as shown in Fig. 2-13. Click "OK".

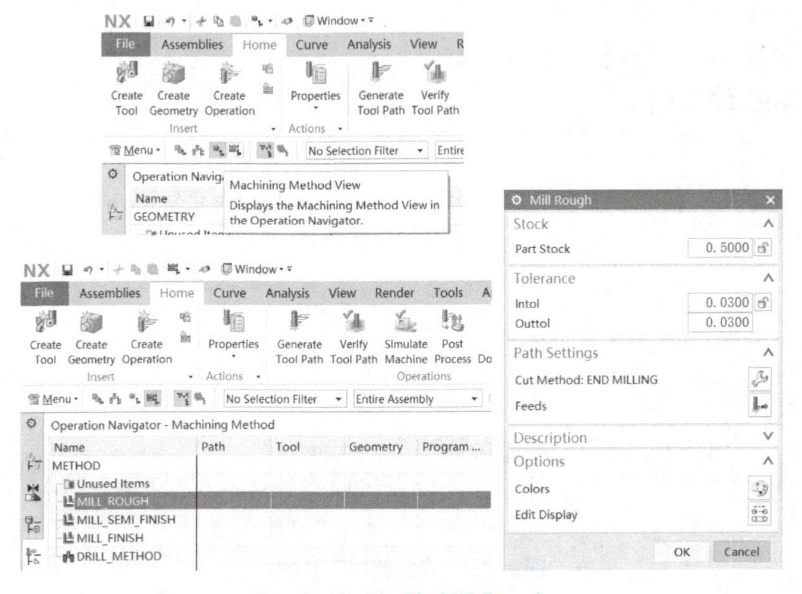

Fig. 2-13 Modify Mill Rough

（14）双击【MILL_SEMI_FINISH】，打开【铣削半精加工】对话框，将【部件余量】设为【0.1】，【内公差】设为【0.01】，【外公差】设为【0.01】，如图 2-14 所示，单击【确定】。

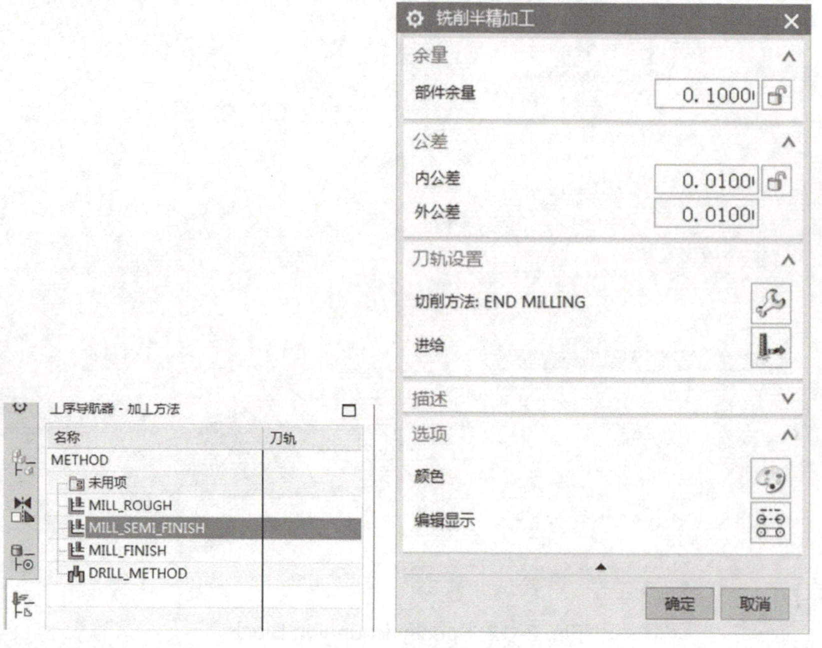

图 2-14　设置半精加工参数

（15）双击【MILL_FINISH】，打开【铣削精加工】对话框，设置【部件余量】为【0】，【内公差】为【0.003】，【外公差】为【0.003】，如图 2-15 所示，单击【确定】。

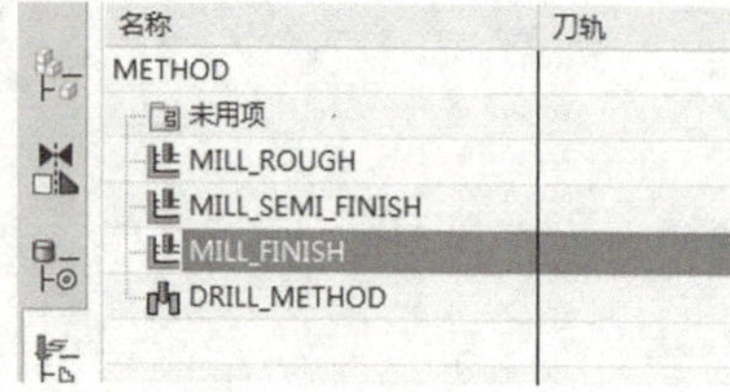

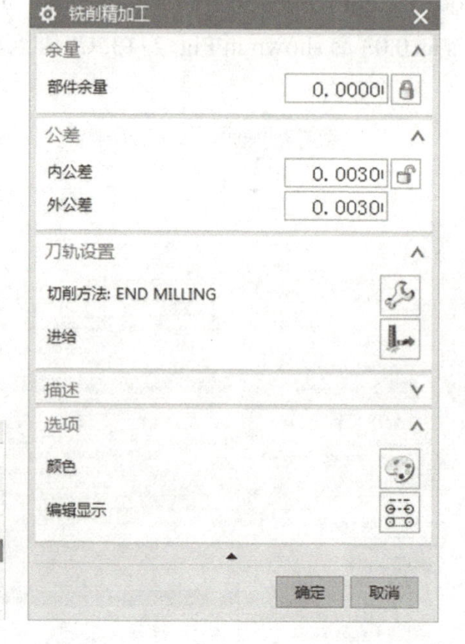

图 2-15　设置精加工参数

（14）Double click "MILL_SEMI_FINISH" to modify "Part Stock" value as 0.1, both "Intol Tolerance" and "Outtol Tolerance" values as 0.01 as shown in Fig. 2-14, click "OK" to close the window.

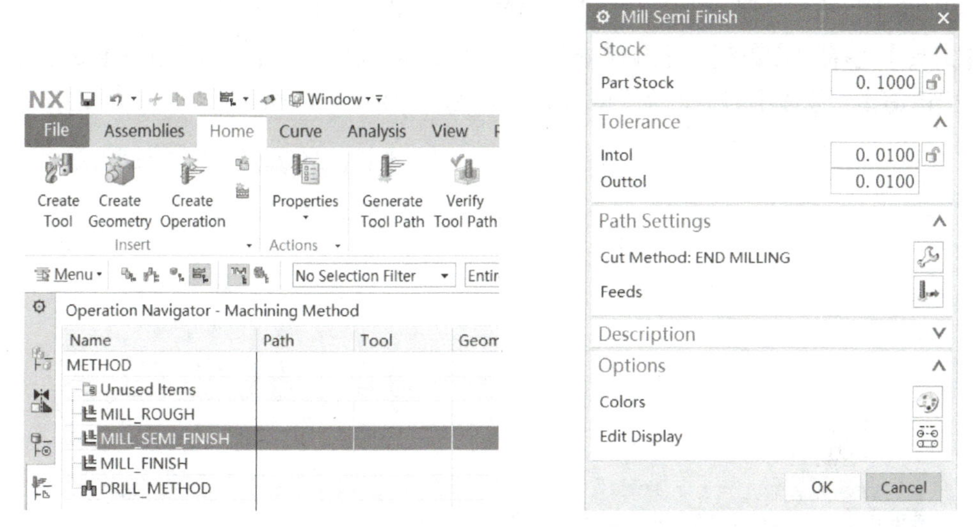

Fig. 2-14　Modify Mill Semi Finish

（15）Double click "MILL_ FINISH" to modify "Part Stock" as 0 and both "Intol Tolerance" and "Outtol Tolerance" values as 0.003 as shown in Fig. 2-15. Click "OK".

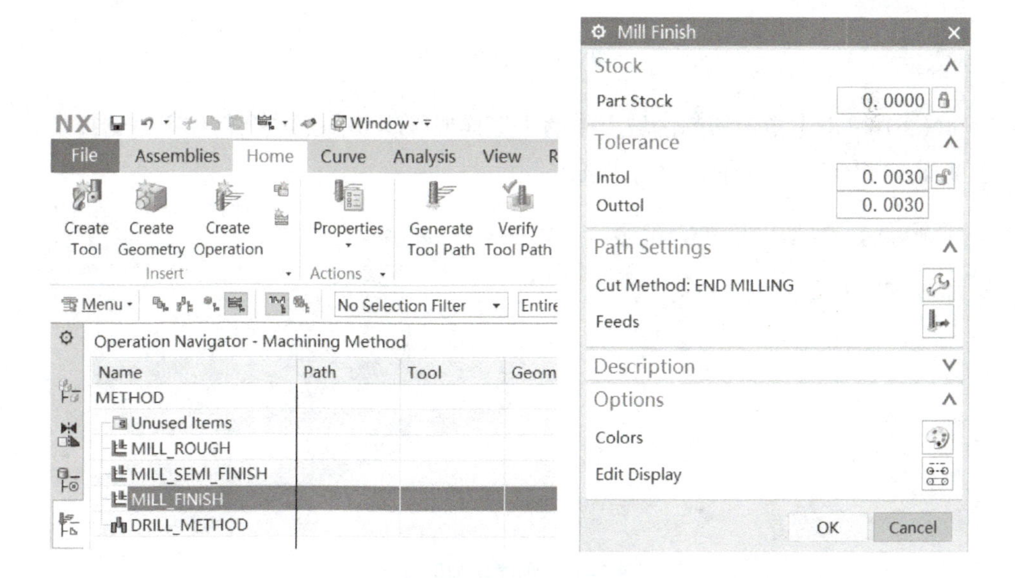

Fig. 2-15　Modify Mill Finish

三、加工程序编制

1. 型面粗加工 -1

（1）单击【创建工序】，打开【创建工序】对话框，【类型】选择【mill_contour】，【工序子类型】选择【型腔铣】，【程序】选择【型面粗加工 -1】，【刀具】选择【D25R5】，【几何体】选择【WORKPIECE】，【方法】选择【MILL_ROUGH】，修改【名称】为【Rough_mill-1】，如图 2-16 所示，单击【确定】。

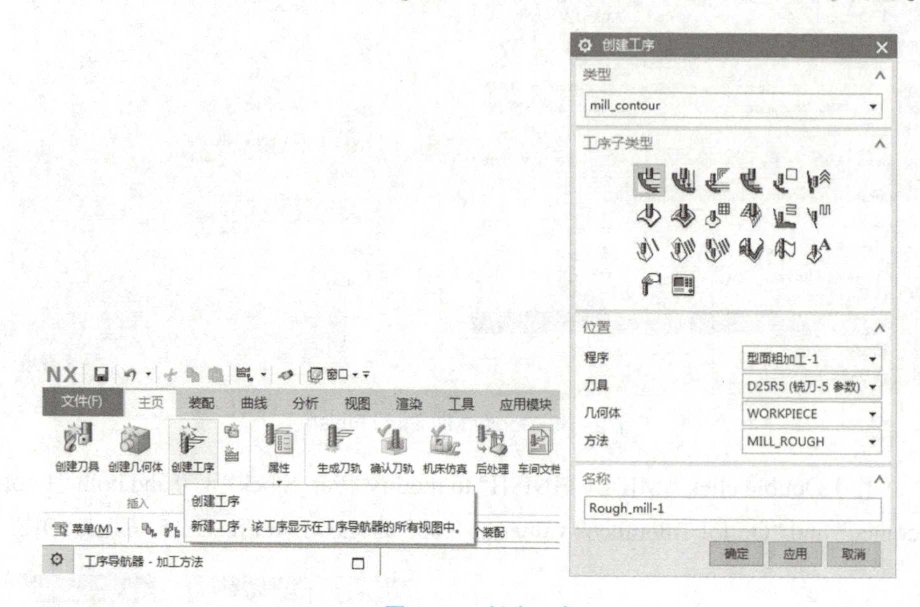

图 2-16　创建工序

（2）单击【指定切削区域】中的【选择或编辑切削区域几何体】，选择切削曲面，如图 2-17 所示，单击【确定】。

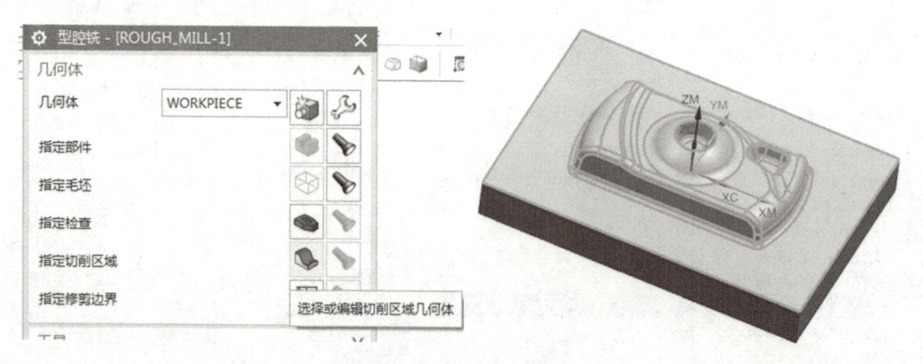

图 2-17　创建切削区域

（3）设置【切削模式】为【跟随周边】，【步距】为【刀具平直百分比】，【平面直径百分比】为【60】，【公共每刀切削深度】为【恒定】，【最大距离】为【1.5 mm】，如图 2-18 所示。

Ⅲ. Programming

1. Rough Machining-1

（1）As shown in Fig. 2-16, click "Create Operation", and then select "mill_contour" in pull-down box of "Type". Select operation subtype as "Cavity Mill", program as "ROUGH MILL-1", tool as "D25R5", geometry as "WORKPIECE" and method as "MILL_Rough". Modify name as "ROUGH_MILL-1", and then click "OK".

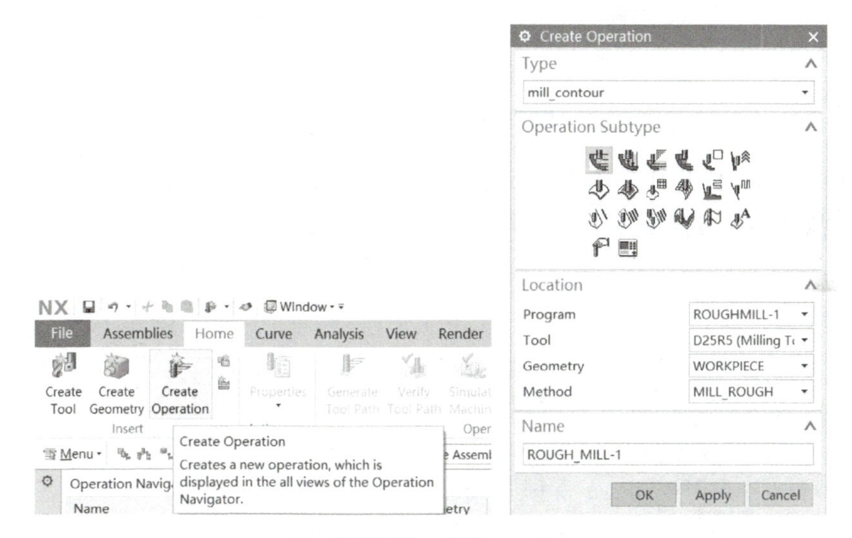

Fig. 2-16 Create Operation

（2）Click button "Specify Cut Area" and select the surface as shown in Fig. 2-17. Click "OK".

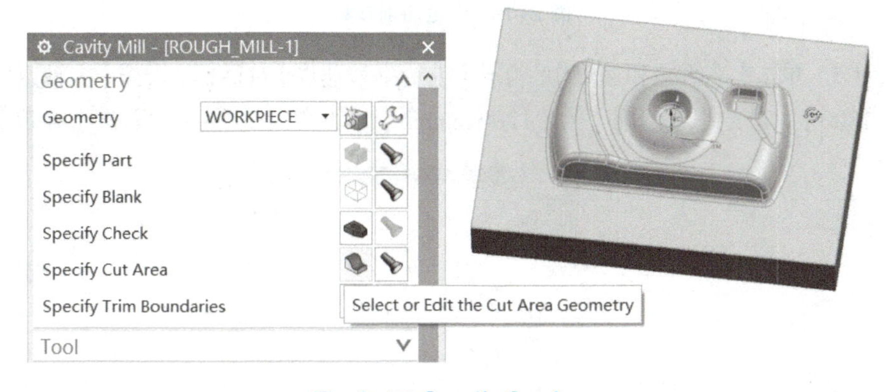

Fig. 2-17 Specify Cut Area

（3）As shown in Fig. 2-18, modify "Cut Pattern" as "Follow Periphery". Modify "Percent of Flat Diameter" as 60, "Common Depth per Cut" as "Constant" and "Maximum Distance" as 1.5 mm.

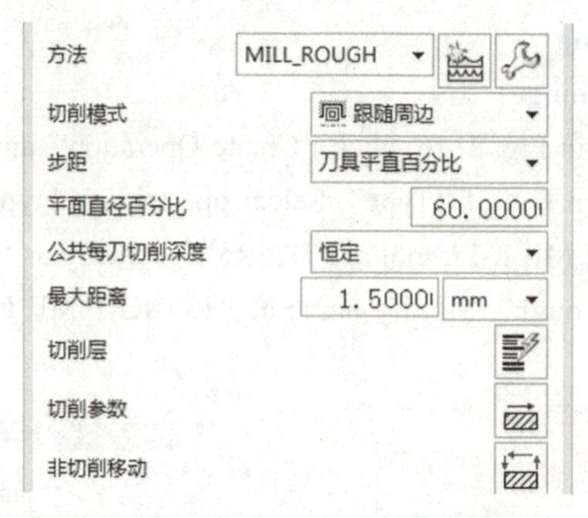

图 2-18　设定切削模式

（4）单击【切削参数】，打开【切削参数】对话框，在【策略】选项卡中，设置【刀路方向】为【向内】，如图 2-19 所示，单击【确定】。

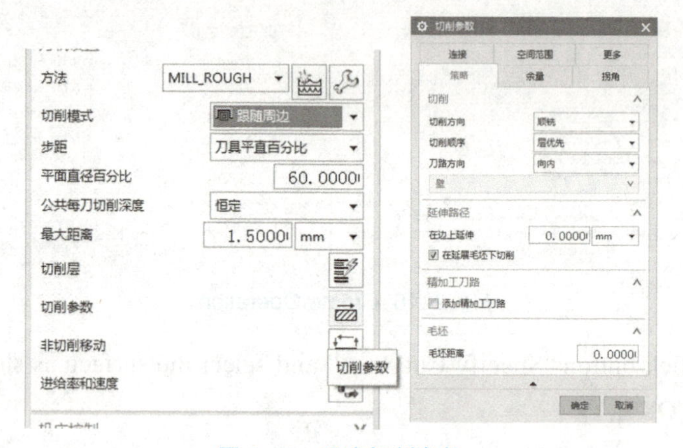

图 2-19　设定切削参数

（5）单击【进给率和速度】，打开【进给率和速度】对话框，设置【主轴速度】为【1800】，进给率【切削】为【1200 mmpm】，如图 2-20 所示，单击【确定】。

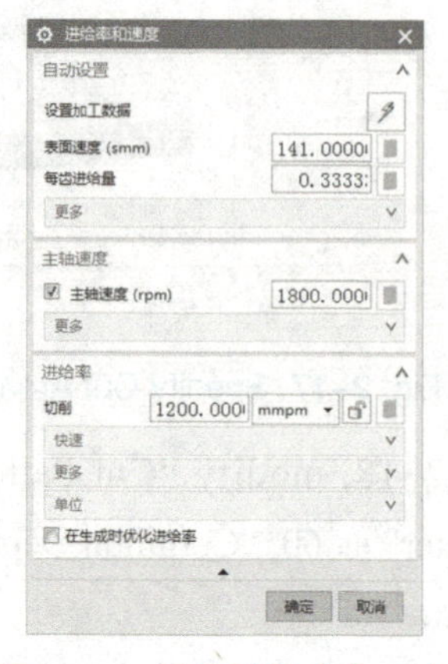

图 2-20　设置进给率和速度

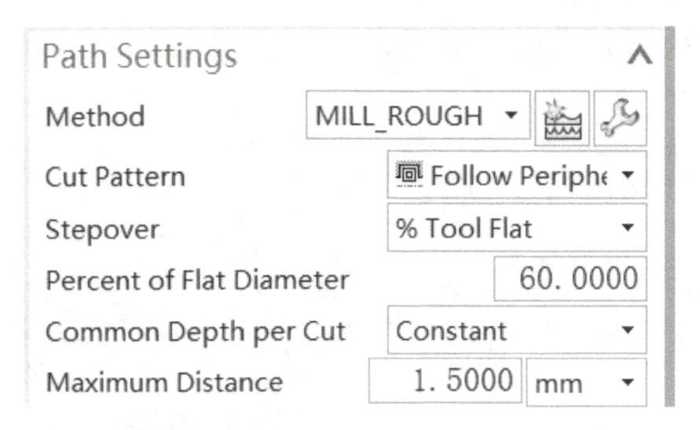

Fig. 2-18 Modify Cut Pattern

(4) As shown in Fig. 2-19, click "Cutting Parameters" button, in "Strategy" tab, select "Inward" as "Pattern Direction". Click "OK".

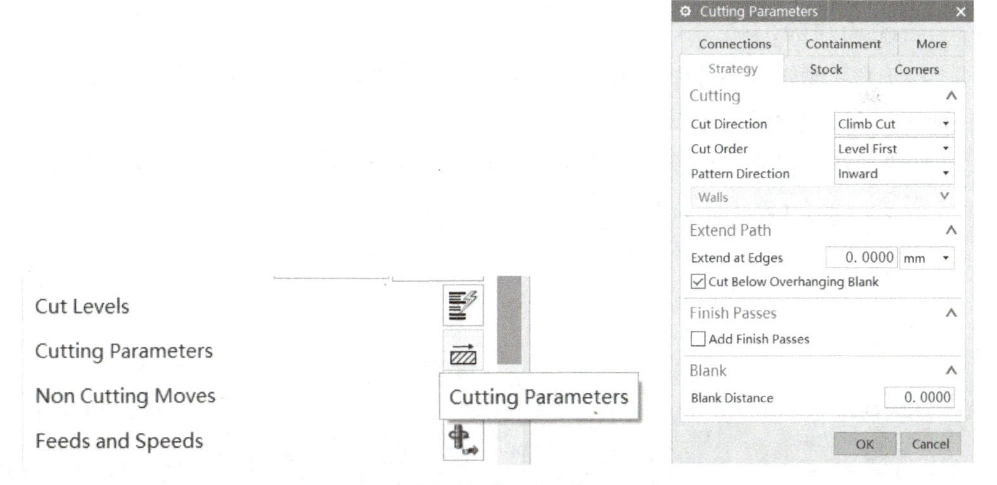

Fig. 2-19 Modify Cutting Parameters

(5) Click "Feeds and Speeds". Modify "Spindle Speed" as 1,800 and "Feed Rates" as 1,200 mmpm as shown in Fig. 2-20. Click "OK".

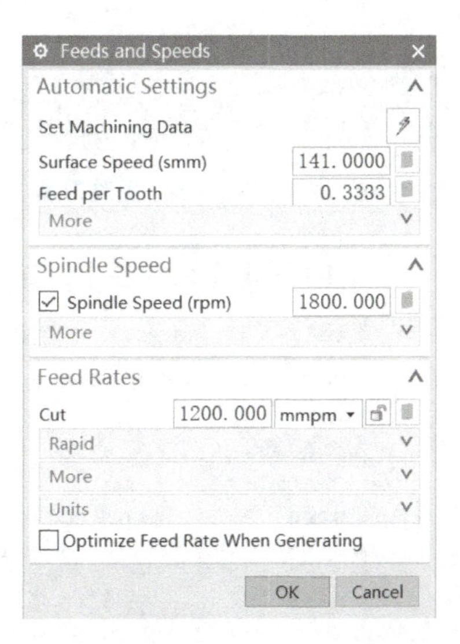

Fig. 2-20 Modify Feeds and Speeds

（6）单击【生成】，查看生成的刀具轨迹，如图 2-21 所示，单击【确定】。

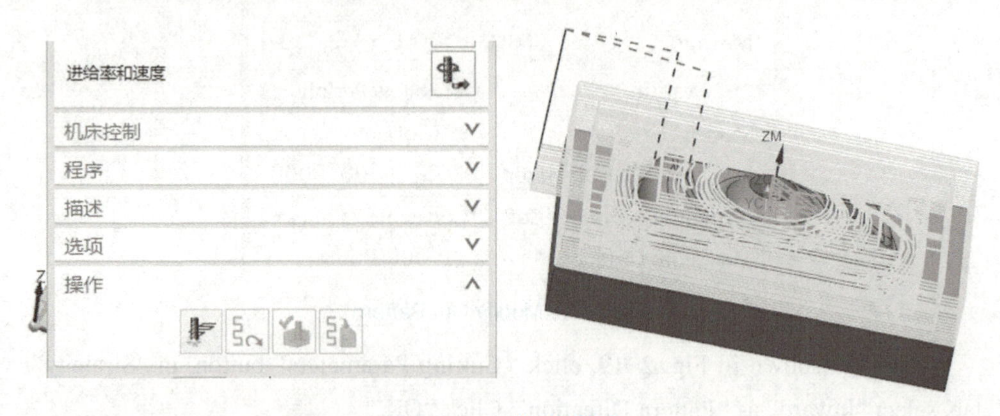

图 2-21 生成刀具轨迹

2. 型面粗加工-2

（1）复制工序名【ROUGH_MILL-1】，在【型面粗加工 -2】下内部粘贴，更改工序名为【ROUGH_MILL-2】。

（2）双击程序，将【刀具】更改为【D10R1-ROUGH】牛鼻刀，如图 2-22 所示。

（3）修改【最大距离】为【0.8 mm】，如图 2-23 所示。

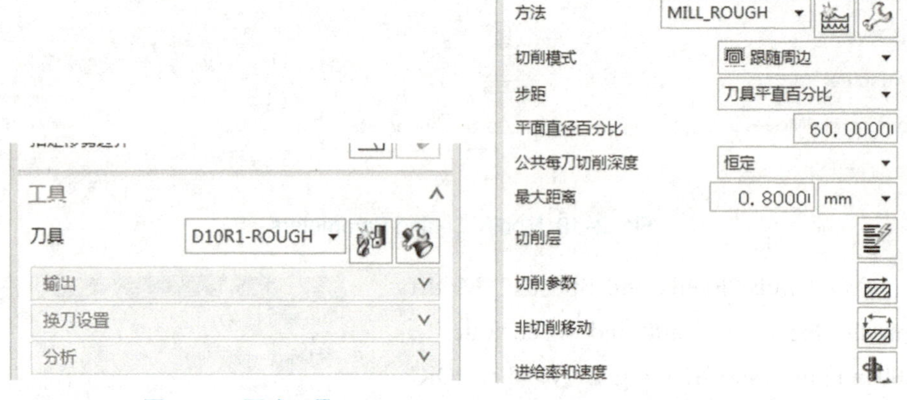

图 2-22 更改刀具　　图 2-23 修改最大距离

（4）单击【切削层】，查看切削层分布，如图 2-24 所示，单击【确定】。

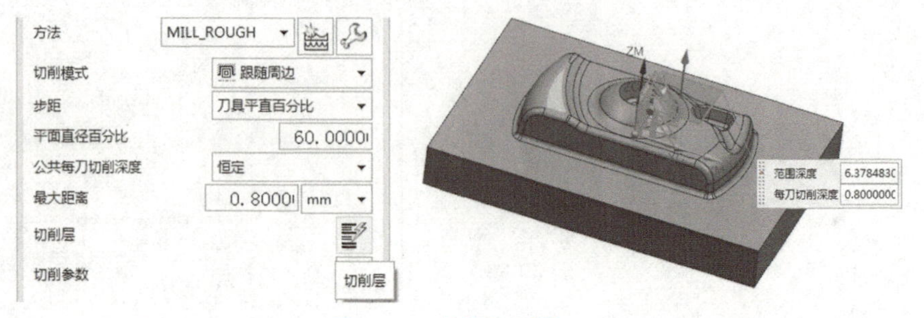

图 2-24 设置切削层

(6) Click "Generate" to view the tool path as shown in Fig. 2-21 and then click "OK".

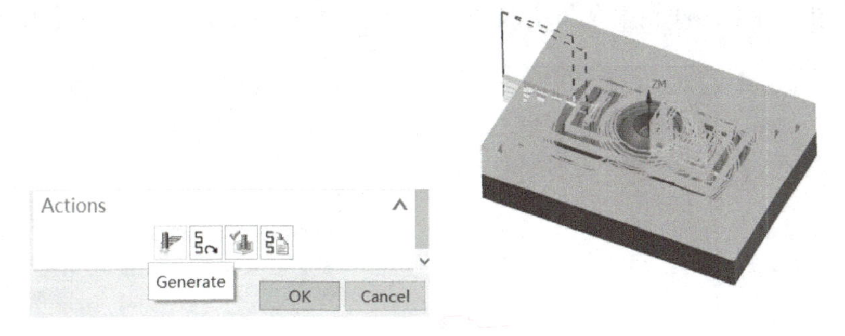

Fig. 2-21 Generate Tool Path

2. Rough Machining-2

(1) Copy program "ROUGH_MILL-1" program and paste it under program "Rough Machining-2". Modify the pasted file name as "ROUGH_MILL-2".

(2) Double click the program and modify the tool as "D10R1-ROUGH" as shown in Fig. 2-22. Click "OK".

(3) Modify "Maximum Distance" as 0.8 mm as shown in Fig. 2-23.

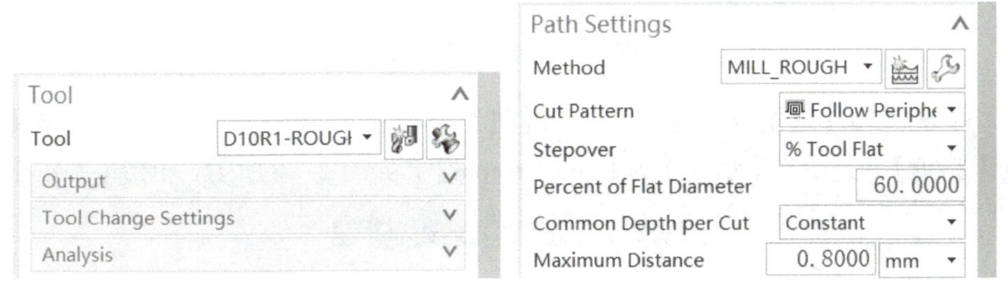

Fig. 2-22 Change Tool　　Fig. 2-23 Modify Maximum Distance

(4) Click "Cut Levels" to check the distribution as shown in Fig. 2-24, and then click "OK".

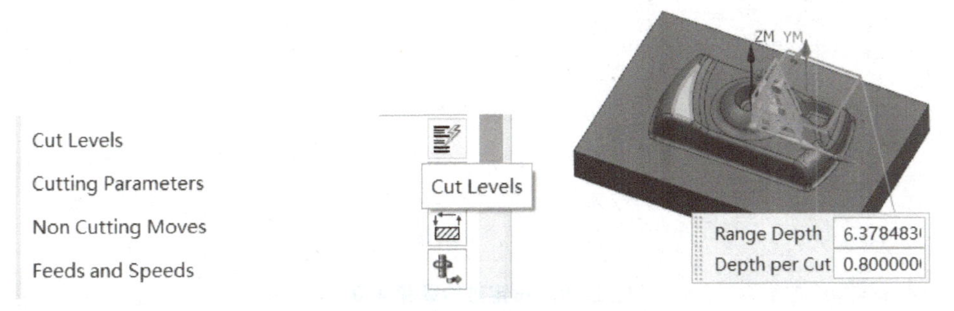

Fig. 2-24 Specify Cut Levels

（5）单击【切削参数】，打开【切削参数】对话框，在【空间范围】选项卡中，【处理中的工件】选择【使用基于层的】，设置【重叠距离】为【1】；在【余量】选项卡中，将【部件侧面余量】改为【0.4】，如图 2–25 所示，单击【确定】。

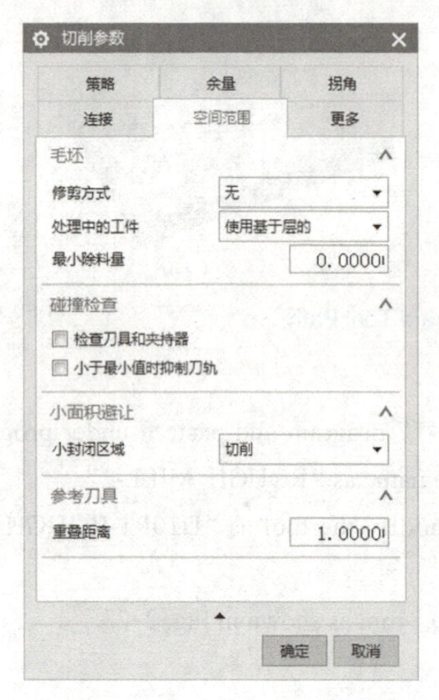

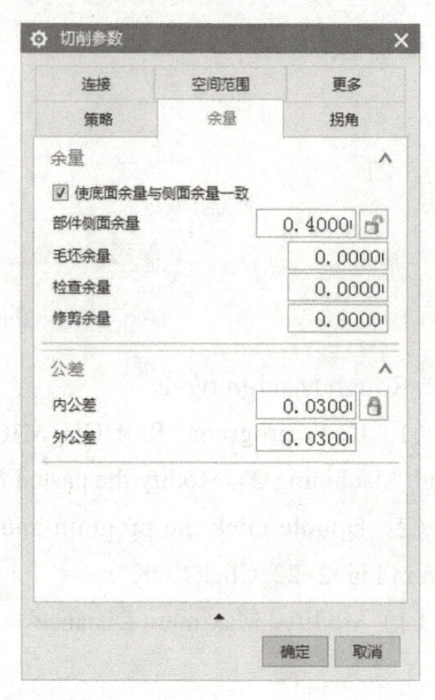

图 2–25　设定切削参数

（6）单击【进给率和速度】，打开【进给率和速度】对话框，设置【主轴速度】为【2600】，进给率【切削】为【1400 mmpm】，如图 2–26 所示，单击【确定】。

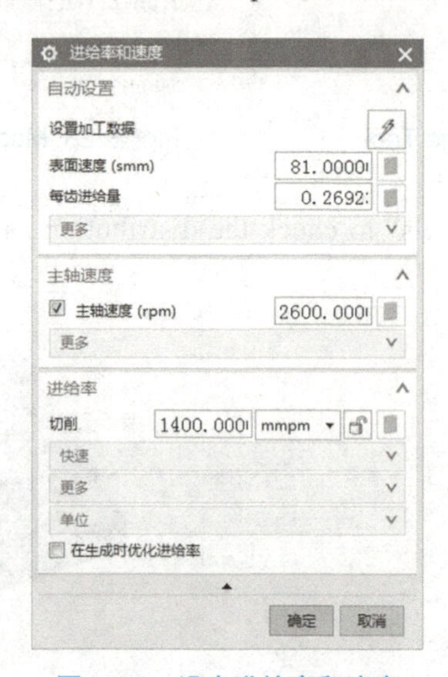

图 2–26　设定进给率和速度

(5) Click "Cutting Parameters". In "Containment" tab, select "Use Level Based" as "In Process Workpiece", modify "Overlap Distance" as 1. In "Stock" tab, modify "Part Side Stock" as 0.4 as shown in Fig. 2-25. Click "OK".

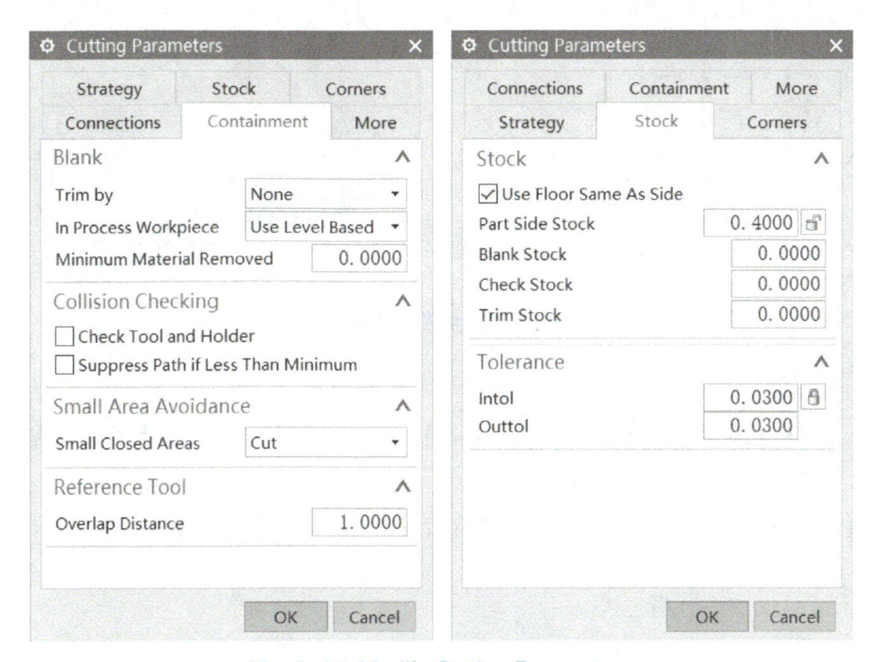

Fig. 2-25 Modify Cutting Parameters

(6) Click "Feeds and Speeds", modify "Spindle Speed" as 2,600 and "Feed Rates" as 1,400 mmpm as shown in Fig. 2-26. Click "OK".

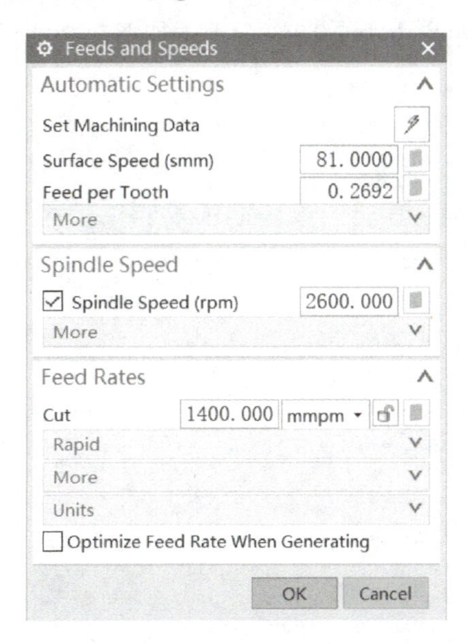

Fig. 2-26 Modify Feeds and Speeds

（7）单击【生成】，查看生成的刀具轨迹，如图 2-27 所示，单击【确定】。

图 2-27　生成刀具轨迹

3. 型面粗加工-3

（1）复制工序名【ROUGH_MILL-2】，在【型面粗加工-3】下内部粘贴，修改工序名为【ROUGH_MILL-3】。

（2）双击程序，刀具更改为【D6R0.5】牛鼻刀，如图 2-28 所示。

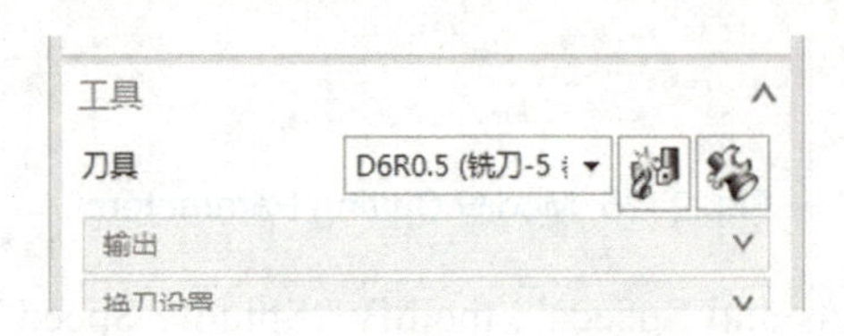

图 2-28　修改刀具

（3）修改【最大距离】为【0.35 mm】，如图 2-29 所示。

图 2-29　修改最大距离

（7）Click “Generate” to view the tool path as shown in Fig. 2-27 and then click “OK”.

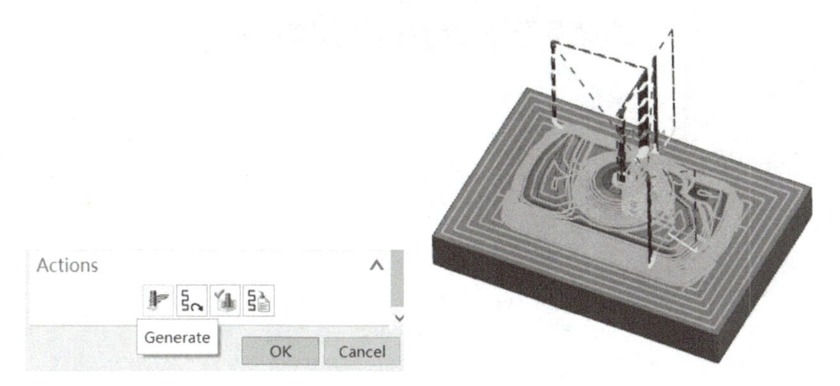

Fig. 2–27 Generate Tool Path

3. Rough Machining–3

（1）Copy program “ROUGH_MILL-2” program and paste it under program “Rough Machining-3”. Modify the pasted file name as “ROUGH_MILL-3”.

（2）Double click the program and modify the tool as “D6R0.5” as shown in Fig. 2-28. Click “OK”.

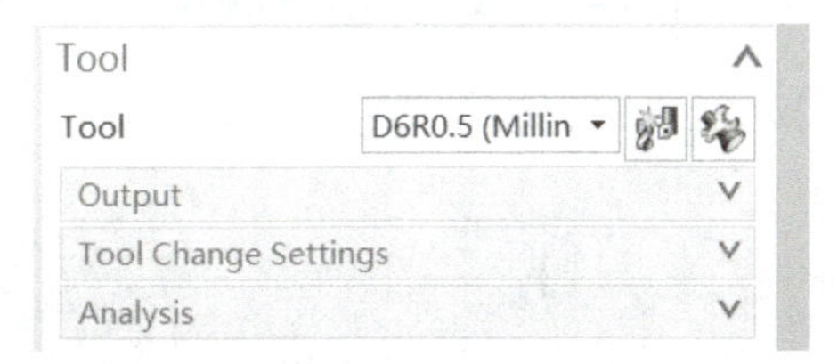

Fig. 2–28 Change Tool

（3）Modify “Maximum Distance” as 0.35 mm as shown in Fig. 2-29.

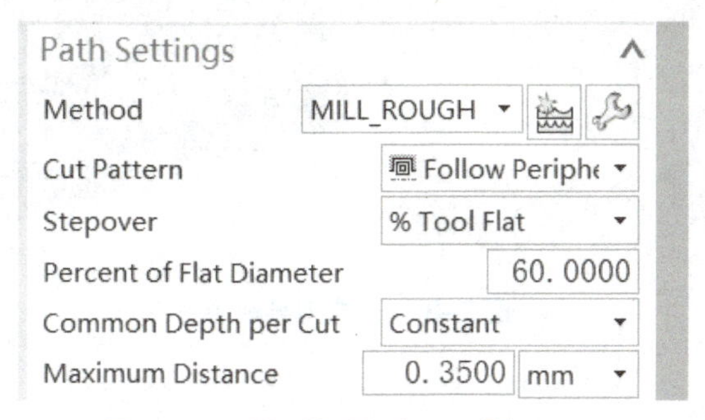

Fig. 2–29 Modify Maximum Distance

（4）单击【进给率和速度】，打开【进给率和速度】对话框，设置【主轴速度】为【3200】，进给率【切削】为【1000 mmpm】，如图 2-30 所示，单击【确定】。

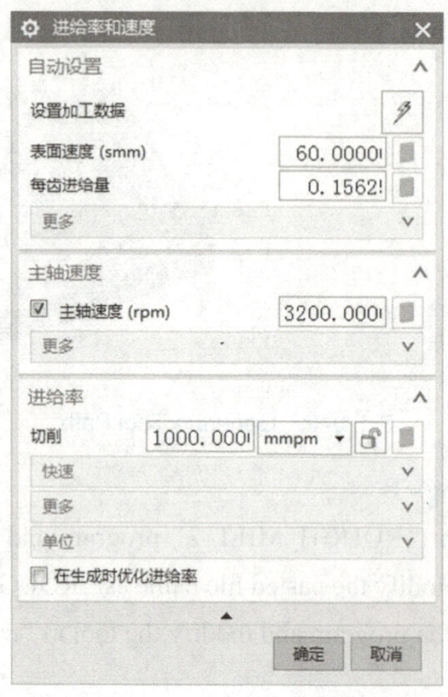

图 2-30　设置进给率和速度

（5）单击【生成】，查看生成的刀具轨迹，如图 2-31 所示，单击【确定】。

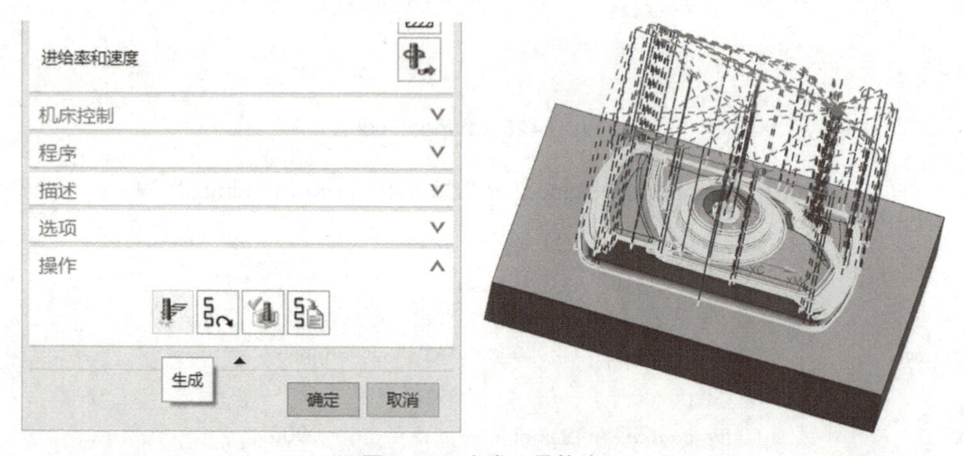

图 2-31　生成刀具轨迹

(4) Click "Feeds and Speeds", modify "Spindle Speed" as 3,200 and "Feed Rates" as 1,000 mmpm as shown in Fig. 2-30. Click "OK".

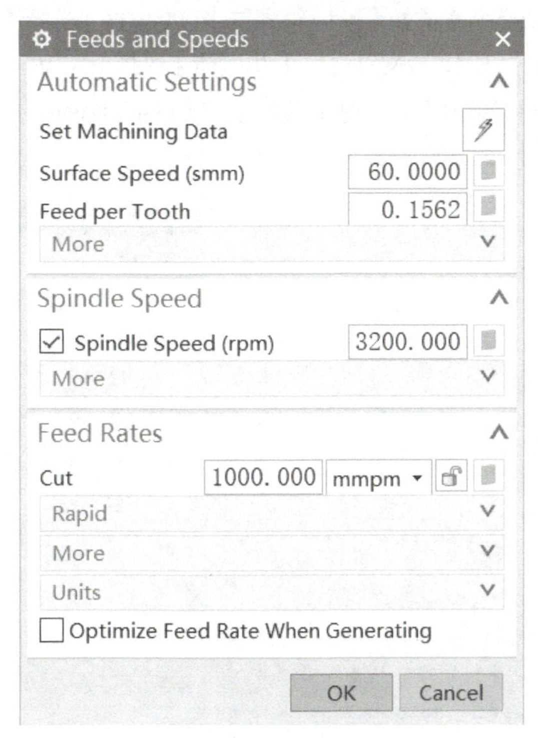

Fig. 2-30 Modify Feeds and Speeds

(5) Click "Generate" to view the tool path as shown in Fig. 2-31, and then click "OK".

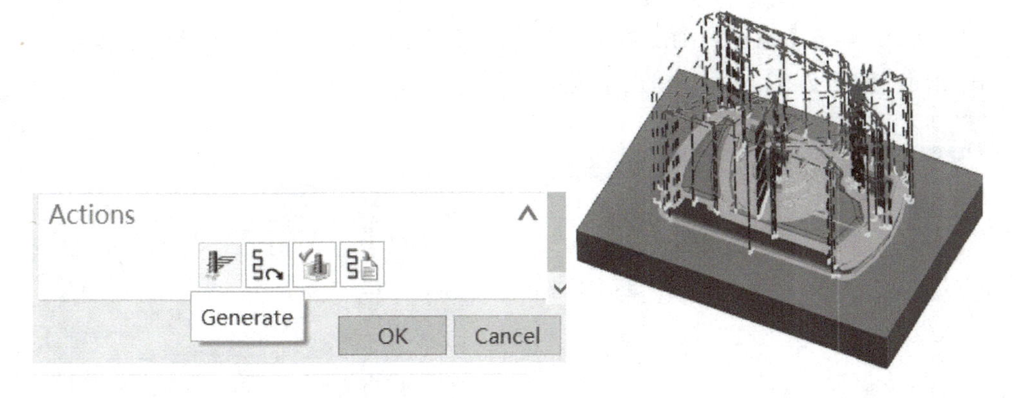

Fig. 2-31 Generate Tool Path

4. 型面半精加工

（1）单击【创建工序】，打开【创建工序】对话框，【类型】选择【mill_contour】，【工序子类型】选择【区域轮廓铣】，【程序】选择【型面半精加工】，【刀具】选择【D6R3-SEMI-FINISH】，【几何体】选择【WORKPIECE】，【方法】选择【MILL_SEMI_FINISH】，修改【名称】为【Semi_finish_mill-1】，如图 2-32 所示，单击【确定】。

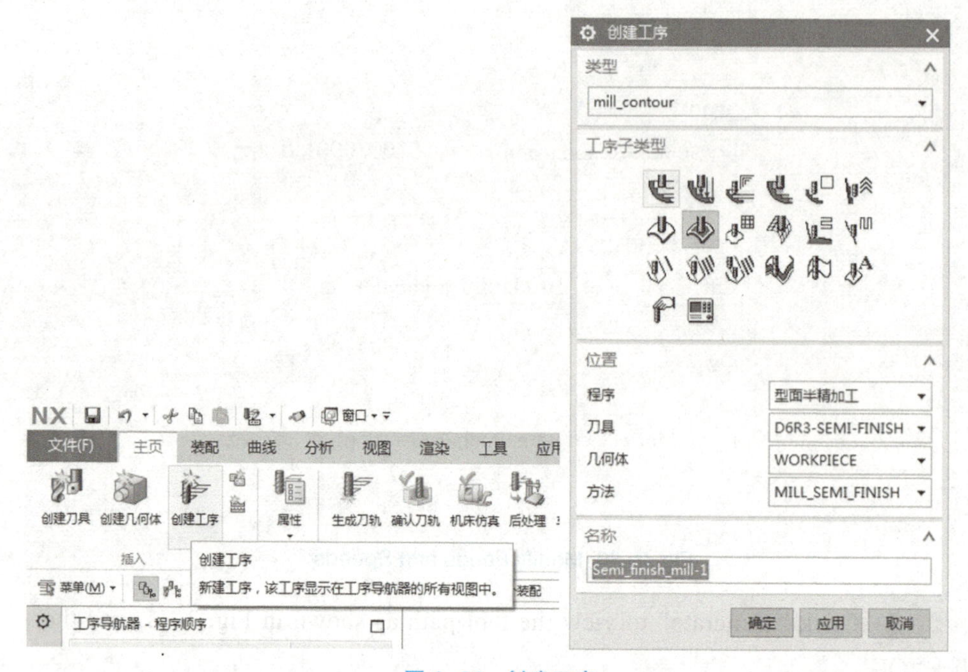

图 2-32 创建工序

（2）单击【指定切削区域】中的【选择或编辑切削区域几何体】，选择切削曲面，如图 2-33 所示，单击【确定】。

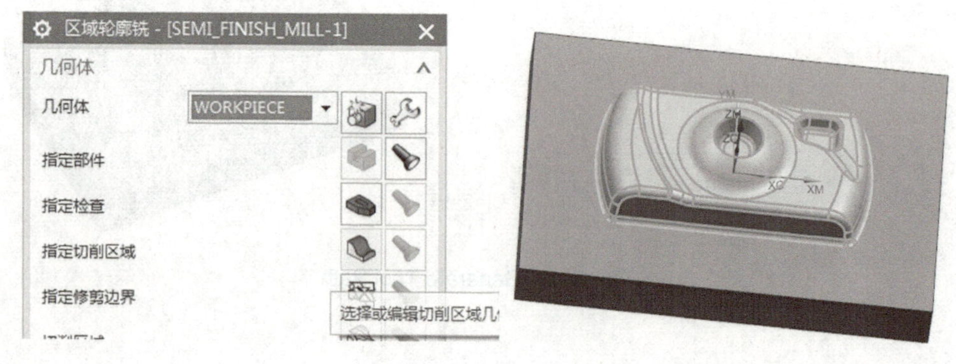

图 2-33 指定切削区域

4. Semi-Finishing

(1) As shown in Fig. 2-32, click "Create Operation", and then select "mill_contour" in pull-down box of "Type". Select operation subtype as "Contour Area", program as "SEMI-FINISHING", tool as "D6R3-SEMI-FINISH", geometry as "WORKPIECE" and method as "MILL_SEMI_FINISH". Modify name as "SEMI_FINISH_MILL-1", and then click "OK".

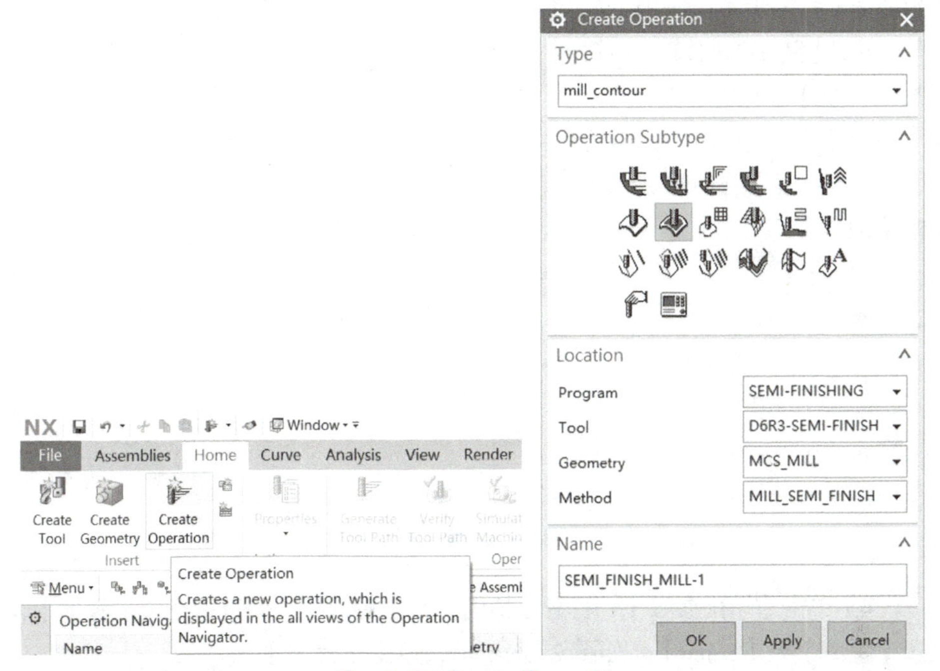

Fig. 2-32 Create Operation

(2) Click "Specify Cut Area" and select surface as shown in Fig. 2-33. Click "OK".

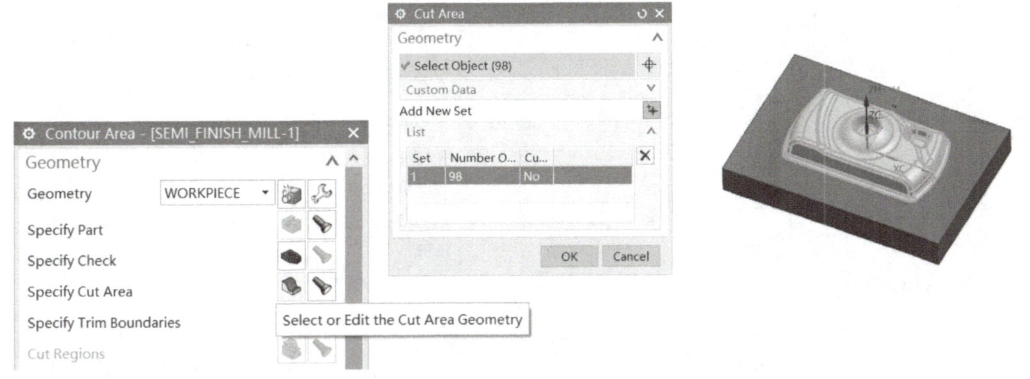

Fig. 2-33 Specify Cut Area

（3）编辑【区域铣削】方法，在打开的【区域铣削驱动方法】对话框中勾选【为平的区域创建单独的区域】，设置【非陡峭切削模式】为【跟随周边】，【刀路方向】为【向外】，【步距】为【恒定】，【最大距离】为【0.5 mm】。【步距已应用】选择【在部件上】，【陡峭切削模式】选择【深度加工往复】，设置【深度切削层】为【恒定】，【深度加工每刀切削深度】为【0.5 mm】，如图 2-34 所示，单击【确定】。

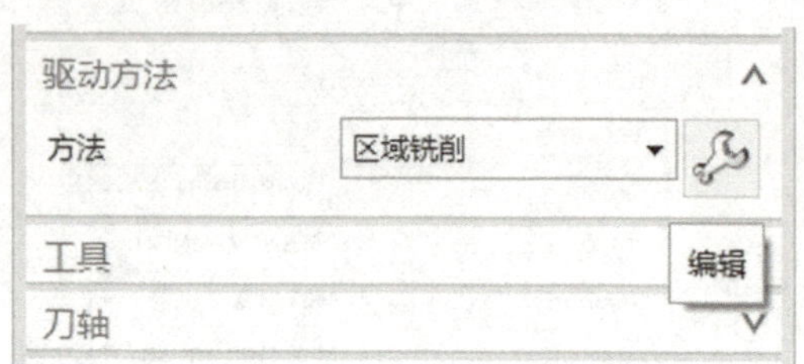

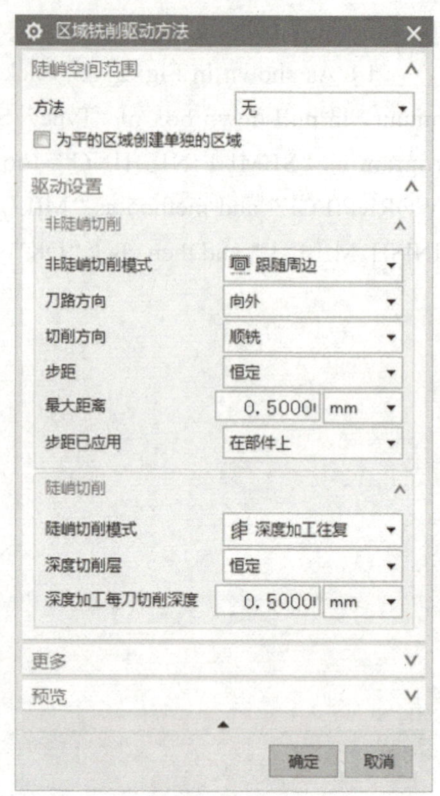

图 2-34 编辑驱动方式

（4）单击【切削参数】，打开【切削参数】对话框，勾选【在边上延伸】，设置【距离】为【1 mm】，如图 2-35 所示，单击【确定】。

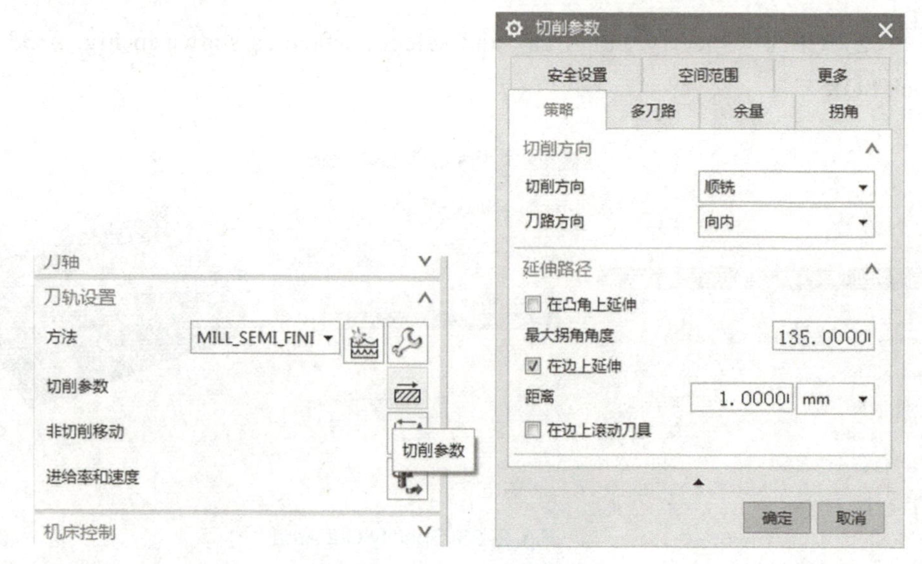

图 2-35 设定切削参数

(3) Click "Edit" to set parameters of drive method. Tick the option "Create Separate Regions For Flat Areas". Select "Follow Periphery" as "Non-steep Cut Pattern", "Outward" as "Pattern Direction", "Constant" as "Stepover", 0.5 mm as "Maximum Distance", "On Part" as "Stepover Applied", "Zlevel Zig Zag" as "Steep Cut Pattern", "Constant" as "Zlevel Cut Levels" , 1 mm as "Zlevel Depth per Cut" as shown in Fig. 2-34. Click "OK".

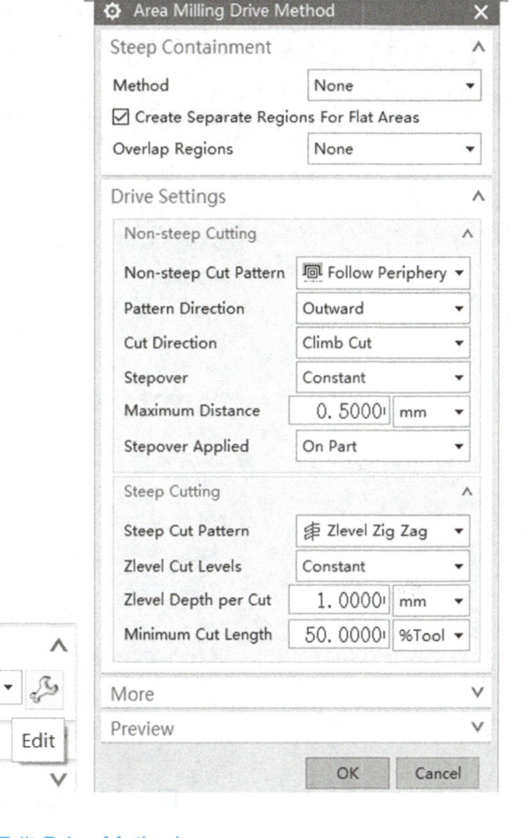

Fig. 2-34 Edit Drive Method

(4) As shown in Fig. 2-35, click "Cutting Parameters" button, in "Strategy" tab, tick the option "Extend at Edges" and modify distance as 1 mm. Click "OK".

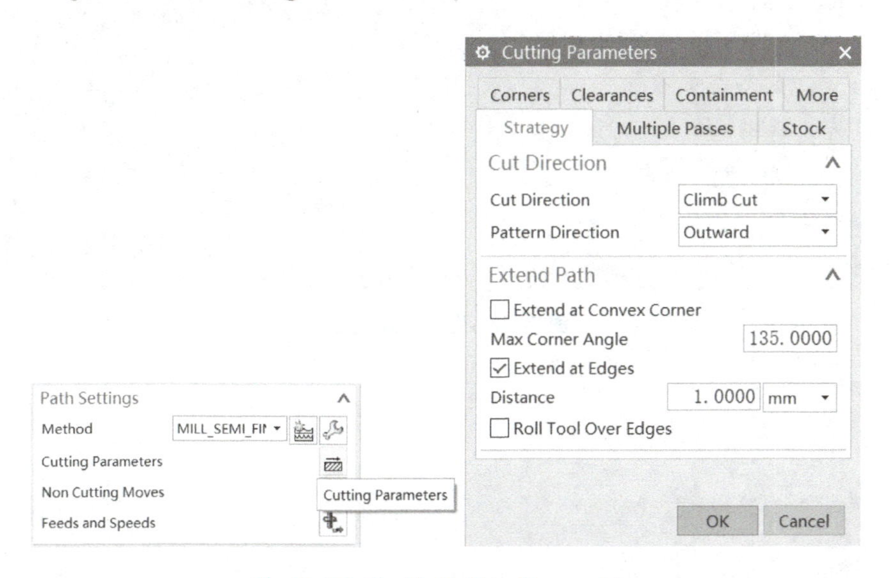

Fig. 2-35 Modify Cutting Parameters

（5）单击【进给率和速度】，打开【进给率和速度】对话框，设置【主轴速度】为【3500】，进给率【切削】为【1600 mmpm】，如图 2-36 所示，单击【确定】。

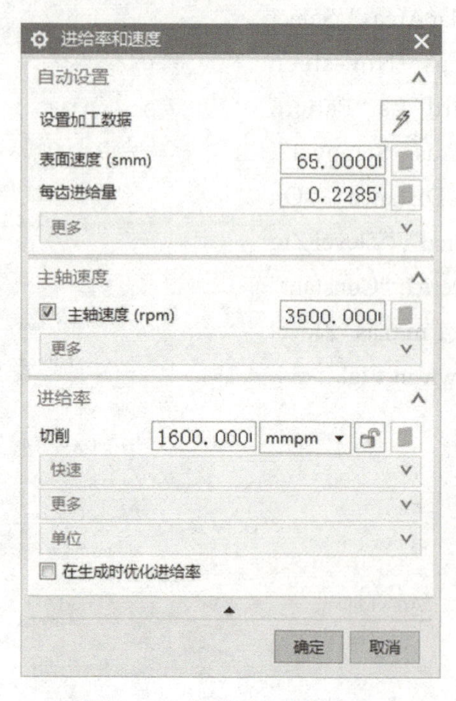

图 2-36　设置进给率和速度

（6）单击【生成】，查看生成的刀具轨迹，如图 2-37 所示，单击【确定】。

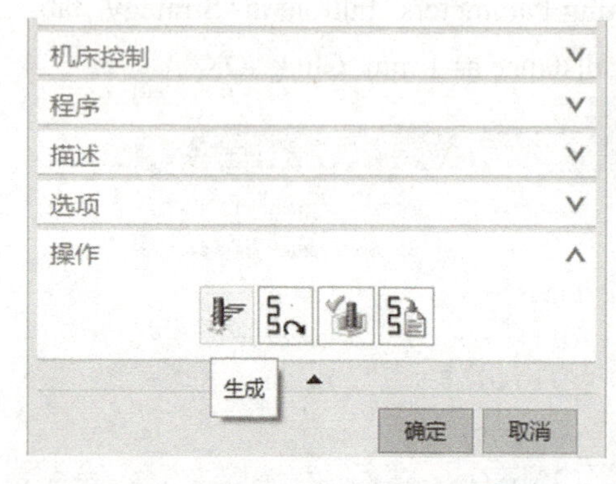

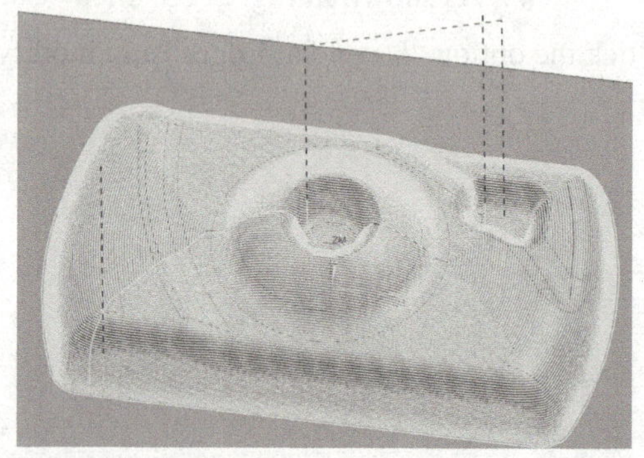

图 2-37　生成刀具轨迹

(5) Click "Feeds and Speeds", modify "Spindle Speed" as 3,500 and "Feed Rates" as 1,600 mmpm as shown in Fig. 2-36. Click "OK".

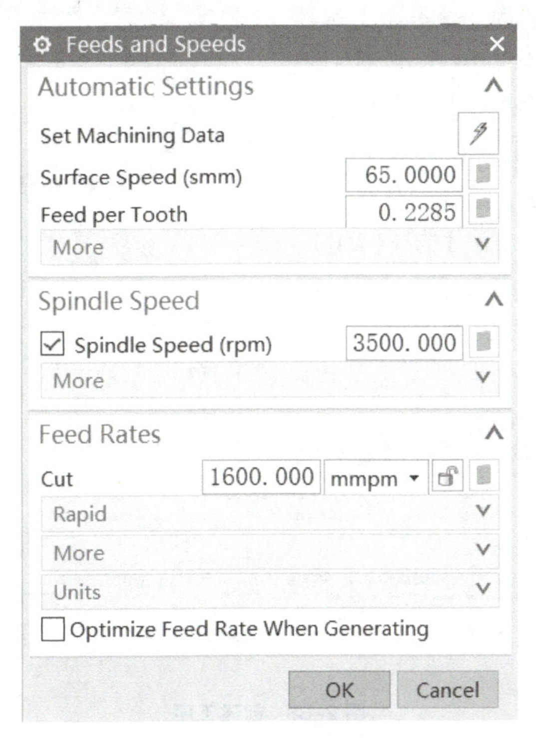

Fig. 2-36 Modify Feeds and Speeds

(6) Click "Generate" to view the tool path as shown in Fig. 2-37, and then click "OK".

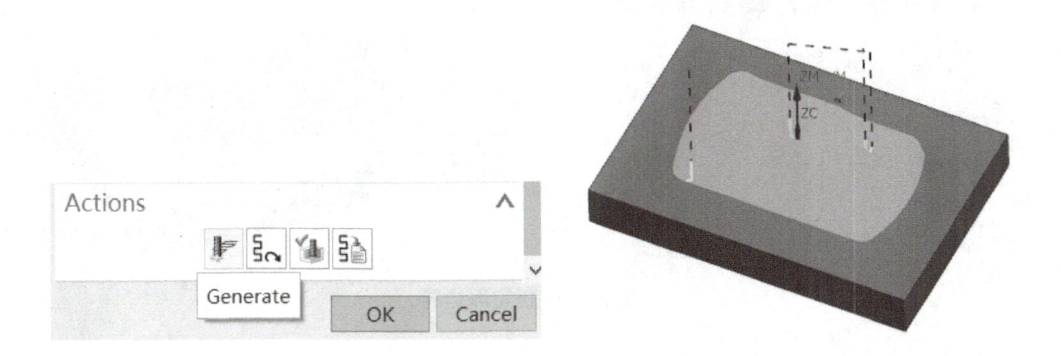

Fig. 2-37 Generate Tool Path

5. 精铣平面

（1）单击【创建工序】，打开【创建工序】对话框，【类型】选择【mill_planar】，【工序子类型】选择【使用边界面铣削】，【程序】选择【精铣平面】，【刀具】选择【D10R1-FINISH】，【几何体】选择【WORKPIECE】，【方法】选择【MILL_FINISH】，修改【名称】为【FACE_MILLING-1】，如图 2-38 所示，单击【确定】。

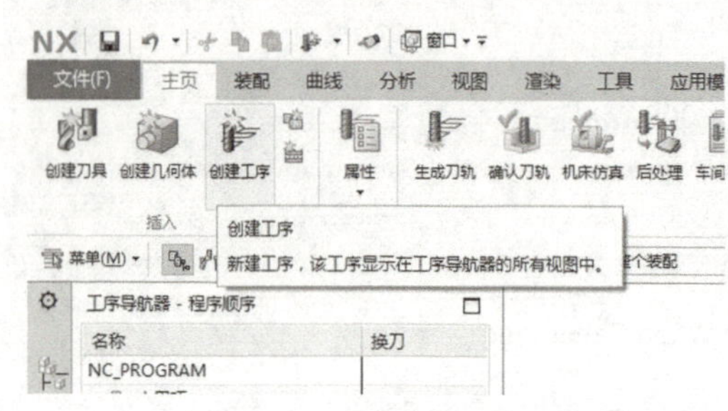

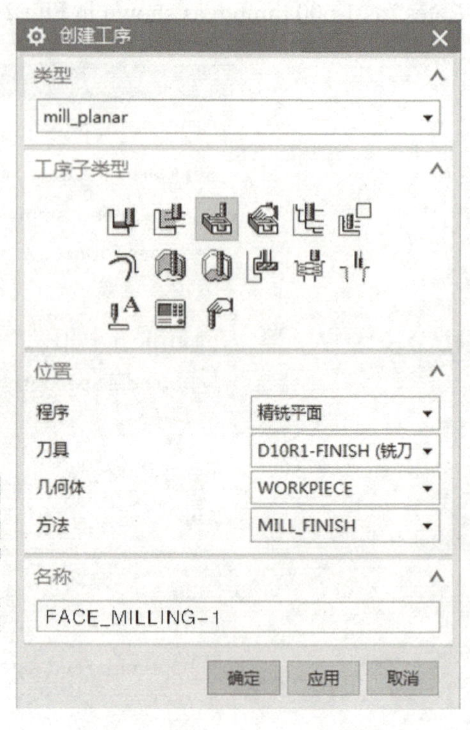

图 2-38　创建工序

（2）单击【指定面边界】中的【选择或编辑面几何体】，选择平面，如图 2-39 所示，单击【确定】。

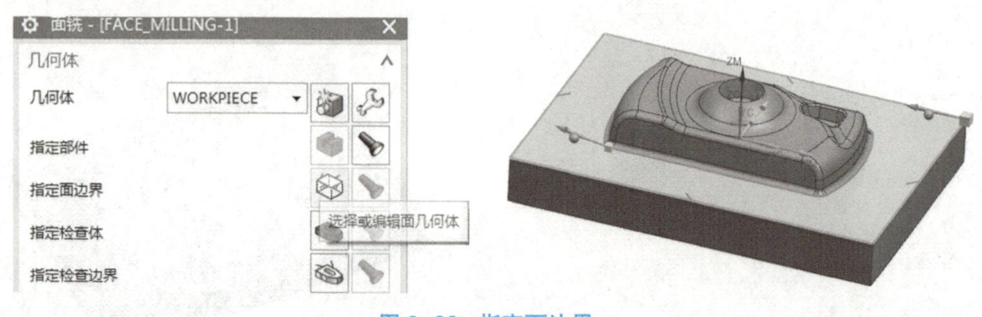

图 2-39　指定面边界

5. Finishing Face Milling

(1) As shown in Fig. 2-38, click "Create Operation", and then select "mill_planar" in pull-down box of "Type". Select operation subtype as "Face Milling with Boundaries", program as "FACE FINISH MILLING", tool as "D10R1-FINISH", geometry as "WORKPIECE" and method as "MILL_FINISH". Modify name as "FACE_MILLING-1", and then click "OK".

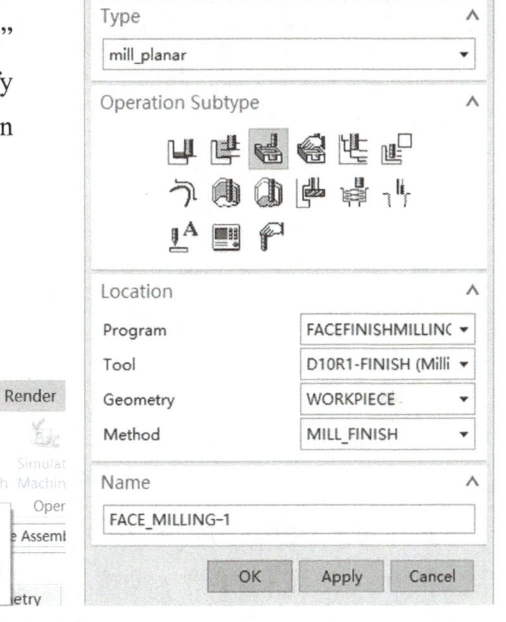

Fig. 2-38 Create Operation

(2) Click "Specify Face Boundaries" and select surface as shown in Fig. 2-39, and click "OK".

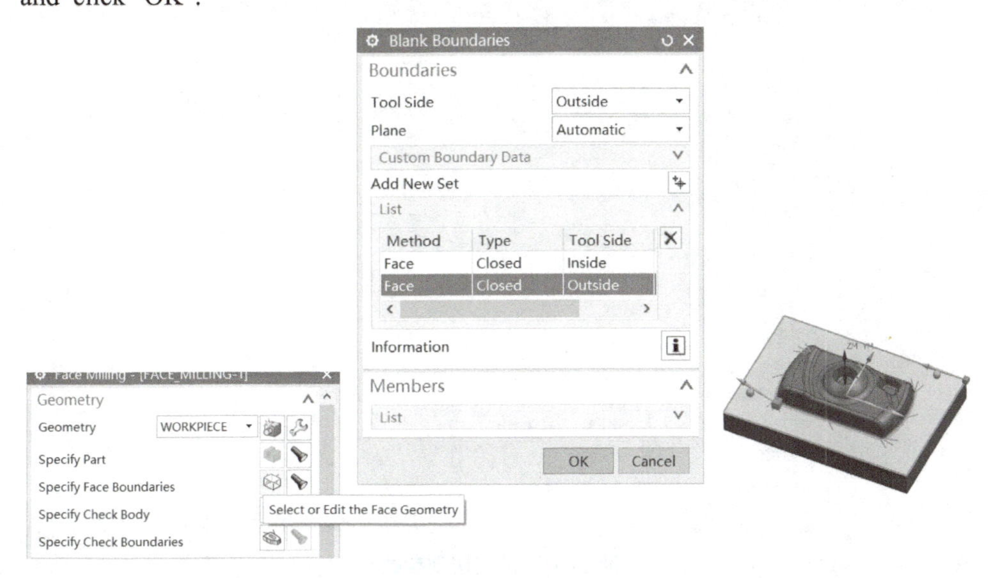

Fig. 2-39 Specify Face Boundaries

（3）【切削模式】选择【跟随周边】，【步距】选择【刀具平直百分比】；设置【平面直径百分比】为【60】，【毛坯距离】为【0.2】，【每刀切削深度】为【0.2】，【最终底面余量】为【0.2】，如图 2-40 所示。

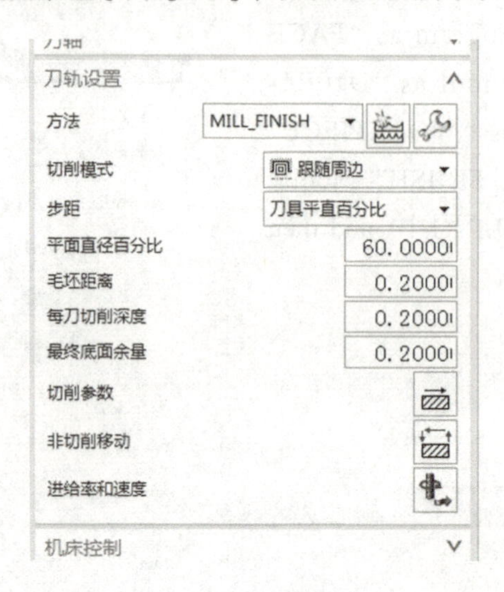

图 2-40　设定切削模式

（4）单击【切削参数】，打开【切削参数】对话框，在【策略】选项卡中，【刀路方向】选择【向内】，勾选【岛清根】；在【余量】选项卡中，设置【壁余量】为【1】，如图 2-41 所示，单击【确定】。

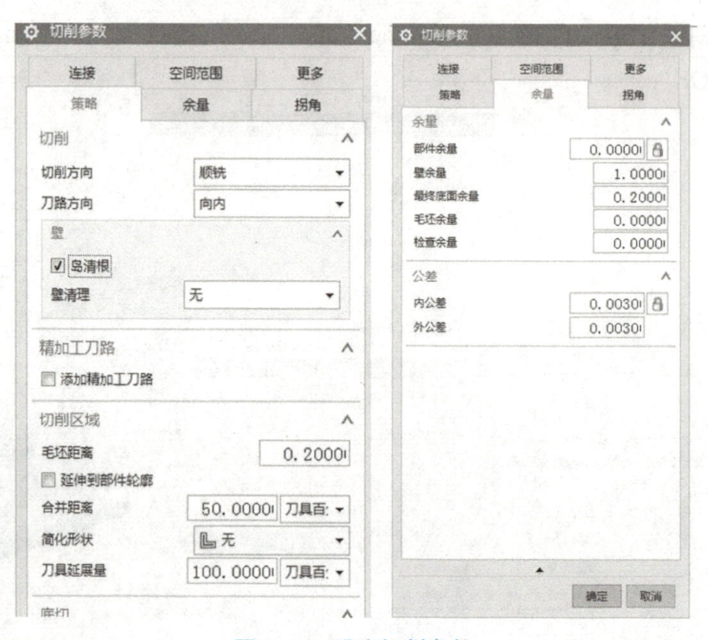

图 2-41　设定切削参数

(3) Modify "Cut Pattern" as "Follow Periphery", select "% Tool Flat" as "Stepover", modify "Percent of Flat Diameter" as 60, "Blank Distance" as 0.2, "Depth Per Cut" as 0.2 and "Final Floor Stock" as 0.2 as shown in Fig. 2-40.

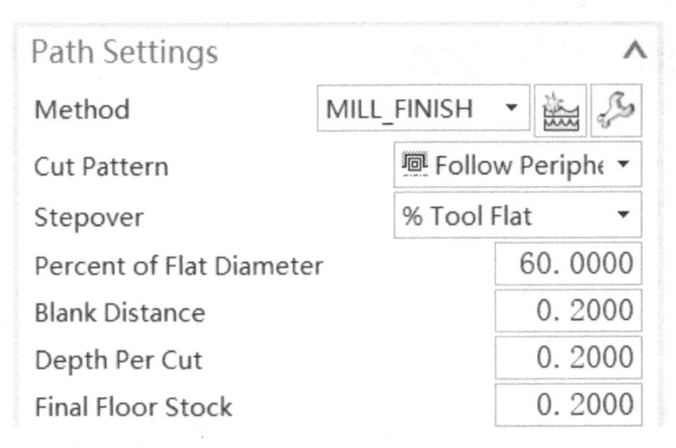

Fig. 2-40　Modify Cut Pattern

(4) As shown in Fig. 2-41, click "Cutting Parameters" button, in "Strategy" tab, select "Inward" as "Pattern Direction" and tick the option "Island Cleanup". In "Stock" tab, modify "Wall Stock" as 1. Click "OK".

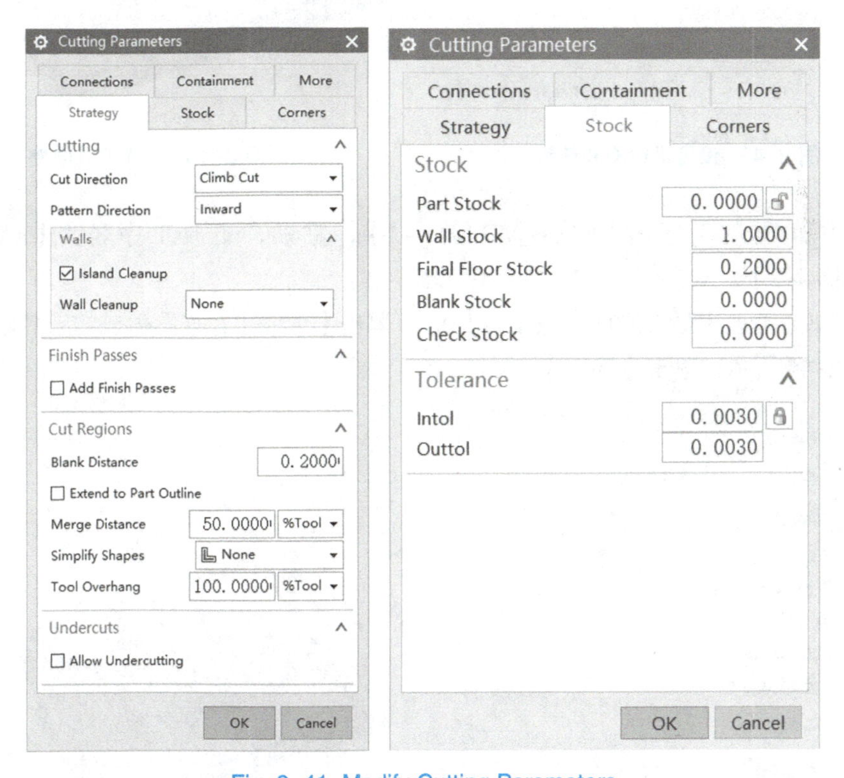

Fig. 2-41　Modify Cutting Parameters

（5）单击【进给率和速度】，打开【进给率和速度】对话框，设置【主轴速度】为【2600】，进给率【切削】为【1400 mmpm】，如图 2-42 所示，单击【确定】。

（6）单击【生成】，查看生成的刀具轨迹，如图 2-43 所示，单击【确定】。

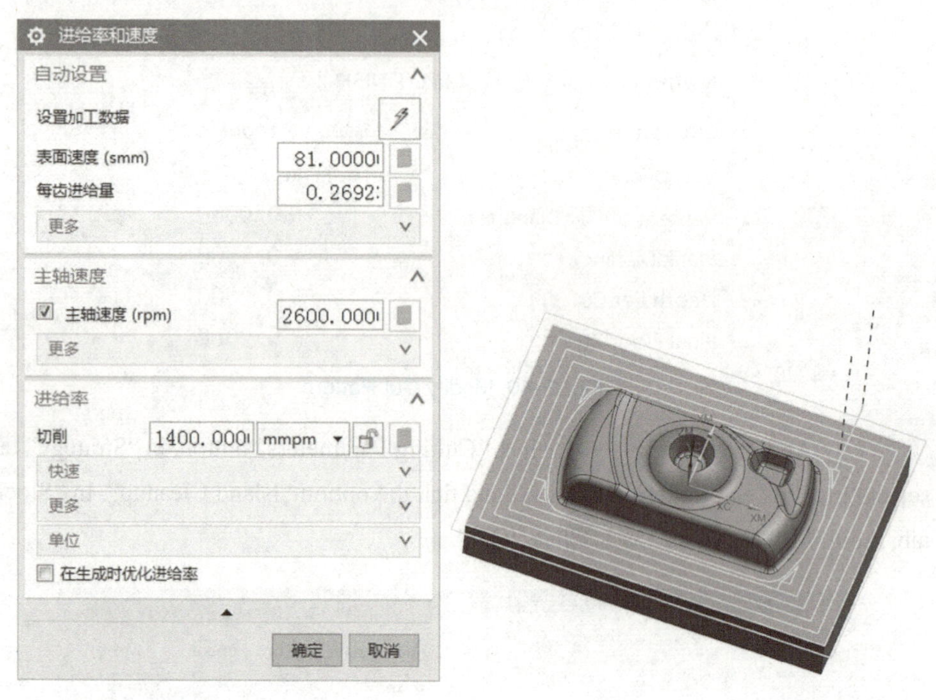

图 2-42 设置进给率和速度　　图 2-43 生成刀具轨迹

（7）复制工序名【FACE_MILLING-1】，粘贴，更改工序名为【FACE_MILLING-2】，双击打开【FACE_MILLING-2】。

（8）修改【最终底面余量】为【0】，单击【生成】，查看生成的刀具轨迹，如图 2-44 所示，单击【确定】。

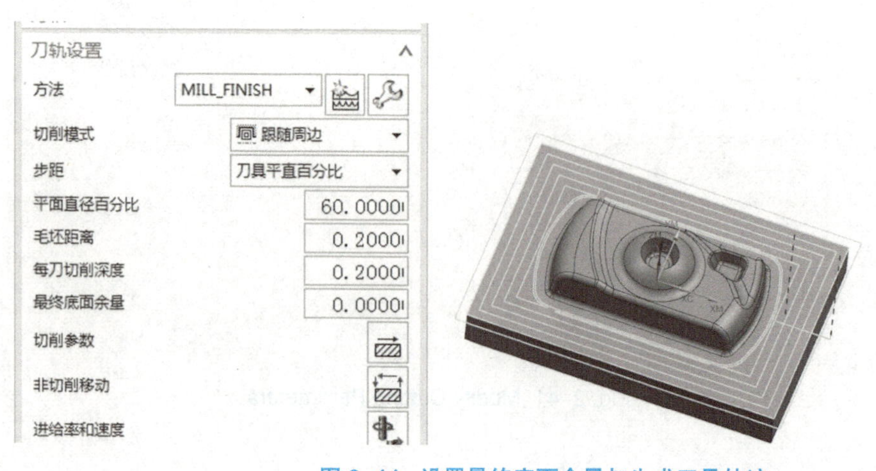

图 2-44 设置最终底面余量与生成刀具轨迹

(5) Click "Feeds and Speeds", modify "Spindle Speed" as 2,600 and "Feed Rates" as 1,400 mmpm as shown in Fig. 2-42. Click "OK".

(6) Click "Generate" to view the tool path as shown in Fig. 2-43, and then click "OK".

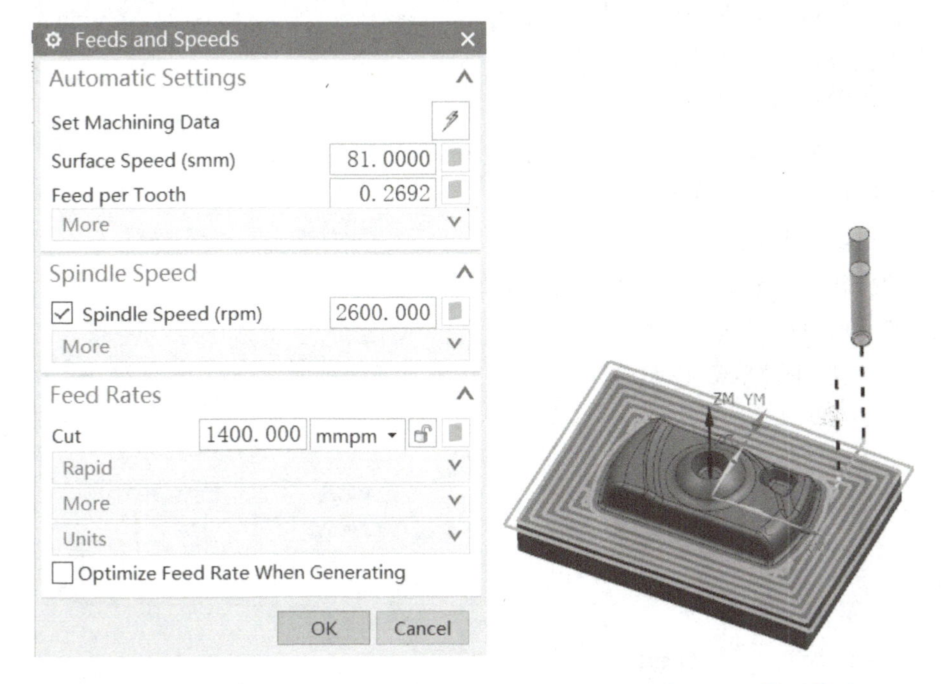

Fig. 2-42 Modify Feeds and Speeds

Fig. 2-43 Generate Tool Path

(7) Copy program "FACE_MILLING-1" and paste it. Modify the pasted file name as "FACE_MILLING-2". Double click the program.

(8) Modify "Final Floor Stock" as 0. Click "Generate" to view the tool path as shown in Fig. 2-44, and then click "OK".

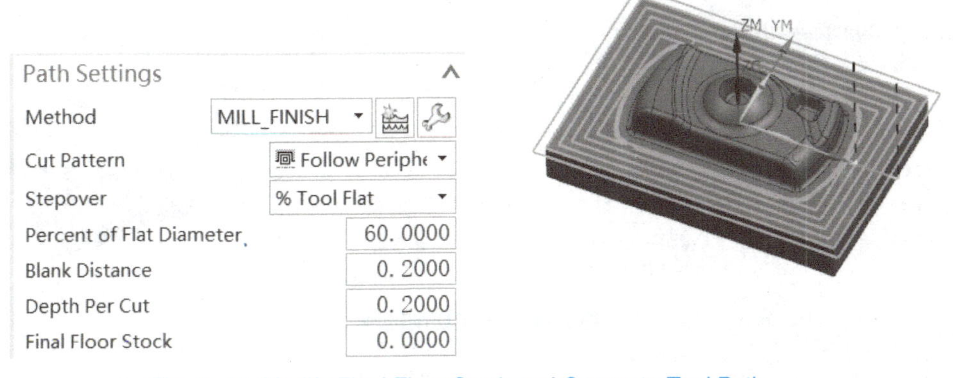

Fig. 2-44 Modify Final Floor Stock and Generate Tool Path

6. 型面精加工 -1

（1）接着对零件进行精加工，复制半精加工程序【SEMI_FINISH_MILL-1】，在型面精加工中选择【内部粘贴】，修改【名称】为【Finish_mill-1】，双击打开【FINISH_MILL-1】。修改【刀具】为【D6R3-FINISH】，【方法】为【MILL_FINISH】。

（2）编辑【区域铣削】方法，更改【最大距离】为【0.2 mm】，【步距已应用】选择【在部件上】，【陡峭切削模式】选择【深度加工往复】，【深度切削层】选择【恒定】，设置【深度加工每刀切削深度】为【0.2】，如图 2-45 所示，单击【确定】。

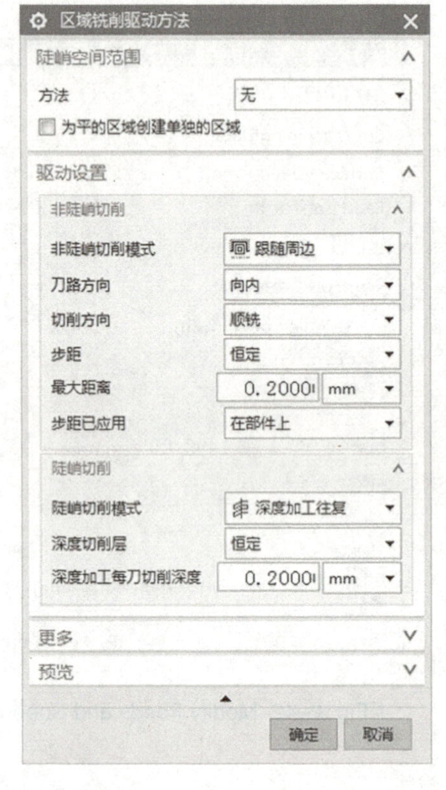

图 2-45　设定驱动方式

（3）单击【指定切削区域】中的【选择或编辑切削区域几何体】，删除中心位置，如图 2-46 所示。

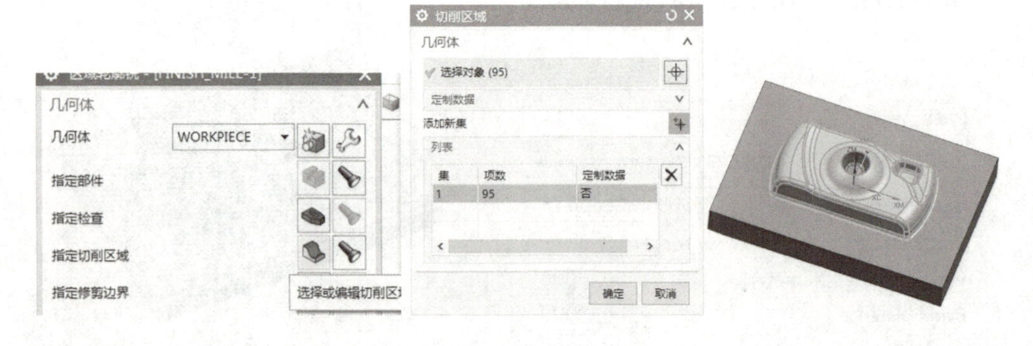

图 2-46　设定切削区域

6. Finishing-1

(1) Copy program "SEMI_FINISH_MILL-1" program and paste it under program "Finishing". Modify the pasted file name as "FINISH_MILL-1". Double click the program, modify the tool as "D6R3-FINISH" and method as "MILL_FINISH".

(2) Click "edit" to set parameters of drive method. Set 0.2 mm as "Maximum Distance", "On Part" as "Stepover Applied", "Zlevel Zig Zag" as "Steep Cut Pattern", "Constant" as "Zlevel Cut Levels", 0.2 mm as "Zlevel Depth per Cut" as shown in Fig. 2-45. Click "OK".

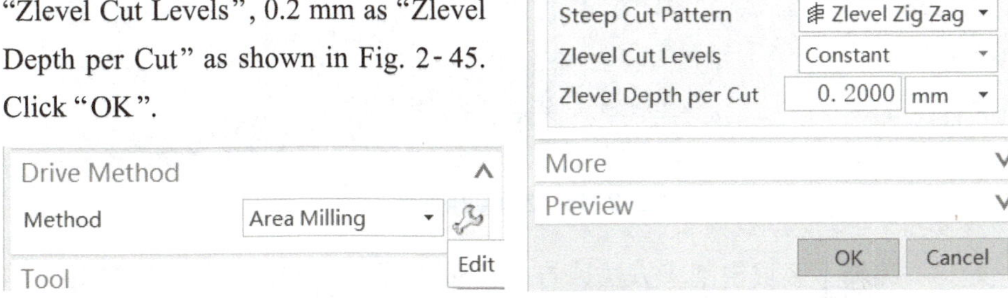

Fig. 2-45 Edit Drive Method

(3) Click "Specify Cut Area" and delete the center part as shown in Fig. 2-46. Click "OK".

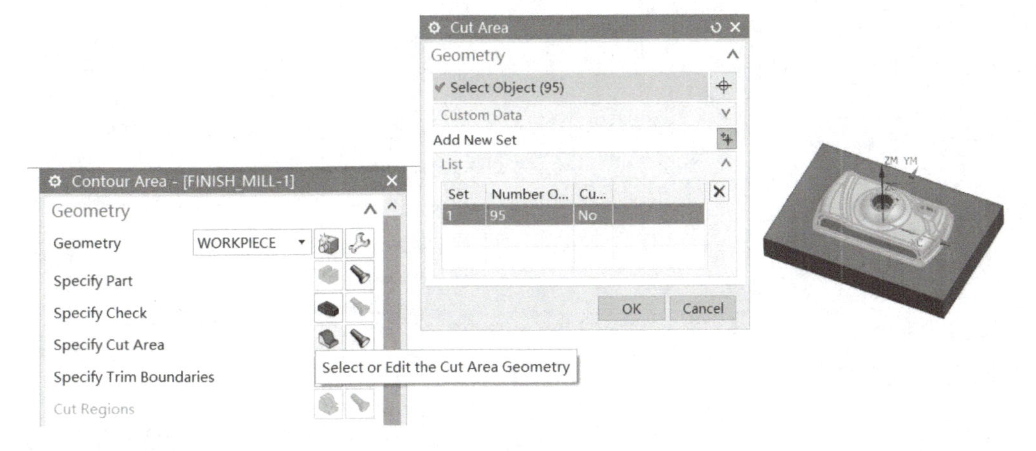

Fig. 2-46 Specify Cut Area

（4）单击【生成】，查看生成的刀具轨迹，如图 2-47 所示，单击【确定】。

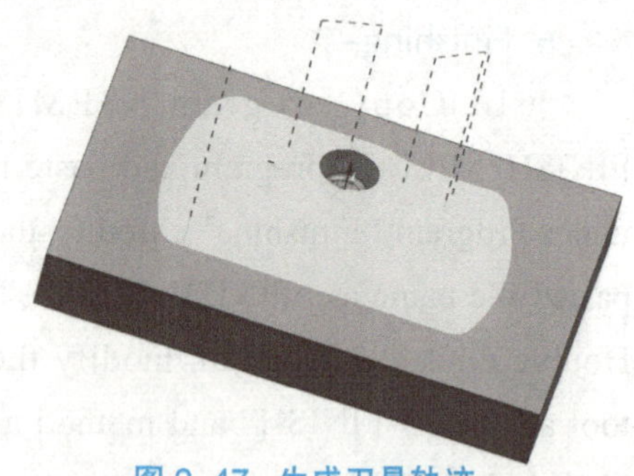

图 2-47 生成刀具轨迹

7. 型面精加工 -2

（1）复制精加工程序【FINISH_MILL-1】，在【型面精加工中 -2】中选择【内部粘贴】，将名称改为【FINISH_MILL-2】，双击打开【FINISH_MILL-2】。

（2）单击【切削区域】，删除已选择面，添加中心曲面，如图 2-48 所示。

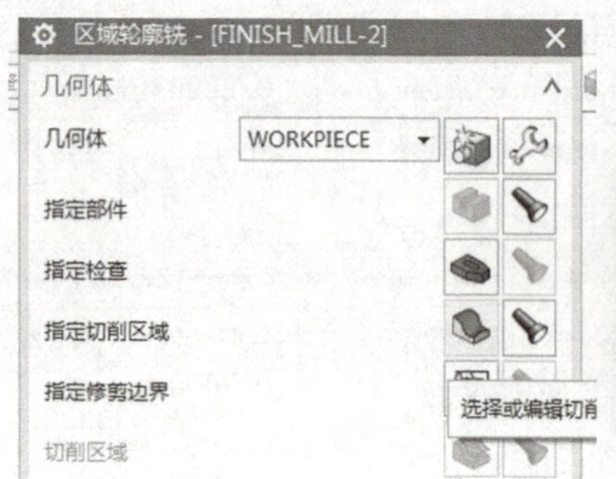

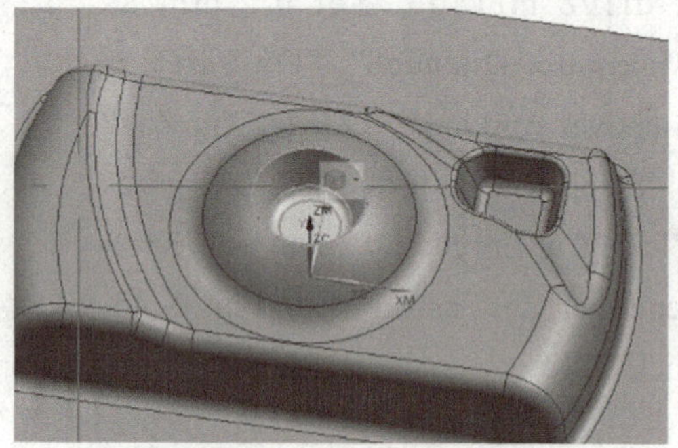

图 2-48 选择切削区域

（3）编辑【区域铣削】方法，打开【区域铣削驱动方法】对话框，更改【刀路方向】为【向内】，如图 2-49 所示，单击【确定】。

图 2-49 设定驱动方式

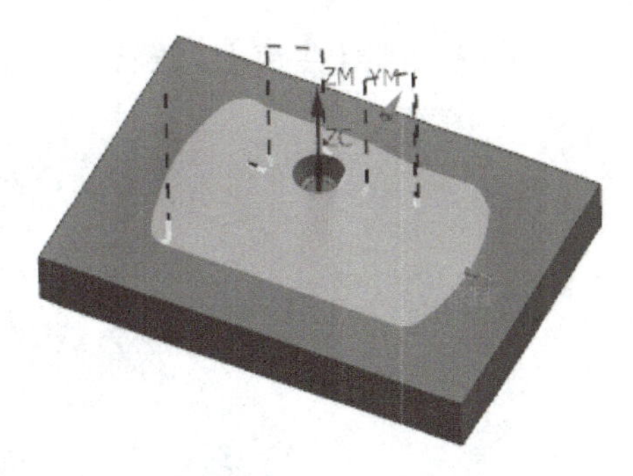

Fig. 2-47　Generate Tool Path

(4) Click "Generate" to view the tool path as shown in Fig. 2-47, and then click "OK".

7. Finishing-2

(1) Copy program "FINISH_MILL-1" program and paste it under program "Finishing-2". Modify the pasted file name as "FINISH_MILL-2". Double click it to open the operation.

(2) Click "Specify Cut Area", delete selected part and reselect part as shown in Fig. 2-48, and then click "OK".

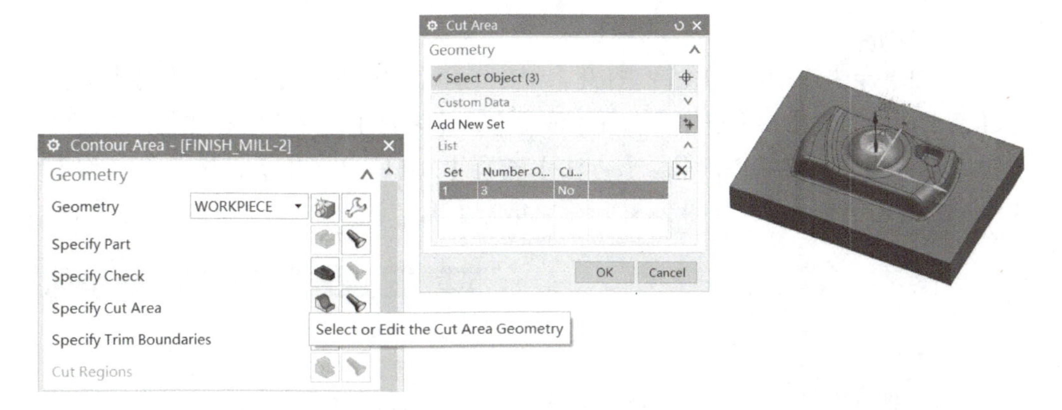

Fig. 2-48　Specify Cut Area

(3) Click "Edit" to set parameters of "Area Milling Drive Method", modify the "Pattern Direction" as "Inward" as shown in Fig. 2-49. Click "OK".

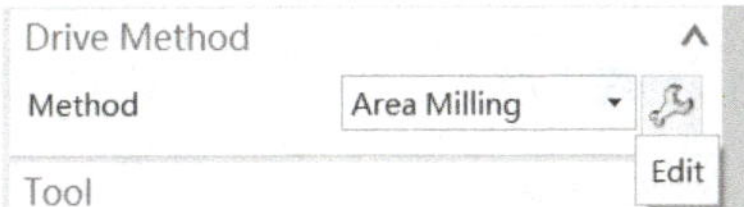

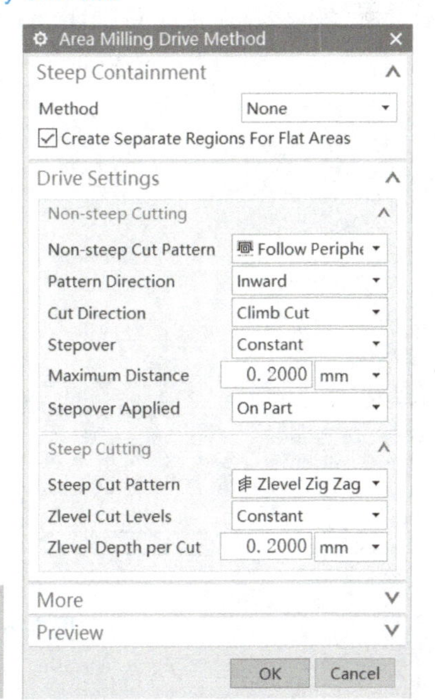

Fig. 2-49　Edit Drive Method

（4）单击【生成】，查看生成的刀具轨迹，如图 2–50 所示，单击【确定】。

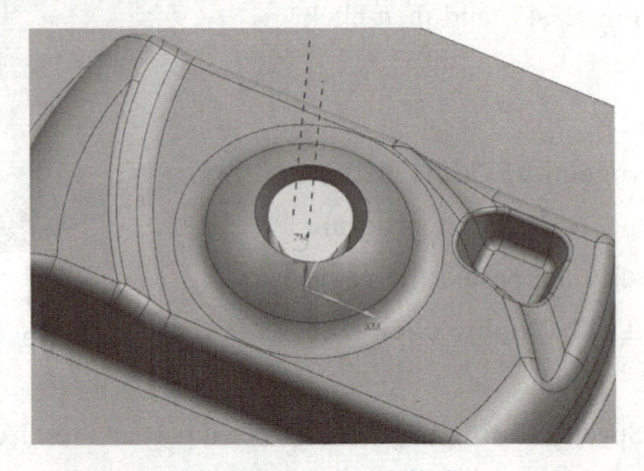

图 2–50　生成刀具轨迹

8. 清根

（1）单击【创建工序】，打开【创建工序】对话框，【工序子类型】选择【清根参考刀具】，【程序】选择【清根】，【刀具】选择【D3R1.5】，【几何体】选择【WORKPIECE】，【方法】选择【MILL_FINISH】，修改【名称】为【FLOWCUT–1】，如图 2–51 所示，单击【确定】。

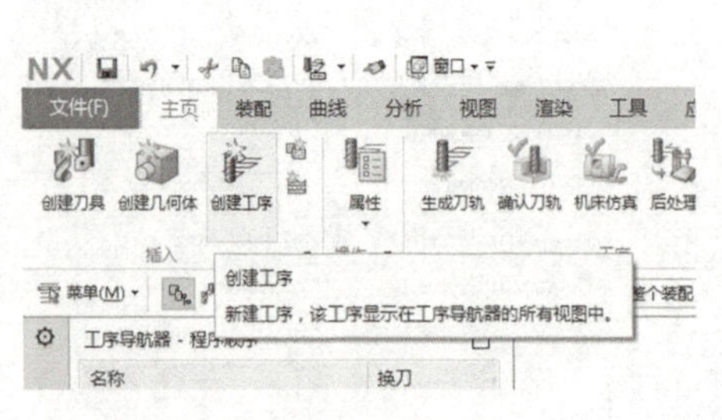

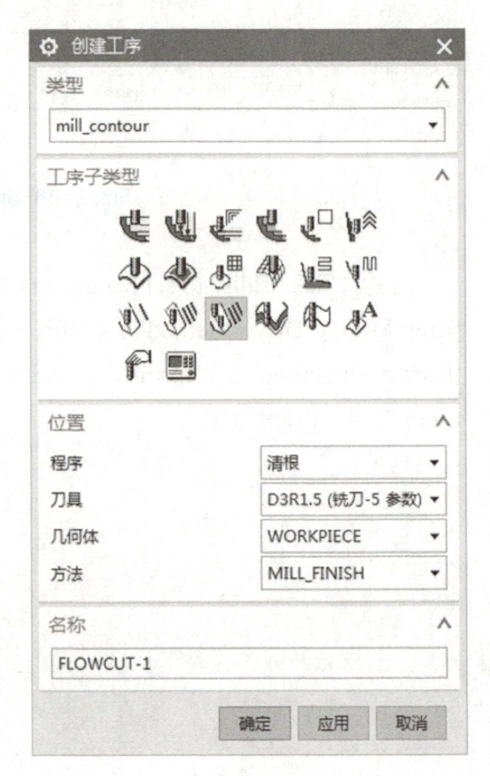

图 2–51　创建工序

(4) Click "Generate" to view the tool path as shown in Fig. 2-50 and then click "OK".

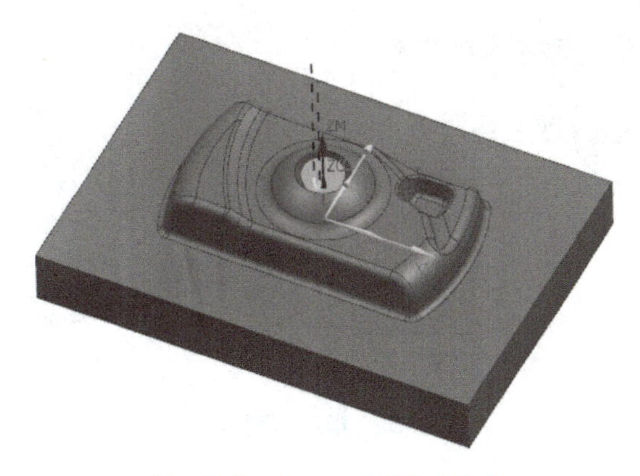

Fig. 2-50 Generate Tool Path

8. Flowcut

(1) As shown in Fig. 2-51, click "Create Operation", and then select "mill_contour" in pull-down box of "Type". Select operation subtype as "Flowcut reference tool", program as "FLOWCUT", tool as "D3R1.5", geometry as "WORKPIECE" and method as "MILL_FINSIH". Modify name as "FLOWCUT-1", and then click "OK".

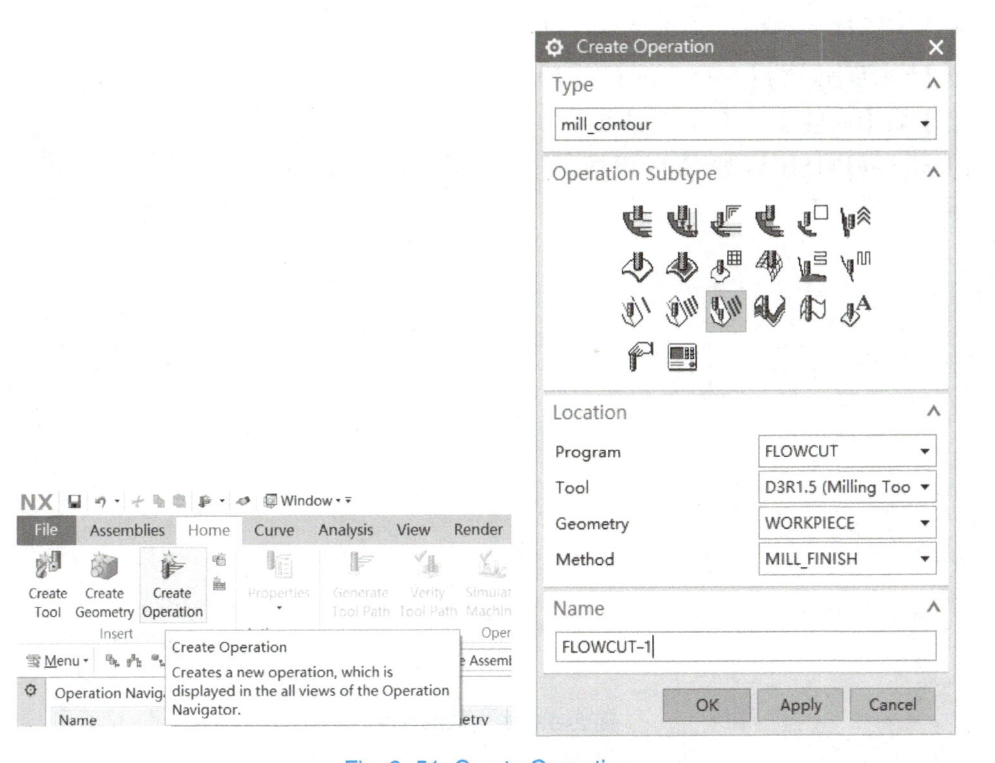

Fig. 2-51 Create Operation

（2）单击【指定切削区域】中的【选择或编辑切削区域几何体】，选择切削曲面，如图 2-52 所示，单击【确定】。

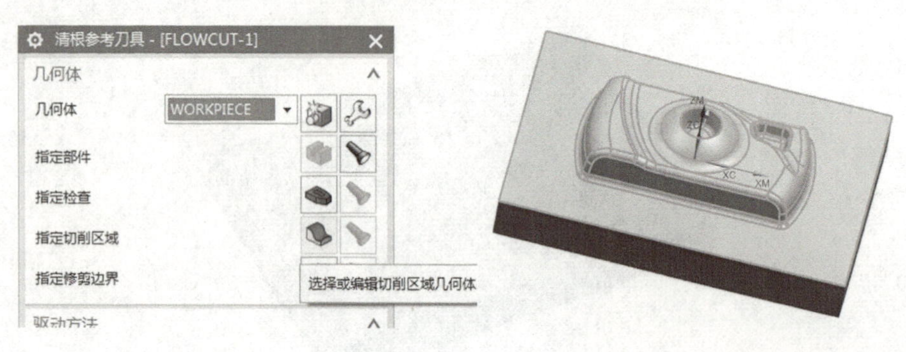

图 2-52 指定切削区域

（3）编辑【清根】方法，打开【清根驱动方法】对话框，【清根类型】选择【参考刀具偏置】；设置【非陡峭切削模式】为【往复】，【切削方向】为【混合】，【步距】为【0.15 mm】，【顺序】选择【先陡】，【陡峭切削模式】选择【往复】，【步距】为【0.15 mm】，【顺序】为【先陡】，【参考刀具】为【D6R3-FINISH】，如图 2-53 所示，单击【确定】。

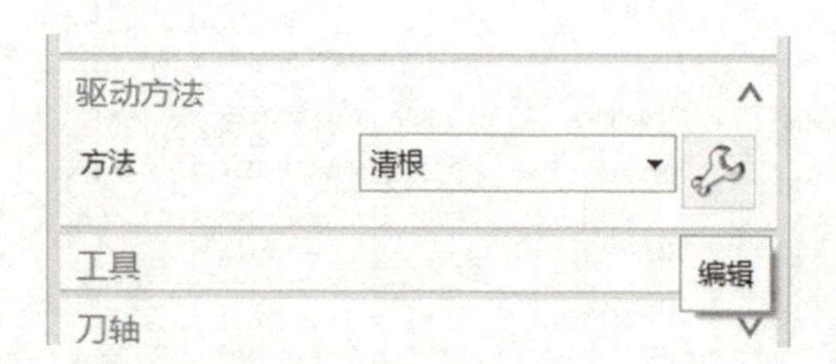

图 2-53 设定切削方法

(2) Click "Specify Cut Area" and select surface as shown in Fig. 2-52. Click "OK".

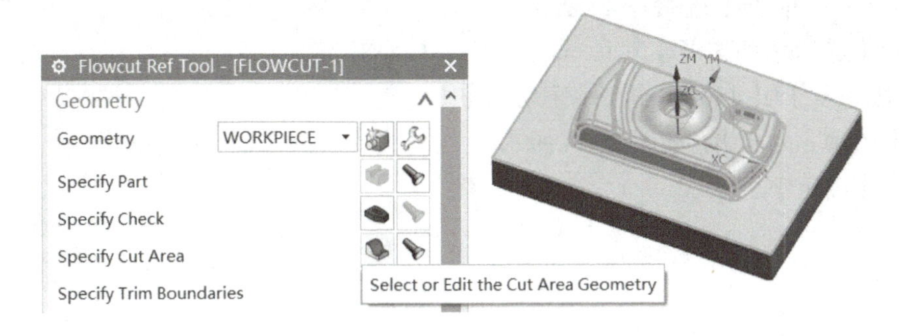

Fig. 2-52　Specify Cut Area

(3) Click "Edit" to set parameters of "Flow Cut Drive Method". Select "Reference Tool Offsets" as flowcut type. Select "Zig Zag" as non-steep cut pattern, "Mixed" as cut direction, 0.15 mm as stepover, "Steep First" as sequencing, "Zig Zag" as steep cut pattern, 0.15 mm as stepover, "Steep First" as sequencing. "D6R3-FINISH" as reference tool as shown in Fig. 2-53. Click "OK".

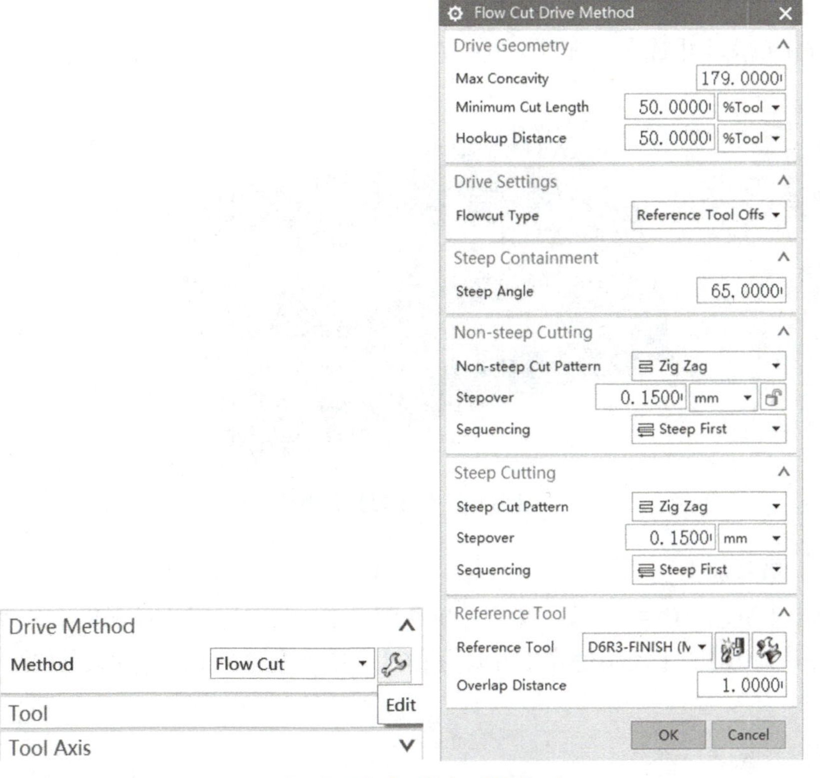

Fig. 2-53　Set Drive Method

（4）单击【切削参数】，采用默认参数，单击【确定】。

（5）单击【进给率和速度】，打开【进给率和速度】对话框，设置【主轴速度】为【4500】，进给率【切削】为【1000 mmpm】，如图 2-54 所示，单击【确定】。

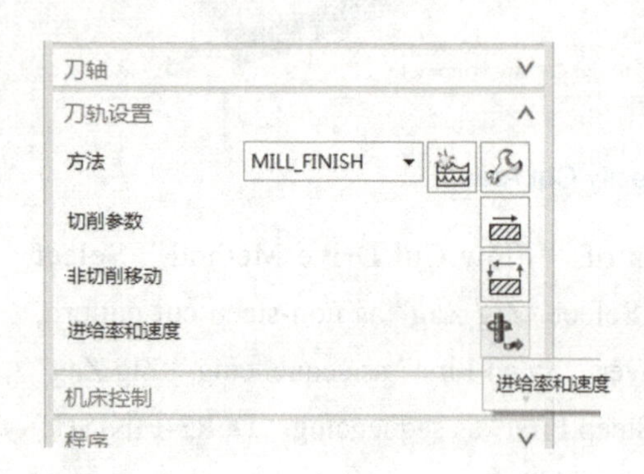

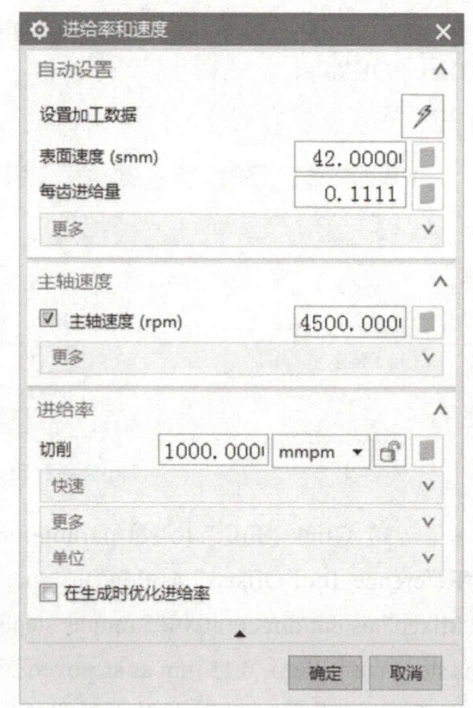

图 2-54　设置进给率和速度

（6）单击【生成】，查看生成的刀具轨迹，如图 2-55 所示，单击【确定】。

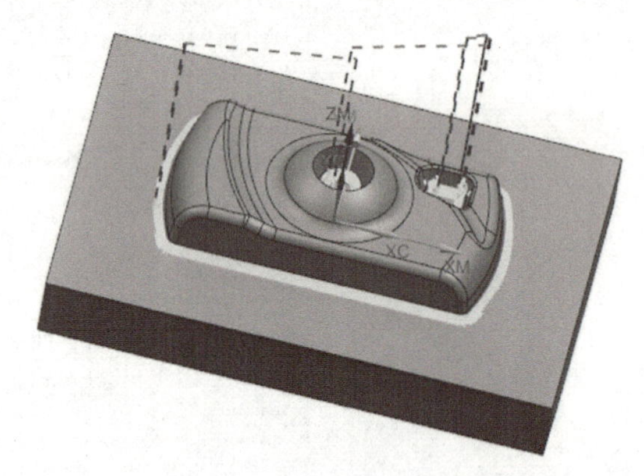

图 2-55　生成刀具轨迹

四、仿真加工

单击【NC_PROGRAM】，单击【确认刀轨】，打开【刀轨可视化】对话框，选择【3D 动态】，【动画速度】可调为【6】，单击【播放】。仿真结果如图 2-56 所示。

(4) Keep Cutting Parameters default.

(5) Click "Feeds and Speeds", modify "Spindle Speed" as 4,500 and "Feed Rates" as 1,000 mmpm as shown in Fig. 2-54. Click "OK".

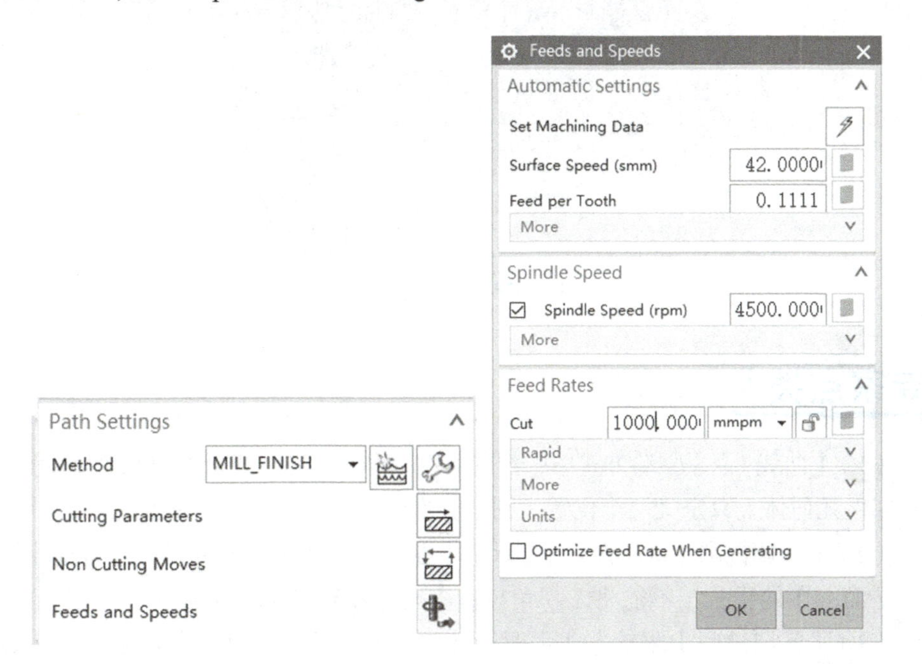

Fig. 2-54　Modify Feeds and Speeds

(6) Click "Generate" to view the tool path as shown in Fig. 2-55, and then click "OK".

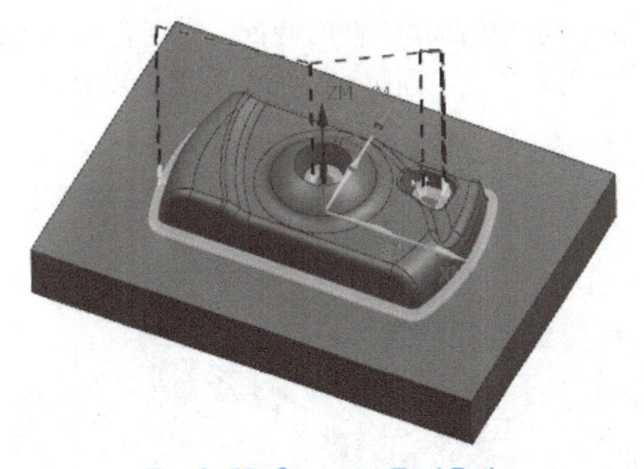

Fig. 2-55　Generate Tool Path

Ⅳ. Simulation Machining

Click "NC_PROGRAM", and then "Verify Tool Path" button. Select "3D Dynamic" and click "Play" button to view the simulation result. The simulation result is shown in Fig. 2-56.

图 2-56 仿真结果

专家点拨

（1）【曲线 / 点】驱动方式中，若只指定一个驱动点，或者指定几个驱动点，使得部件几何体上只定义一个位置，则不会生成刀轨且会显示出错消息。

（2）【区域铣削】驱动方式中，计算【在部件上】的步距所需的时间比计算【在平面上】的更长，不能将【拐角控制】与【在部件上】选项一起使用。

（3）【区域铣削】驱动方式主要设计用于使用【在部件上】选项时的精加工刀路，且不支持多个深度。

课后训练

根据图 2-57 所示的壳体凸模零件的特征，制定合理的工艺路线，设置必要的加工参数，生成刀具轨迹，通过相应的后处理生成数控加工程序，并运用机床加工零件。

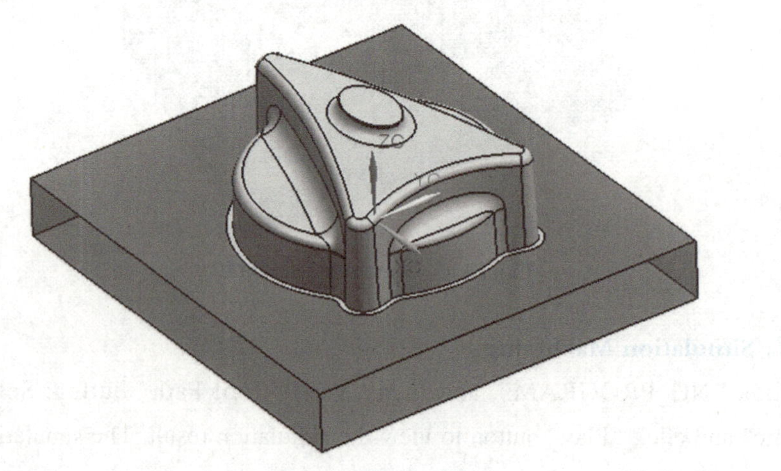

图 2-57 壳体凸模零件

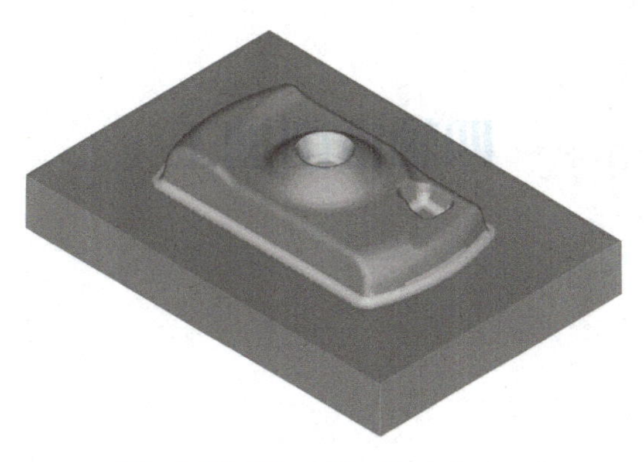

Fig. 2-56 Simulation Result

Expert Reviews

（1） In curve / point driving mode, if only one driving point is specified, or several driving points are specified so that only one position is defined on the part geometry, the tool path will not be generated and an error message will be displayed.

（2）In area milling drive mode, it takes longer to calculate the step "on part" than "on plane", and corner control cannot be used with the "on part" option.

（3）The area milling drive method is mainly designed for the finishing tool path when using the "on part" option, and does not support multiple levels.

Practice

According to the characteristics of disc parts as shown in Fig. 2-57, make a reasonable processing technic, set necessary parameters, generate tool path, generate NC program through corresponding post-processor, and use machine tools to machine parts.

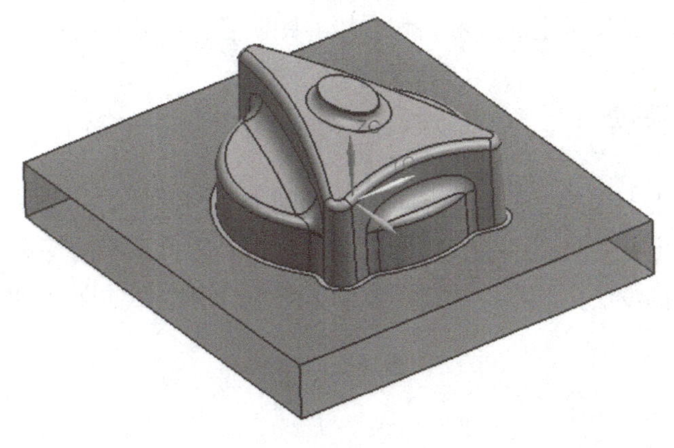

Fig. 2-57 Shell Punch Parts

模块 2 四轴铣削加工

四轴铣削加工通常是指四轴联动加工，就是在 3 个线性轴（X、Y、Z）的基础上增加了一个旋转轴或者摆动轴。相较于传统的三轴铣削加工，四轴铣削加工改变了加工模式，提高了加工能力、零件的复杂度和精度，解决了很多复杂零件的加工难题。

NX CAM 的可变轴轮廓铣是用于精加工由轮廓曲面形成的区域的加工方法。它可以通过精确控制刀轴和投影矢量，使刀轨沿着非常复杂的曲面的复杂轮廓移动。该种加工方法常用于四、五轴铣削编程。

项目 3 滚轴的铣削编程与仿真加工

学习目标

能力目标：能运用 NX 软件完成滚轴的铣削编程与仿真加工。

知识目标：掌握可变轴铣削几何体设置；

掌握【曲线 / 边】驱动方法设置；

掌握投影矢量方法设置；

掌握刀轴的设置。

素质目标：激发学生的学习兴趣，培养团队合作和创新精神。

项目导读

滚轴是印染上用的一种零件，是四轴铣削加工中常见的一类零件。这类零件在圆柱面上开出一些具有一定规律的图案，由于其零件形状的特殊性，采用车削或者三轴铣削都没法完成零件加工。零件上一般会有内凹曲面、圆弧面等特征。

Module 2 4-axis Milling

4-axis milling usually refers to 4-axis linkage machining, that is, a rotating axis or swing axis is added on the basis of three linear axes (X, Y and Z). Compared with the traditional 3-axis milling, 4-axis machining changes the machining mode, enhances the machining ability, improves the complexity and accuracy of parts, and solves many machining problems of complex parts.

Variable axis contour milling of NX CAM is a machining method for Finishing areas formed by contour surfaces. It can make the tool path move along the complex contour of very complex surface by accurately controlling the tool axis and projection vector. It is commonly used for 4-and 5-axis milling programming.

Project 3 Milling Programming and Simulation Machining of Roller

Learning Objectives

Capacity Objective: Complete model programming and simulation machining with NX software.

Knowledge Objective: Master variable axis milling geometry settings;
Master curve / edge driving method settings;
Master the method and setting of projection vector;
Master the setting of tool axis.

Quality objective: Stimulate students' interest in learning and cultivate the spirit of teamwork as well as innovation.

Project Guidance

Roller is a kind of part used in printing and dyeing. It is a common kind of part in 4-axis milling. This kind of part have some regular patterns on the cylindrical surface. Due to the particularity of its part shape, it is impossible to complete the part processing by turning or 3-axis milling. Parts generally have concave surfaces, arc surfaces and other features.

任务描述

学生以企业制造部门 MC 数控程序员的身份进入 NX CAM 功能模块，根据滚轴零件的特征，制定合理的工艺路线，创建型腔铣、可变轴轮廓铣等加工操作，创建必要的参考几何体，设置必要的加工参数，生成刀具轨迹，检验刀具轨迹是否正确合理，并对操作过程中存在的问题进行研讨和交流，通过相应的后处理生成数控加工程序，并运用机床加工零件。

项目实施

按照零件加工要求，制定滚轴的加工工艺；编制滚轴加工程序；完成滚轴的仿真加工，后处理得到数控加工程序，完成零件加工。

一、制定加工工艺

1. 滚轴零件分析

该零件形状比较简单，主要由曲面、圆弧面、端面、圆柱面等组成，主要加工内容为星型曲面、过渡面，经过对零件的分析，可知零件上最小的内凹圆弧半径为 2 mm，所以精加工时刀具半径不能大于 2 mm。

2. 毛坯选用

零件材料为 45# 钢圆棒，尺寸为 ϕ100 mm × 2 500 mm。零件长度、直径尺寸已经精加工到位，无须再次加工。

3. 加工工序卡制定

零件选用立式四轴联动机床加工（立式加工中心，带有绕 X 轴旋转的回转台），三爪卡盘夹持，遵循先粗后精加工原则，粗加工采用“3+1”轴型腔铣方式，精加工采用四轴联动加工。制定的加工工序卡如表 3-1 所示。

Task Description

Students operate NX CAM functional modules as MC programmers in enterprise manufacturing department. According to the characteristics of model parts, establish a reasonable processing route, create machining operations such as cavity milling and variable axis contour milling, create necessary reference geometry, and set necessary machining parameters, generate tool path and check the generated tool path. Besides, students should discuss the problems occurred in the operation. By selecting the corresponding post-processor to generate NC machining programs, students import them into machine tools to complete the parts processing.

Project Implementation

Firstly, formulate processing technic under processing requirements. Secondly, program for machining model. Finally, complete simulation machining, and then import the NC machining program obtained by post processor to the machine tool to complete machining.

I. Formulate Processing Technic

1. Structural analysis of connecting block

The structure of the part is relatively simple, mainly composed of curved surface, end face, cylindrical surface, etc. The main processing content is the star curved surface and the excessive surface. After the analysis of the parts, it is known that the radius of the smallest concave arc on the part is 2 mm, so the tool radius can not be greater than 2 mm when finishing.

2. Blank selection

The part material is 45# steel round bar, and the size is ϕ100 mm × 2,500 mm. The length and diameter of the parts have been finished in place and no machining required.

3. Formulation of processing procedure card

The parts are processed by vertical 4-axis linkage machine tool (vertical machining center with rotary table rotating around X-axis), clamped by three jaw chuck, following the principle of rough machining first and then finish machining. The rough machining adopts "3+1" -axis cavity milling method, and the finish machining adopts 4-axis linkage machining. The processing procedure is shown in Table 3-1.

表 3-1 加工工序卡

零件号：27024003		工序名称：滚轴铣削加工		工艺流程卡_工序单
材料：45#	页码：1	工序号：01		版本号：0
夹具：三爪卡盘	工位：MC	数控程序号：		

刀具及参数设置				
加工内容	刀具号	刀具规格	主轴转速 /rpm	进给速度 /mmpm
型面粗加工	T01	D6R0	1800	1200
型面侧壁精加工	T02	D3R0	2600	1400
型面底面精加工	T03	B4	3200	1000

更改号	更改内容	批准	日期
拟制：　日期：	审核：　日期：	批准：	日期：

二、加工准备

（1）启动 NX，单击【打开】，打开【打开】对话框，选择【滚轴 .prt】文件，如图 3-1 所示，单击【OK】，打开零件模型。

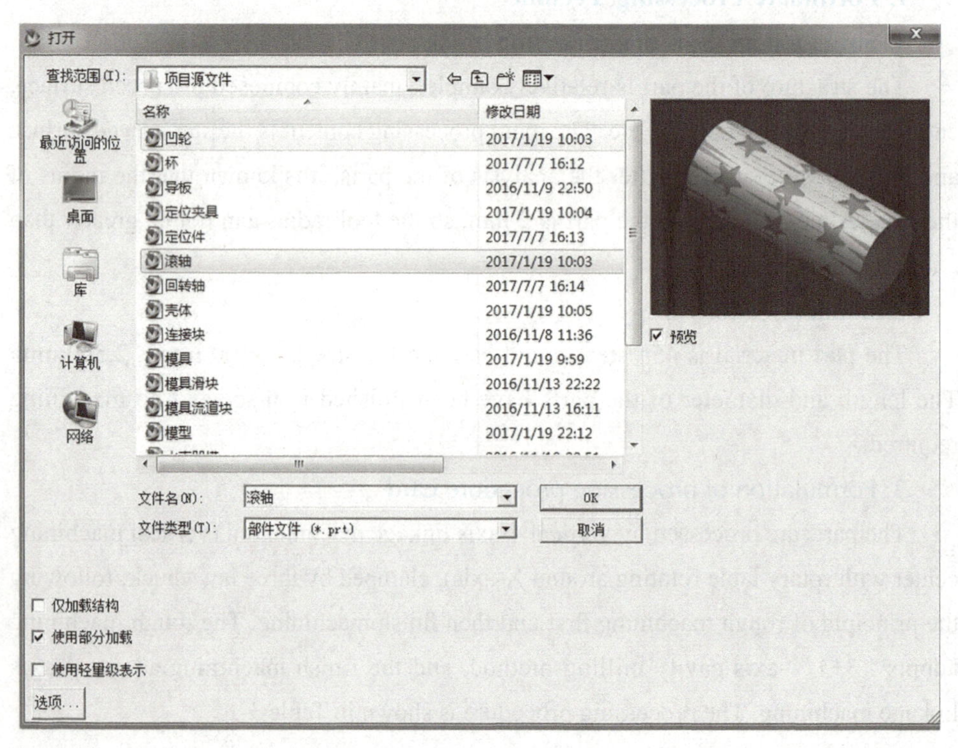

图 3-1 打开文件

Table 3-1 Processing Procedure Card

Part number: 27024003		Name of process: Milling of roller		Process card - Process sheet	
Material: 45#	Page number: 1			Procedure number: 01	Version number: 0
Fixture: Three-jaw chuck	Work station: MC			CNC program number:	
Tool and parameter setting					
Tool number	Tool specification	Processing content	Spindle speed /rpm	Feed speed /mmpm	
T01	D6R0	Rough machining	1800	1200	
T02	D3R0	Side finishing	2600	1400	
T03	B4	Bottom finishing	3200	1000	
02					
01					
Change number	Change content	Approve	Date		
Draws:	Date: / review: / Date:	Approve:	Date:		

Ⅱ. Preparation for Processing

（1）As shown in Fig. 3-1, start NX, click “OPEN”, select model “Roller” (.prt file), and click “OK”. Then, the part model can be seen in WCS.

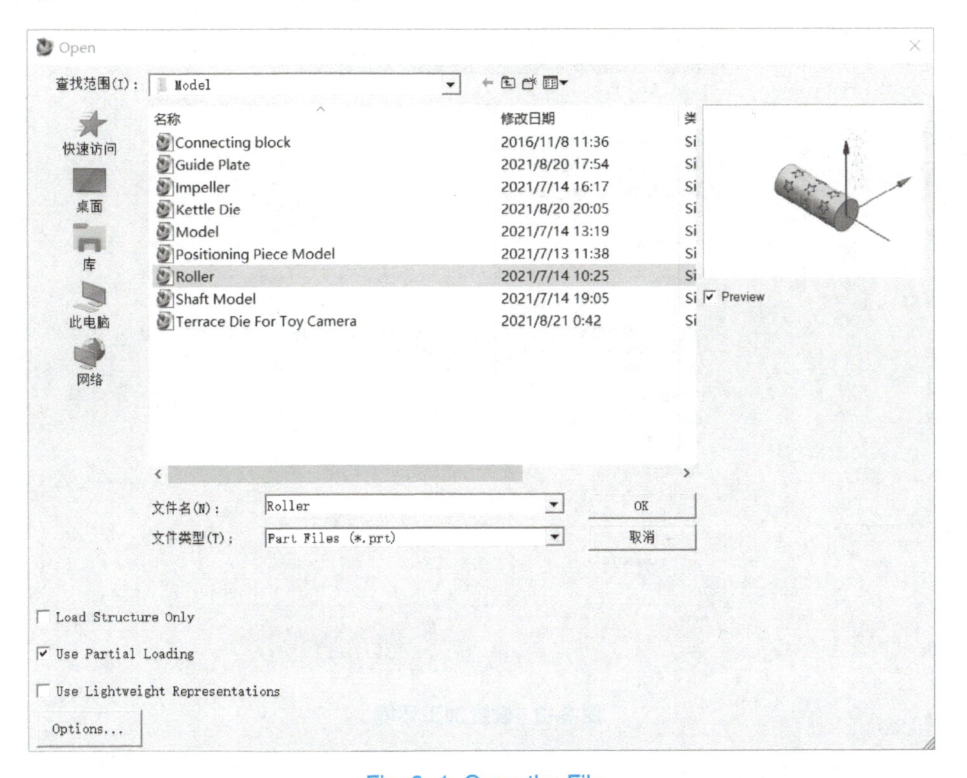

Fig. 3-1 Open the File

（2）选择【应用模块】选项卡，单击【加工】，如图 3-2 所示，进入加工环境。

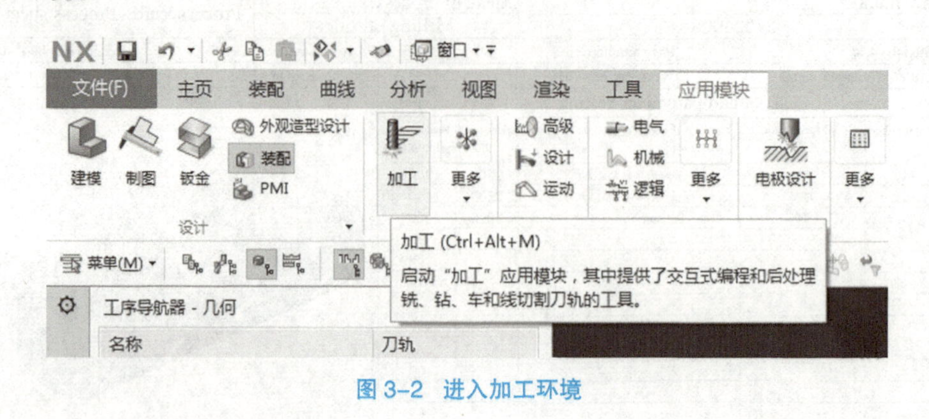

图 3-2 进入加工环境

（3）在弹出的【加工环境】对话框中，在【CAM 会话配置】选项中选择【cam_general】，在【要创建的 CAM 设置】选项中选择【mill_planar】，如图 3-3 所示，单击【确定】。

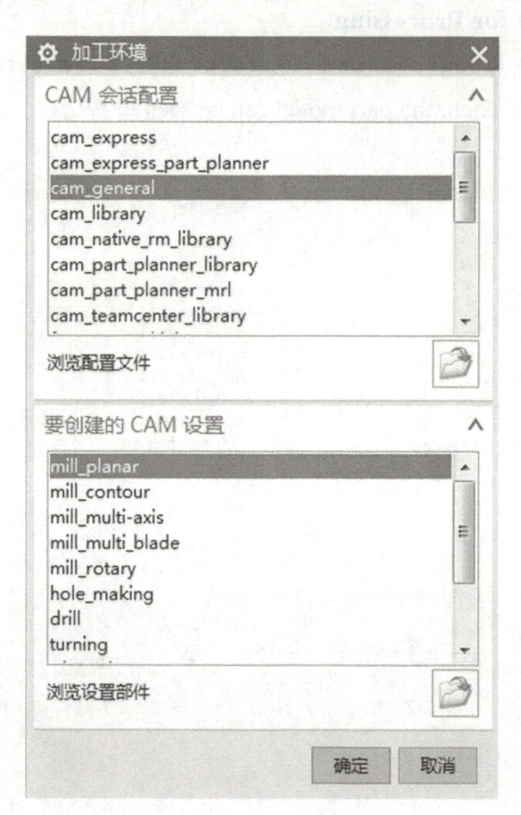

图 3-3 设置加工环境

（2）Select "Application", and click "Manufacturing". The manufacturing menu can be seen as shown in Fig. 3-2.

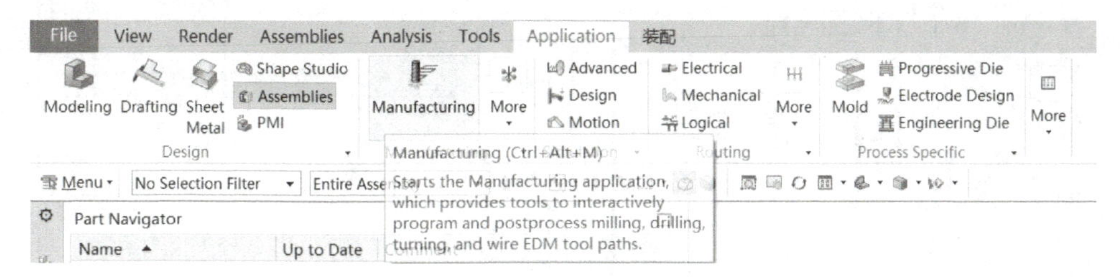

Fig. 3-2 Enter Manufacturing Environment

（3）As shown in Fig. 3-3, in "Machining Environment" window, select "cam_general" in "CAM Session Configuration" tab, and then, select "mill_planar" in "CAM Setup to Create" tab. Finally, click "OK".

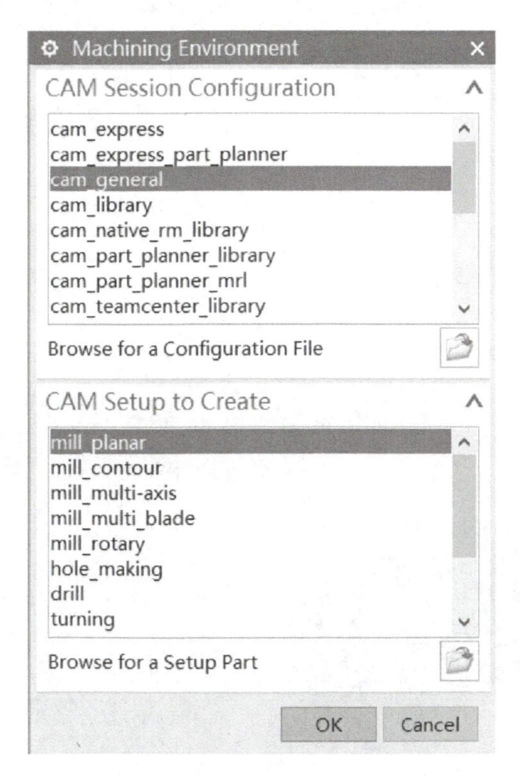

Fig. 3-3 Set Machining Environment

（4）单击【几何视图】，双击【MCS_MILL】，如图 3-4 所示。

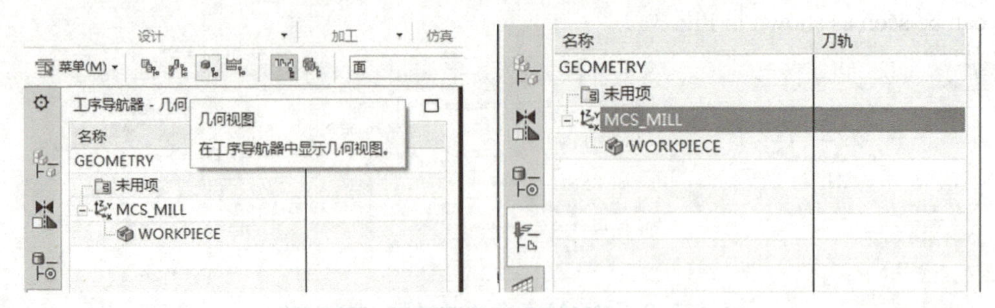

图 3-4　几何视图

（5）将加工坐标系移动到【圆柱面】的端部，同时把【旋转轴】设置成 X 方向旋转，如图 3-5 所示。

（6）安全几何体采用【圆柱】来代替，同时指定点，指定矢量，半径为【80】，如图 3-6 所示。

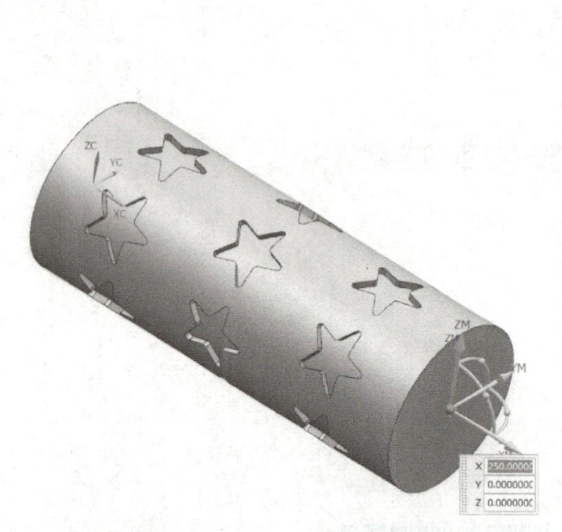

图 3-5　加工坐标系

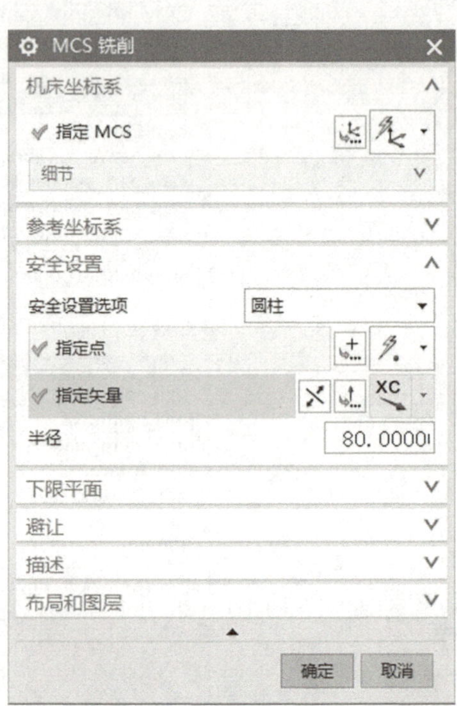

图 3-6　设置机床坐标系

(4) As shown in Fig. 3-4, click "Geometry". Then double click "MCS_MILL".

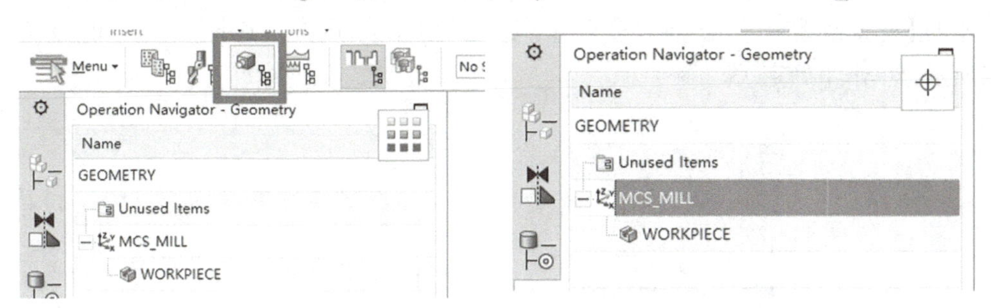

Fig. 3–4 Geometry View

(5) In "MCS Mill" window, click "CSYS Dialog" and then migrate MCS to the end of cylinder and set X-axis as the rotation axis as shown in Fig. 3-5, and then click "OK" to close the window.

(6) As shown in Fig. 3-6, select "Cylinder" as "Clearance Option", specify point and vector and modify "Radius" as 80.

Fig. 3–5 Set MCS

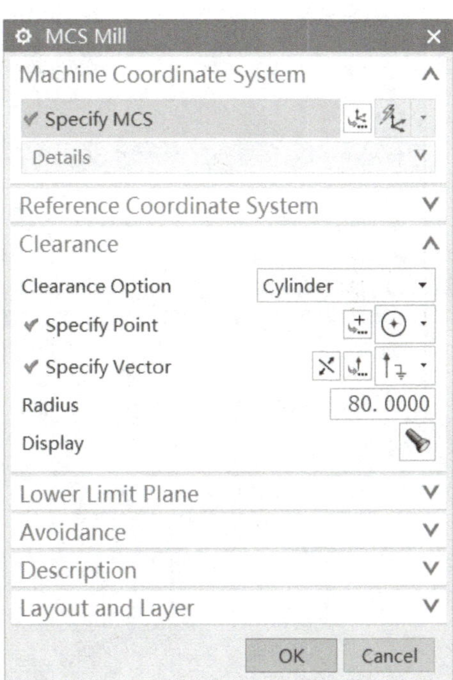

Fig. 3–6 Specify MCS

（7）双击【WORKPIECE】选项，弹出【工件】对话框，单击【指定毛坯】中的【选择或编辑毛坯几何体】弹出【毛坯几何体】对话框，选择毛坯，如图 3-7 所示，单击【确定】。

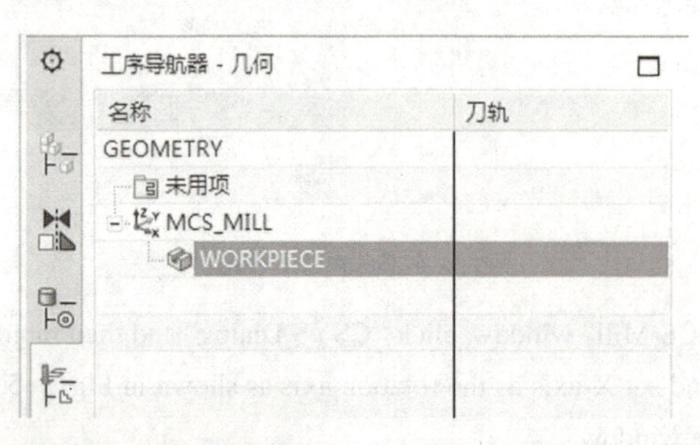

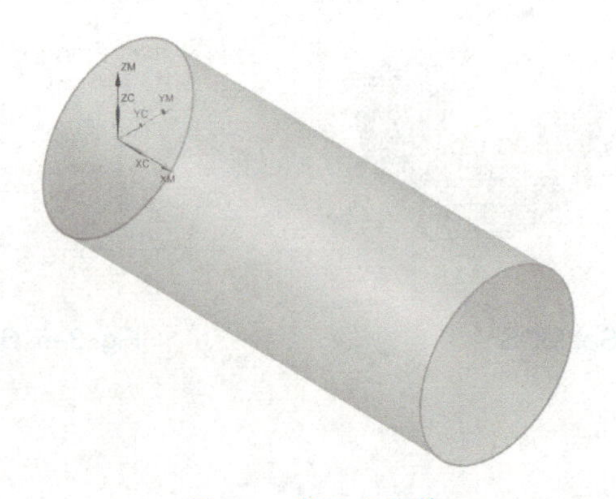

图 3-7 指定几何体

（7）Double click “WORKPIECE” and click “Specify Blank” in “Workpiece” window. In “Blank Geometry” window, select the part as shown in Fig. 3-7. Click “OK”.

Fig. 3-7 Specify Blank Geometry

（8）单击【指定部件】中的【选择或编辑部件几何体】，弹出【部件几何体】对话框，选择【面】来定义，选择凹腔面，如图 3-8 所示。

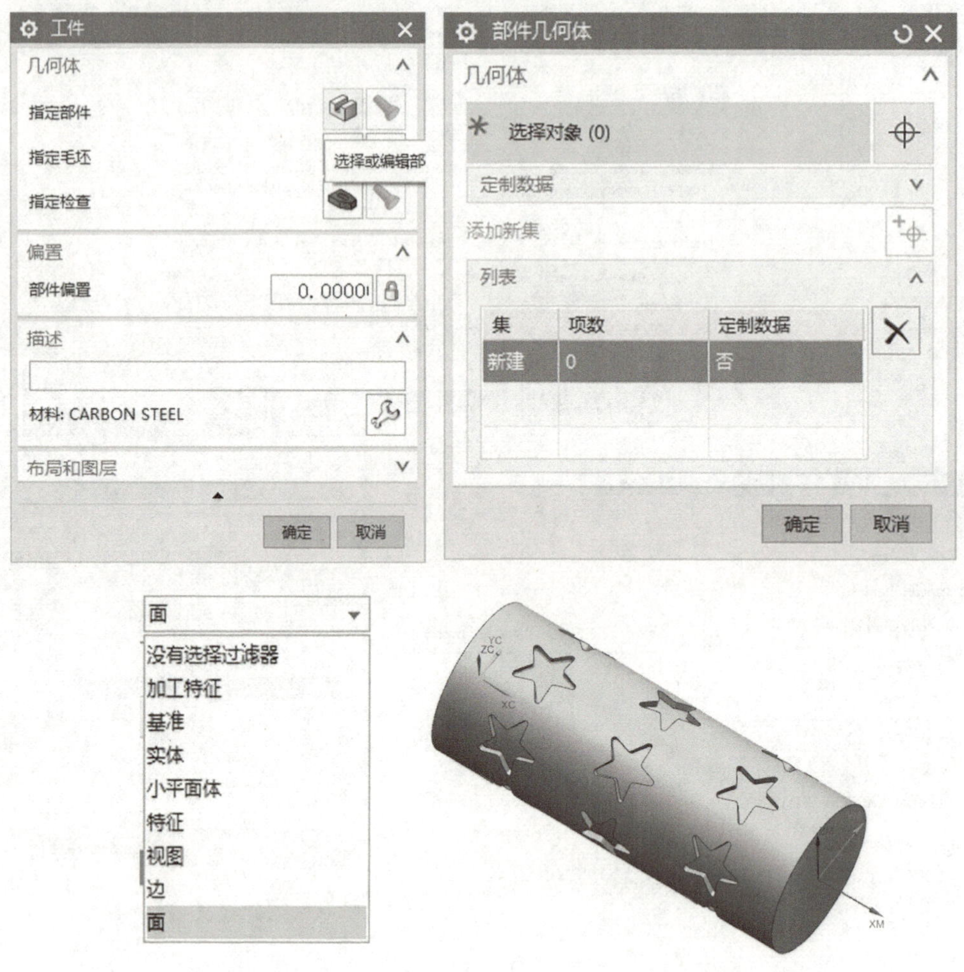

图 3-8　指定部件

（9）单击【机床视图】，单击【创建刀具】，如图 3-9 所示。

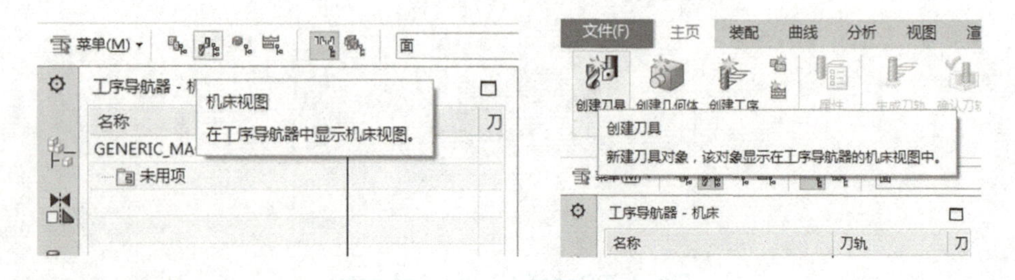

图 3-9　创建刀具

(8) Click "Specify Part" in "Workpiece" window. In "Part Geometry" window, select the surface as shown in Fig. 3-8. Click "OK" to close the window.

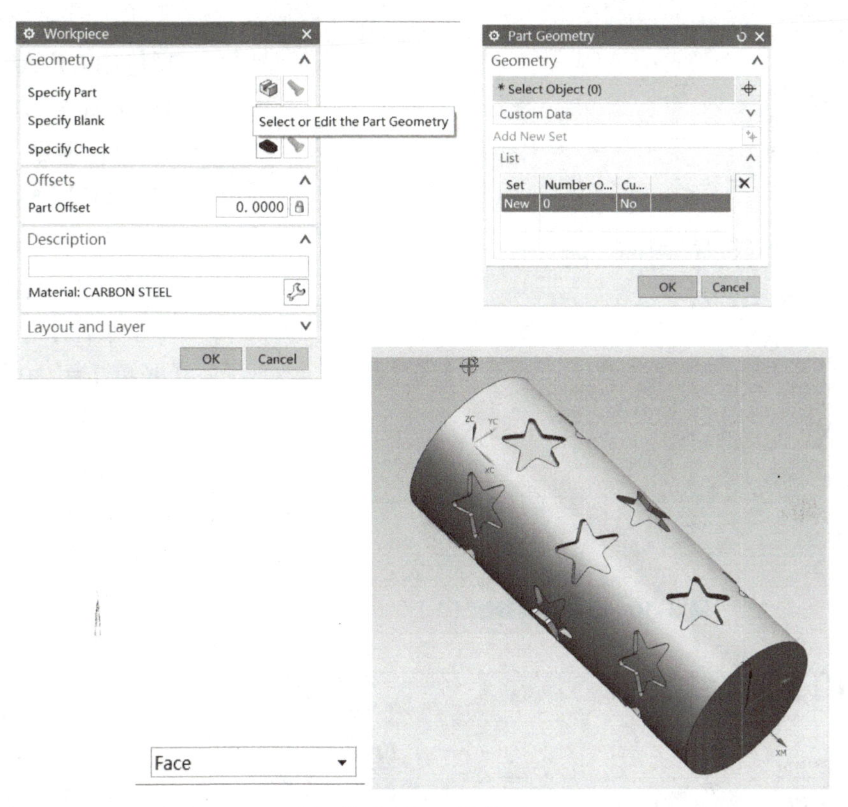

Fig. 3-8　Specify Part

(9) Click "Machine Tool View", and then click "Create Tool" as shown in Fig. 3-9.

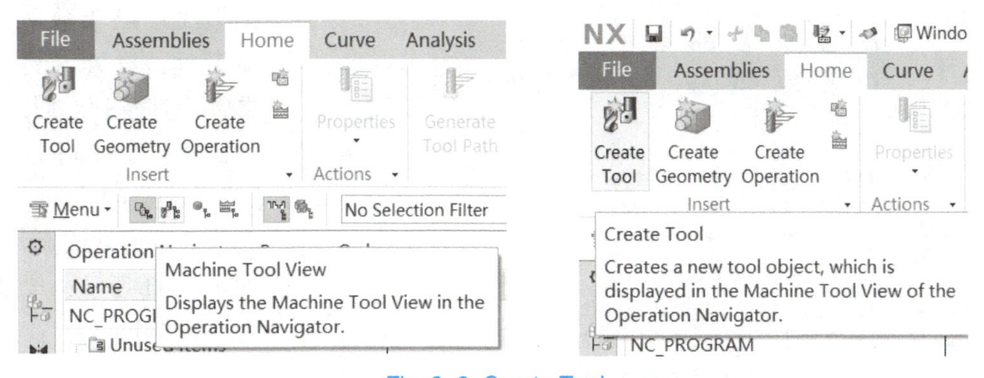

Fig. 3-9　Create Tool

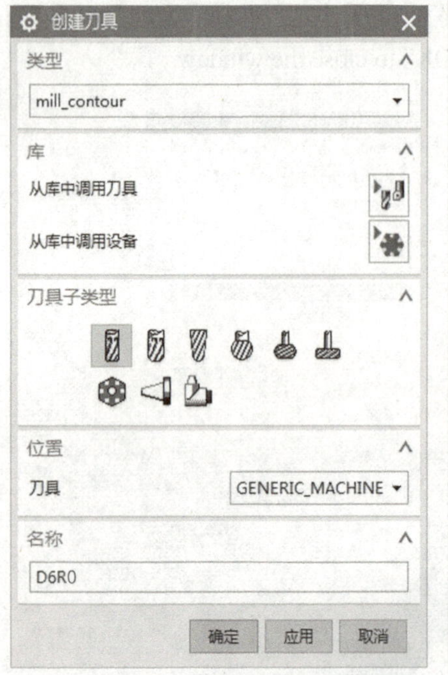

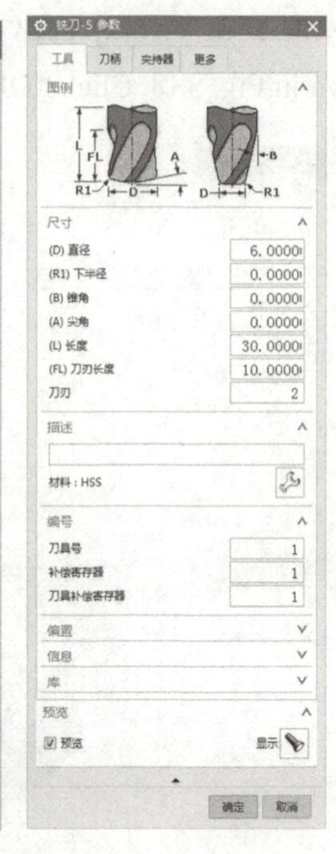

图 3-10　设置刀具参数（1）

（10）在【创建刀具】对话框中，【刀具子类型】选择【MILL】，修改【刀具】名称为【D6R0】，单击【应用】。弹出【铣刀-5 参数】对话框，设置刀具参数：【直径】为【6】，【下半径】为【0】，【长度】为【30】，【刀刃长度】为【10】，其他参数为默认值，如图 3-10 所示，单击【确定】。

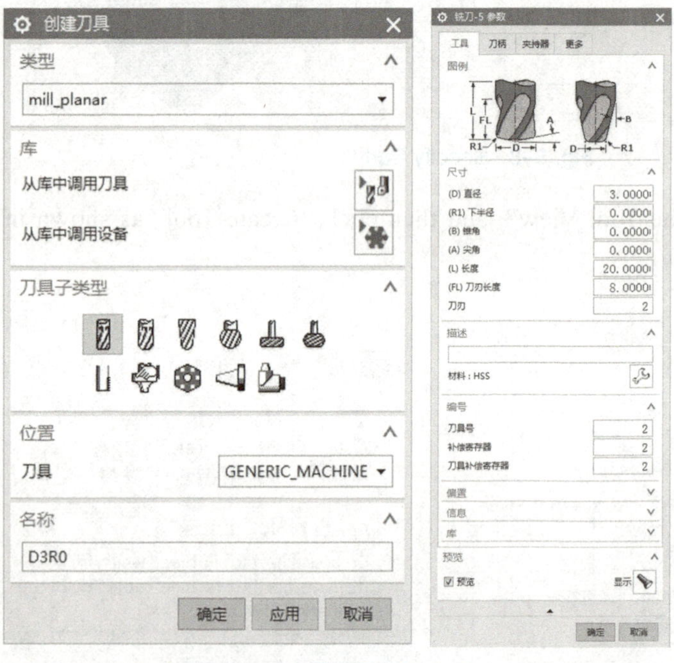

图 3-11　设置刀具参数（2）

（11）回到【创建刀具】对话框，【刀具子类型】选择【MILL】，修改【名称】为【D3R0】，单击【应用】。弹出【铣刀-5 参数】对话框，设置刀具参数：【直径】为【3】，【长度】为【20】，【刀刃长度】为【8】，其他参数为默认值，如图 3-11 所示，单击【确定】。

(10) As shown in Fig. 3-10, in the "Create Tool" window, select "mill" in "Tool Subtype" and modify "Name" as "D6R0", then click "OK". In "Milling Tool-5 Parameters" window, "Diameter" value is 6 and "Lower Radius" value is 0. Other parameters default, click "OK" to close the window.

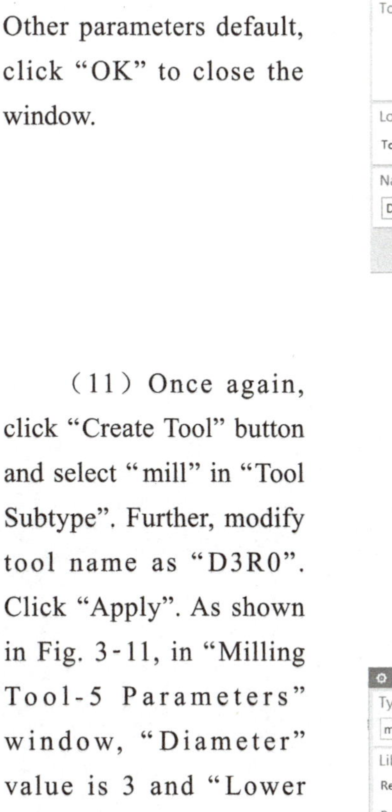

Fig. 3-10 Modify Tool Parameters (1)

(11) Once again, click "Create Tool" button and select "mill" in "Tool Subtype". Further, modify tool name as "D3R0". Click "Apply". As shown in Fig. 3-11, in "Milling Tool-5 Parameters" window, "Diameter" value is 3 and "Lower Radius" value is 0. Other parameters default, click "OK" to close the window.

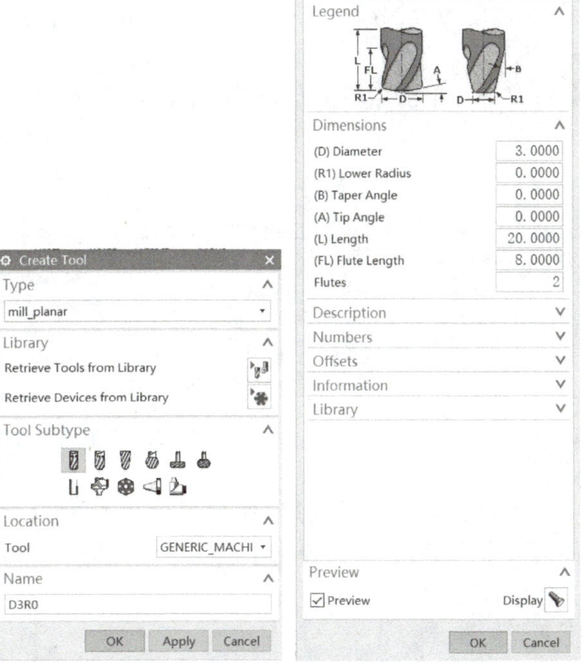

Fig. 3-11 Modify Tool Parameters (2)

（12）回到【创建刀具】对话框，【刀具子类型】选择【BALL-MILL】，修改【名称】为【BALL4】，单击【应用】。弹出【铣刀－球头铣】对话框，设置刀具参数：【球直径】为【4】，【长度】为【15】，其他参数为默认值如图3-12所示。

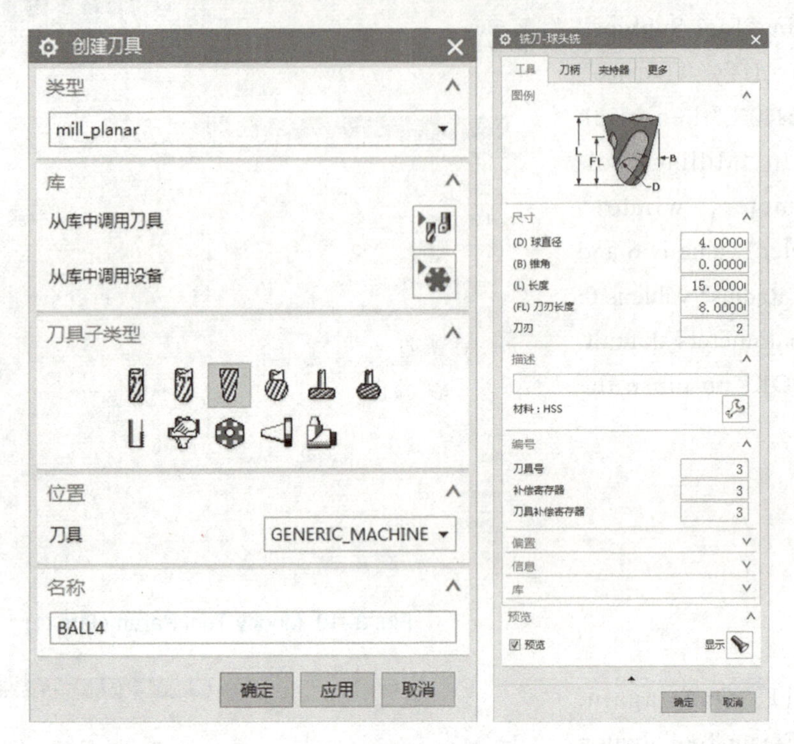

图3-12 设置刀具参数（3）

（13）单击【加工方法视图】，如图3-13所示。

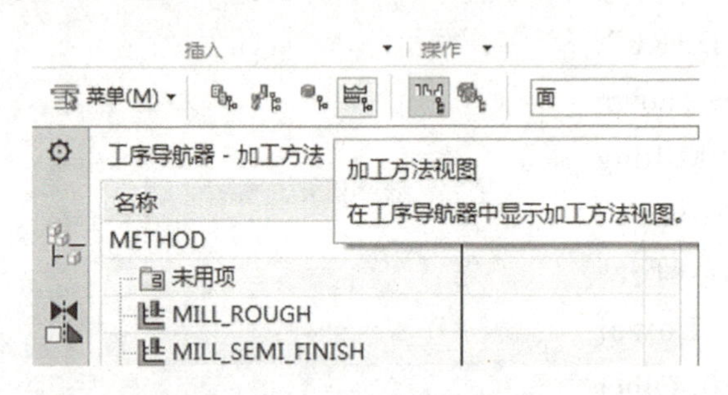

图3-13 设置加工方法

（14）双击【MILL_SEMI_FINISH】，弹出【铣削半精加工】对话框，修改【部件余量】为【0.15】，单击【确定】，如图3-14所示。

(12) Click "Create Tool" button and select "ball-mill" in "Tool Subtype". Further, modify tool name as "BALL4". Click "Apply". As shown in Fig. 3-12, in "Milling Tool-Ball Mill" window, set "Ball Diameter" value as 4. Click "OK" to close the window.

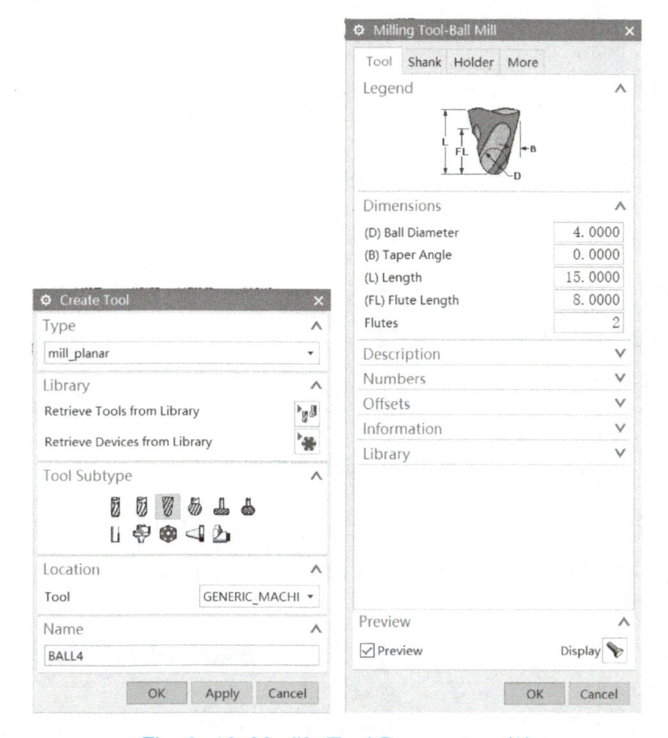

Fig. 3-12 Modify Tool Parameters (3)

(13) Click "Machining Method View" button as shown in Fig. 3-13.

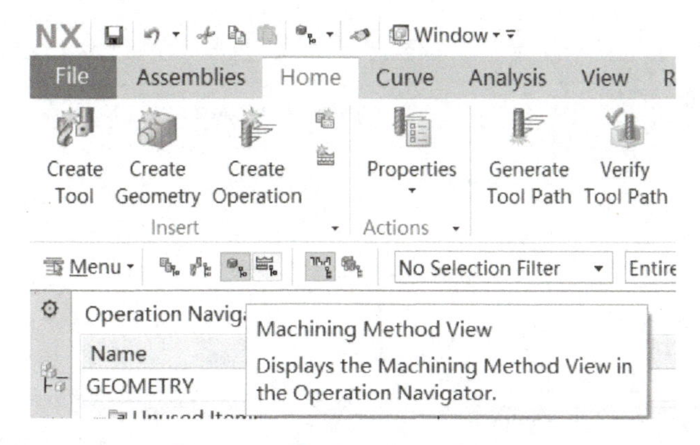

Fig. 3-13 Set Machining Method

(14) Double click "MILL_SEMI_FINISH" to modify "Part Stock" value as 0.15 as shown in Fig. 3-14. Click "OK" to close the window.

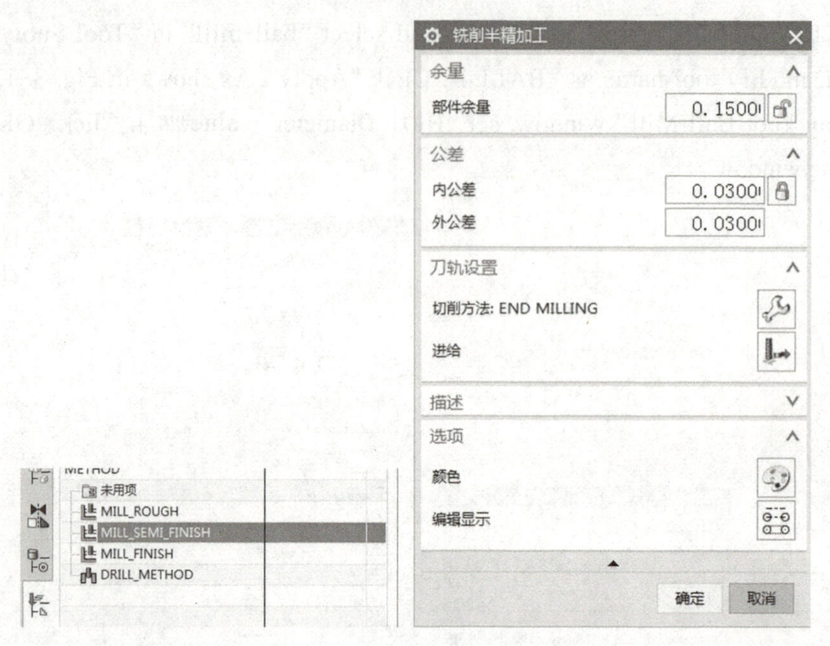

图 3-14 铣削半精加工

（15）双击【MILL_FINISH】，弹出【铣削精加工】对话框，修改【内公差】和【外公差】均为【0.01】，如图 3-15 所示，单击【确定】。

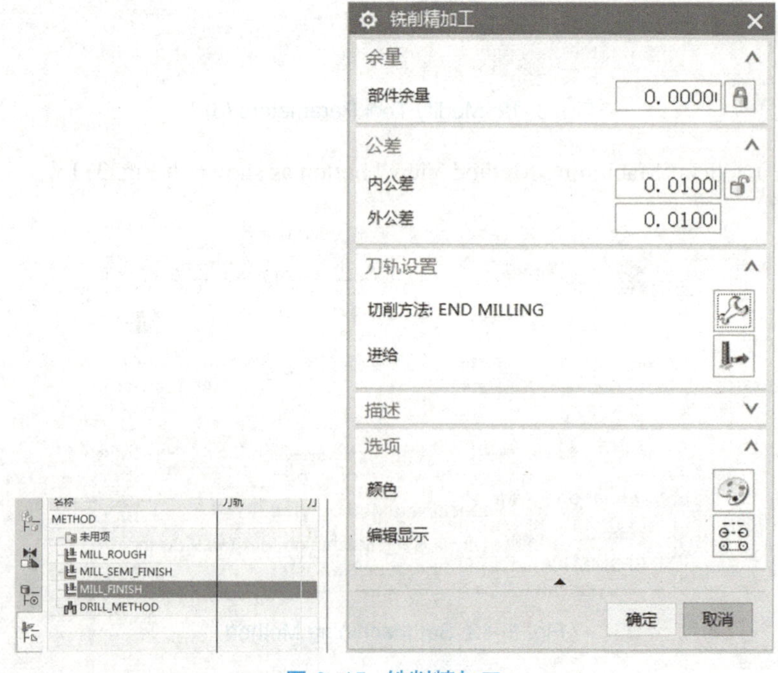

图 3-15 铣削精加工

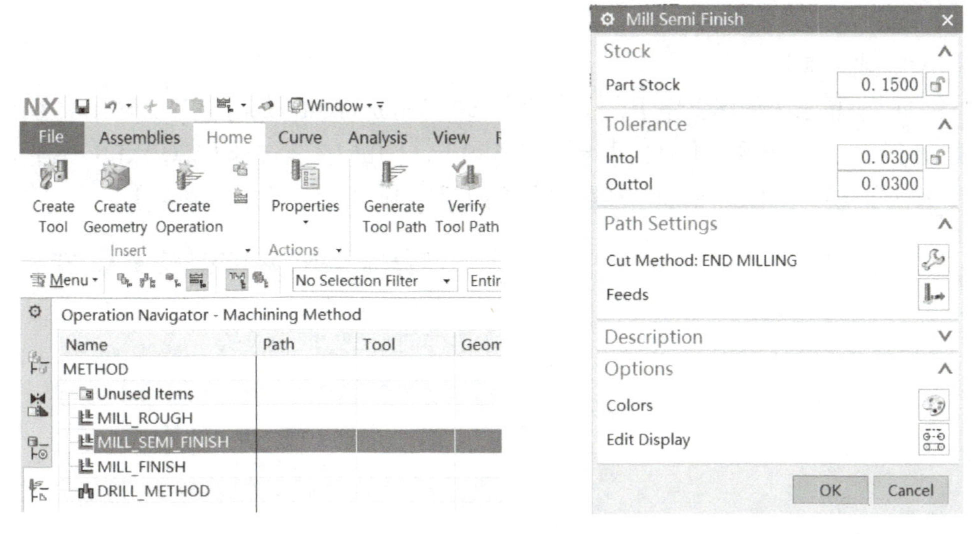

Fig. 3-14 Mill Semi Finish

(15) Double click "MILL_ FINISH" to modify both "Intol Tolerance" and "Outtol Tolerance" values as 0.01 as shown in Fig. 3-15, and then click "OK".

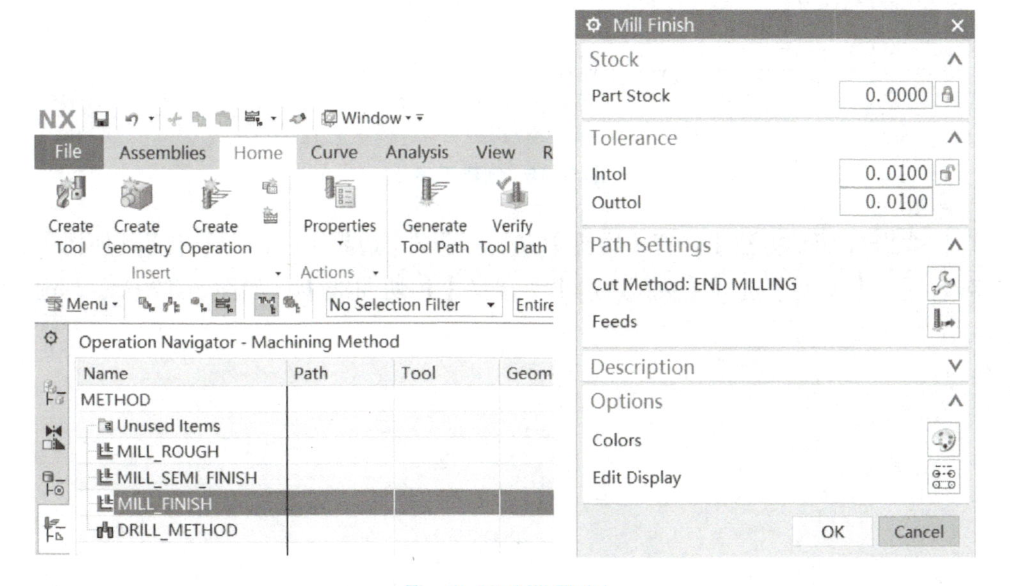

Fig. 3-15 Mill Finish

三、加工程序编制

1. 型面粗加工

（1）单击【创建工序】，【类型】选择【mill_multi-axis】，【工序子类型】选择【可变轮廓铣】，【刀具】选择【D6R0】，【几何体】选择【WORKPIECE】，【方法】选择【MILL_SEMI_FINISH】，如图 3-16 所示，单击【确定】。

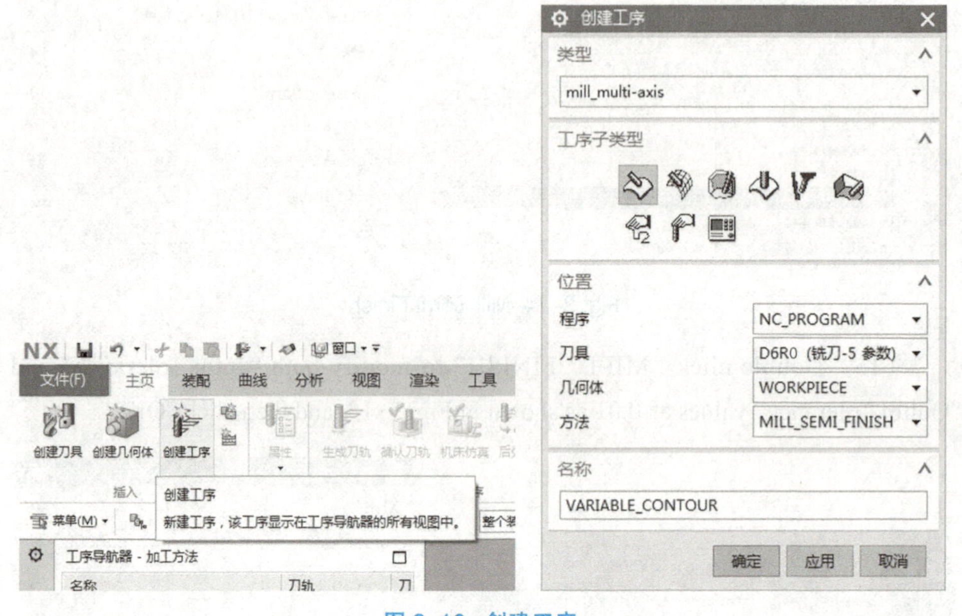

图 3-16 创建工序

（2）【驱动方法】选择【边界】，单击定义【边界】，打开【边界几何体】对话框，【类型】选择【曲线 / 边】。选择【相切曲线】，如图 3-17 所示，选择星型环边。

Ⅲ. Programming

1. Rough Machining

（1）As shown in Fig. 3-16, click “Create Operation”, and then select “mill_multi-axis” in pull-down box of “Type”. Select operation subtype as “variable contour”, tool as “D6R0”, geometry as “WORKPIECE” and method as “MILL_SEMI_FINISH”, and then click “OK”.

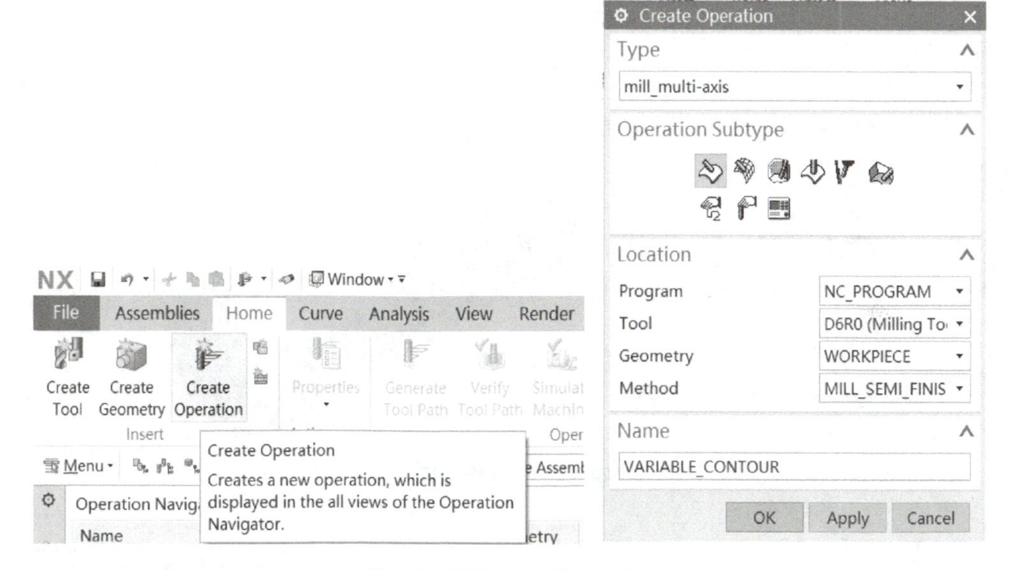

Fig. 3-16 Create Operation

（2）Select “Boundary” as “Drive Method”, click “Define Boundary”, select “Curves/Edges” as “Mode”. Then select “Tangent Curves” and curves as shown in Fig. 3-17.

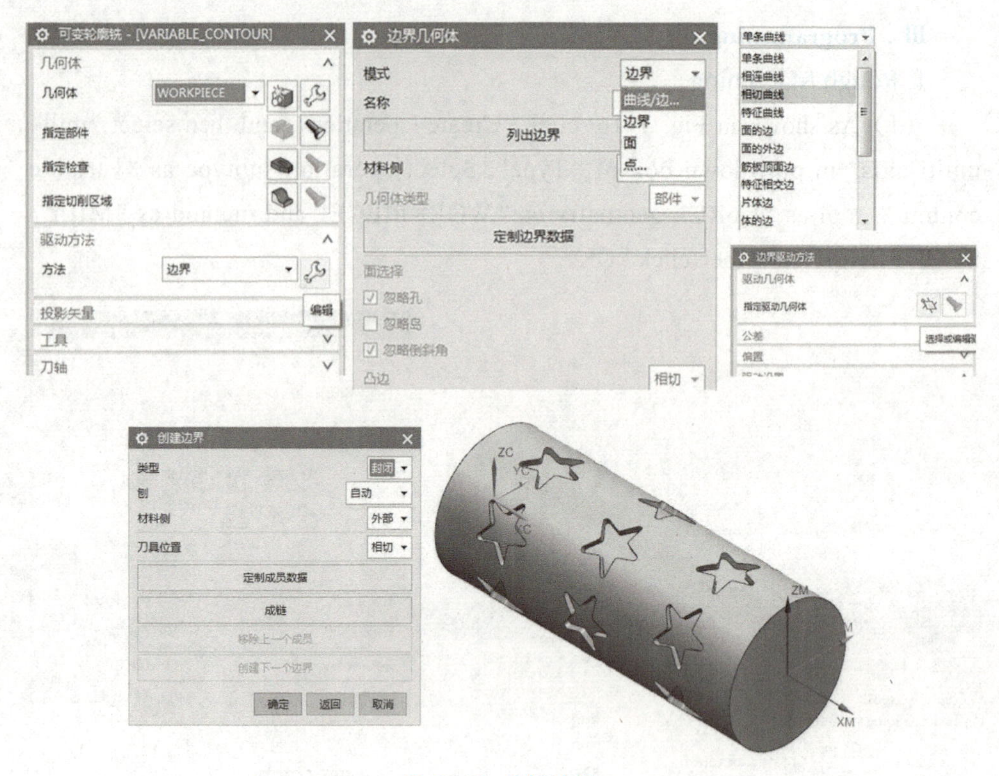

图 3-17 选择边界

（3）单击【偏置】，设置【边界偏置】为【0.5】。将【驱动设置】中的【切削模式】改为【跟随周边】，【刀路方向】改为【向外】，如图 3-18 所示。

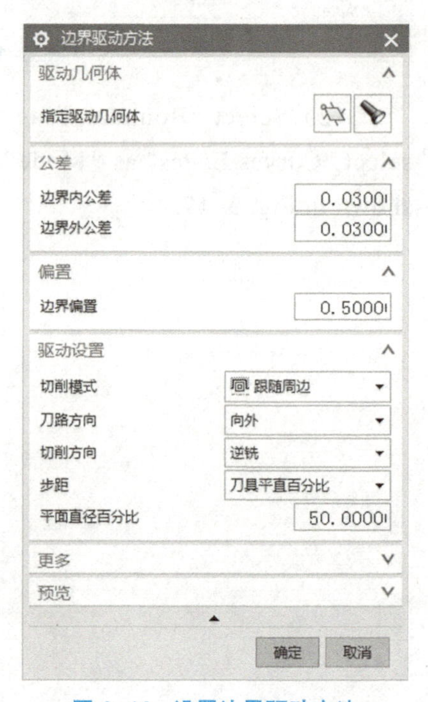

图 3-18 设置边界驱动方法

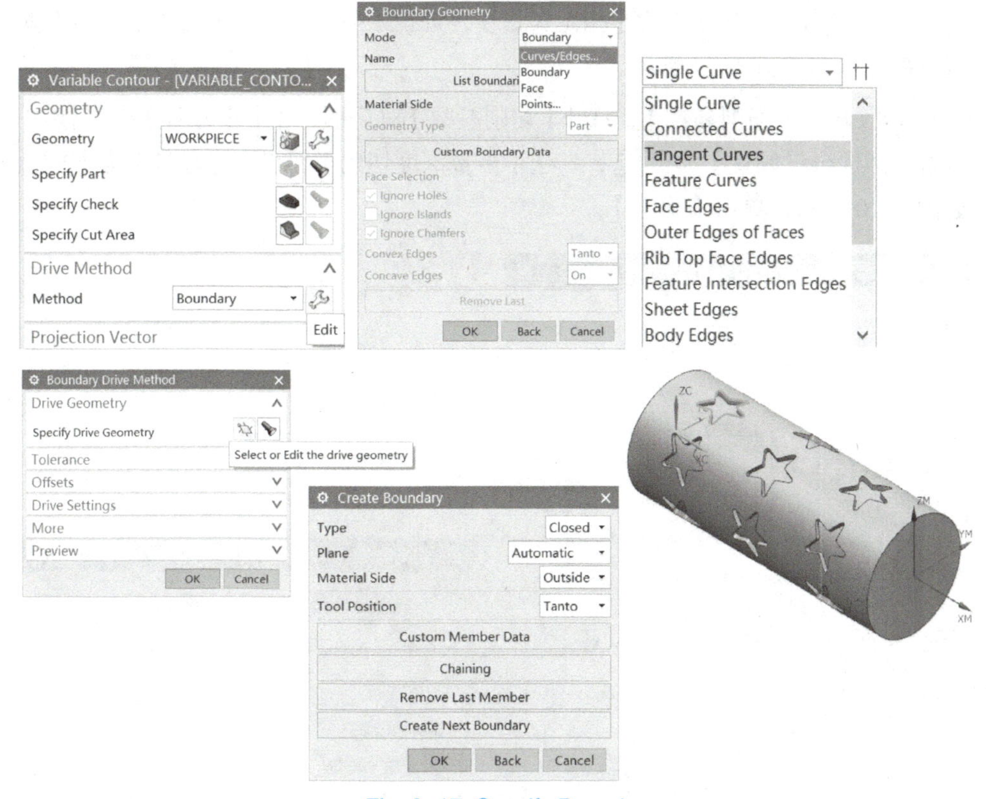

Fig. 3-17　Specify Boundary

（3）Set “Boundary Offset” as 0.5. Modify “Cut Pattern” as “Follow Periphery” and “Pattern Direction” as “Outward” as shown in Fig. 3-18.

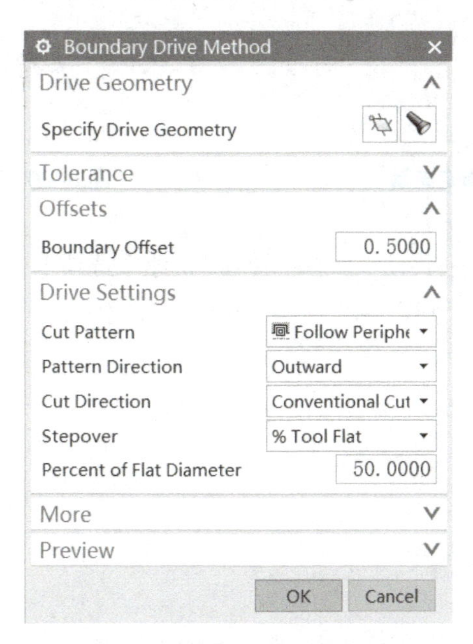

Fig. 3-18　Modify Boundary Drive Method

（4）【投影矢量】设为【指定矢量】，选择【-ZC 轴】，【刀轴】设为【垂直于部件】，如图 3-19 所示。

（5）单击【切削参数】，打开【切削参数】对话框，在【多刀路】选项卡中，设置【部件余量偏置】为【4】，勾选【多重深度切削】，设置【增量】为【1】，如图 3-20 所示，单击【确定】。

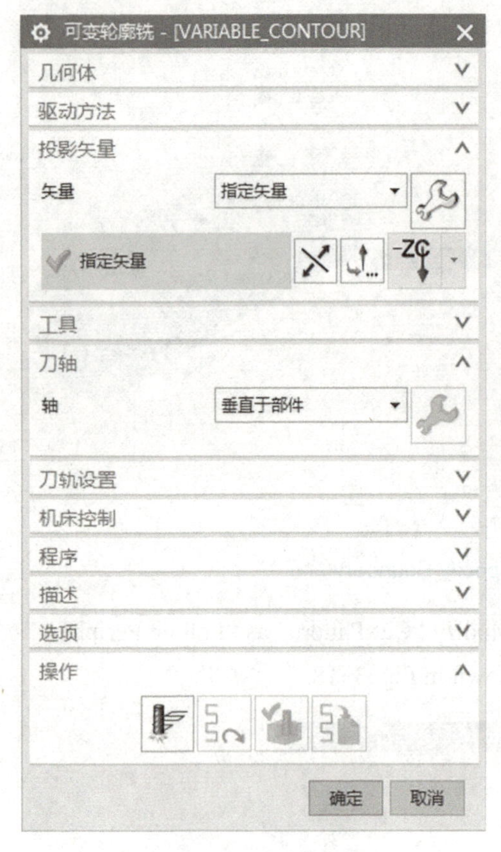

图 3-19　指定矢量

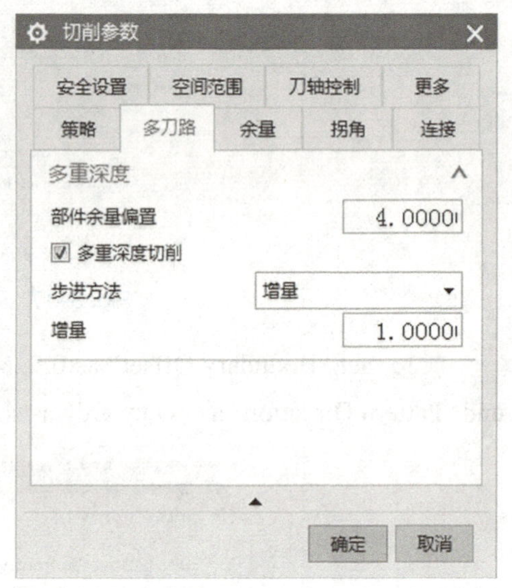

图 3-20　设置切削参数

（6）单击【非切削移动】，在【转移 / 快速】选项卡中，【安全设置选项】选择【使用继承的】，如图 3-21 所示，单击【确定】。

(4) As shown in Fig. 3-19, select “Specify Vector” as projection vector, and select “-ZC axis” as the vector. Select “Normal to Part” as tool axis.

(5) As shown in Fig. 3-20, click “Cutting Parameters” button, and in “Multiple Passes” tab, modify “Part Stock Offset” as 4. Tick the option “Multi-Depth Cut” and modify “Increment” as 1. Click “OK”.

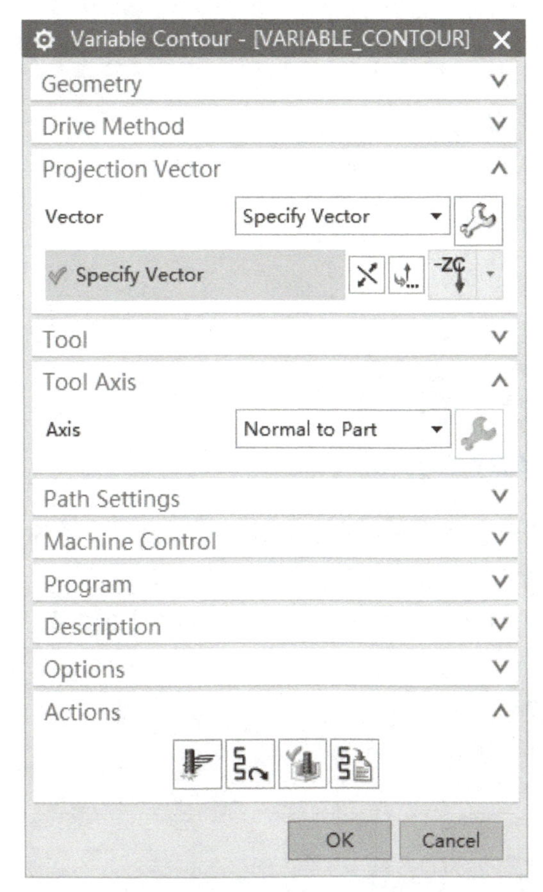

Fig. 3-19 Specify Vector

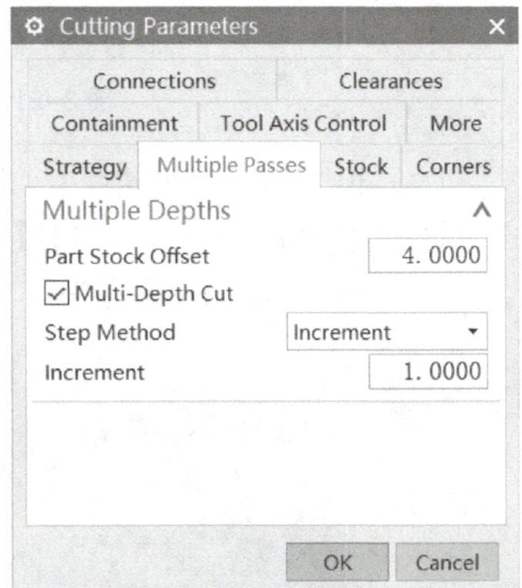

Fig. 3-20 Modify Cutting Parameters

(6) As shown in Fig. 3-21, click “Non Cutting Moves” button, and in “Transfer/Rapid” tab, select “Use Inherited” as “Clearance Option”. Click “OK”.

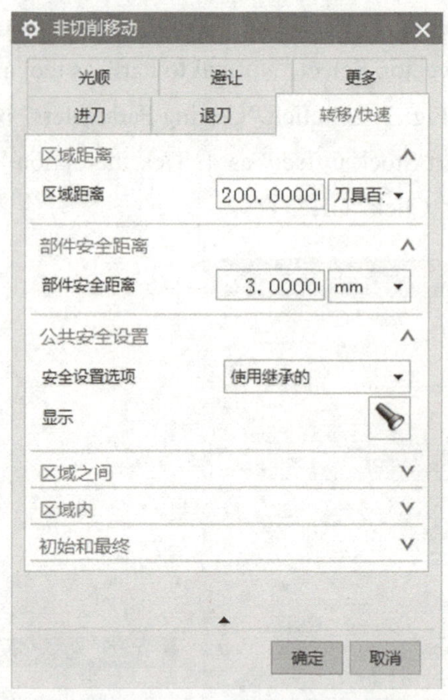

图 3-21 设置非切削移动

（7）单击【进给率和速度】，打开【进给率和速度】对话框，设置【主轴速度】为【1800】，进给率【切削】为【1200 mmpm】，单击【确定】。

（8）单击【生成】，查看生成的刀具轨迹，如图 3-22 所示，单击【确定】。

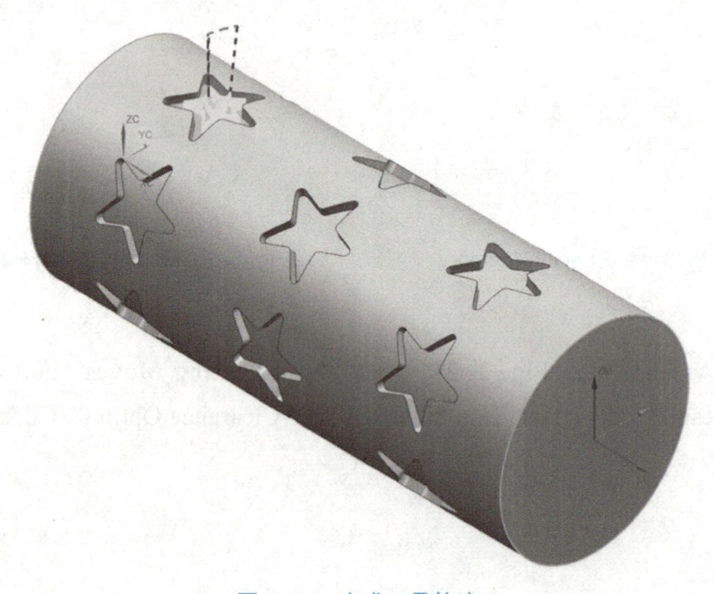

图 3-22 生成刀具轨迹

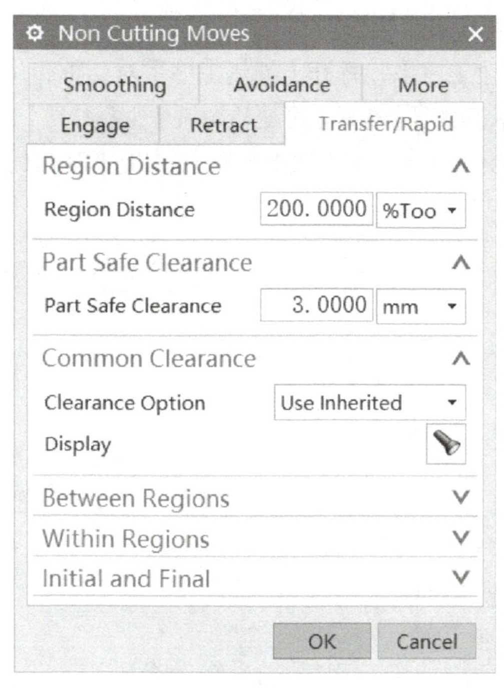

Fig. 3–21 Modify Non Cutting Moves

(7) Click "Feeds and Speeds", modify "Spindle Speed" as 1,800 and "Feed Rates" as 1,200 mmpm. Click "OK".

(8) Click "Generate" to view the tool path as shown in Fig. 3-22, and then click "OK".

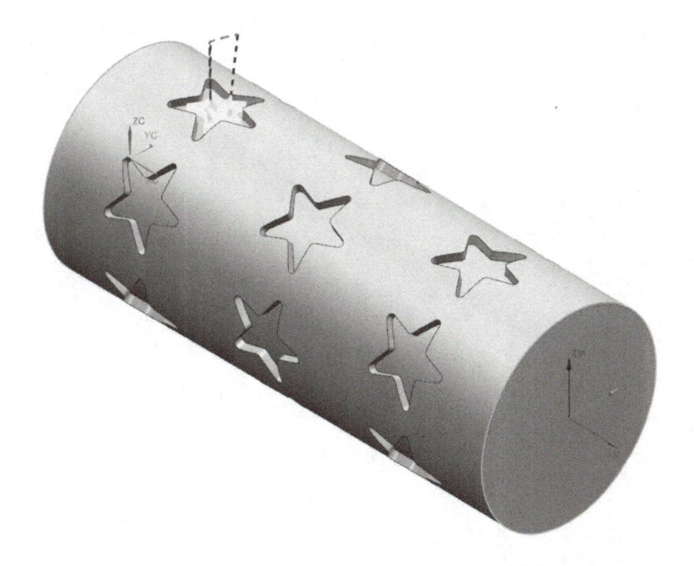

Fig. 3–22 Generate Tool Path

2. 型面侧面壁精加工

（1）单击【创建工序】，打开【创建工序】对话框，【类型】选择【mill_multi-axis】，【工序子类型】选择【可变轮廓铣】，【刀具】选择【D3R0】，【几何体】选择【WORKPIECE】，【方法】选择【MILL_FINISH】，如图 3-23 所示，单击【确定】。

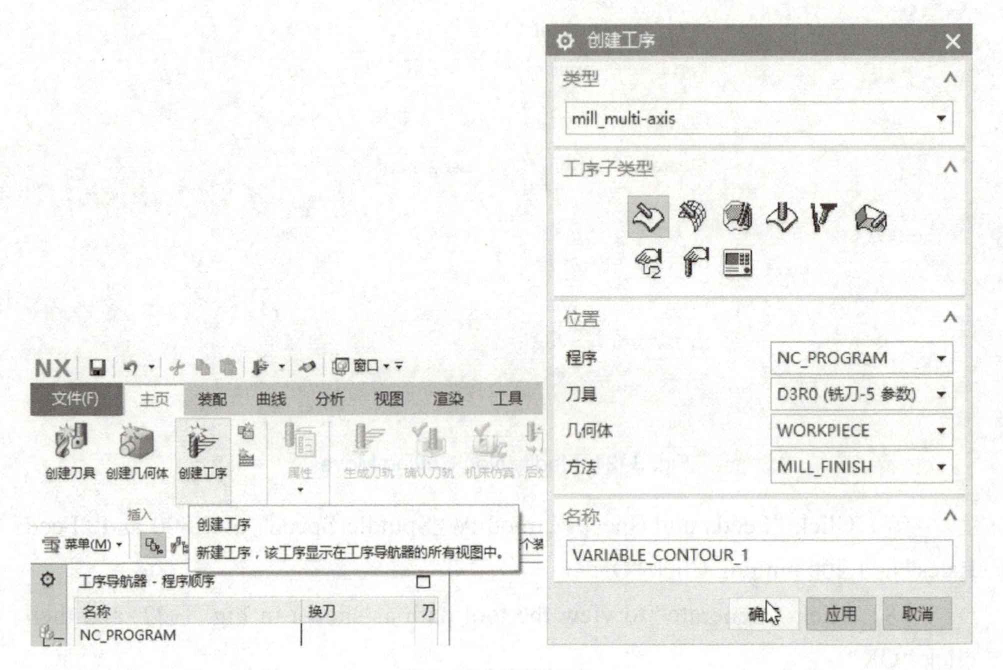

图 3-23 创建工序

2. Side Finishing

(1) As shown in Fig. 3-23, click "Create Operation", and then select "mill_multi-axis" in pull-down box of "Type". Select operation subtype as "variable contour", tool as "D3R0", geometry as "WORKPIECE" and method as "MILL_FINISH", and then click "OK".

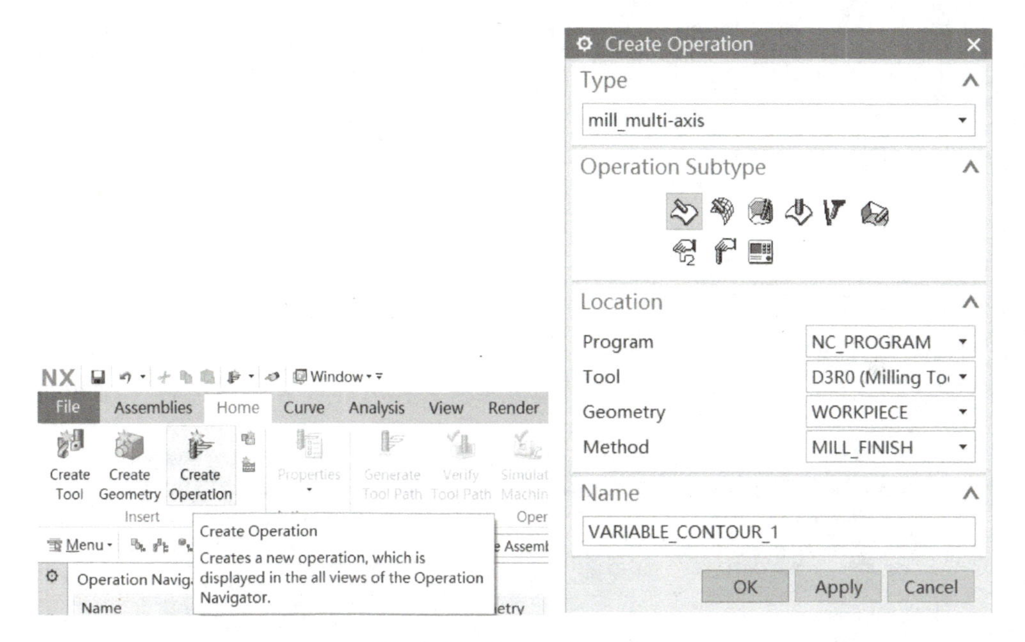

Fig. 3-23　Create Operation

（2）【驱动方法】选择【曲线 / 点】，选择星型曲线，单击【驱动设置】，设置【左偏置】为【1.5 mm】，【切削步长】为【公差】，【公差】为【0.01】，如图 3-24 所示，单击【确定】。

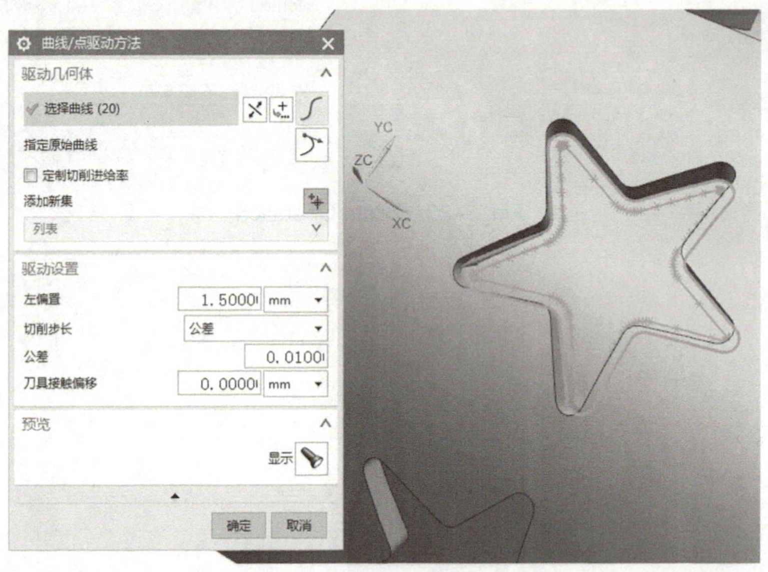

图 3-24 设置驱动方式

（2）Select "Curve/Point" as "Drive Method", select curves of stars as shown in Fig. 3-24. Modify "Offset left" as 1.5 mm, set "Tolerance" as "Cut Step" and modify it as 0.01 mm.

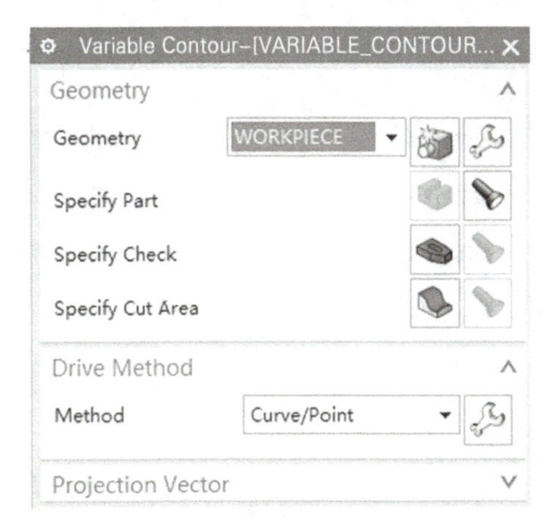

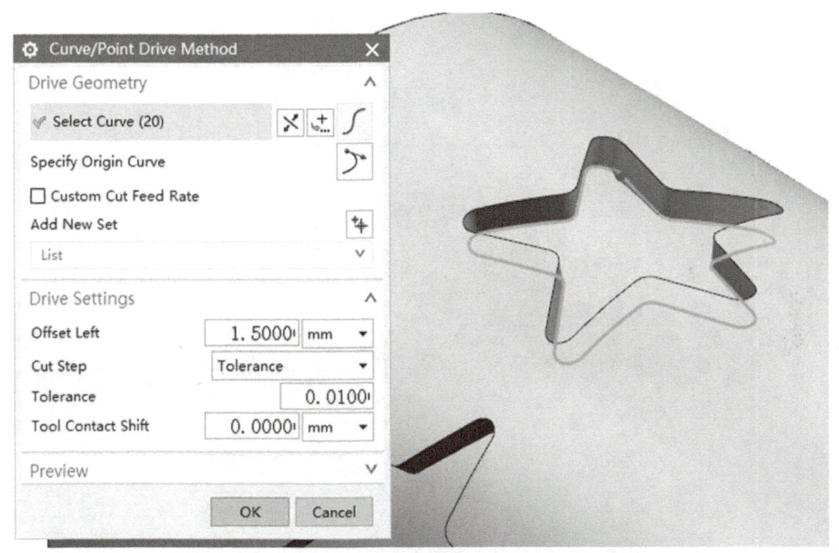

Fig. 3-24 Modify Drive Method

（3）单击【投影矢量】，设置【矢量】为【刀轴】；单击【刀轴】，设置【轴】为【垂直于部件】，如图 3-25 所示。

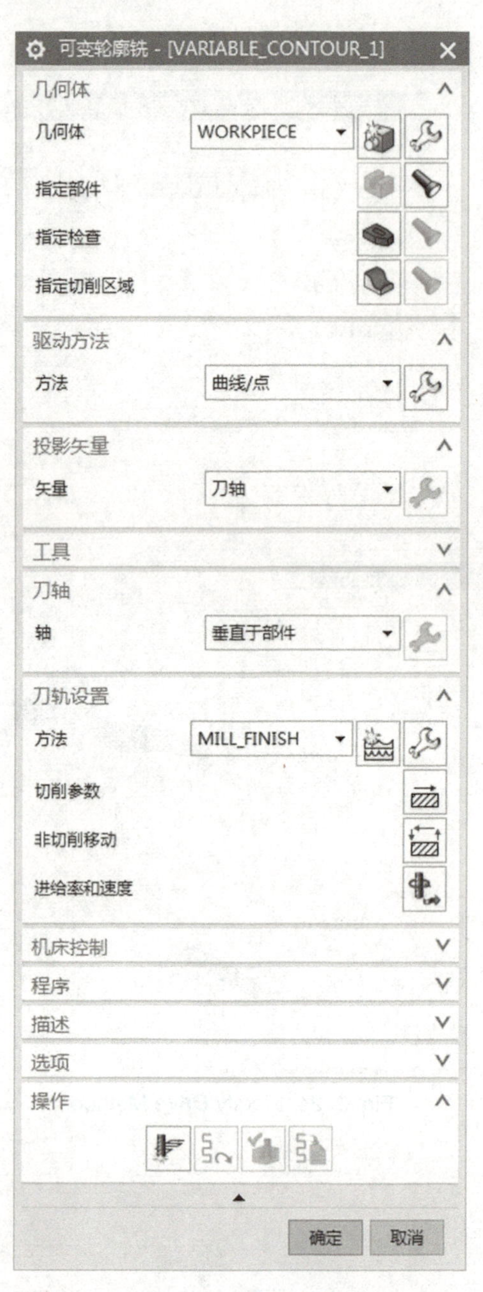

图 3-25　设置参数

（3）Select "Tool Axis" as "Vector" and "Normal to Part" as "Tool Axis" as shown in Fig. 3-25.

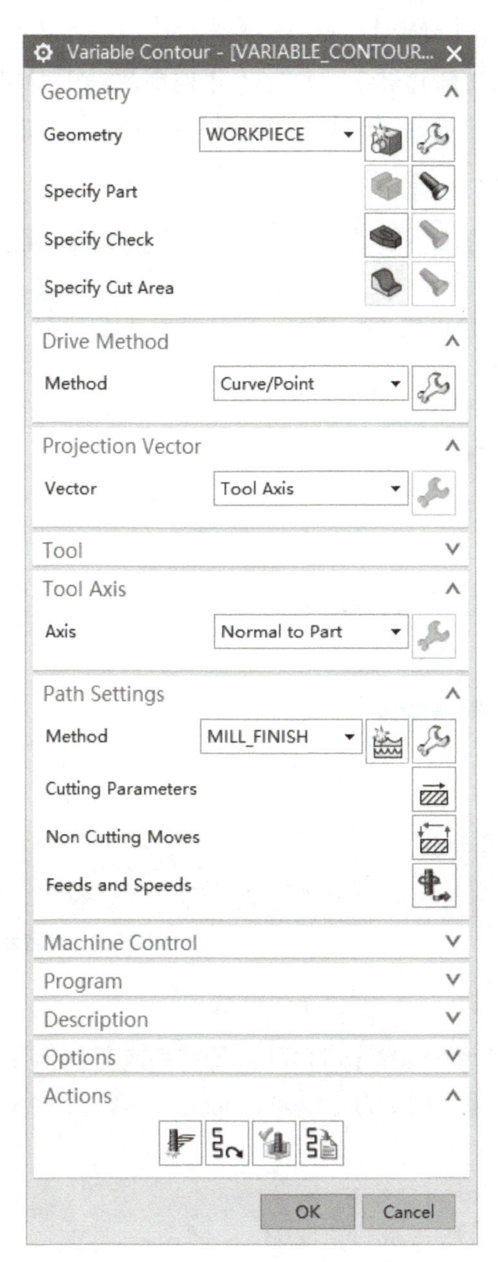

Fig. 3-25　Select Vector

（4）单击【切削参数】，打开【切削参数】对话框，在【多刀路】选项卡中，设置【部件余量偏置】为【4】，勾选【多重深度切削】，设置【增量】为【0.5】，如图 3-26 所示，单击【确定】。

（5）单击【非切削移动】，打开【非切削移动】对话框，在【转移 / 快速】选项卡中，【安全设置选项】选择【使用继承的】，如图 3-27 所示，单击【确定】。

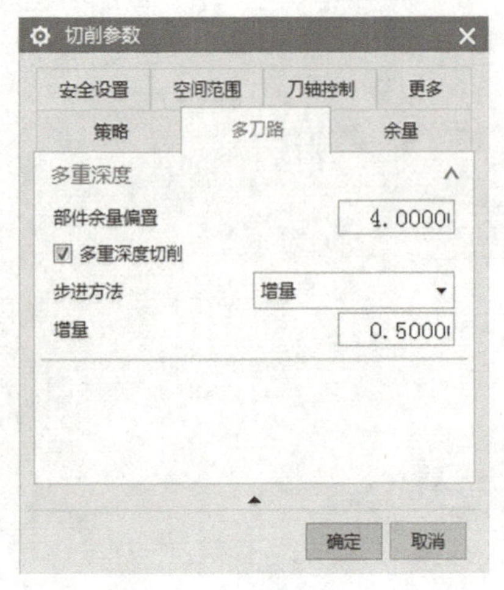

图 3-26　设置切削参数

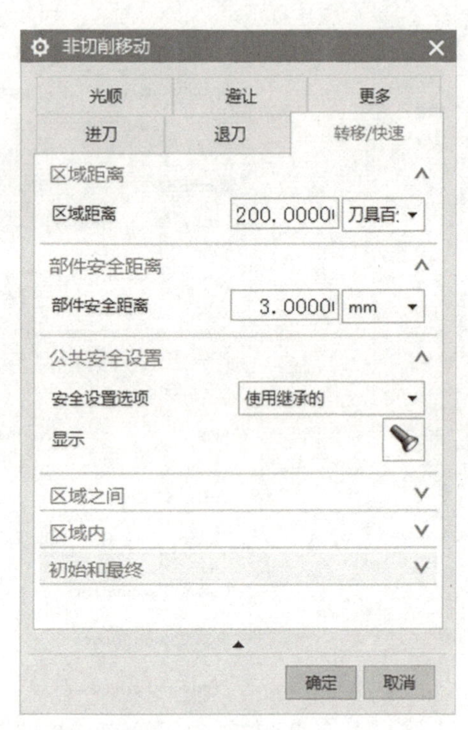

图 3-27　设置非切削移动

（6）单击【进给率和速度】，打开【进给率和速度】对话框，设置【主轴速度】为【2600】，进给率【切削】为【1400 mmpm】，单击【确定】。

（4）As shown in Fig. 3-26, Click "Cutting Parameters" button, and in "Multiple Passes" tab, modify "Part Stock Offset" as 4. Tick the option "Multi-Depth Cut" and modify "Increment" as 0.5. Click "OK".

（5）As shown in Fig. 3-27, click "Non Cutting Moves" button, and in "Transfer/Rapid" tab, select "Use Inherited" as "Clearance Option". Click "OK".

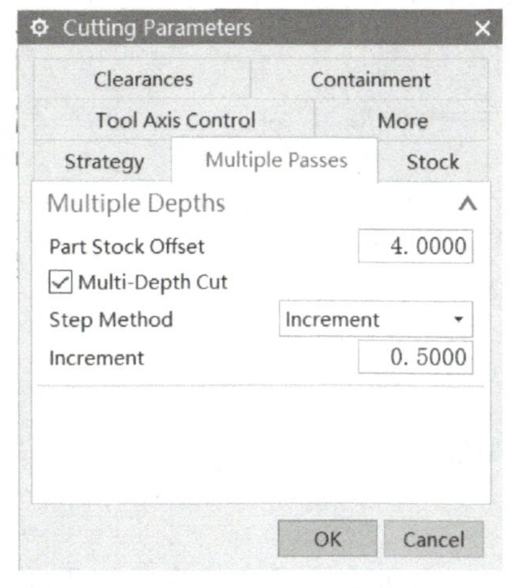

Fig. 3-26 Modify Cutting Parameters

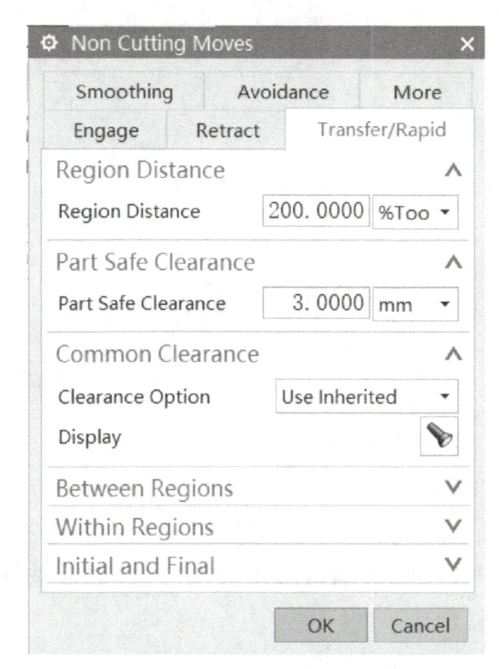

Fig. 3-27 Modify Non Cutting Moves

（6）Click "Feeds and Speeds", modify "Spindle Speed" as 2,600 and "Feed Rates" as 1,400 mmpm. Click "OK".

（7）单击【生成】，查看生成的刀具轨迹，如图 3-28 所示，单击【确定】。

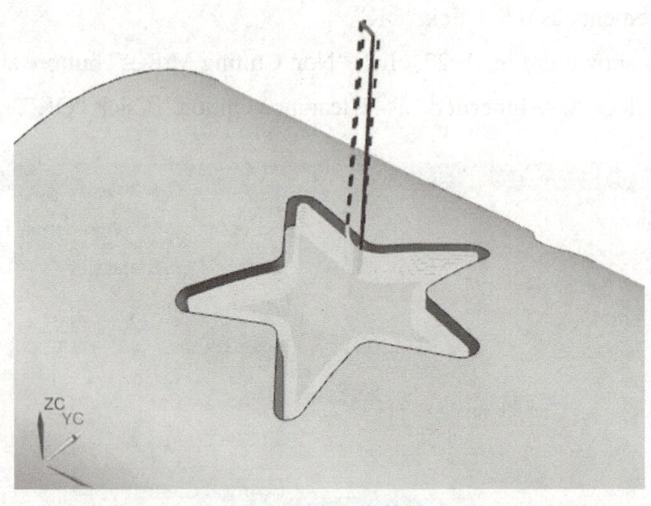

图 3-28 生成刀具轨迹

3. 型面底面精加工

（1）单击【创建工序】，打开【创建工序】对话框，【类型】选择【mill_multi-axis】，【工序子类型】选择【可变轮廓铣】，【刀具】选择【BALL4】，【几何体】选择【WORKPIECE】，【方法】选择【MILL_FINISH】，如图 3-29 所示，单击【确定】。

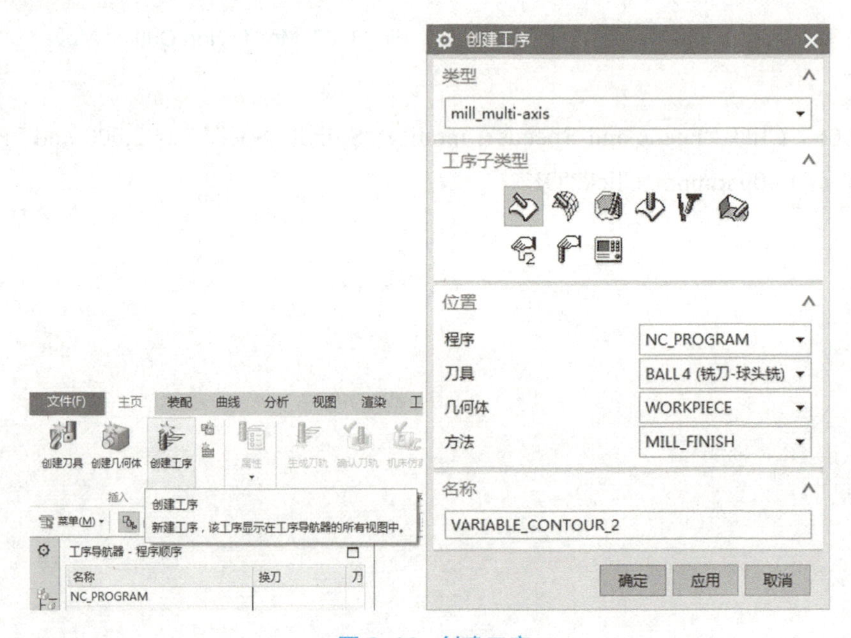

图 3-29 创建工序

(7) Click "Generate" to view the tool path as shown in Fig. 3-28, and then click "OK".

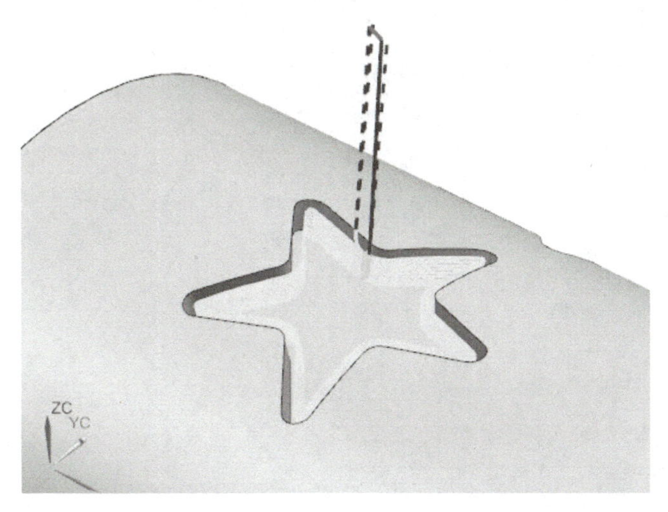

Fig. 3–28 Generate Tool Path

3. Bottom Finishing

(1) As shown in Fig. 3-29, click "Create Operation", and then select "mill_multi-axis" in pull-down box of "Type". Select operation subtype as "variable contour", tool as "BALL4", geometry as "WORKPIECE" and method as "MILL_FINISH", and then click "OK".

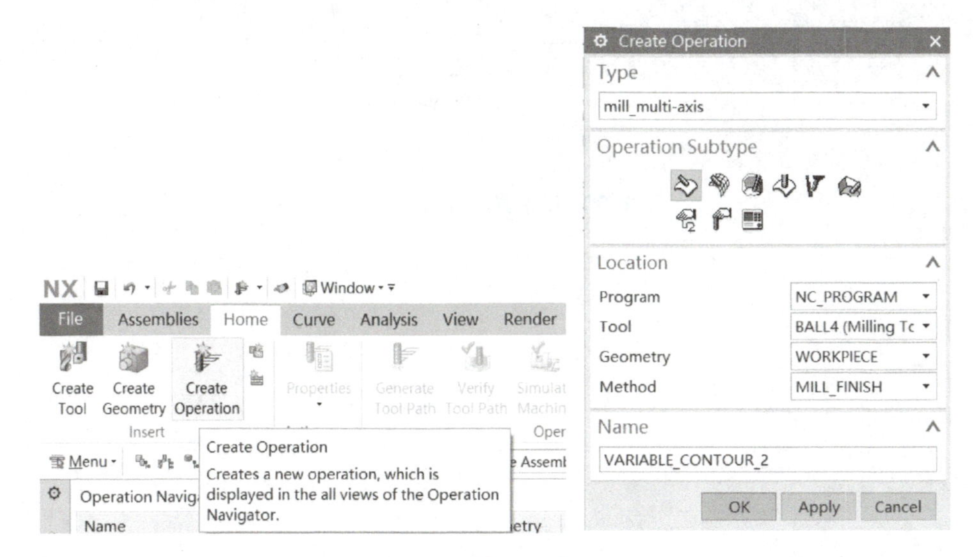

Fig. 3–29 Create Operation

（2）【驱动方法】选择【边界】，单击定义【边界】，【类型】选择【曲线/边】，选择【相切曲线】，选择星型环边，如图 3-30 所示。

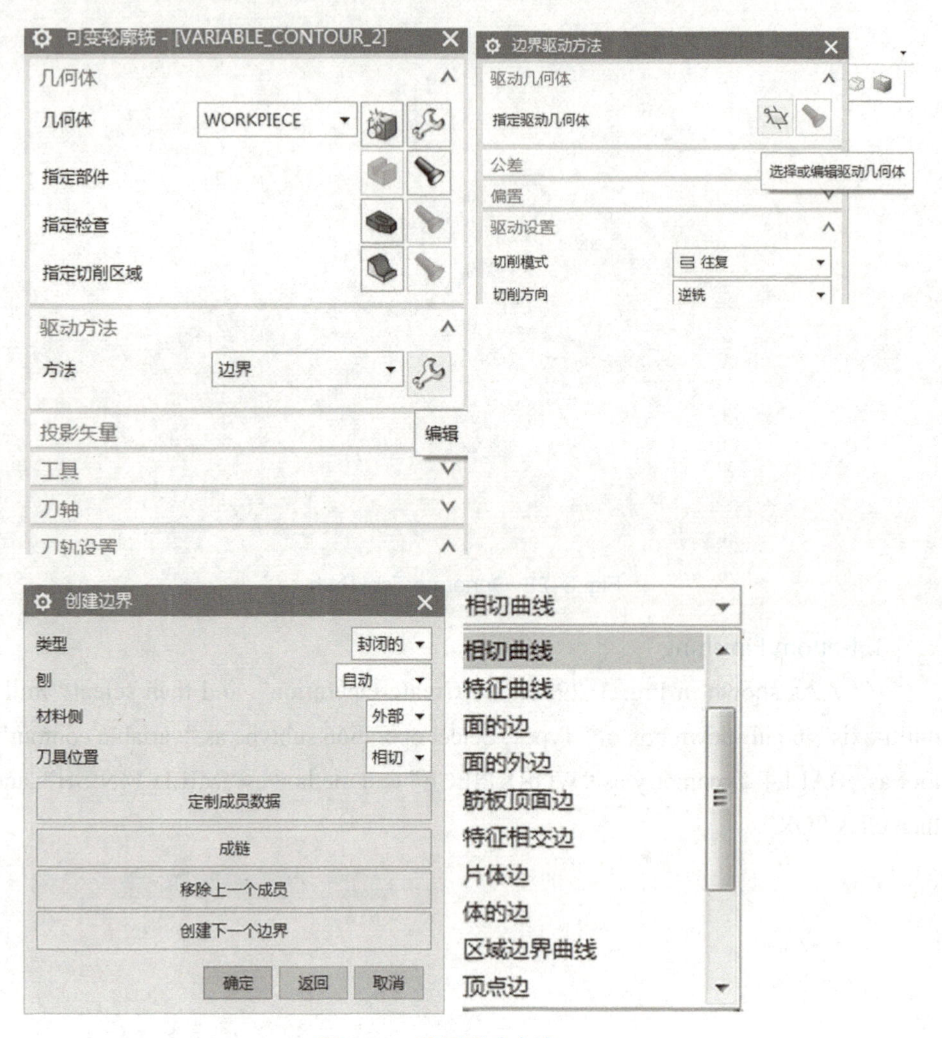

图 3-30 设置驱动方法

（2）Select boundary as drive method, click “Define Boundary”, select “Curves/Edges” as mode, then select “Tangent Curves” and curves shown as in Fig. 3-30.

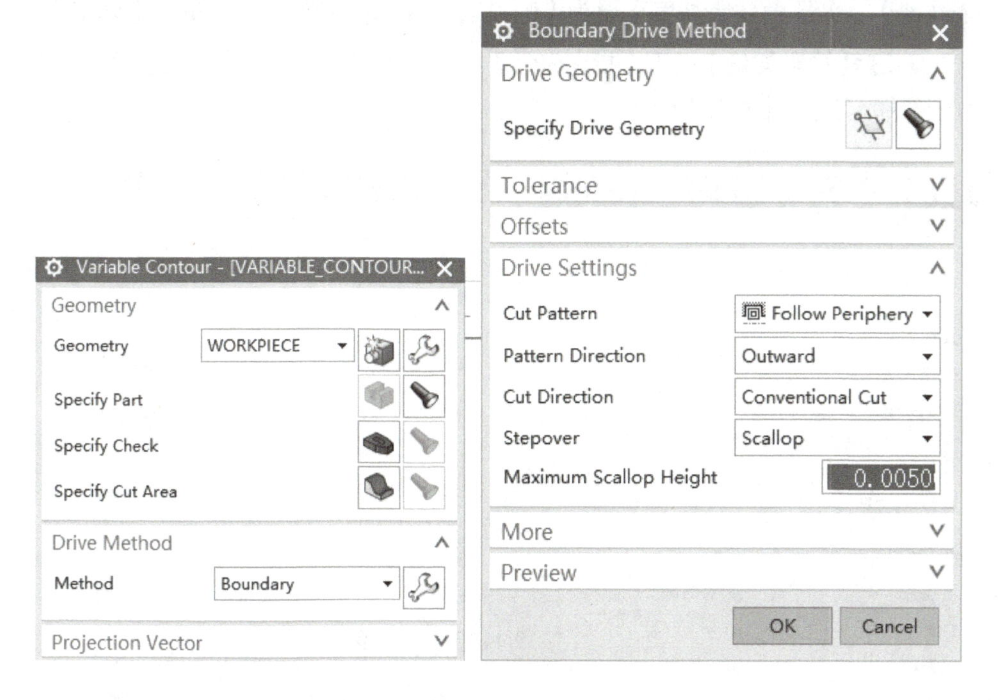

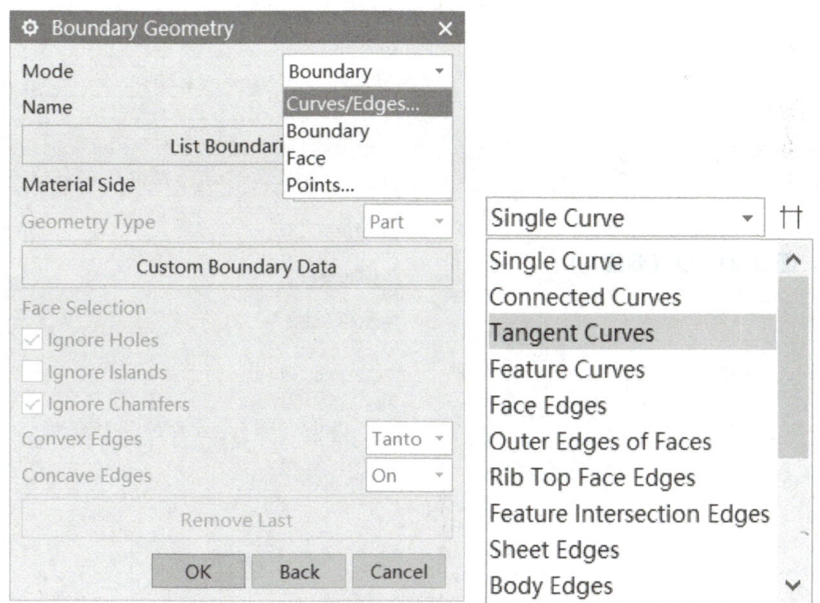

Fig. 3-30 Modify Drive Method

（3）单击【偏置】，设置【边界偏置】为【0.3】。在【驱动设置】中将【切削模式】改为【跟随周边】，【刀路方向】改为【向外】，【步距】改为【残余高度】，【最大残余高度】设为【0.005】，如图3-31所示，单击【确定】。

（4）【投影矢量】设为【指定矢量】，选择【-ZC轴】，【刀轴】设为【垂直于部件】，如图3-32所示。

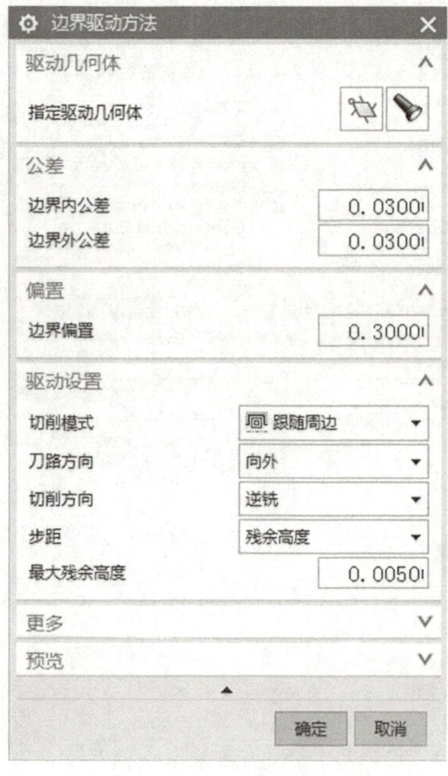

图3-31 设置参数（1）

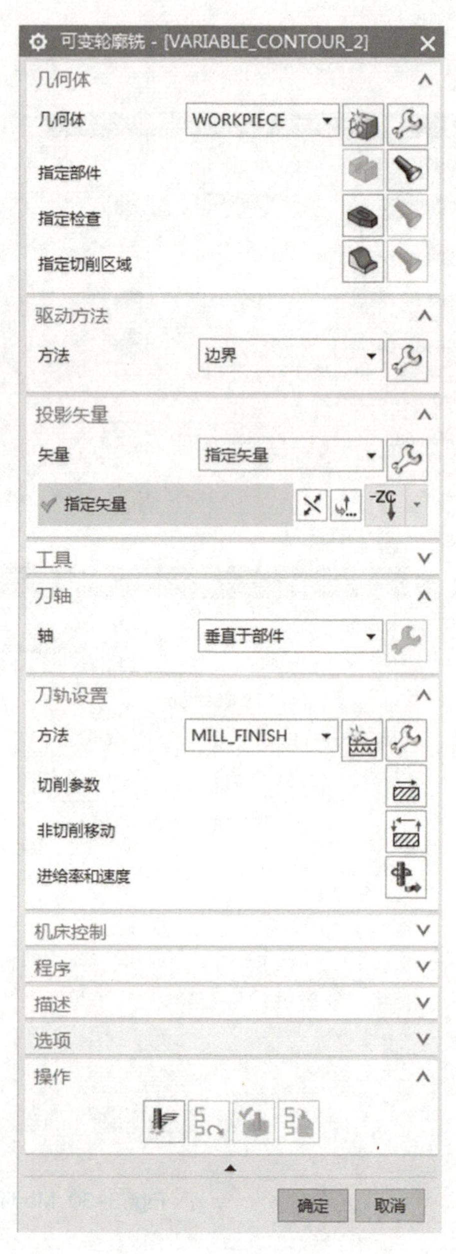

图3-32 设置参数（2）

（3）Set “Boundary Offset” as 0.3. Modify “Cut Pattern” as “Follow Periphery” and “Pattern Direction” as “Outward”, select “Scallop” as “Stepover” and modify it as 0.005 as shown in Fig. 3-31. Click “OK”.

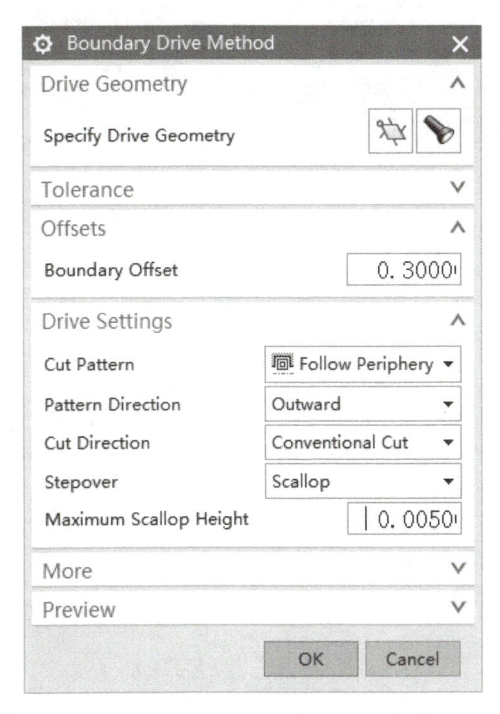

Fig. 3-31 Specify Parameters (1)

（4） As shown in Fig. 3-32, select “Specify Vector” as projection vector, and select “-ZC axis” as the vector. Select “Normal to Part” as tool axis.

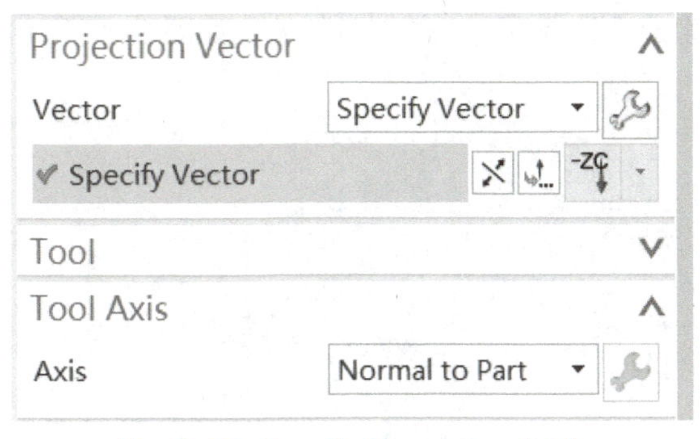

Fig. 3-32 Specify Parameters (2)

（5）单击【非切削移动】，打开【非切削移动】对话框，在【转移/快速】选项卡中，【安全设置选项】选择【使用继承的】，如图 3-33 所示，单击【确定】。

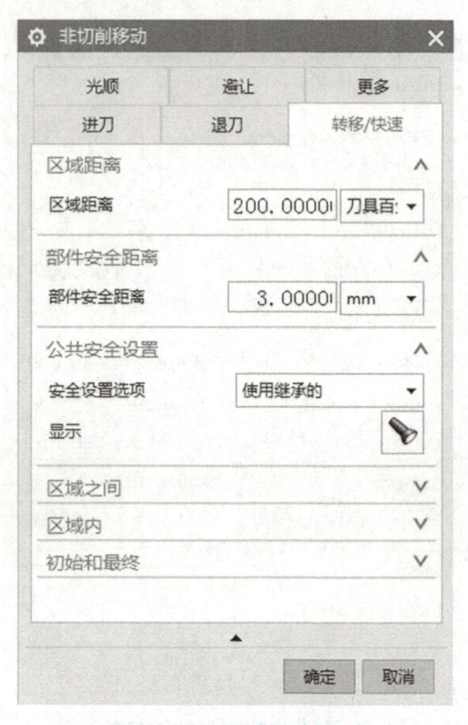

图 3-33　设置非切削移动

（6）单击【进给率和速度】，打开【进给率和速度】对话框，设置【主轴速度】为【3200】，进给率【切削】为【1000 mmpm】，单击【确定】。

（7）单击【生成】，查看生成的刀具轨迹，如图 3-34 所示，单击【确定】。

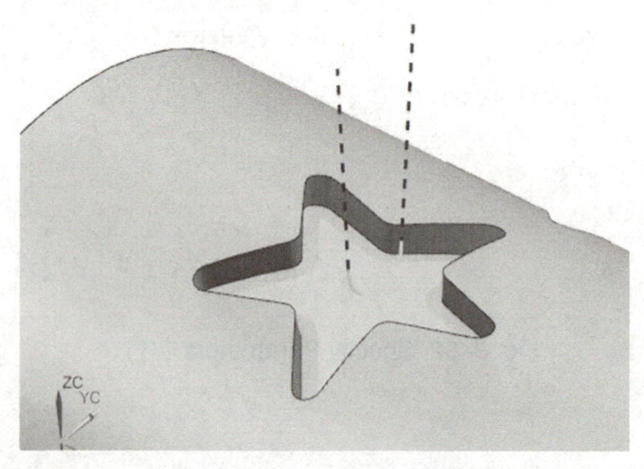

图 3-34　生成刀具轨迹

(5) As shown in Fig. 3-33, click "Non Cutting Moves" button, and in "Transfer/Rapid" tab, select "Use Inherited" as "Clearance Option". Click "OK".

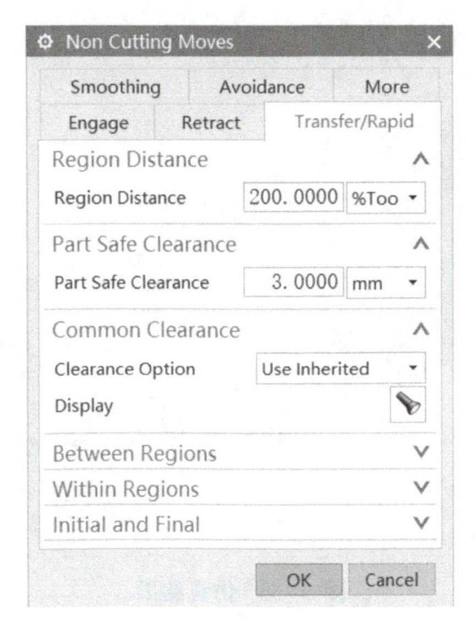

Fig. 3-33 Modify Non Cutting Moves

(6) Click "Feeds and Speeds", modify "Spindle Speed" as 3,200 and "Feed Rates" as 1,000 mmpm. Click "OK".

(7) Click "Generate" to view the tool path as shown in Fig. 3-34, and then click "OK".

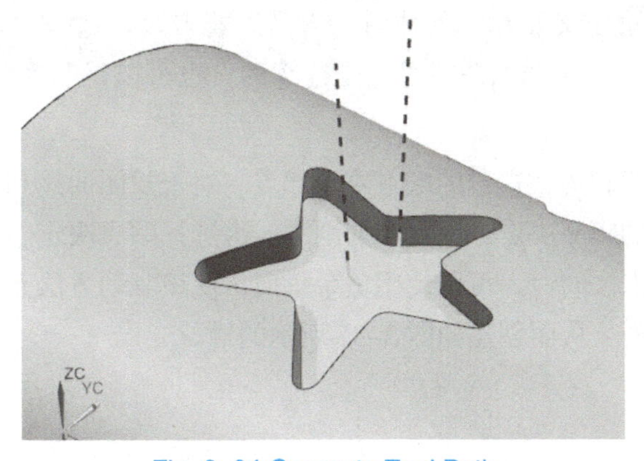

Fig. 3-34 Generate Tool Path

四、仿真加工

单击【NC_PROGRAM】，单击【确认刀轨】，选择【3D 动态】，单击【播放】，仿真结果（3D 动态）如图 3-35 所示。

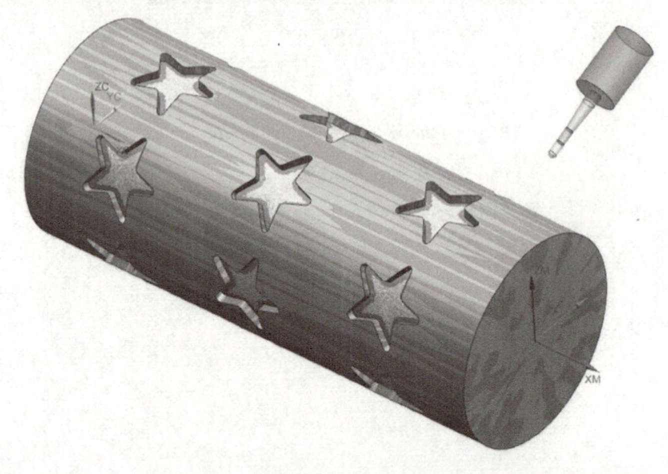

图 3-35 仿真结果

专家点拨

使用【曲线 / 点】驱动方法通过指定点和选择曲线或面边缘定义驱动几何体。指定点后，驱动轨迹创建为指定点之间的线段；指定曲线或边时，沿选定曲线和边生成驱动点。驱动几何体投影到部件几何体上，然后在此生成刀具轨迹。曲线可以是开放的或封闭的、连续的或非连续的及平面的或非平面的。

当由曲线或边定义驱动几何体时，刀具沿着刀轨按选择的顺序从一条曲线或边运动至下一条。所选的曲线可以是连续的，也可以是非连续的，如图 3-36 所示。

对于开放曲线和边，选定的端点决定起点。对于封闭曲线和边，起点和切削方向由选择线段的顺序决定。原点和切削方向由选择顺序决定。可以用指定原点曲线命令修改原点。同时，可以使用负余量值，使该驱动方法允许刀具只在低于选定部件表面切削，从而创建如图 3-37 所示的槽。

Ⅳ. Simulation Machining

Click "NC_PROGRAM", and then "Verify Tool Path" button. Select "3D Dynamic" and click "Play" button to view the simulation result. The simulation result is shown in Fig. 3-35.

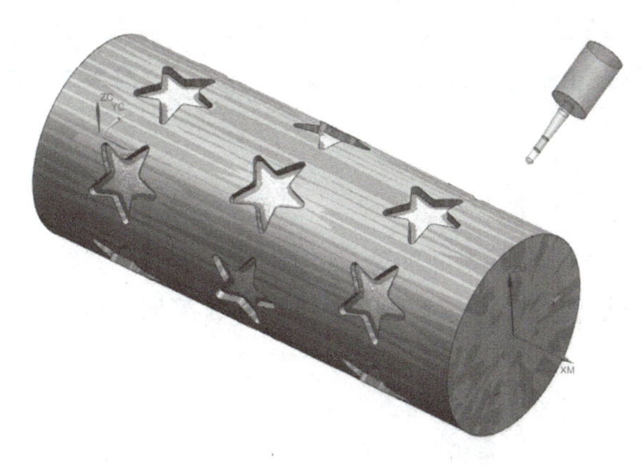

Fig. 3-35 Simulation Result

Use the curve / point drive method to define drive geometry by specifying points and selecting curve or face edges. After specify points, the drive track is created as a segment between the specified points; When a curve or edge is specified, drive points are generated along the selected curve and edge. The drive geometry is projected onto the assembly geometry, where the tool path is generated. Curves can be open or closed, continuous or discontinuous, and planar or non-planar.

When the drive geometry is specified by a curve or edge, the tool moves from one curve to the next in the selected order along the tool path. Selected curves can be continuous or discontinuous as shown in Fig. 3-36.

For open curves and edges, the selected endpoint determines the starting point. For closed curves and edges, the starting point and cutting direction are determined by the order in which curves are selected. The base point and cutting direction are determined by the selection order. The base point can be modified with the specify point/curve command. Meanwhile, negative margin values can be used, this drive method allows the tool to cut only below the surface of the selected part as shown in Fig. 3-37.

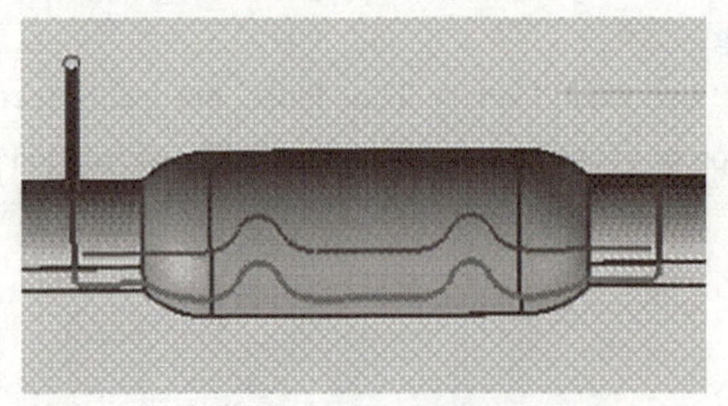

图 3-36　由曲线定义的驱动几何体

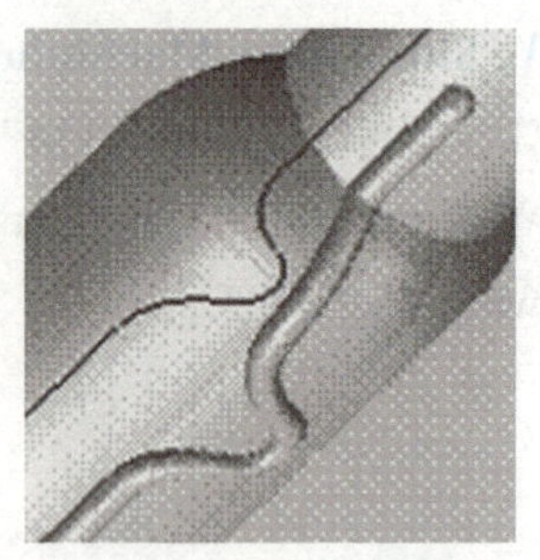

图 3-37　负余量槽

课后训练

根据图 3-38 所示的异形零件的特征，制定合理的工艺路线，设置必要的加工参数、生成刀具轨迹、通过相应的后处理生成数控加工程序，并运用机床加工零件。

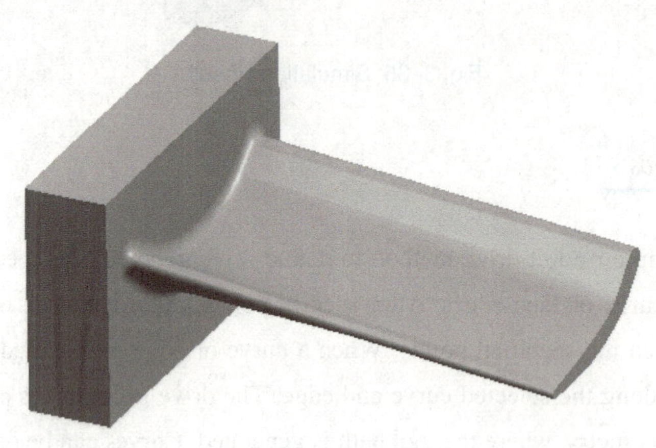

图 3-38　异形零件

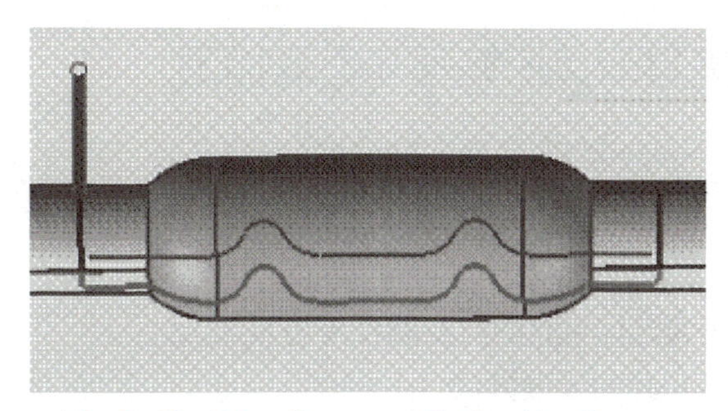

Fig. 3-36　Drive Geometry Defined by Curves

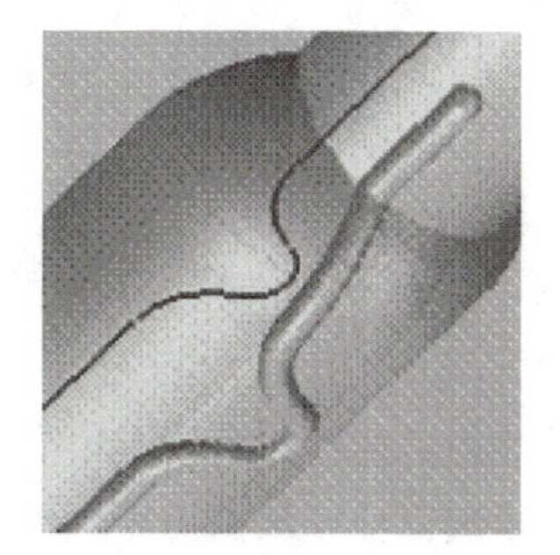

Fig. 3-37　Negative Margin Slot

Practice

According to the characteristics of disc parts as shown in Fig. 3-38, make a reasonable processing technic, set necessary parameters, generate tool path, generate NC program through corresponding post-processor, and use machine tools to machine parts.

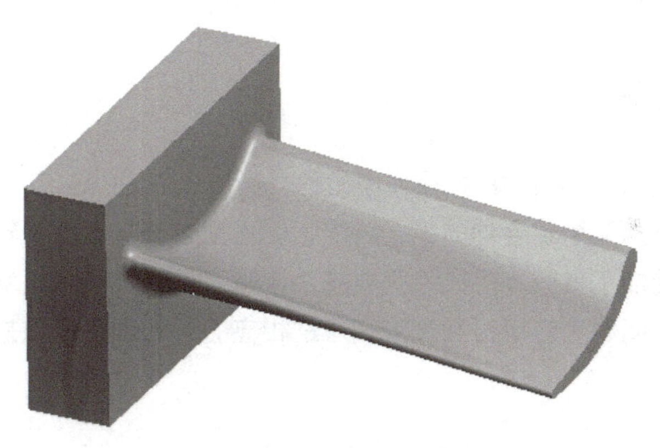

Fig. 3-38　Special Shaped Parts

项目 4 定位夹具的数控编程与仿真加工

学习目标

能力目标： 能运用 NX 软件完成定位夹具的数控编程与仿真加工。
知识目标： 掌握可变轴铣削几何体的设置方法；
掌握旋转底面精加工的设置方法；
掌握刀具轴的设置方法。
素质目标： 激发学生的学习兴趣，培养团队合作和创新精神。

项目导读

定位夹具是机械结构中常见的一类零件。这类零件的特点是结构比较简单，零件整体外形成块状，零件上一般会有台阶、圆弧角、配合孔等特征。在编程与加工过程中要特别注意台阶面的精度和配合孔的精度。

任务描述

学生以企业制造部门 MC 数控程序员的身份进入 NX CAM 功能模块，根据定位夹具零件的特征，制定合理的工艺路线，创建型腔铣、可变轴轮廓铣等加工操作，设置必要的加工参数，生成刀具轨迹，检验刀具轨迹是否正确合理，并对操作过程中存在的问题进行研讨和交流，通过相应的后处理生成数控加工程序，并运用机床加工零件。

Project 4 CNC Programming and Simulation Machining of Positioning Fixture

Capacity Objective: Complete model programming and simulation machining with NX software.

Knowledge Objective: Master variable axis milling geometry settings;
Master the setting method of finish machining of rotating bottom surface;
Master the setting of tool axis.

Quality objective: Stimulate students' interest in learning and cultivate the spirit of teamwork as well as innovation.

Positioning fixture is a kind of common parts in mechanical structure. The characteristics of this kind of parts are that the structure is relatively simple, the overall shape of the parts is block, and the parts generally have the characteristics of steps, arc angles, matching holes and so on. In the process of programming and machining, special attention should be paid to the accuracy of step surface and matching hole.

Students operate NX CAM functional modules as MC programmers in enterprise manufacturing department. According to the characteristics of model parts, establish a reasonable processing route, create machining operations such as cavity milling and variable axis contour milling, create necessary reference geometry, and set necessary machining parameters, generate tool path and check the generated tool path. Besides, students should discuss the problems occurred in the operation. By selecting the corresponding post-processor to generate NC machining programs, students import them into machine tools to complete the parts processing.

项目实施

按照零件加工要求，制定定位夹具加工工艺；编制定位夹具加工程序；完成定位夹具的仿真加工，通过相应的后处理生成数控加工程序，完成零件加工。

一、制定加工工艺

1. 定位夹具件结构分析

该定位夹具结构比较简单，主要由台阶、圆弧过渡面、孔等组成，主要加工内容为外形、台阶、孔。

2. 毛坯选用

零件材料由厚度为 90 mm 的 45# 钢板切割而成，尺寸为 255 mm × 228 mm × 90 mm。零件四周单边最小余量为 3 mm，零件厚度方向为了保证零件的装夹，余量为 7 mm。

3. 加工工序卡制定

零件选用立式四轴联动机床加工，平口钳夹持，遵循先粗后精、先面后孔的加工原则。制定的加工工序卡如表 4–1 所示。

表 4–1 加工工序卡

零件号：357902-1			工序名称：定位夹具铣削加工		工艺流程卡_工序单
材料：45#		页码：1	工序号：01		版本号：0
夹具：平口钳		工位：MC	数控程序号：		
刀具及参数设置					
刀具号	刀具规格	加工内容	主轴转速/rpm	进给速度/mmpm	
T01	D20R0.4	型面粗加工	1800	1200	
T02	B16	型面精加工-1	2600	1400	
T03	B8	型面精加工-2	2600	1400	
T03	B8	型面精加工-3	2600	1400	
T03	ZXZ12	中心钻	800	100	
T04	D17.8	钻孔	800	100	
T05	D18	铰孔	800	100	
02					
01					
更改号	更改内容		批准	日期	
拟制：	日期：	审核：	日期：	批准：	日期：

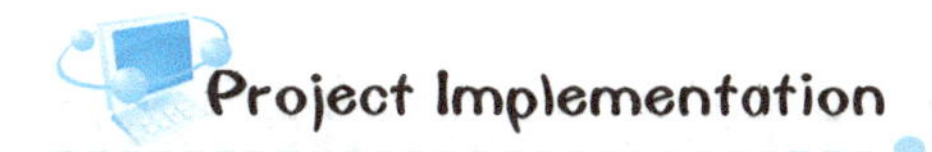

Project Implementation

Firstly, formulate processing technic under processing requirements. Secondly, program for machining model. Finally, complete simulation machining, and then import the NC machining program obtained by post processor to the machine tool to complete machining.

Ⅰ. Formulate Processing Technic

1. Structural analysis of positioning fixture

The structure of the positioning fixture is relatively simple. It is mainly composed of steps, arc transition surface, holes and other features. The main processing contents are shape, steps and holes.

2. Blank selection

The material of the part is cut from 45# steel plate with a thickness of 90 mm, and the size is 255 mm × 228 mm × 90 mm. The minimum allowance on one side around the part is 3 mm, and the allowance in the thickness direction of the part is 7 mm to ensure the clamping of the part.

3. Formulation of processing procedure card

The parts are processed by vertical 4-axis linkage machine tool, clamped by flat pliers, and follow the processing principle of rough before fine, face before hole. The processing procedure is shown in Table 4-1.

Table 4-1 Processing Procedure Card

Part number: 357902-1		Name of process: Milling of positioning fixture		Process card - Process sheet
Material: 45#	Page number: 1		Procedure number: 01	Version number: 0
Fixture: Parallel-jaw vice	Work station: MC		CNC program number:	
Tool and parameter setting				
Tool number	Tool specification	Processing content	Spindle speed /rpm	Feed speed /mmpm
T01	D20R0.4	Rough machining	1800	1200
T02	B16	Finish machining-1	2600	1400
T03	B8	Finish machining-2	2600	1400
T03	B8	Finish machining-3	2600	1400
T03	ZXZ12	Drill center hole	800	100
T04	D17.8	Drill	800	100
T05	D18	Reaming	800	100
02				
01				
Change number	Change content	Approve	Date	
Draws:	Date: / review: Date:	Approve:	Date:	

二、加工准备

（1）启动NX，单击【打开】，在弹出的【打开】对话框中选择【定位夹具.prt】文件，如图4-1所示，单击【OK】，打开零件模型。

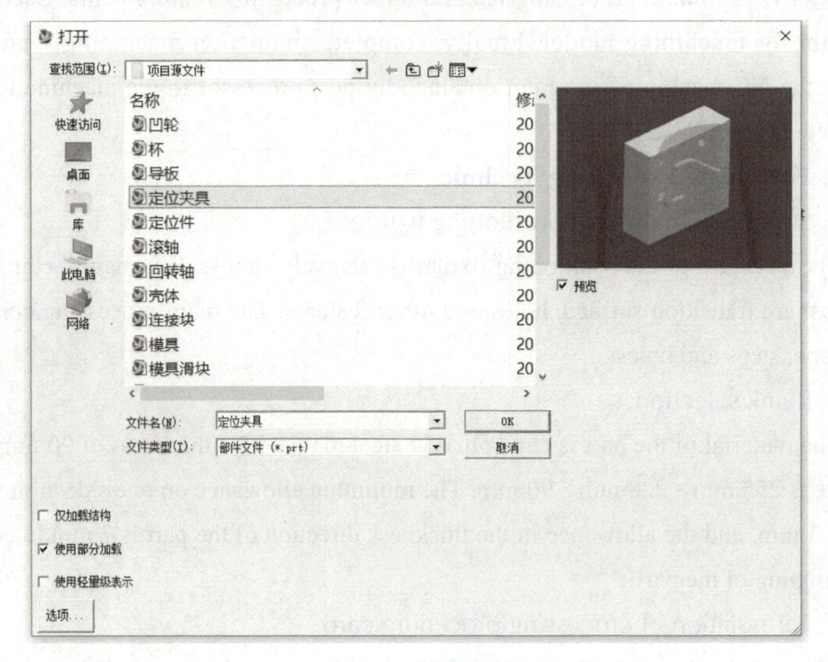

图4-1 打开文件

（2）选择【应用模块】选项卡，单击【加工】，如图4-2所示，进入加工环境。

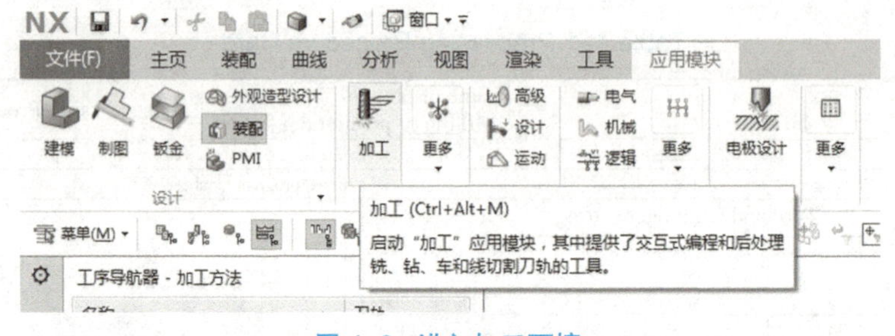

图4-2 进入加工环境

（3）在弹出的【加工环境】对话框中，在【CAM会话配置】选项中选择【cam_general】，在【要创建的CAM设置】选项中选择【mill_contour】，如图4-3所示，单击【确定】。

Ⅱ. Preparation for Processing

（1）As shown in Fig. 4-1, start NX, click “OPEN”, select model “Positioning Fixture” (.prt file), and click “OK”. Then, the part model can be seen in WCS.

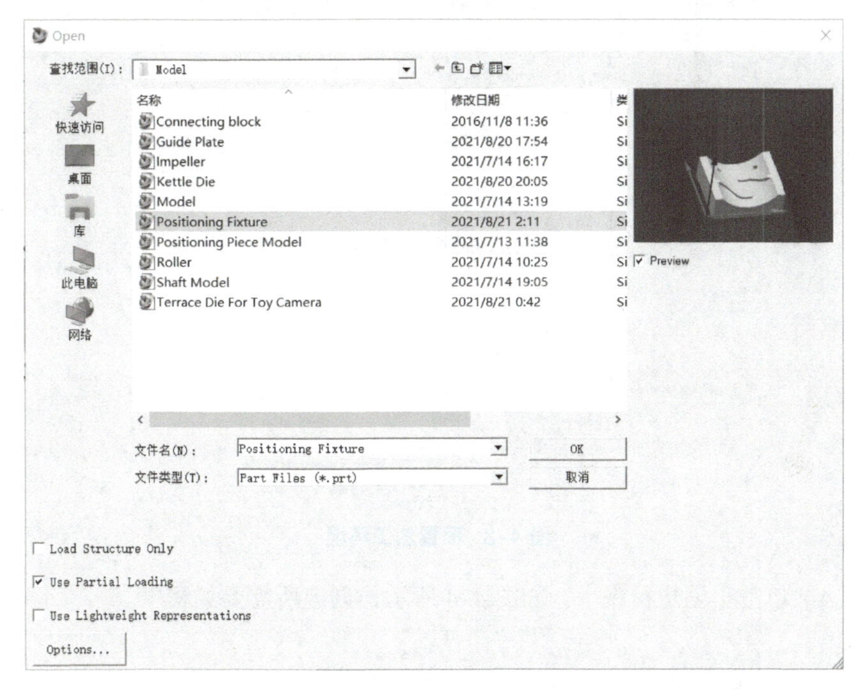

Fig. 4-1 Open the File

（2）Select “Application”, and click “Manufacturing”. The manufacturing menu can be seen as shown in Fig. 4-2.

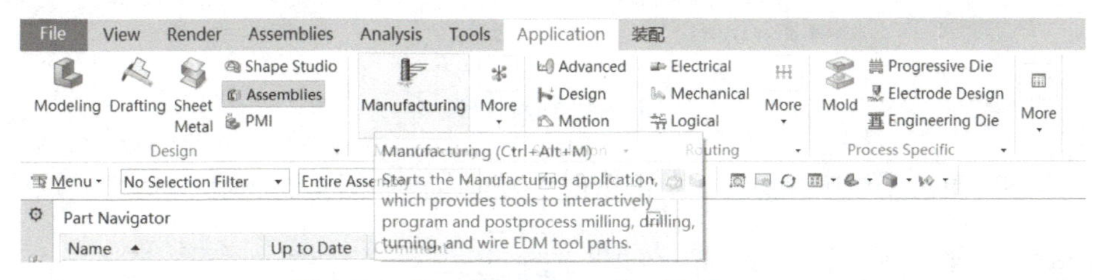

Fig. 4-2 Enter Manufacturing Environment

（3）As shown in Fig. 4-3, in “Machining Environment” window, select “cam_general” in “CAM Session Configuration” tab, and then, select “mill_contour” in “CAM Setup to Create” tab. Finally, click “OK”.

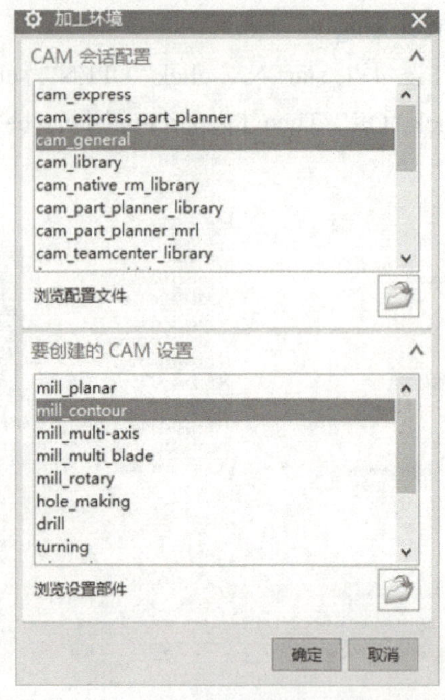

图 4–3　配置加工环境

（4）单击【创建程序】，如图 4–4 所示，创建所需要的程序。

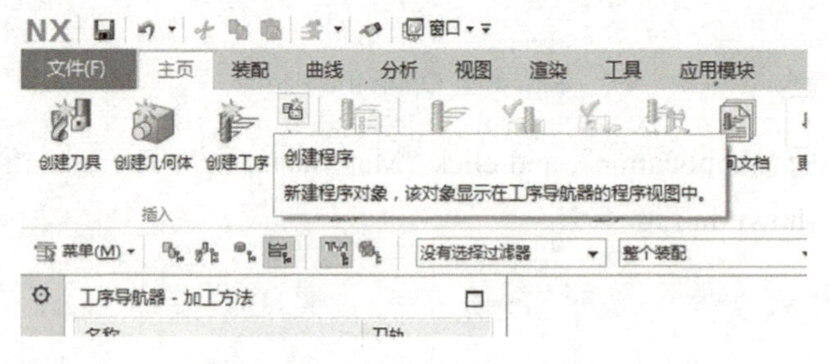

图 4–4　创建程序

（5）单击【机床视图】，单击【创建刀具】，如图 4–5 所示。

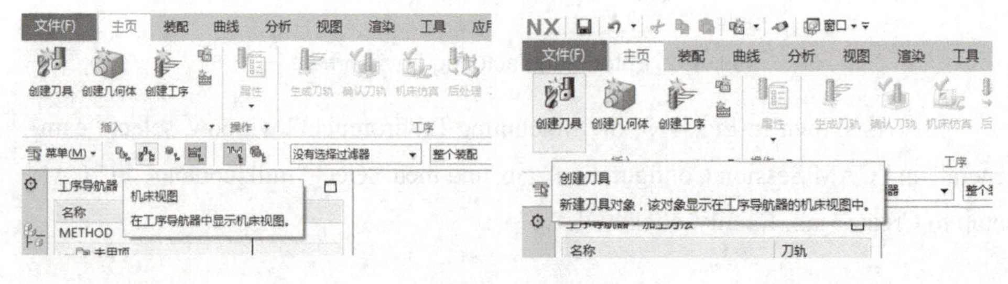

图 4–5　创建刀具

（6）在【创建刀具】对话框中，【刀具子类型】选择【MILL】，修改【名称】为【D20R0.4】，如图 4–6 所示，单击【应用】。弹出【铣刀 –5 参数】对话框，

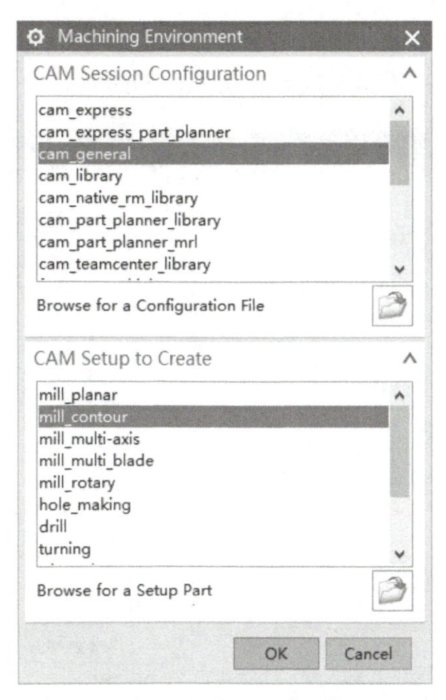

Fig. 4–3 Set Machining Environment

(4) As shown in Fig. 4-4, click "Create Program" to create programs needed.

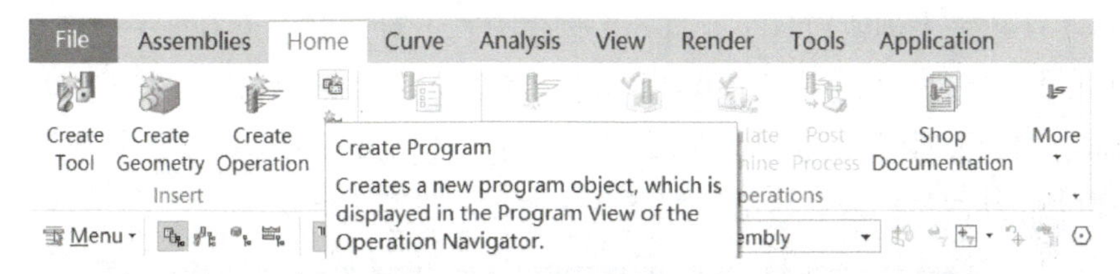

Fig. 4–4 Create Program

(5) Click "Machine Tool View", and then click "Create Tool" as shown in Fig. 4-5.

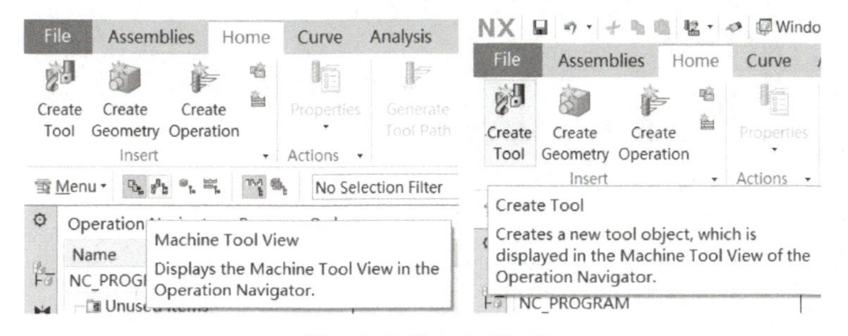

Fig. 4–5 Create Tool

(6) As shown in Fig. 4-6, in the "Create Tool" window, select "mill" in "Tool Subtype" and modify "Name" as "D20R0.4", and click "OK". As Shown in Fig. 4-7,

设置刀具参数：【直径】为【20】，【下半径】为【0.4】，【长度】为【45】，【刀刃长度】为【20】，其他参数为默认值，如图 4–7 所示，单击【确定】。

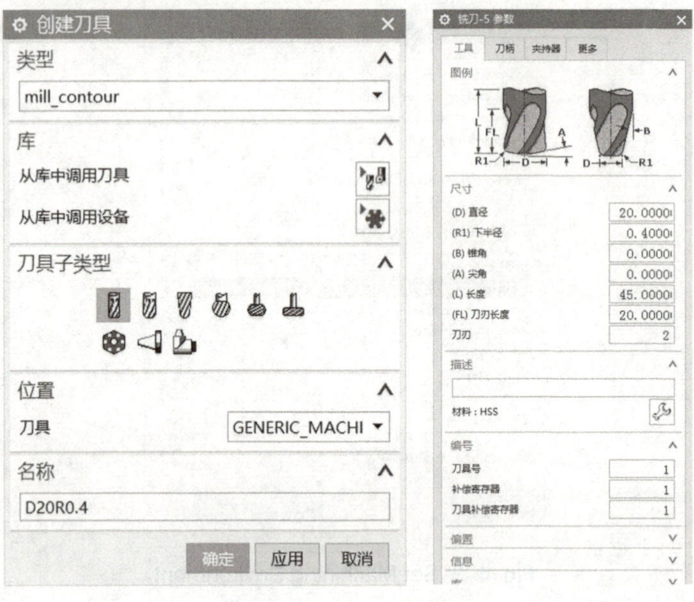

图 4–6　创建刀具　　　图 4–7　设置刀具参数（1）

（7）单击【创建刀具】，打开【创建刀具】对话框，【刀具子类型】选择【球头刀】，修改【名称】为【BALL16】，单击【应用】。弹出【铣刀 – 球头铣】对话框，设置刀具参数：【球直径】为【16】，【长度】为【40】，【刀刃长度】为【20】，其他参数为默认值，如图 4–8 所示，单击【确定】。

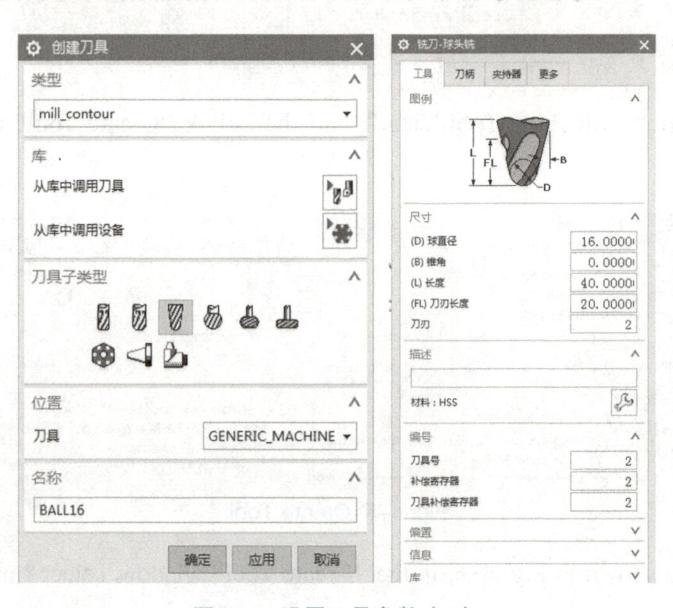

图 4–8　设置刀具参数（2）

in "Milling Tool-5 Parameters" window, "Diameter" value is 20 and "Lower Radius" value is 0.4. Other parameters default, click "OK" to close the window.

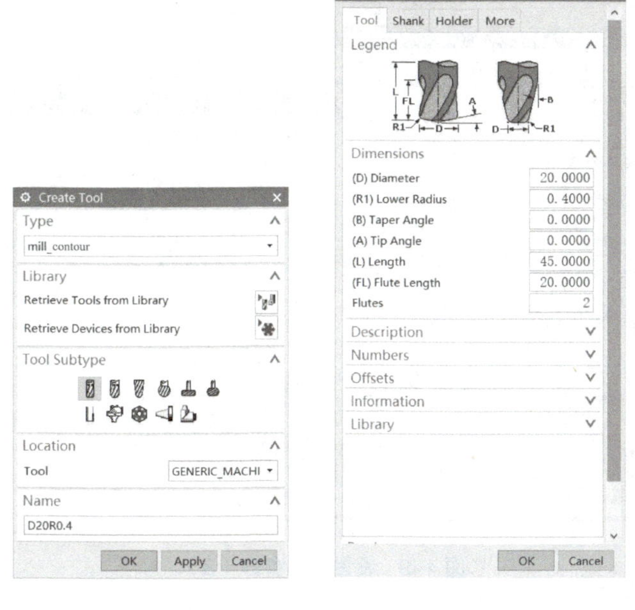

Fig. 4–6 Create Tool   Fig. 4–7 Modify Tool Parameters (1)

(7) As shown in Fig. 4-8, in the "Create Tool" window, select "ball-mill" in "Tool Subtype", modify "Name" as "BALL16", and click "OK". In "Milling Tool-Ball Mill" window, "Ball Diameter" value is 16. Other parameters default, click "OK" to close the window.

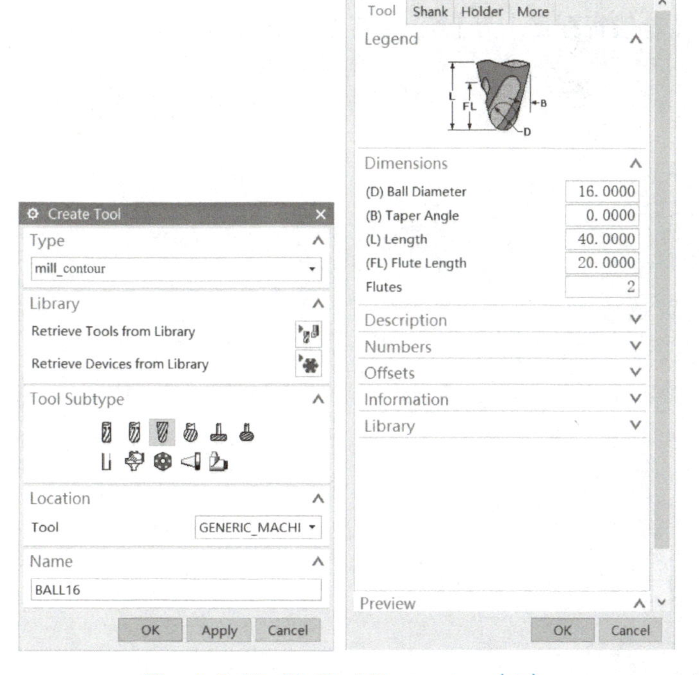

Fig. 4–8 Modify Tool Parameters (2)

（8）单击【创建刀具】，【刀具子类型】选择【球头刀】，【刀具名称】改为【BALL8】，单击【应用】。弹出【铣刀－球头铣】对话框，设置刀具参数：【球直径】为【8】，其他参数为默认值，如图 4-9 所示，单击【确定】。

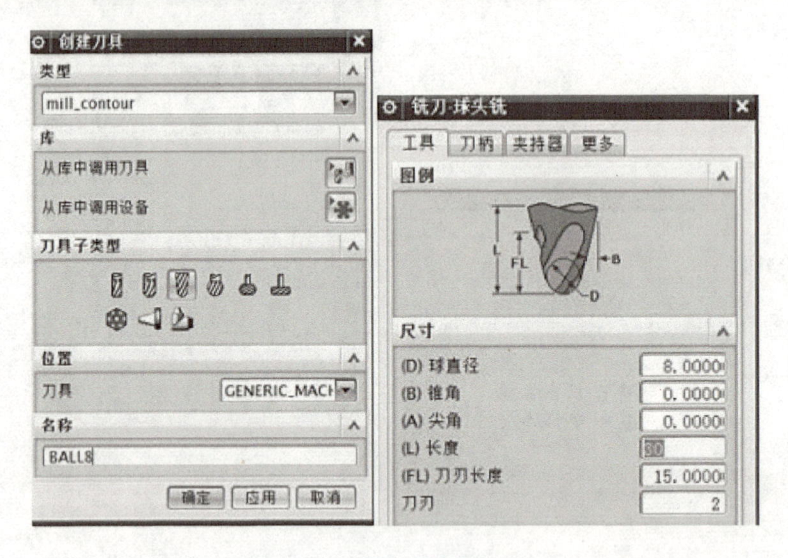

图 4-9 设置刀具参数（3）

（9）单击【创建刀具】，【类型】选择【drill】，【刀具子类型】选择【中心钻】，【刀具名称】改为【ZXZ12】，单击【确定】。弹出【钻刀】对话框，设置刀具参数：【直径】为【12】，【刀尖角度】为【100.3888】，【刀具号】为【4】，【补偿寄存器】为【4】，其他参数为默认值，如图 4-10 所示，单击【确定】。

(8) As shown in Fig. 4-9, in the "Create Tool" window, select "ball-mill" in"Tool Subtype", modify "Name" as "BALL8", and then click "Apply". In "Milling Tool-Ball Mill" window, "Ball Diameter" value is 8. Other parameters default, click "OK" to close the window.

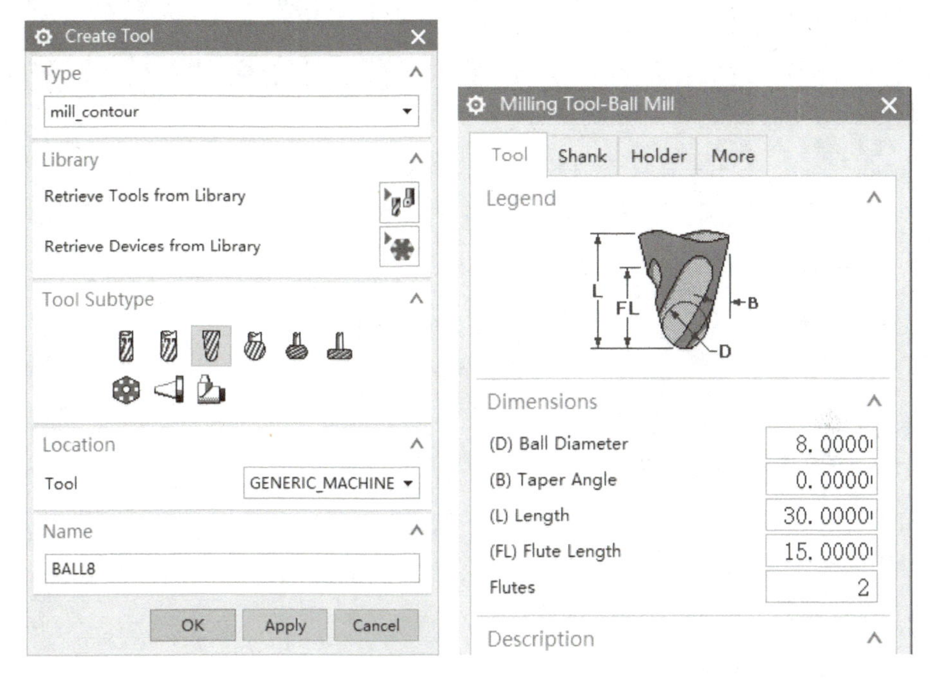

Fig. 4-9 Modify Tool Parameters (3)

(9) As shown in Fig. 4-10, click "Create Tool" window, tool type select "drill", select "spotdrilling tool" in tool subtype and modify "Name" as "ZXZ12", and then click "OK". In "Drilling Tool" window, "Diameter" value is 12, "Point Angle" value is 100.3888, "Tool Number" is 4, "Adjust Register" is 4. Other parameters default, click "OK" to close the window.

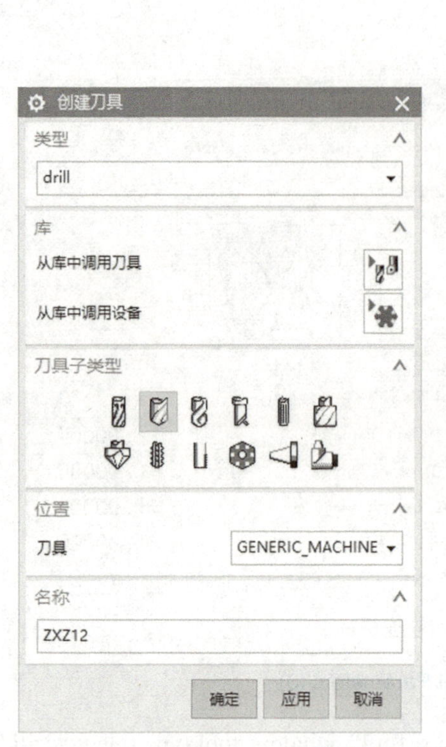

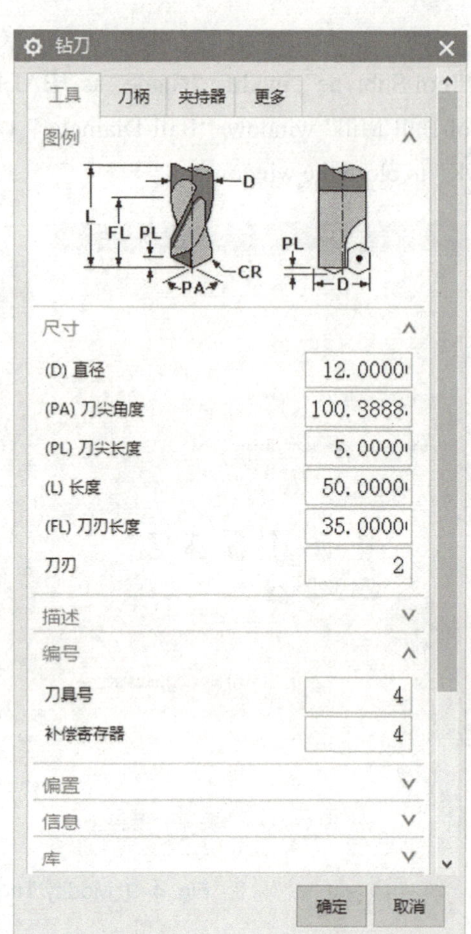

图 4-10 设置刀具参数（4）

（10）单击【创建刀具】，刀具子类型选择【球头刀】，刀具名称【D17.8Drill】，单击【确定】。弹出【钻刀】对话框，设置刀具参数：【直径】为【17.8】，【刀具号】为【5】，【补偿寄存器】为【5】，其他参数为默认值，如图 4-11 所示，单击【确定】。

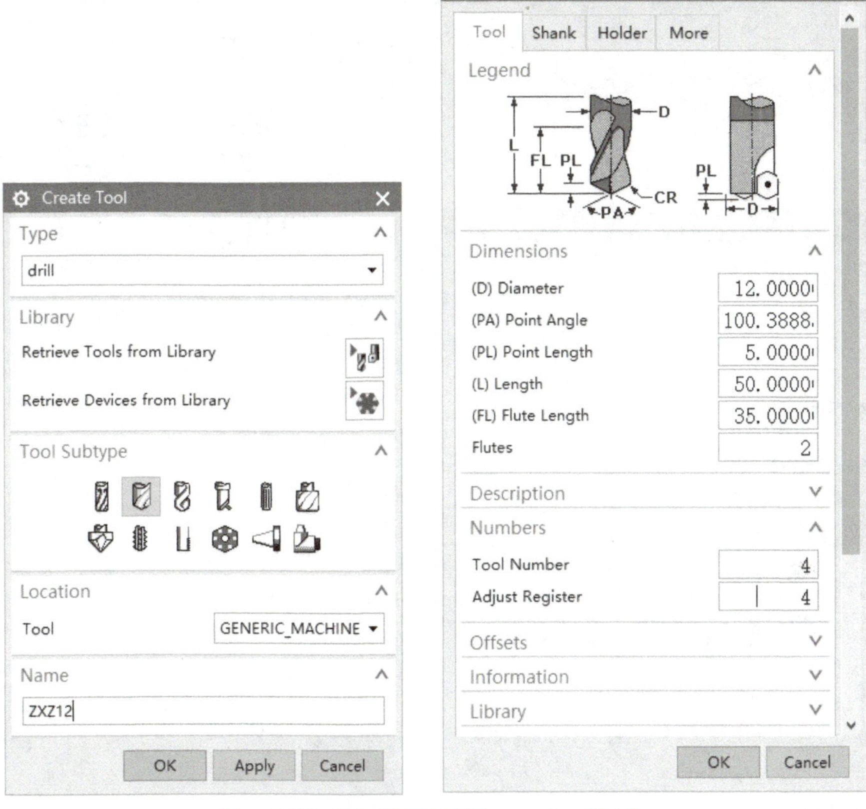

Fig. 4-10　Modify Tool Parameters (4)

(10) As shown in Fig. 4-11, click "Create Tool" window, select "ball-mill" in "Tool Subtype" and modify "Name" as " D17.8Drill ", and then click "Apply". In "Drilling Tool " window, "Diameter" value is 17.8, "Tool Number" is 5, "Adjust Register" is 5. With other parameters default, click "OK" to close the window.

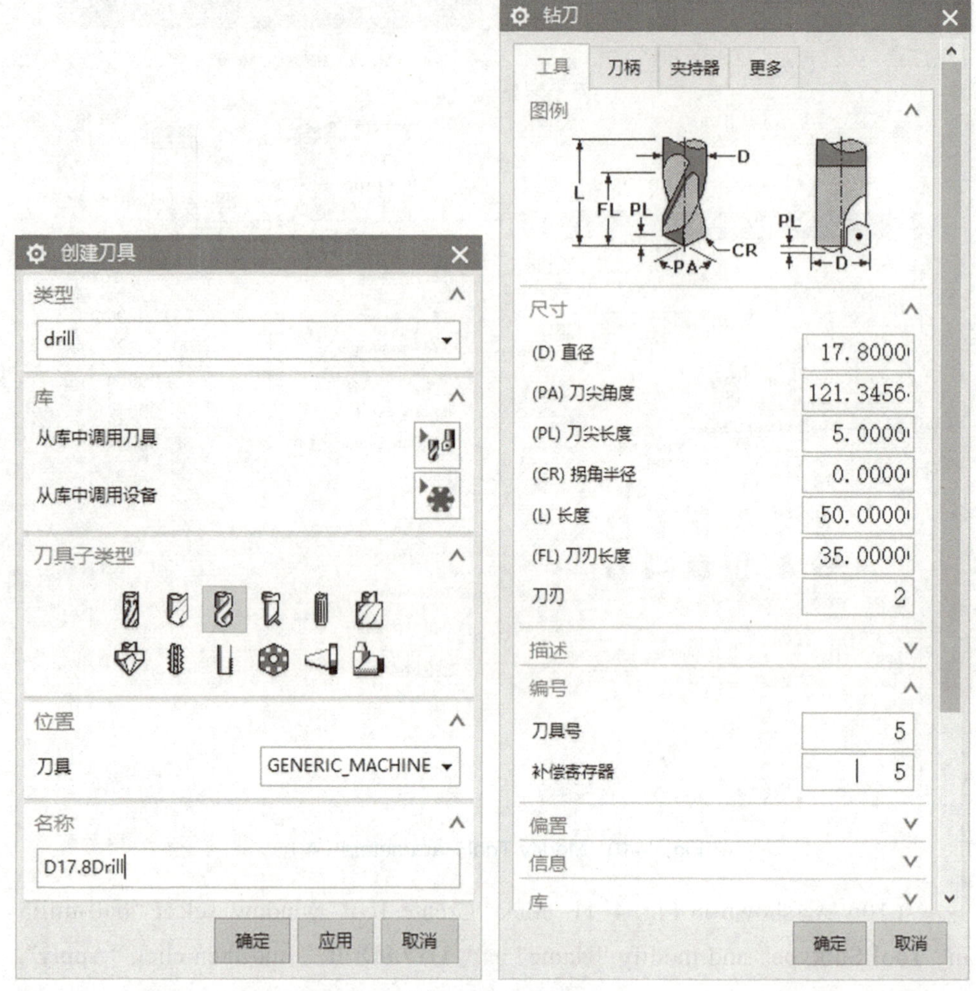

图 4-11 设置刀具参数（5）

（11）单击【创建刀具】，刀具子类型选择【铰刀】，刀具名称【D18REAMER】，单击【确定】。弹出【钻刀】对话框，设置刀具参数：【直径】为【18】，【刀具号】为【6】，【补偿寄存器】为【6】，其他参数为默认值，如图 4-12 所示，单击【确定】。

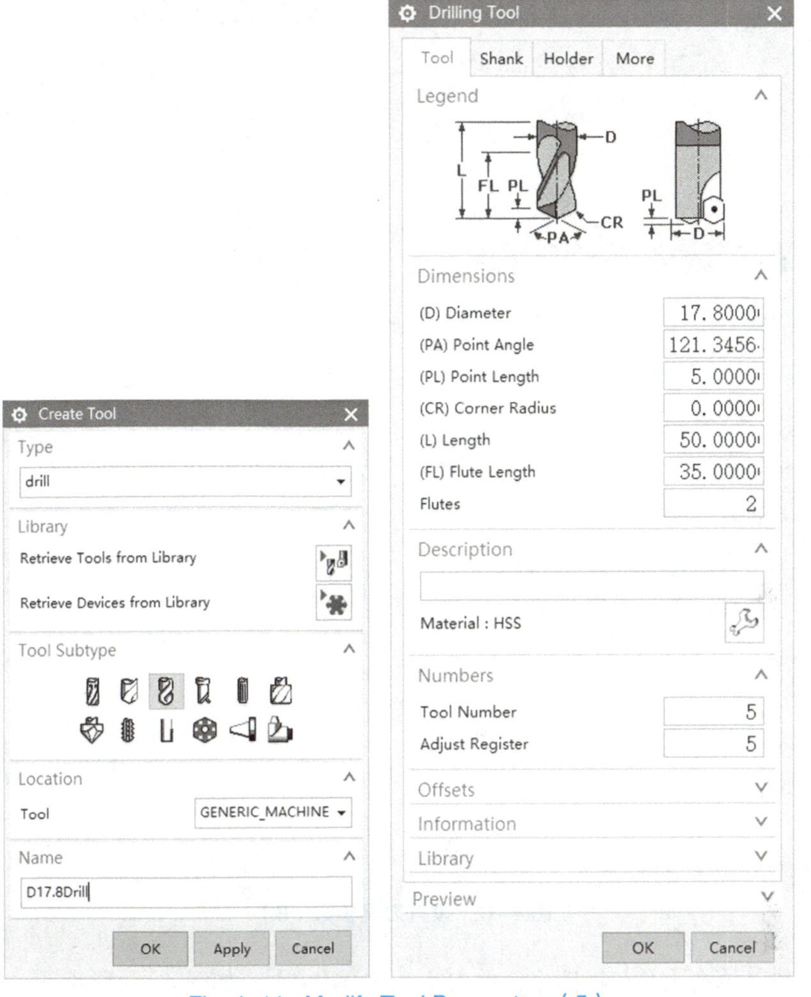

Fig. 4–11　Modify Tool Parameters (5)

(11) As shown in Fig. 4-12, click "Create Tool" window, select "reamer" in "Tool Subtype" and modify "Name" as "D18REAMER ", and then click "OK". In "Drilling Tool" window, "Diameter" value is 18, "Tool Number" is 6, "Adjust Register" is 6. With other parameters default, click "OK" to close the window.

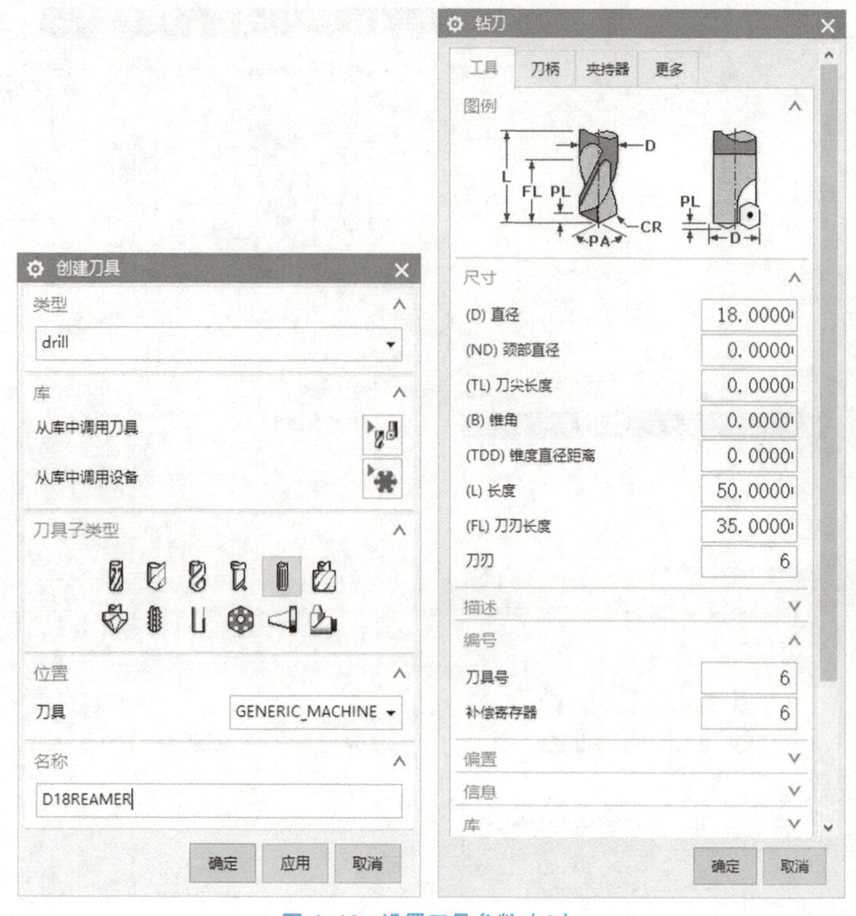

图 4-12　设置刀具参数（6）

（12）单击【几何视图】，单击【MCS_MILL】，如图 4-13 所示。

图 4-13　几何视图

（13）单击【工序导航器 - 几何】中的【WORKPIECE】，打开【工件】对话框，单击【指定部件】中的【选择或编辑部件几何体】，如图 4-14 所示。

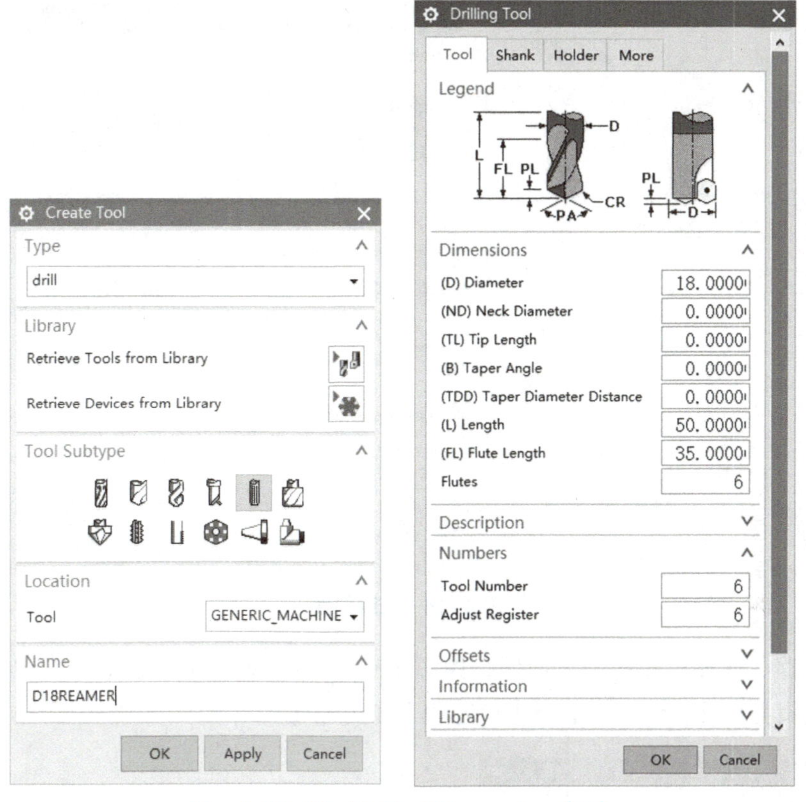

Fig. 4–12　Modify Tool Parameters (6)

（12）Click "Geometry View" and then click "MCS_MILL" as shown in Fig. 4-13.

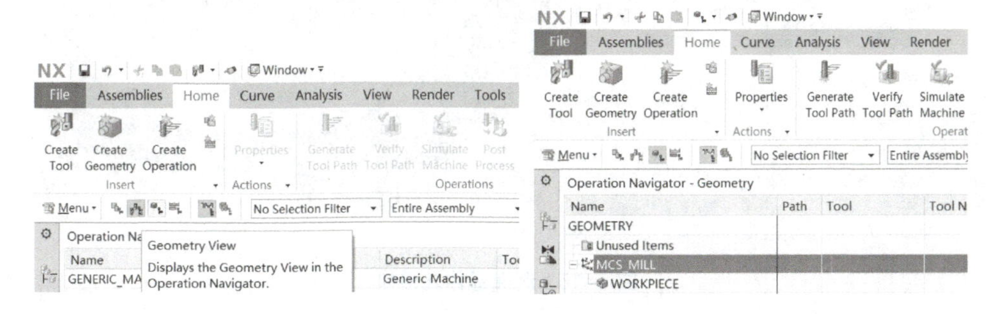

Fig. 4–13　Geometry View

（13）Click the "WORKPIECE" in the "Operation Navigator-Geometry" and click "Specify Part" in "Workpiece" window as shown in Fig. 4-14.

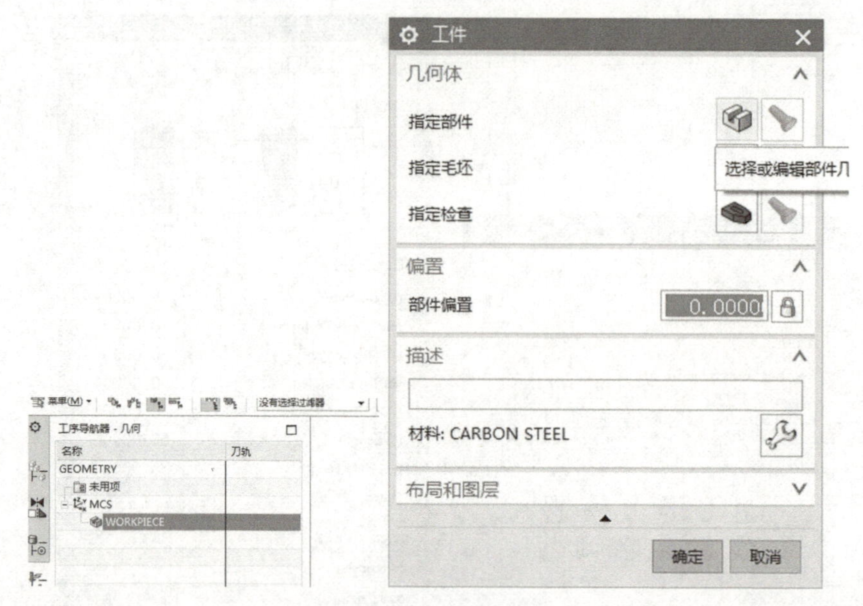

图 4-14　指定部件

（14）选择部件作为几何体，如图 4-15 所示，单击【指定毛坯】中的【选择或编辑毛坯几何体】，选择【包容块】作为毛坯。

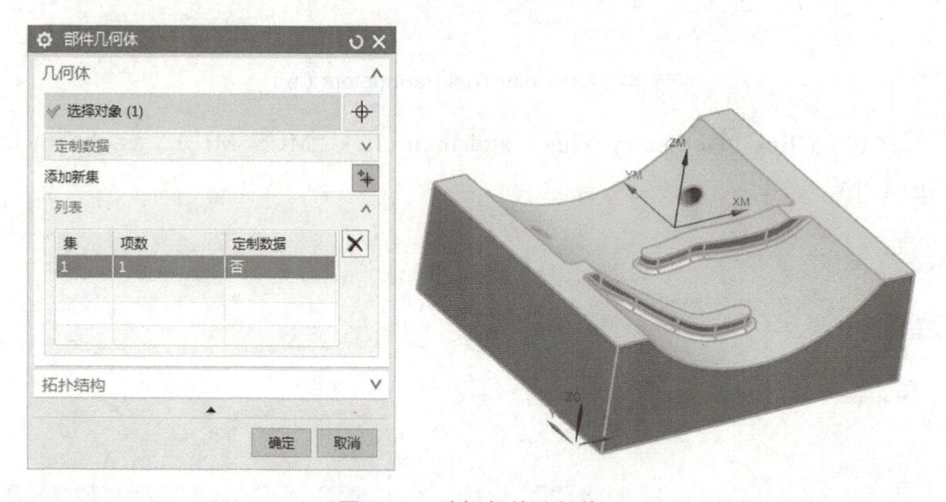

图 4-15　选择部件几何体

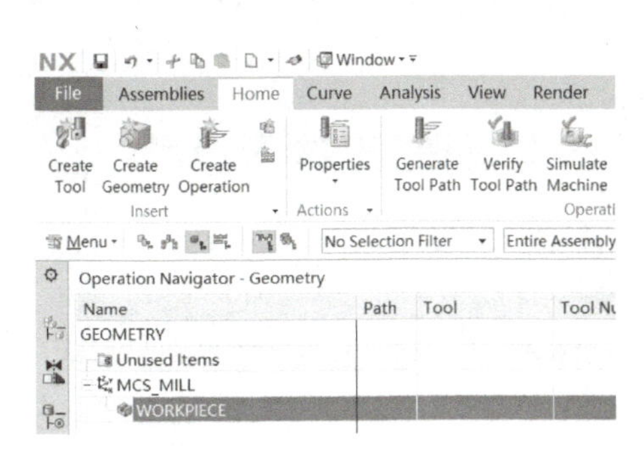

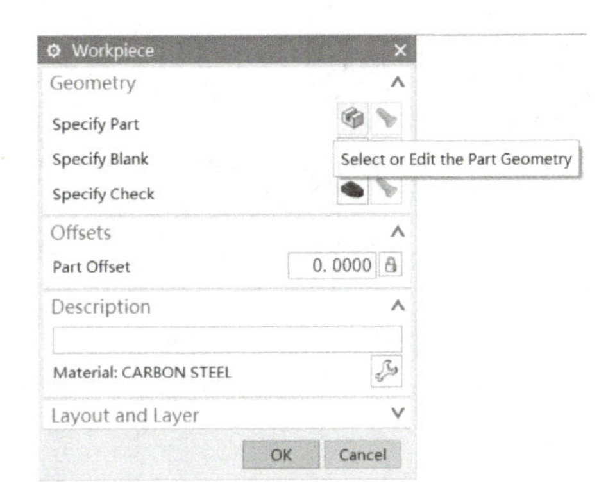

Fig.4-14 Specify Part

(14) In "Part Geometry" window, select the solid part as Fig. 4-15. Click "Specify Blank" in "Workpiece" window. In "Blank Boundaries" window, select " Bounding Block" as type. Click "OK".

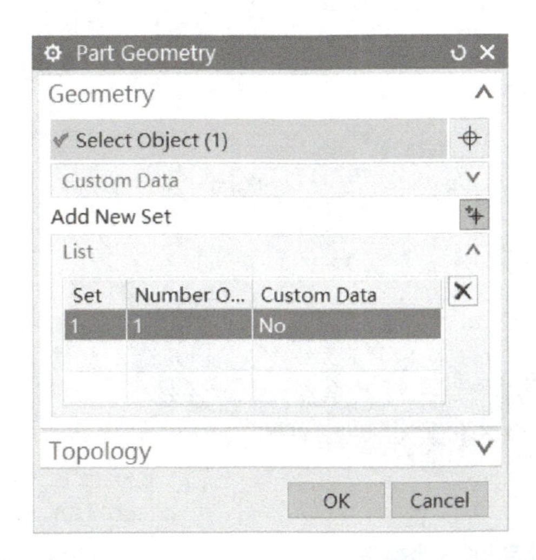

Fig. 4-15 Select Part Geometry

（15）单击【MCS】窗口，单击下拉菜单，选择【指定 MCS】，如图 4-16 所示。

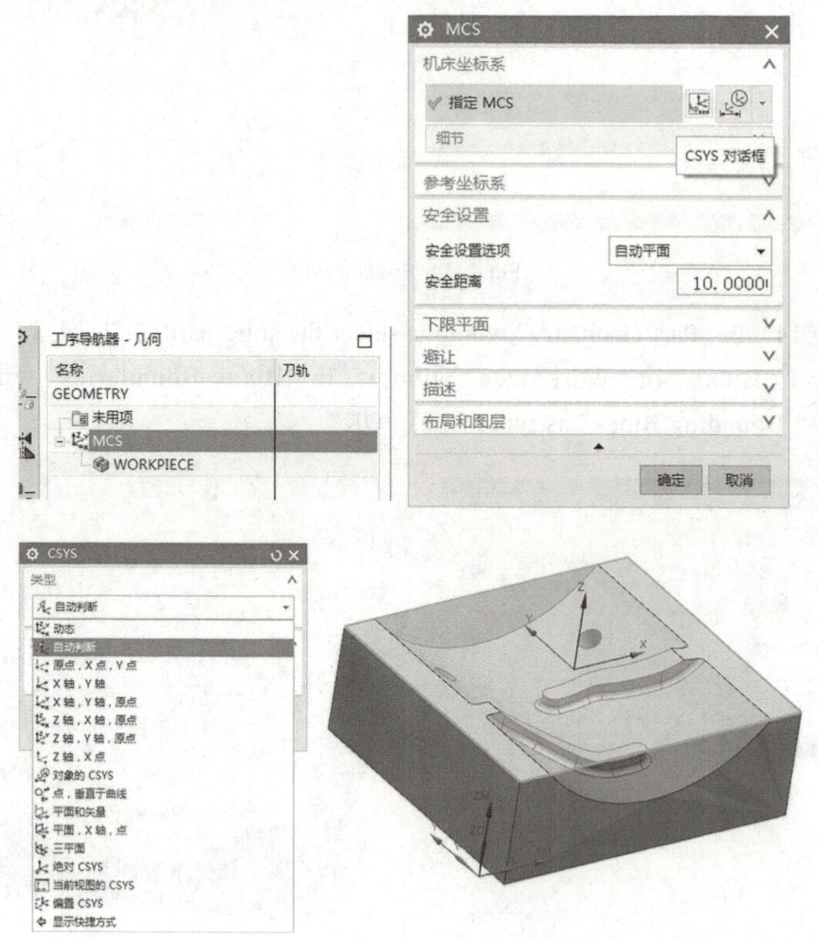

图 4-16　设定 MCS 坐标系

（16）更改【安全距离】为【30】，如图 4-17 所示，单击【确定】。

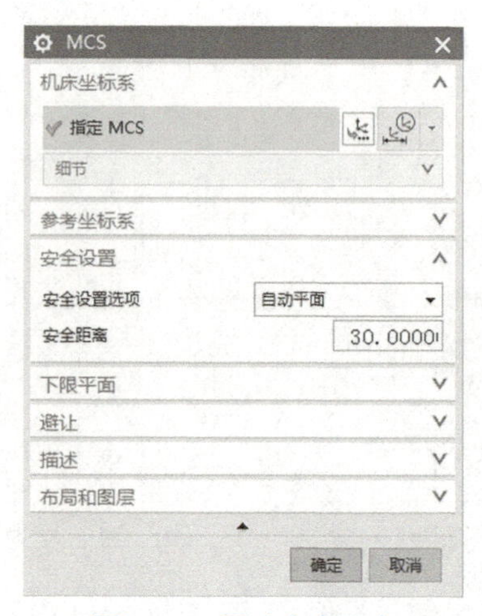

图 4-17　更改安全距离

（15）Click “MCS MILL”. In “MCS Mill” window, click “CSYS Dialog”, and specify MCS as shown in Fig. 4-16.

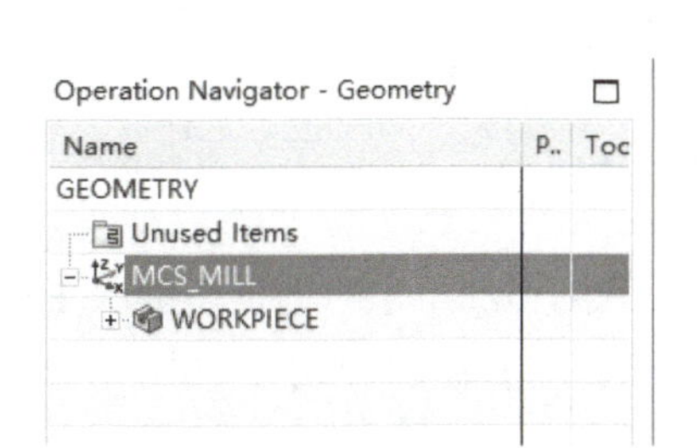

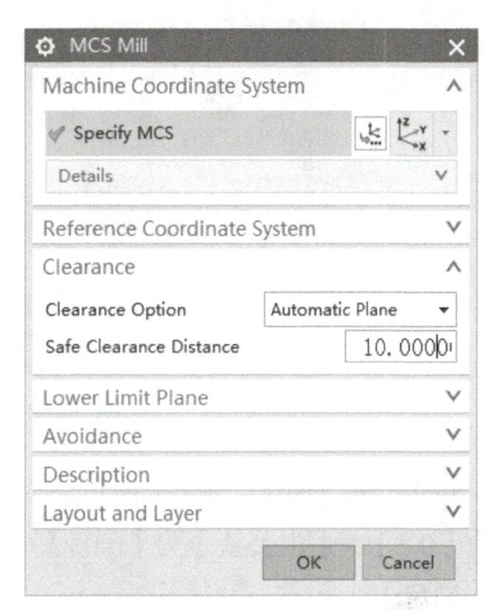

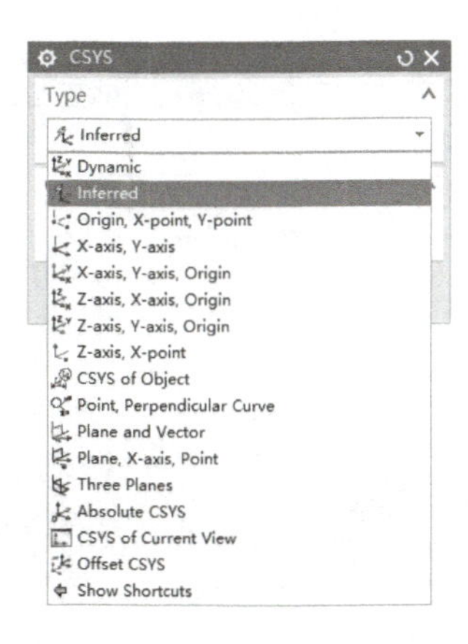

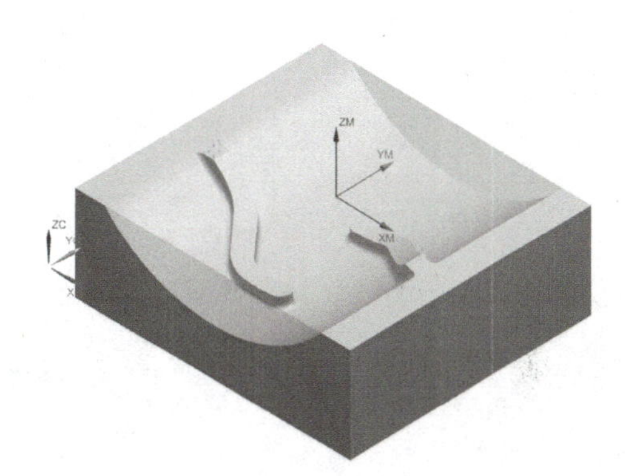

Fig. 4-16　Set MCS

（16）Change “Safe Clearance Distance” to 30 as shown in Fig. 4-17, and then click “OK” to close the window.

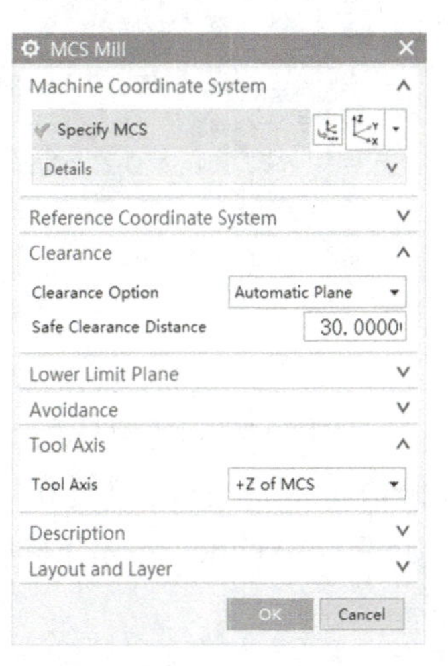

Fig. 4-17　Modify Safe Clearance Distance

（17）单击【加工方法】，如图 4–18 所示。

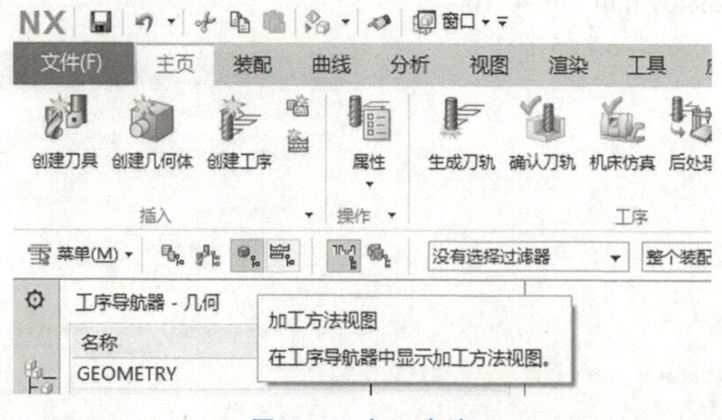

图 4–18　加工方法

（18）双击【MILL_ROUGH】，弹出【铣削粗加工】对话框，修改【部件余量】为【0.5】，【内公差】为【0.03】，【外公差】为【0.03】，如图 4–19 所示，单击【确定】。

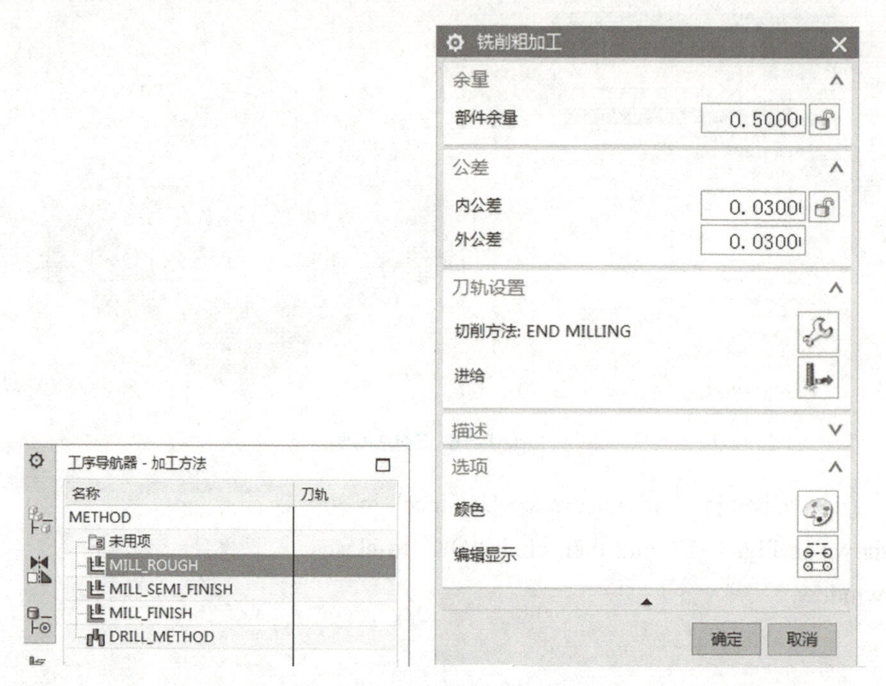

图 4–19　设定铣削粗加工参数

（19）双击【MILL_SEMI_FINISH】，弹出【铣削半精加工】对话框，修改【部件余量】为【0.1】，【内公差】为【0.01】，【外公差】为【0.01】，如图 4–20 所示，单击【确定】。

(17) Click "Machining Method View" button as shown in Fig. 4-18.

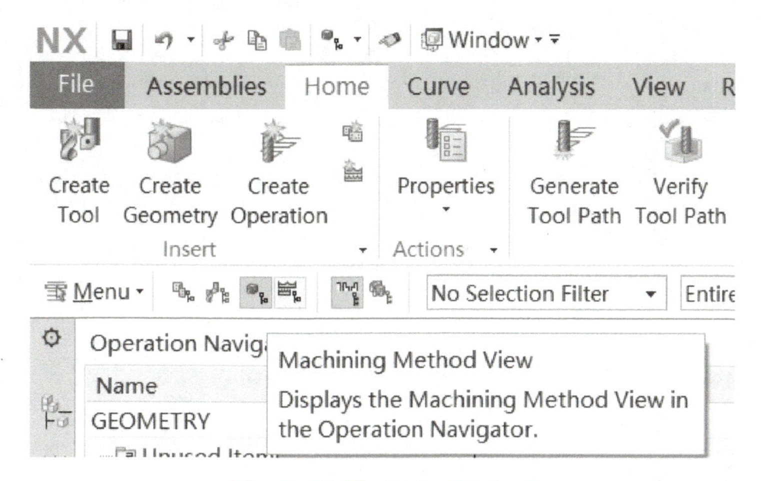

Fig. 4-18 Machining Method

(18) Double click " MILL_ROUGH " to modify "Part Stock" value as 0.5 and both "Intol Tolerance" and "Outtol Tolerance" values as 0.03 as shown in Fig. 4-19. Click "OK".

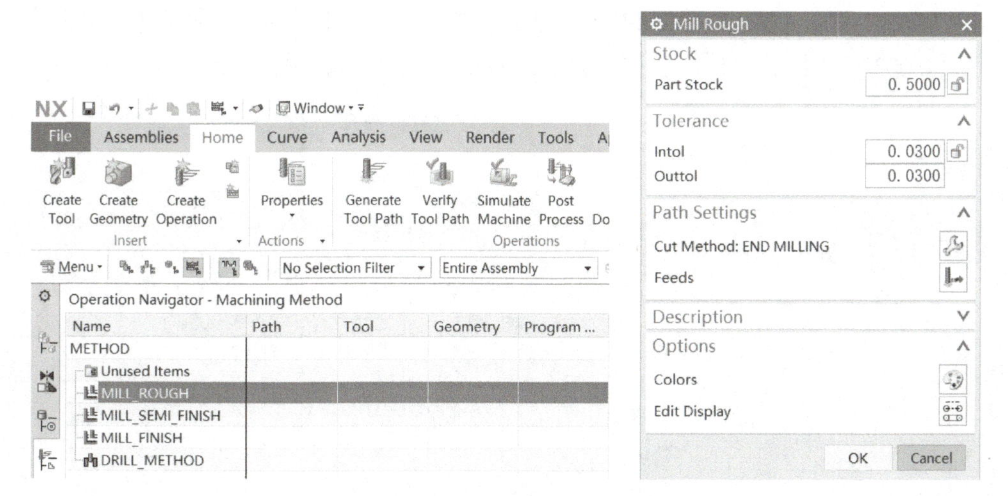

Fig. 4-19 Modify Mill Rough

(19) Double click " MILL_SEMI_FINISH " to modify "Part Stock" value as 0.1, and both "Intol Tolerance" and "Outtol Tolerance" values as 0.01 as shown in Fig. 4-20. Click "OK" to close the window.

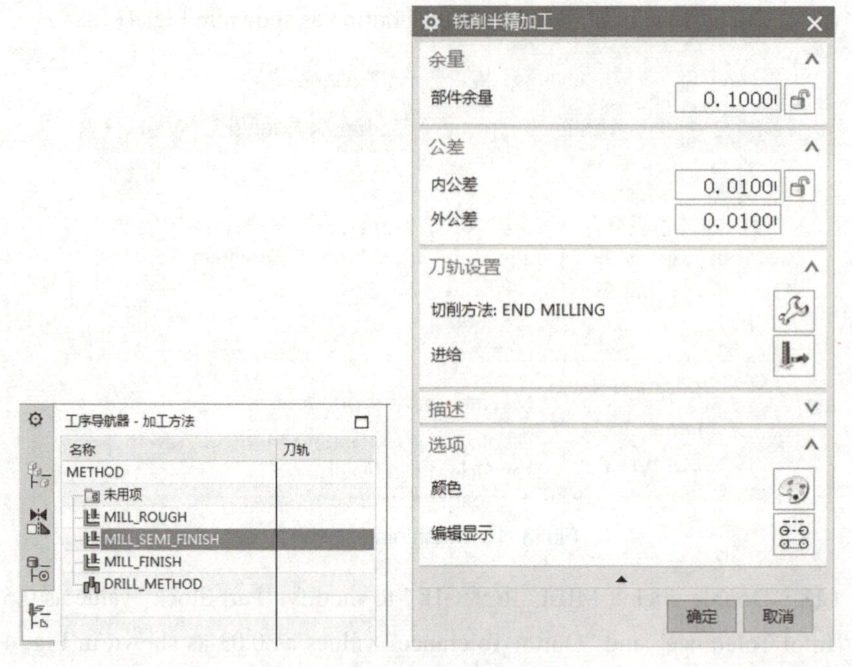

图 4-20　设定铣削半精加工参数

（20）双击【MILL _FINISH】，弹出【铣削精加工】对话框，修改【内公差】为【0.003】，【外公差】为【0.003】，如图 4-21 所示，单击【确定】。

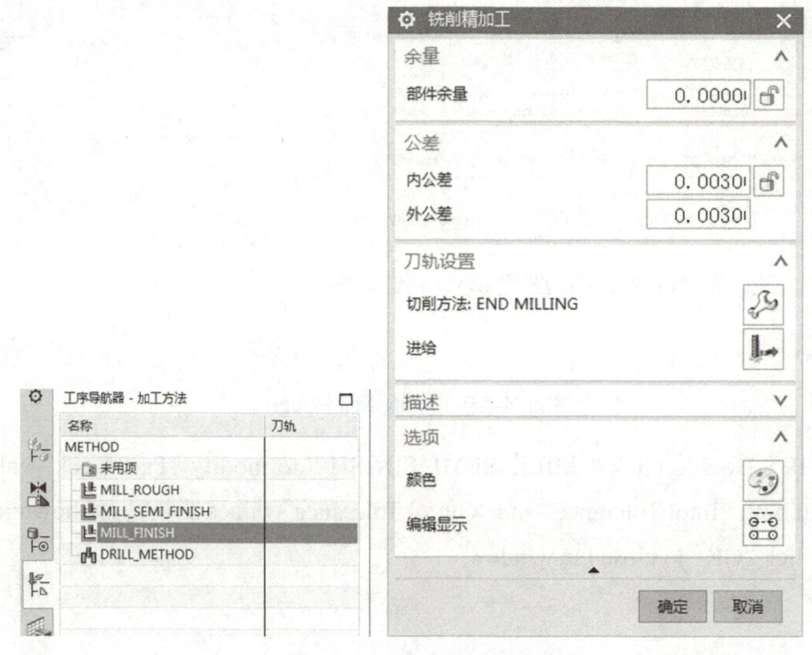

图 4-21　设定铣削精加工参数

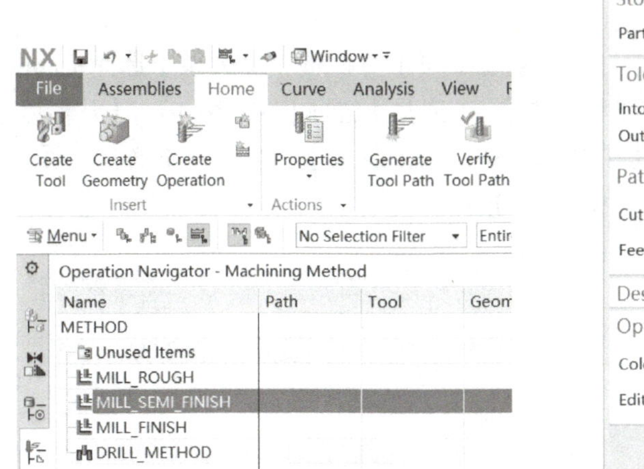

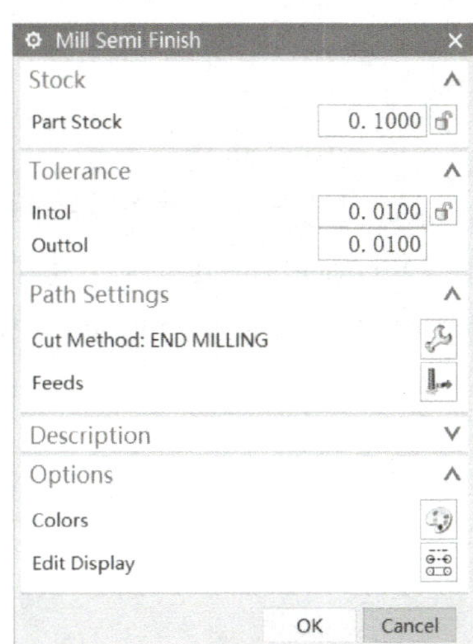

Fig. 4–20 Modify Mill Semi Finish

(20) Double click " MILL_FINISH " to modify both "Intol Tolerance" and "Outtol Tolerance" values as 0.003 as shown in Fig. 4-21. Click "OK".

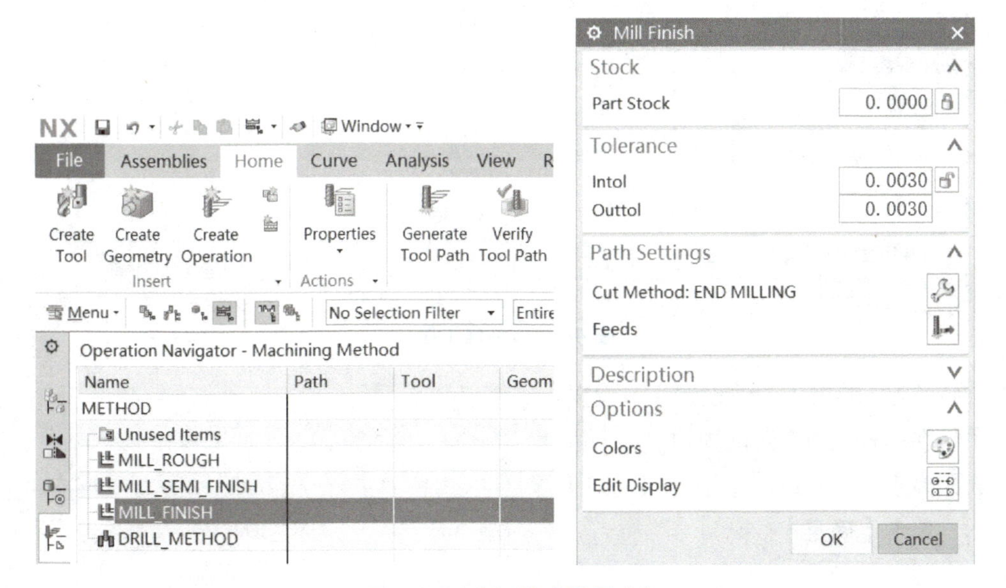

Fig. 4–21 Modify Mill Finish

三、加工程序编制

1. 型面粗加工

（1）单击【创建工序】，打开【创建工序】对话框，【类型】选择【mill_contour】，【工序子类型】选择【型腔铣】，【刀具】选择【D20R0.4】，【几何体】选择【WORKPIECE】，【方法】选择【MILL_ROUGH】，修改【名称】为【型面粗加工】，如图 4–22 所示，单击【确定】。

图 4–22　创建工序

（2）单击【刀轨设置】，设置【切削模式】为【跟随周边】，【最大距离】为【2 mm】，如图 4–23 所示，单击【确定】。

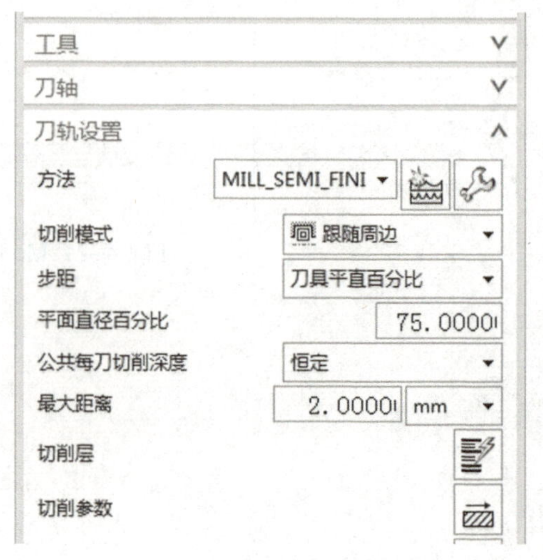

图 4–23　设置刀轨

Ⅲ. Programming

1. Rough Machining

（1）As shown in Fig. 4-22, click “Create Operation”, and then select “mill_contour” in pull-down box of “Type”. Select operation subtype as “Cavity Mill”, tool as “D20R0.4”, geometry as “WORKPIECE” and method as “MILL-ROUGH”. Modify name as “ Rough-mill ”, and then click “OK”.

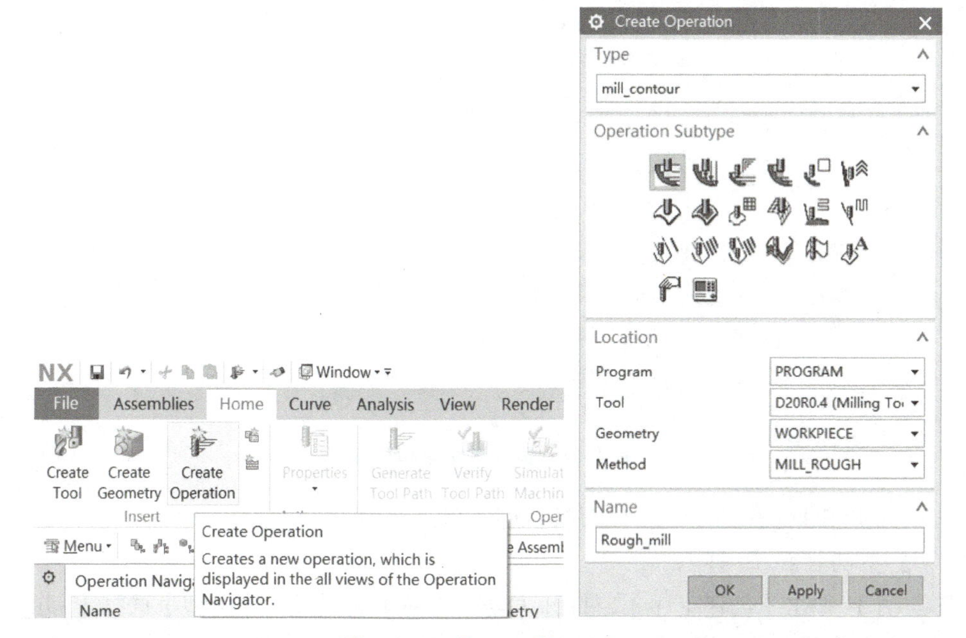

Fig. 4-22 Create Operation

（2）Modify “Maximum Distance” as 2 mm as shown in Fig. 4-23.

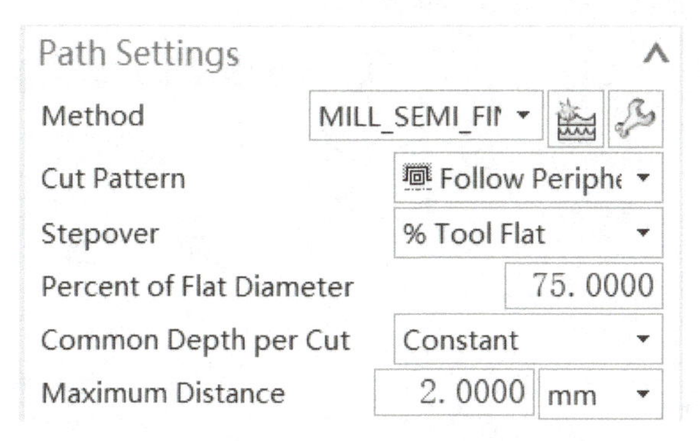

Fig. 4-23 Modify Path Settings

（3）单击【进给率和速度】，设置【主轴速度】为【1800】，进给率【切削】为【1200 mmpm】，如图 4-24 所示，单击【确定】。

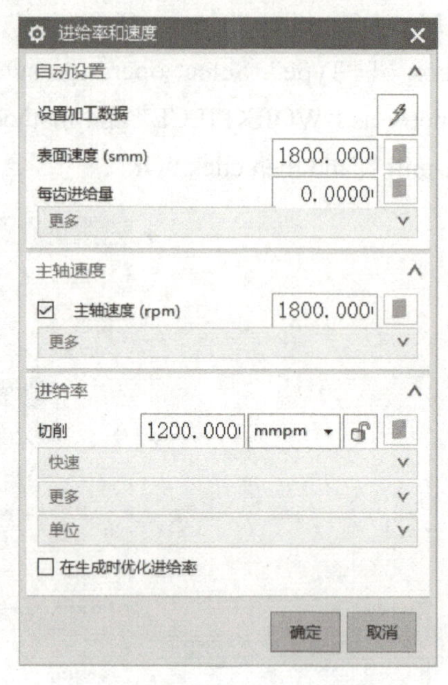

图 4-24 设置进给率和速度

（4）单击【生成】，生成刀具轨迹，单击【确定】。

2. 型面精加工 -1

（1）单击【创建工序】，打开【创建工序】对话框，【类型】选择【mill_rotary】，【工序子类型】选择【旋转底面】，【刀具】选择【BALL16】，【几何体】选择【WORKPIECE】，修改【方法】为【MILL_FINISH】，【名称】为【型面精加工 -1】，如图 4-25 所示，单击【确定】。

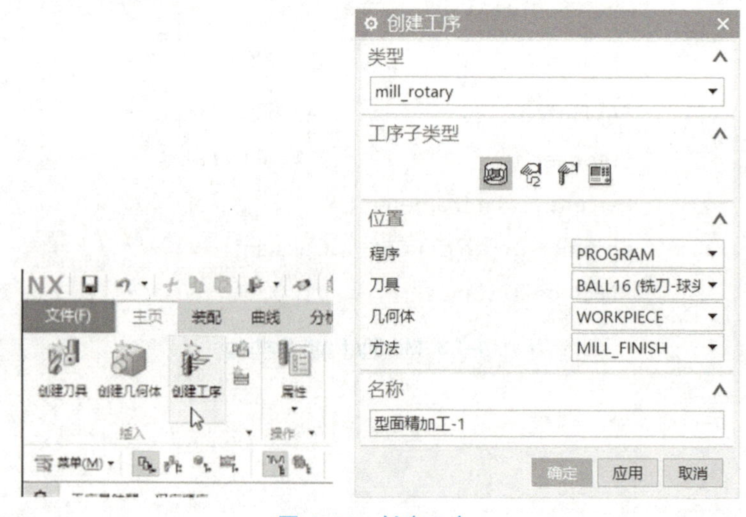

图 4-25 创建工序

（3）Click “Feeds and Speeds”, modify “Spindle Speed” as 1,800 and “Feed Rates” as 1,200 mmpm as shown in Fig. 4-24. Click “OK”.

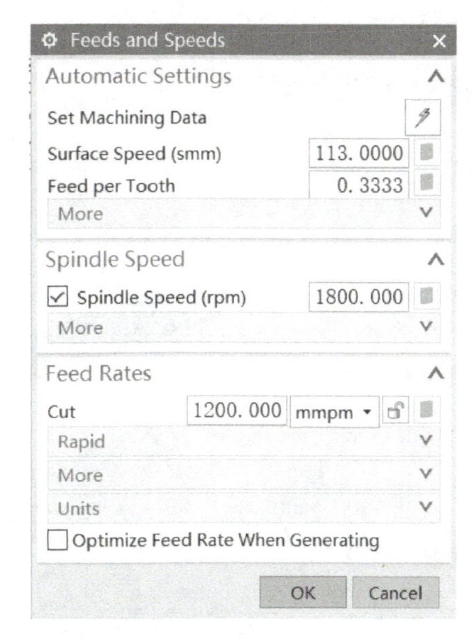

Fig. 4-24 Modify Feeds and Speeds

（4）Click “Generate” to view the tool path and then click “OK”.

2. Finish Machining-1

（1）As shown in Fig. 4-25, click “Create Operation”, and then select “mill_rotary” in pull-down box of “Type”. Select operation subtype as “rotary floor”. Modify name as “Finish_mill-1”, and then click “OK”.

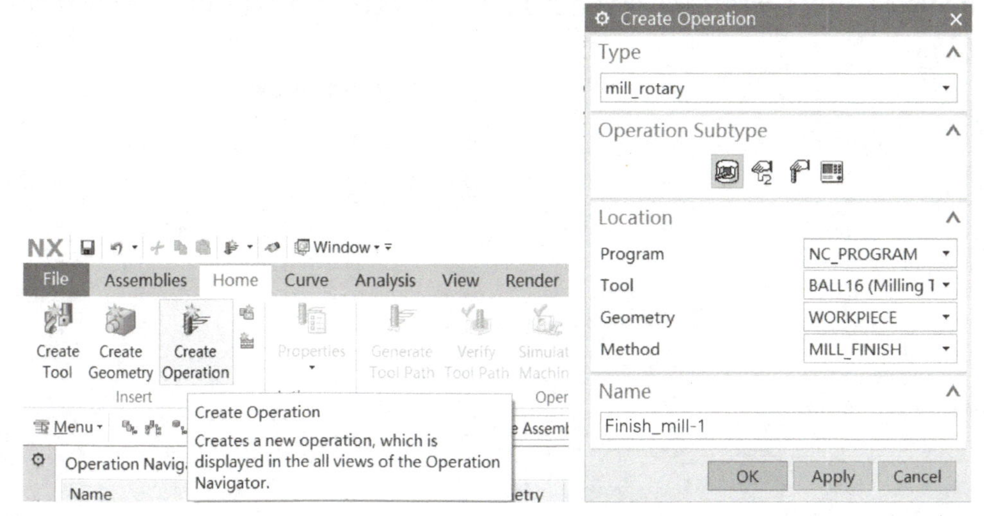

Fig. 4-25 Create Operation

（2）在【旋转底面】对话框中单击【指定底面】，选择图 4–26 上排右图所示的面作为底面，单击【确定】，单击【指定壁】，选择图 4–26 下排右图所示的壁作为壁。

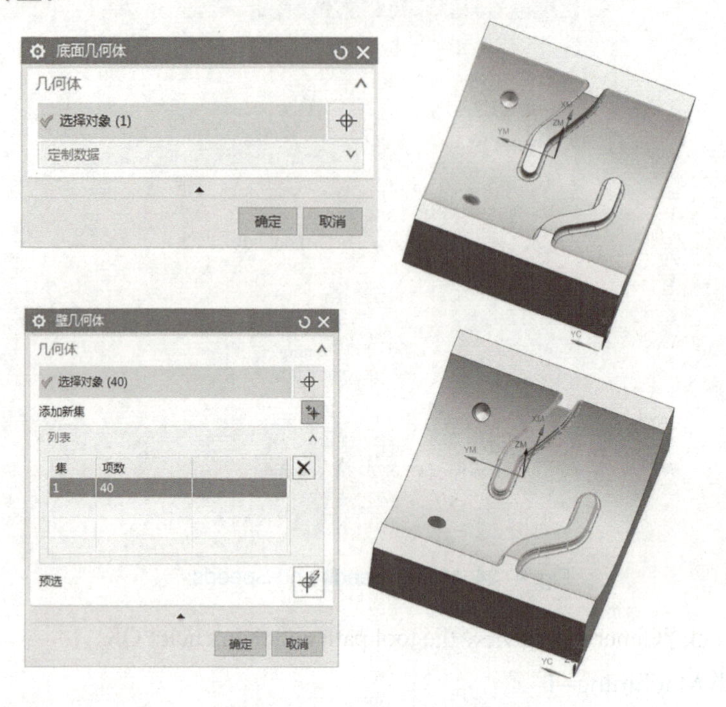

图 4–26 指定底面和壁几何体

（3）单击【驱动方法】，打开【旋转底面精加工驱动方法】对话框，【旋转轴】选择【指定】，【指定矢量】选择【+YC】，【指定点】选择工件的中心点（150，0，165）；设置【步距】为【残余高度】，【最大残余高度】为【0.01】，如图 4–27 所示。

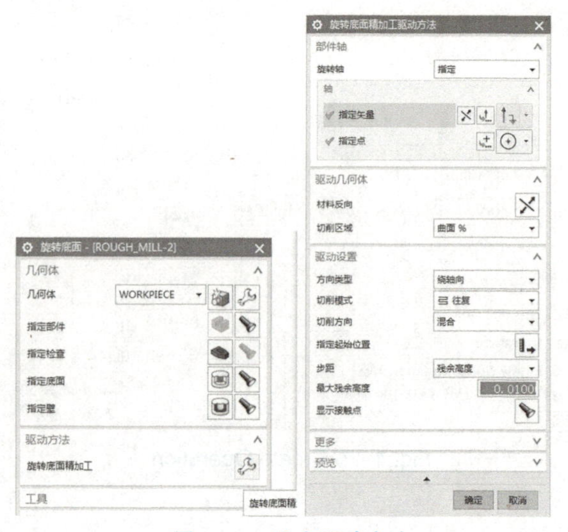

图 4–27 设定驱动方法

(2) Click "Specify Floor" and select the surface as shown in Fig. 4-26 upper right. Click "Specify Walls" and select the surface as shown in Fig. 4-26 lower right.

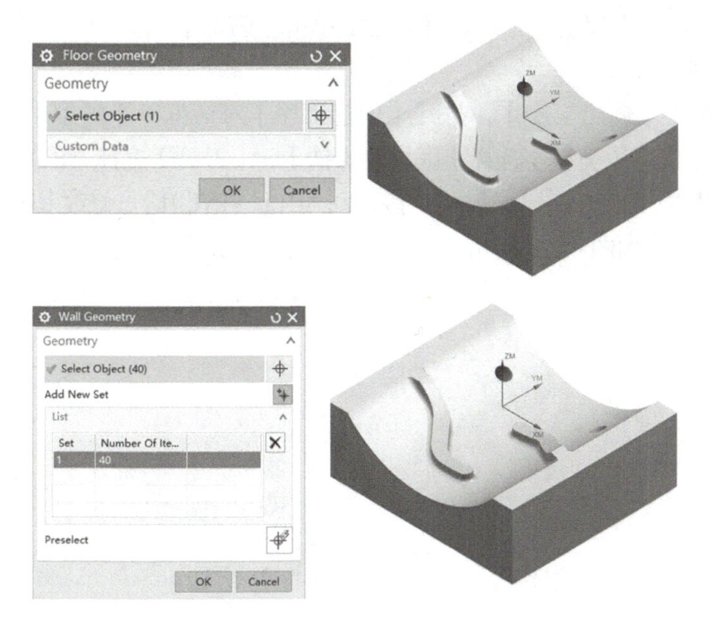

Fig. 4-26 Specify Floor and Walls

(3) Click "Drive Method". In the window of "Rotary Floor Finish Drive Method", select "Specify" as axis of rotation. Select "+YC" as specify vector and select midpoint of part as specify point (150,0,165). Select "Scallop" as stepover and modify "Maximum Scallop Height" as 0.01 as shown in Fig. 4-27.

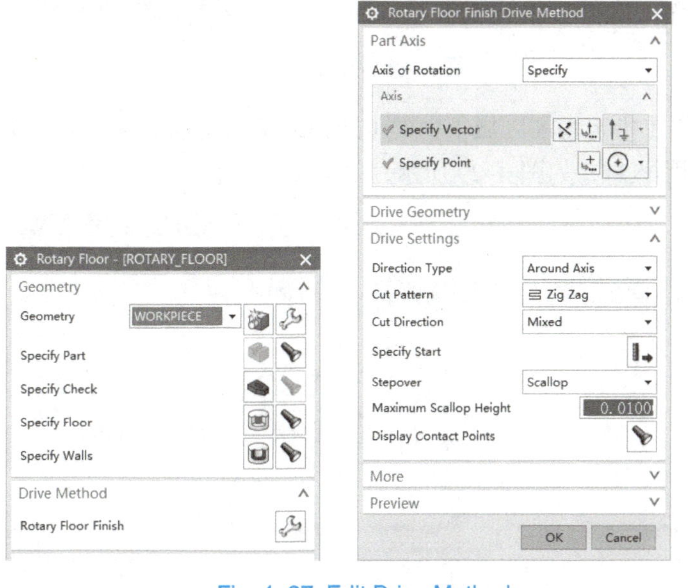

Fig. 4-27 Edit Drive Method

（4）单击【进给率和速度】，打开【进给率和速度】对话框，设置【主轴速度】为【1800】，进给率【切削】为【1200 mmpm】，单击【确定】。

（5）单击【生成】，生成刀具轨迹，单击【确定】。

3. 型面精加工 -2

（1）单击【创建工序】，打开【创建工序】对话框，【类型】选择【mill_contour】，【工序子类型】选择【区域轮廓铣】，【刀具】选择【BALL8】，【几何体】选择【WORKPIECE】，【方法】选择【MILL_FINISH】，修改【名称】为【型面精加工 -2】，如图 4-28 所示，单击【确定】。

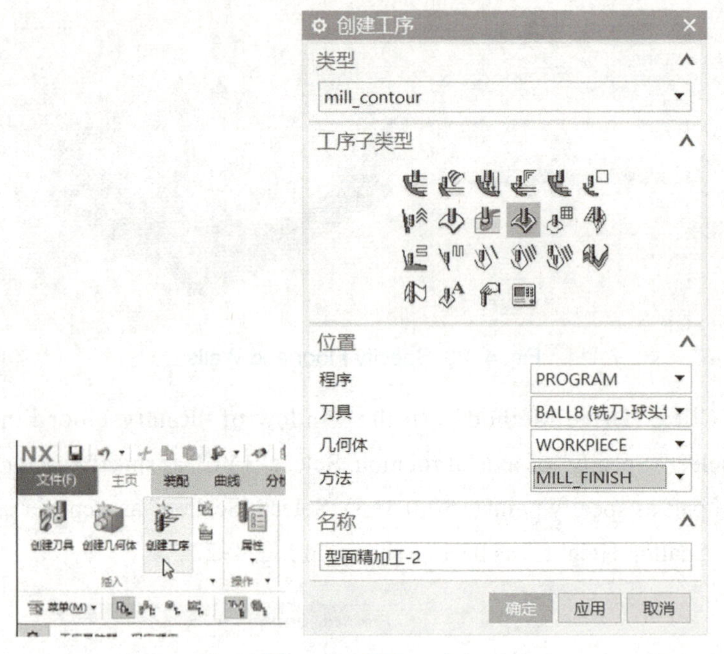

图 4-28 创建工序

（2）单击【指定切削区域】中的【选择或编辑切削区域几何体】，然后在工件上指定切削区域，如图 4-29 所示，单击【确定】。

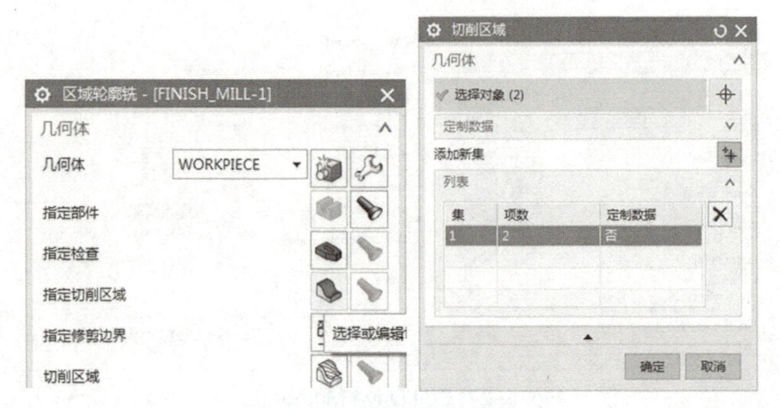

图 4-29 指定切削区域

(4) Click “Feeds and Speeds”, modify “Spindle Speed” as 1,800 and “Feed Rates” as 1,200 mmpm. Click “OK”.

(5) Click “Generate” to view the tool path and then click “OK”.

3. Finish Machining–2

(1) As shown in Fig. 4-28, click “Create Operation”, and then select “mill_contour” in pull-down box of “Type”. Select operation subtype as “contour area”, tool as “BALL8”, geometry as “WORKPIECE” and method as “MILL_FINISH”. Modify name as “Finish_mill-2”, and then click “OK”.

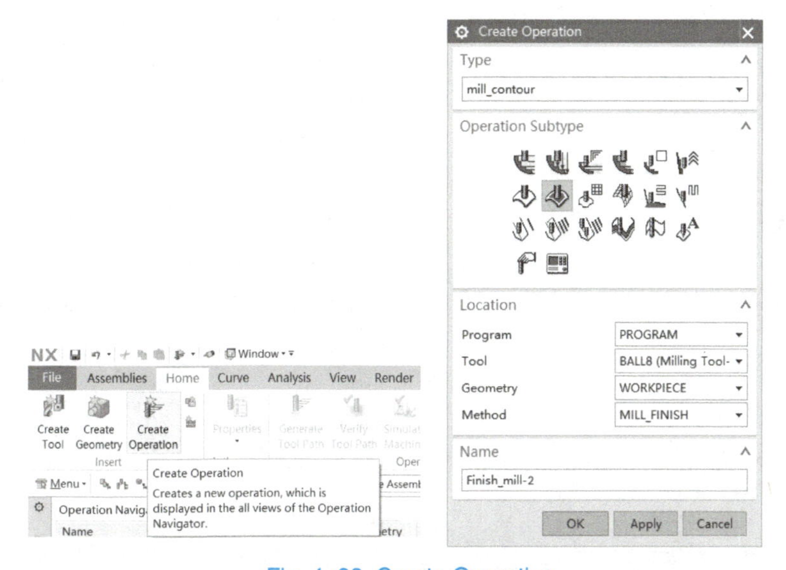

Fig. 4–28 Create Operation

(2) Click “Specify Cut Area” and select part as shown in Fig. 4-29. Click “OK”.

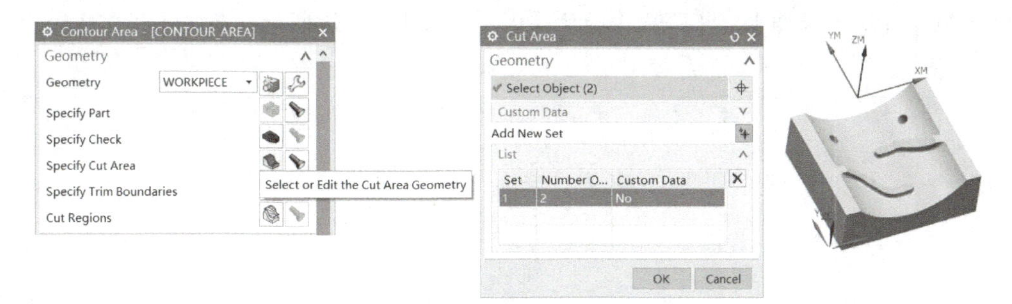

Fig. 4–29 Specify Cut Area

（3）在【区域轮廓铣】中单击【驱动方法】，设置【步距】为【残余高度】，【最大残余高度】为【0.01】，如图 4-30 所示。

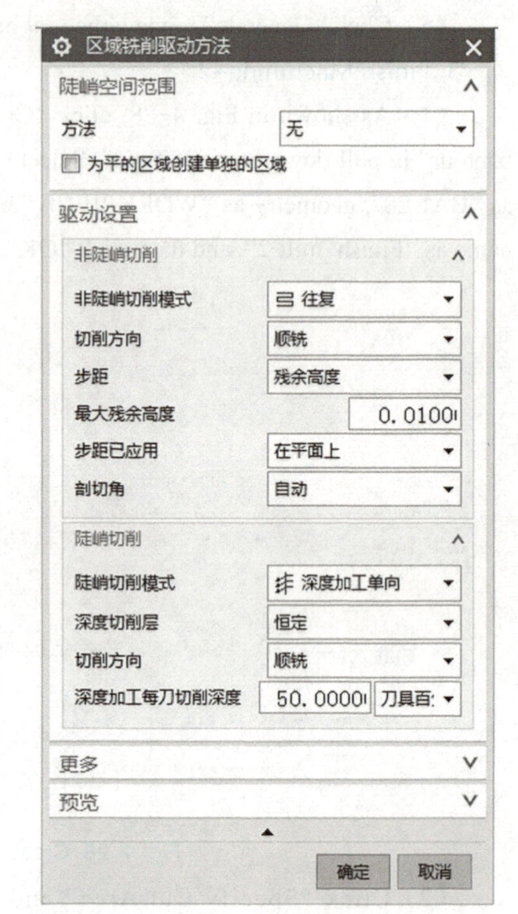

图 4-30　设定驱动参数

（4）单击【切削参数】，打开【切削参数】对话框，在【策略】选项卡中，勾选【在边上延伸】，【距离】改为【0.5】，如图 4-31 所示，单击【确定】。

（3）Click “Edit” to set parameters of drive method. Select “Scallop” as stepover and modify “Maximum Scallop Height” as 0.01 as shown in Fig. 4-30.

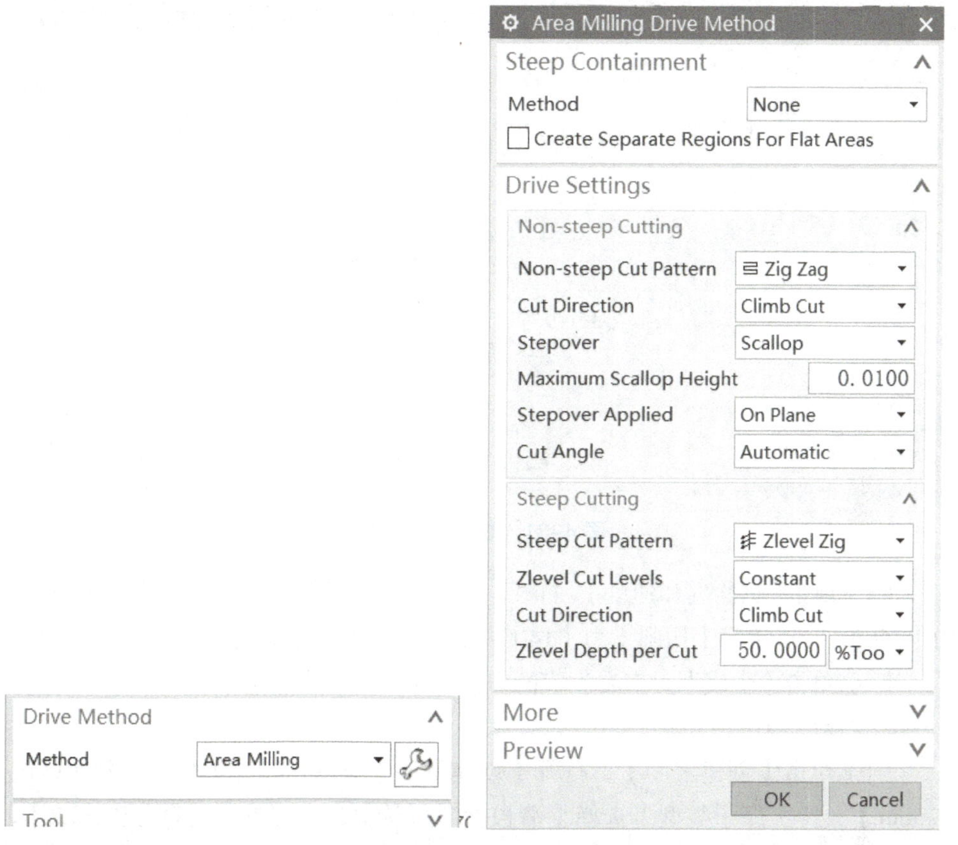

Fig. 4-30 Modify Drive Method Parameters

（4）As shown in Fig. 4-31, click “Cutting Parameters” button, in “Strategy” tab, tick the option “Extend at Edges” and modify distance as 0.5. Click “OK”.

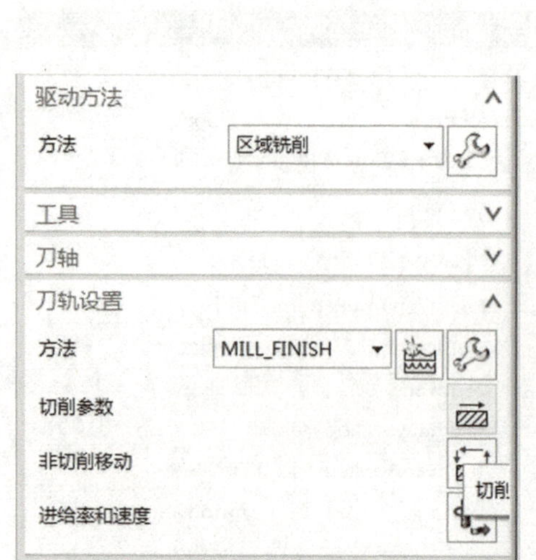

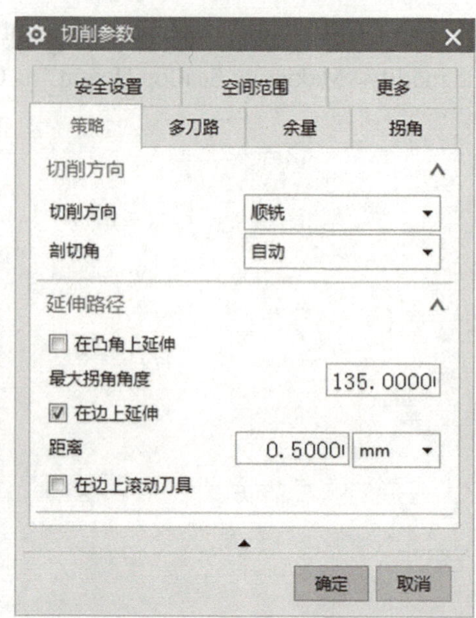

图 4–31　更改切削参数

（5）单击【进给率和速度】，打开【进给率和速度】对话框，设置【主轴速度】为【2600】，进给率【切削】为【1400 mmpm】，单击【确定】。

（6）单击【生成】，查看生成的刀具轨迹，单击【确定】。

4. 型面精加工 –3

（1）单击【创建工序】，打开【创建工序】对话框，【类型】选择【mill_contour】，【工序子类型】选择【清根参考刀具】，【刀具】选择【BALL8】，【几何体】选择【WORKPIECE】，【方法】选择【MILL_FINISH】，修改【名称】为【型面精加工 -3】，如图 4–32 所示，单击【确定】。

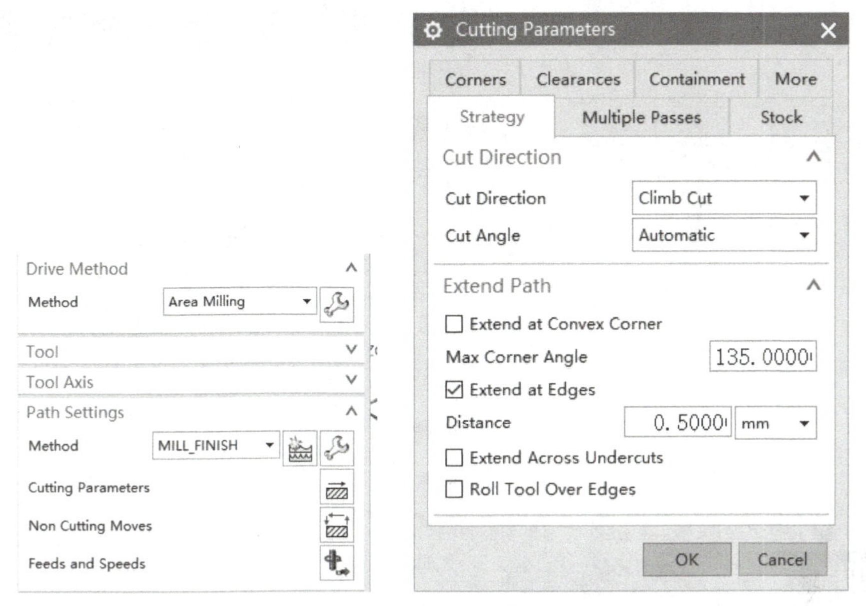

Fig. 4–31 Modify Cutting Parameters

(5) Click "Feeds and Speeds", modify "Spindle Speed" as 2,600 and "Feed Rates" as 1,400 mmpm. Click "OK".

(6) Click "Generate" to view the tool path and then click "OK".

4. Finishing Machining–3

(1) As shown in Fig. 4-32, click "Create Operation", and then select "mill_contour" in pull-down box of "Type" and "Flowcut Reference Tool" in "Operation Subtype". Select tool as "BALL8", geometry as "WORKPIECE" and method as "MILL_FINISH". Modify name as " Finish_mill-3", and then click "OK".

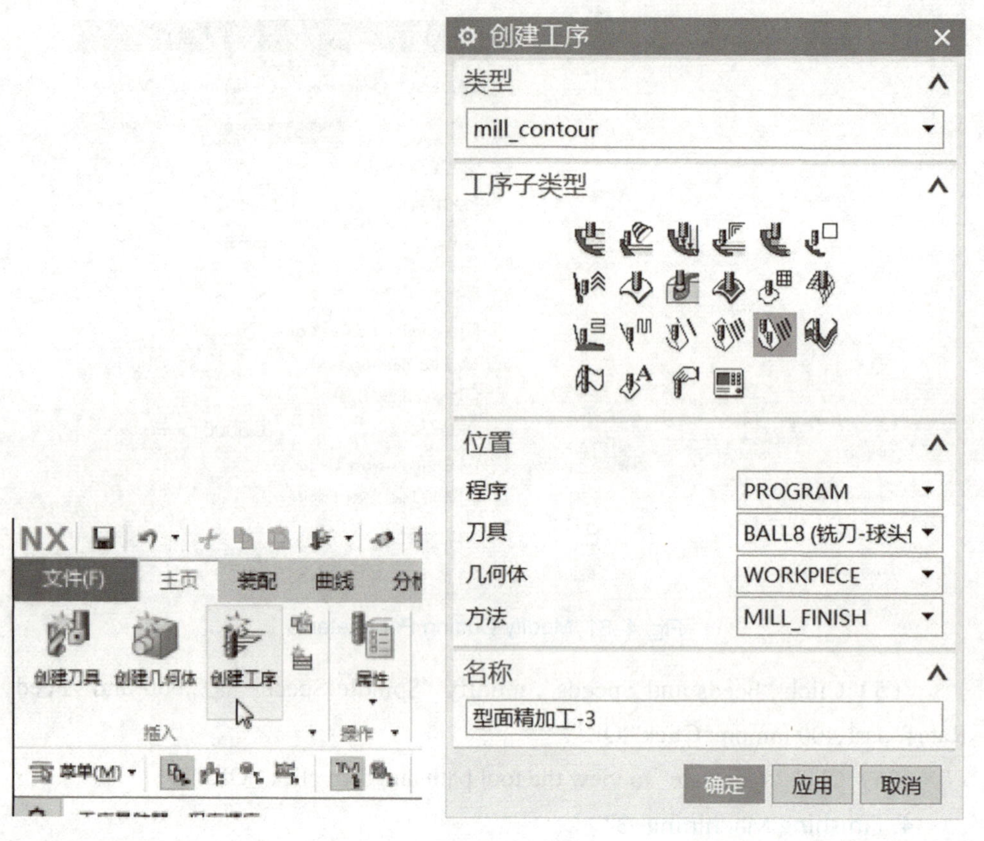

图 4-32 创建工序

（2）单击【驱动方法】，打开【清根驱动方法】对话框，【非陡峭切削】下的【顺序】选择【先陡】，【陡峭切削】下的【陡峭切削模式】选择【同非陡峭】，【参考刀具】选择【BALL16】，【重叠距离】改为【0.3】，单击【确定】，如图 4-33 所示。

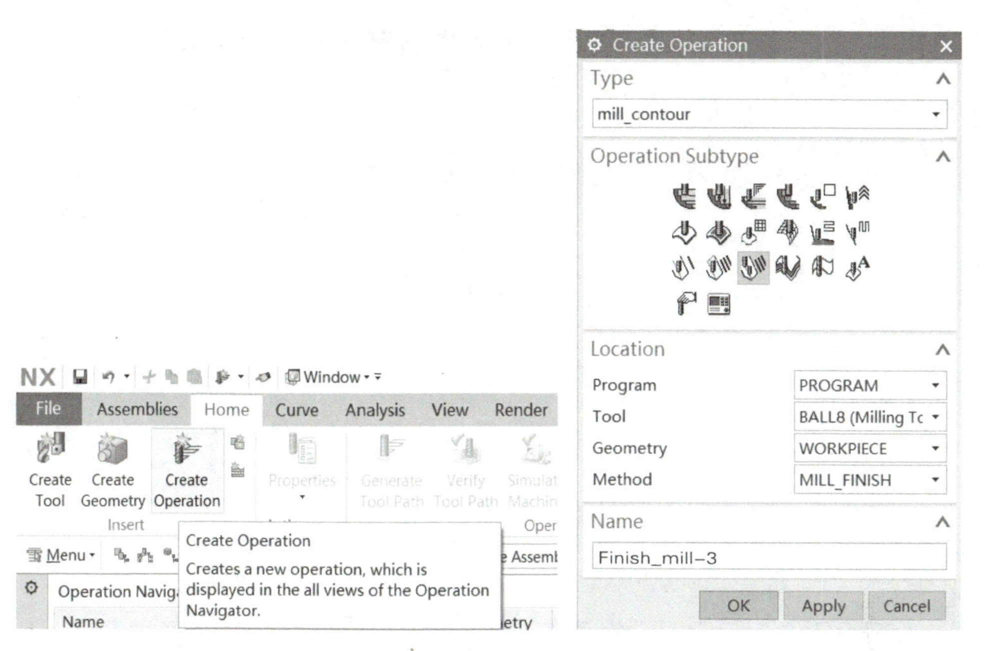

Fig. 4-32 Create Operation

(2) Click "Edit" to set parameters of drive method. Select "Steep First" as sequencing, select "Same as Non-Steep" as steep cut pattern. Select "BALL16" as reference tool as shown in Fig. 4-33. Click "OK".

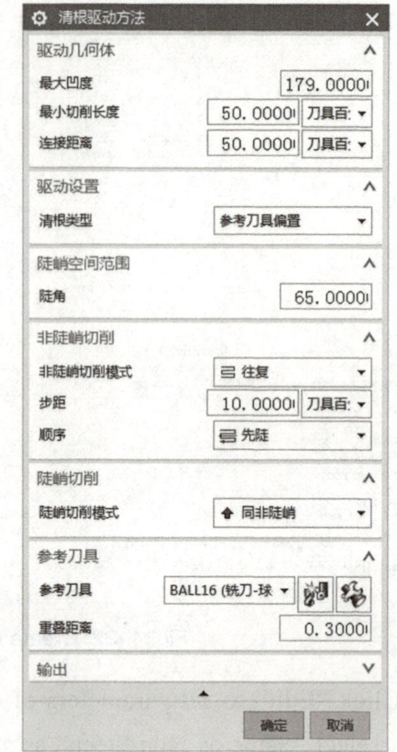

图 4-33 驱动设置

（3）单击【进给率和速度】，打开【进给率和速度】对话框，【主轴速度】设为【2600】，进给率【切削】设为【1400 mmpm】，如图 4-34 所示，单击【确定】。

（4）单击【生成】，创建刀具轨迹，单击【确定】。

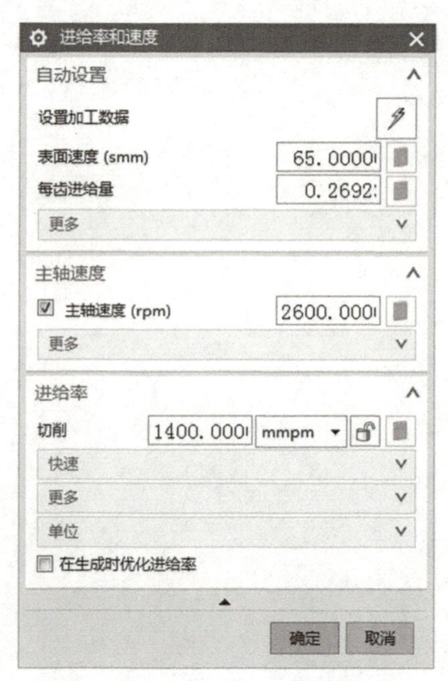

图 4-34 设置进给率和速度

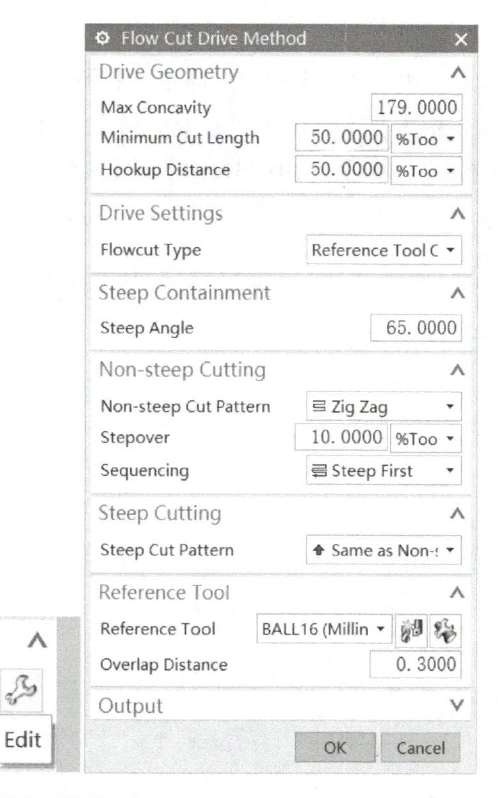

Fig. 4–33 Drive Method

(3) Click "Feeds and Speeds", modify "Spindle Speed" as 2,600 and "Feed Rates" as 1,400 mmpm as shown in Fig. 4-34. Click "OK".

(4) Click "Generate" to view the tool path and then click "OK".

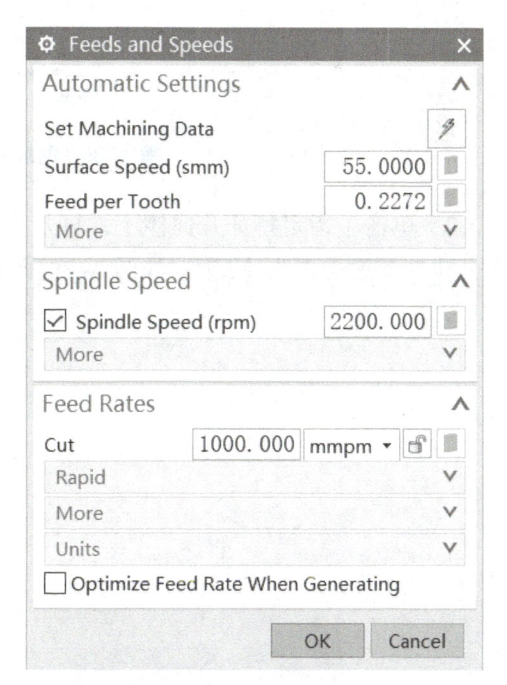

Fig. 4–34 Modify Feeds and Speeds

5. 中心孔

（1）单击【创建工序】，打开【创建工序】对话框，【类型】选择【hole_making】，【工序子类型】选择【DRILLING】，【刀具】选择【ZXZ12】，【几何体】选择【WORKPIECE】，【方法】选择【DRILL_METHOD】，修改【名称】为【中心孔】，如图 4–35 所示，单击【确定】。

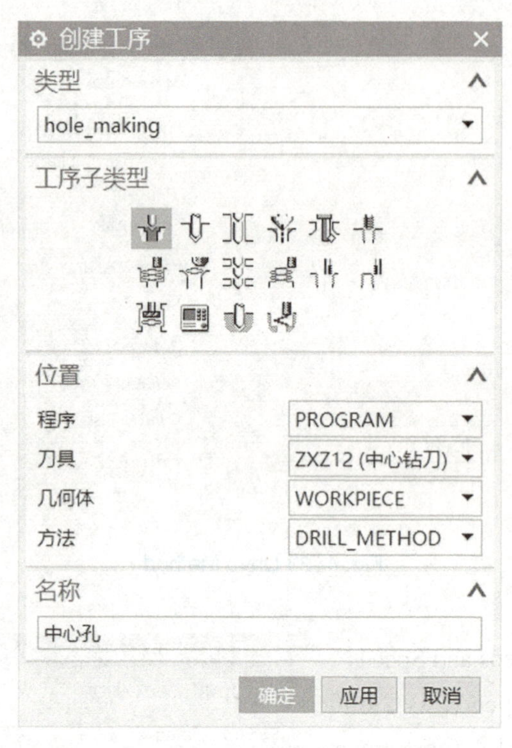

图 4–35 创建工序

（2）单击【指定特征几何体】，打开【特征几何体】对话框，选择模型上的 2 个孔，深度设置为【3】，如图 4–36 所示，单击【确定】。

5. Drill Center Hole

（1）As shown in Fig. 4-35, click “Create Operation”, and then select “hole_making” in pull-down box of “Type”. Select operation subtype as “drilling”, tool as “ZXZ12”, geometry as “ WORKPIECE ” and method as “ DRILL_METHOD ”. Modify name as “ Drill-1”, and then click “OK”.

Fig. 4-35　Create Operation

（2）Click “Specify Feature Geometry”. In the “Feature Geometry” dialog box, select two holes on the model, and set the depth as 3 as shown in Fig. 4-36. Click “OK”.

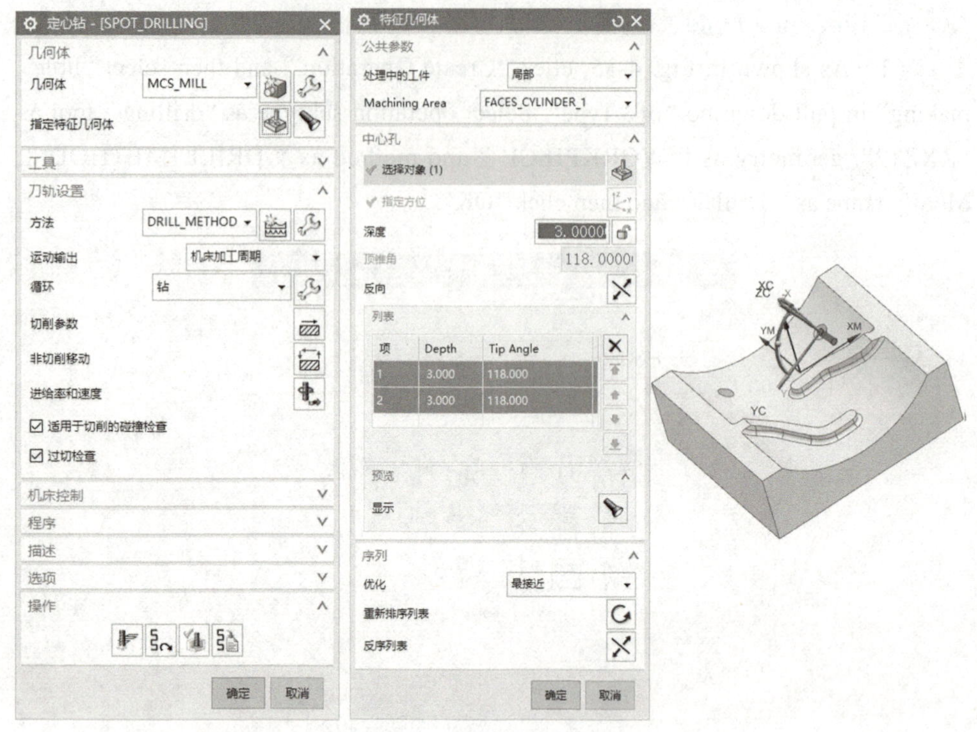

图 4-36 指定特征几何体

（3）单击【进给率和速度】，打开【进给率和速度】对话框，设置【主轴速度】为【800】，进给率【切削】为【100 mmpm】，如图 4-37 所示，单击【确定】。

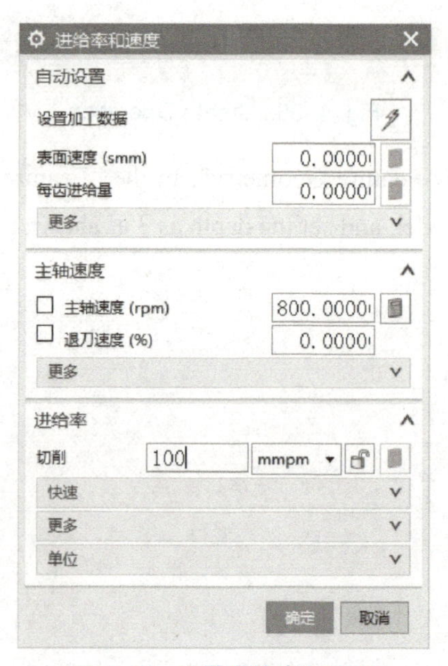

图 4-37 设置进给率和速度

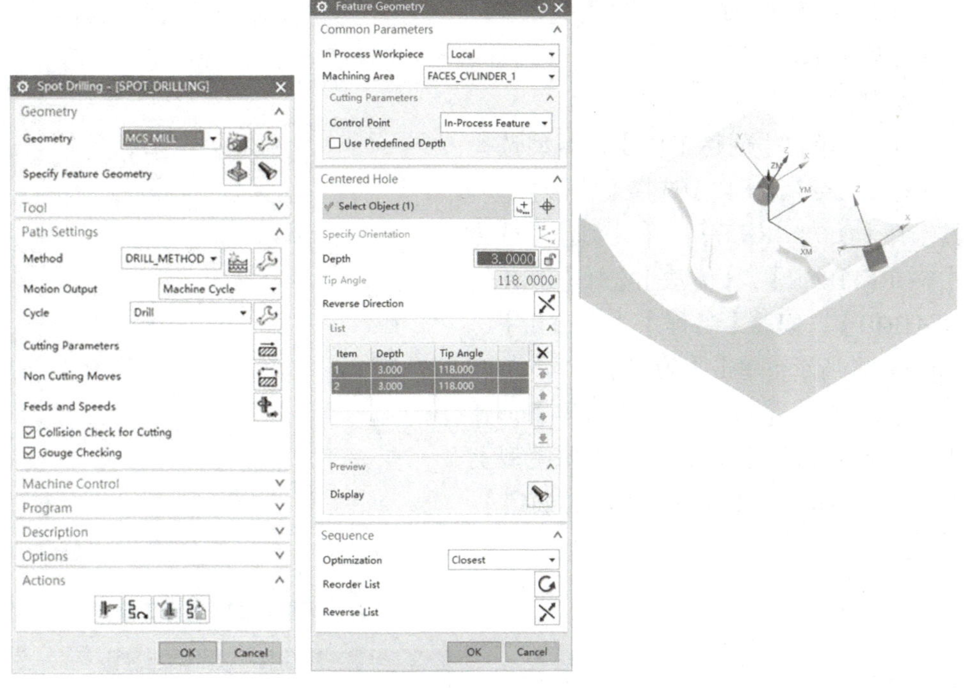

Fig. 4-36　Specify Feature Geometry

（3）Click "Feeds and Speeds". Set "Spindle Speed" as 800, "Feed Rates" as 100 mmpm as shown in Fig. 4-37, and click "OK".

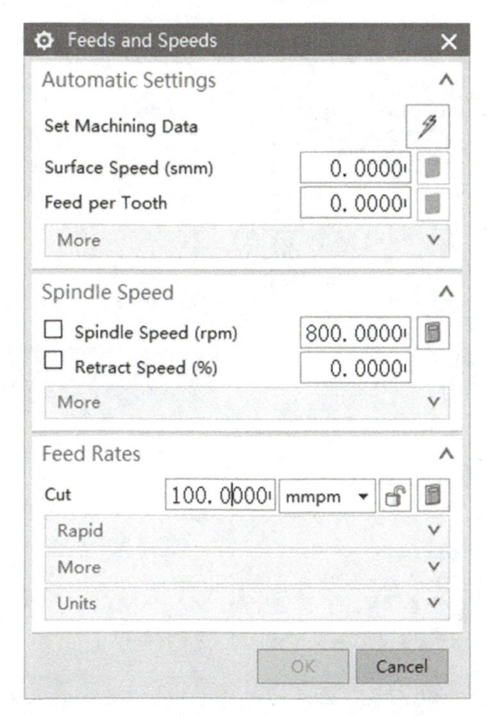

Fig. 4-37　Specify Feeds and Speeds

（4）单击【生成】，创建刀具轨迹，单击【确定】。

6. 钻孔

（1）单击【创建工序】，类型选择【hole_making】，子类型选择【钻孔】，刀具选择【D17.8DRILL】，【几何体】选择【WORKPIECE】，【方法】选择【DRILL_METHOD】，修改【名称】为【钻孔】，如图 4-38 所示，单击【确定】。

（2）单击【指定特征几何体】，弹出【特征几何体】对话框，选择模型上的 2 个孔，如图 4-39 所示，单击【确定】。

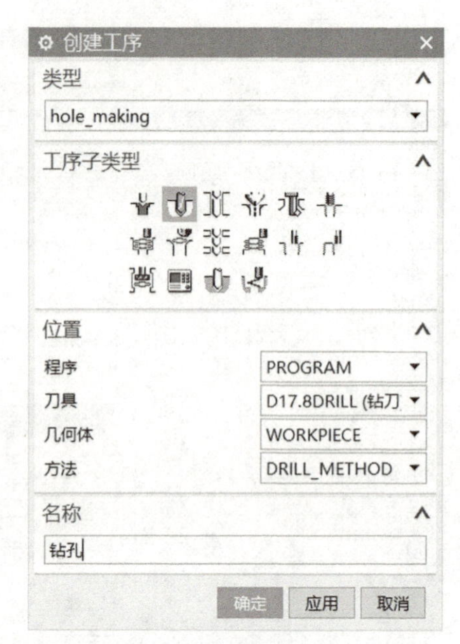

图 4-38　创建工序

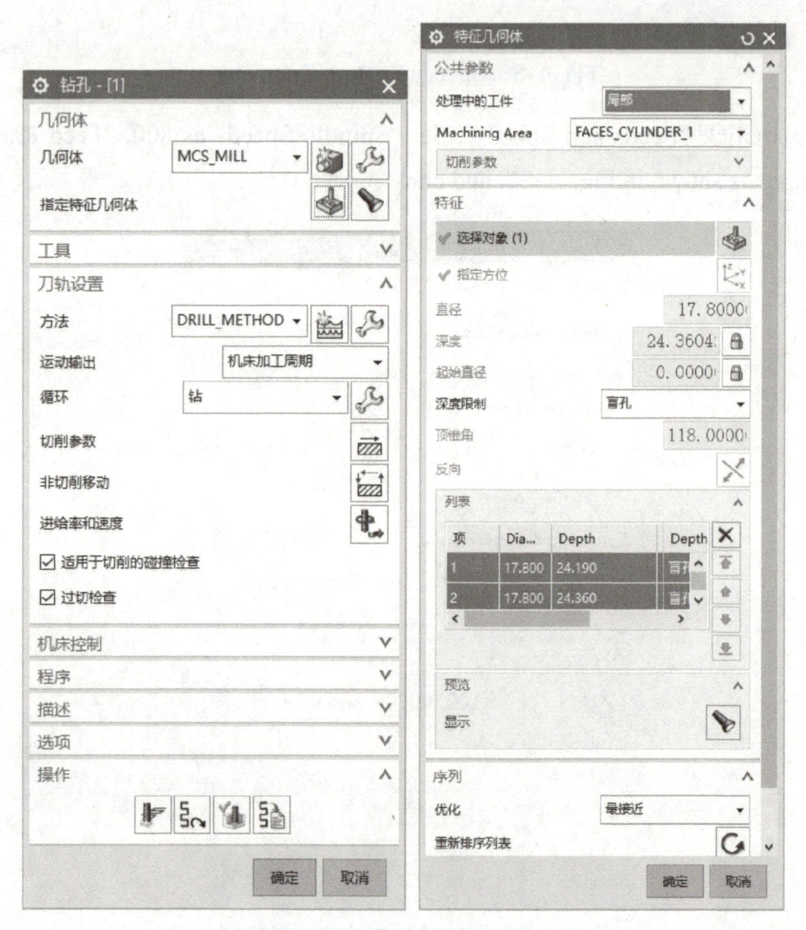

图 4-39　指定特征几何体

(4) Click “Generate Tool Path”, and click “OK”.

6.Drill

(1) Click “Create Operation”, type “hole_making”, subtype “drilling-2”, tool “D17.8_DRILL”, geometry “WORKPIECE”, method “DRILL_METHOD”, name “Drill-2”. As shown in Fig. 4-38, click “OK”.

(2) Click “Specify Feature Geometry”, the “Feature Geometry” dialog box appears, select the two holes on the model as shown in Fig. 4-39, and click “OK”.

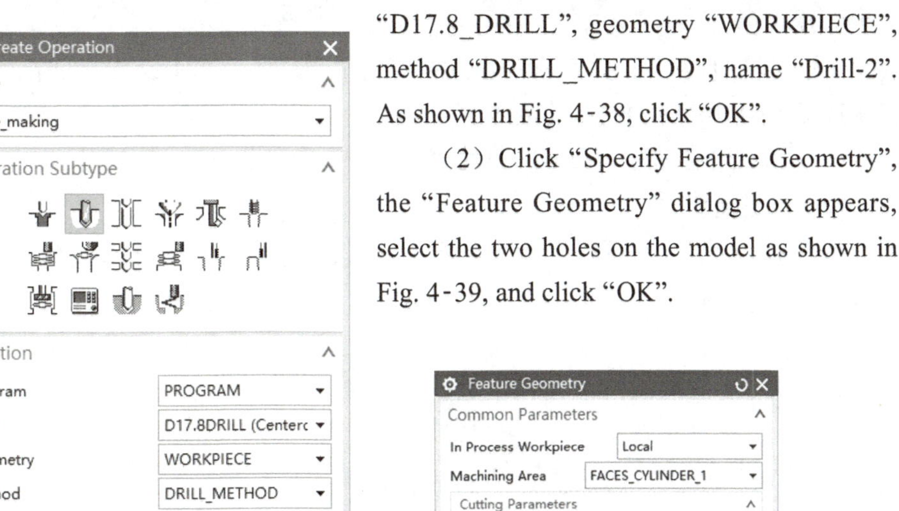

Fig. 4-38 Create Operation

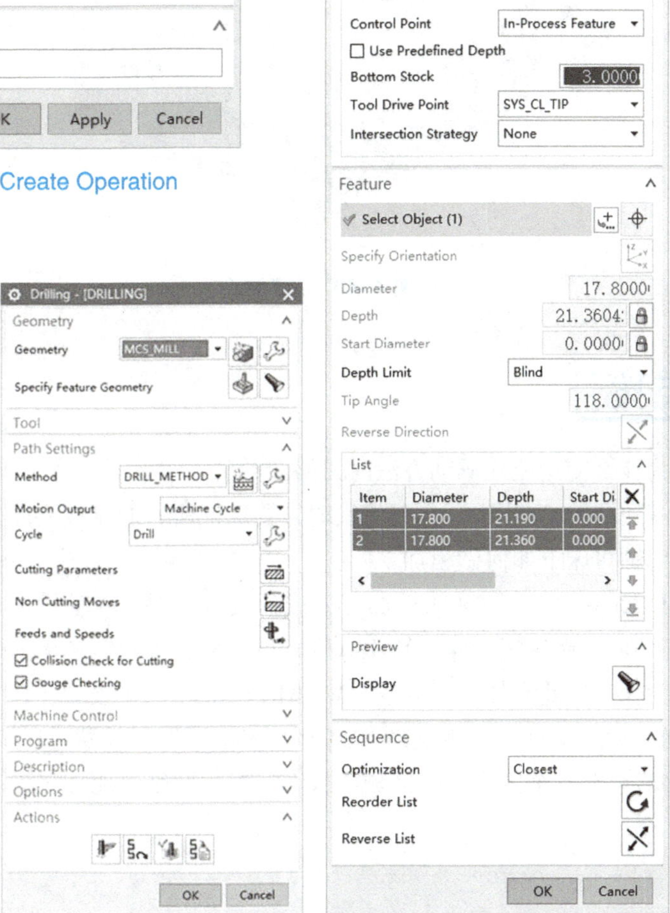

Fig. 4-39 Specify Feature Geometry

（3）单击【进给率和速度】，【主轴速度】为【800 rpm】，进给率【切削】为【100 mmpm】，单击【确定】。

（4）单击【生成】，创建刀具轨迹，单击【确定】。

7. 铰孔

（1）复制【钻孔】工序，修改工序名为【铰孔】，双击【铰孔】工序，【刀具】选择【D18REAMER】，其他参数不变。

（2）单击【生成】，创建刀具轨迹，单击【确定】。

四、仿真加工

单击【NC_PROGRAM】，单击【确认刀轨】，选择【3D 动态】，如图 4-40 所示，单击【播放】。

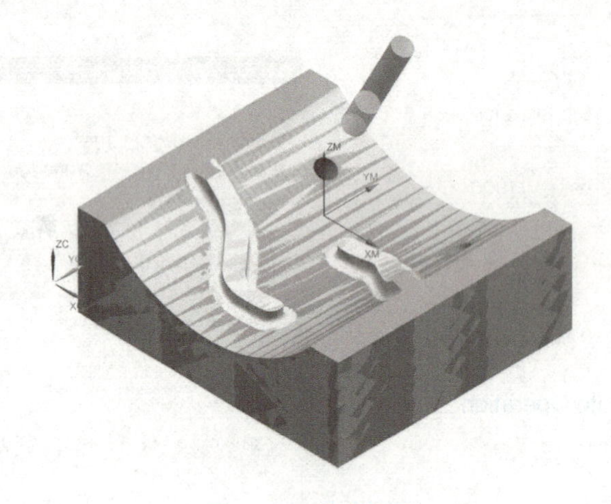

图 4-40 仿真结果

专家点拨

（1）为什么使用流线驱动方法时刀路会混乱？

答：在流线驱动方法中，选择流线时，要保证所有的流线方向一致。

（2）四轴机床上如何做粗加工？

答：可以采用“3+1”轴的粗加工方法，就是先把第 4 轴设为 0 度，用型腔铣做粗加工一面，然后把第 4 轴转 180 度，再用型腔铣做粗加工另外一面。

（3）为什么做此零件是在设置【WORKPIECE】时不设置【PART】呢？

答：做多轴加工时一般在设置【WORKPIECE】时不设置【PART】，因为多轴加工很多时候要一个区域一个区域，或者一组面一组面地加工。

（3）Click “Feeds and Speeds”. Set “Spindle Speed” as 800, “Feed Rates” as 100 mmpm, and click “OK”.

（4）Click “Generate ” to view the tool path, and click “OK”.

7. Reaming

（1）Copy the “Drill-2” process, modify the process named “Drill-3”, double-click the process “Drill-3”, select the tool “D18REAMER”, and keep other parameters unchanged.

（2）Click “Generate ” to view the tool path, and click “OK”.

Ⅳ. Simulation Machining

Click “NC_PROGRAM”, and then “Verify Tool Path” button. Select “3D Dynamic” and click “Play” button to view the simulation result. The simulation result is shown in Fig. 4-40.

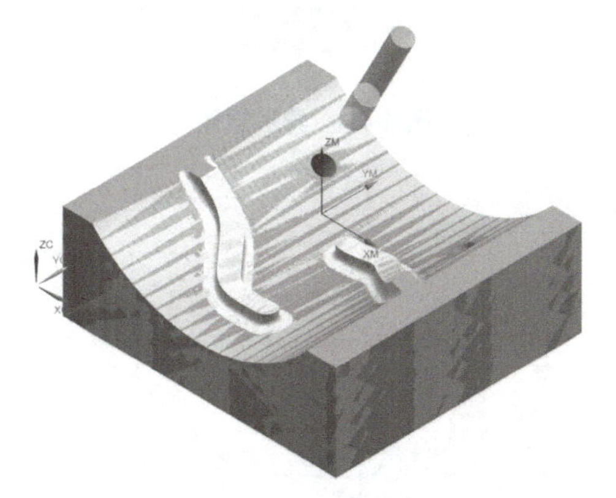

Fig. 4-40 Simulation Result

（1）Q: Why is the knife path confused when using streamline driving method?

A: In the streamline driving method, when selecting the streamline, it is necessary to ensure that all streamline directions are consistent.

（2）Q: How can you do rough machining on 4-axis machine tool?

A: The rough machining method of “3+1”-axis can be adopted. That is, first set the fourth axis to 0 degrees, use cavity milling to rough one side, turn the fourth axis 180 degrees, and then use cavity milling to rough the other side.

（3）Q: Why not specify part when specify workpiece for this model?

A: During multi-axis machining, we generally don’t specify part because multi-axis machining often machining area by area or group of faces by group of faces.

课后训练

根据图 4-41 所示的异形零件的特征，制定合理的工艺路线，设置必要的加工参数，生成刀具轨迹，通过相应的后处理生成数控加工程序，并运用机床加工零件。

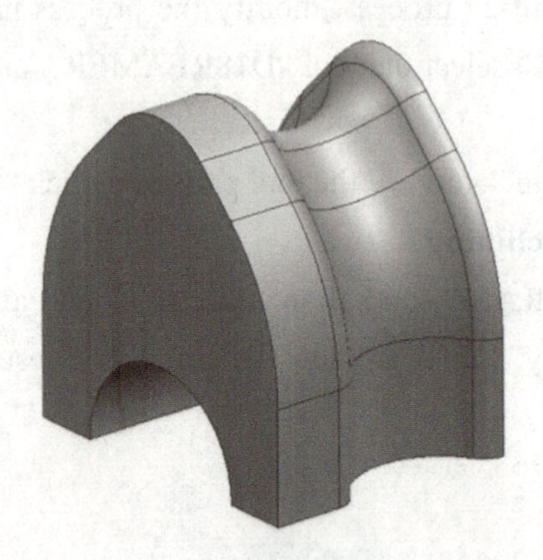

图 4-41　异形零件

Practice

According to the characteristics of disc parts as shown in Fig. 4-41, make a reasonable processing technic, set necessary parameters, generate tool path, generate NC program through corresponding post-processor, and use machine tools to machine parts.

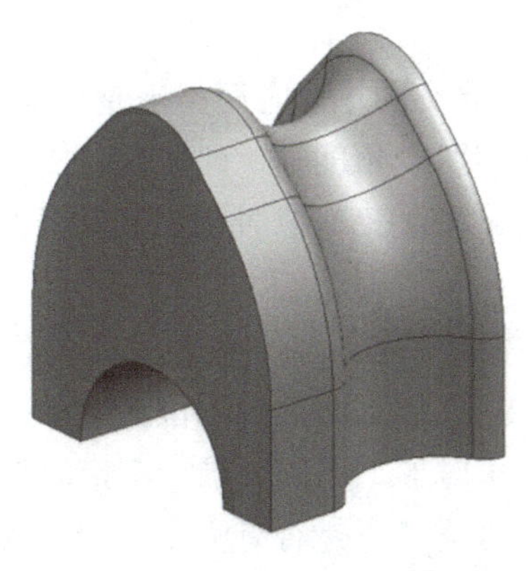

Fig4–41 Special Shaped Parts

模块 3 五轴铣削加工

五轴联动加工技术已经成熟并且应用越来越广泛，从机床制造的角度来看，五轴机床比三轴机床多两个角度轴，即转台或摆头。从五轴加工应用的角度来看，机床的角度轴的配置、CAM 软件的刀具轴线控制、刀具轨迹的后处理是关键技术。

NX CAM 的可变轴曲面轮廓铣为五轴铣削加工提供了很好的解决方案，它常采用驱动面投影方法，生成加工面上的刀位轨迹，这种方法可以使得驱动面和加工面分离，从而降低了对加工面的要求，不论加工面属于单个曲面或者混合曲面，也不论加工面是否连续、是否有突变，NX CAM 都能生成满意的刀位轨迹。

项目 5 壳体的数控编程与仿真加工

学习目标

能力目标：能运用 NX 软件完成壳体的数控编程与仿真加工。

知识目标：掌握五轴加工投影矢量的设置方法；

掌握五轴加工刀轴的设置方法；

掌握五轴加工外形轮廓铣的加工方法。

素质目标：激发学生的学习兴趣，培养团队合作和创新精神。

5-axis Milling

5-axis linkage processing technology has been mature and widely used. From the point of view of machine tool manufacturing. 5-axis machine tool has two more angle axes than 3-axis machine tool, namely turntable or swing head.From the point of view of 5-axis machining application, the key technologies are the configuration of the angle axis of the machine tool, the tool axis control of CAM software and the post-processing of the tool path.

The variable axis surface profile milling of NX CAM provides a good solution for 5-axis milling. It often uses the drive surface projection method to generate the tool path of the machined surface. This method can separate the drive surface from the machined surface, thus reducing the requirement on the machined surface, whether the machined surface is a single surface or a mixed surface. In addition, NX CAM can generate a satisfactory tool position trajectory regardless of whether the machining surface is continuous or not.

Project 5　CNC Programming and Simulation Machining of Shell

Learning Objectives

Capacity Objective: Complete the shell programming and simulation machining with NX software.

Knowledge Objective: Master the projection vector setting of 5-axis machining;

Master the setting method of 5-axis machining tool axis;

Master the contour milling method of 5-axis machining.

Quality objective: Stimulate students' interest in learning and cultivate the spirit of teamwork as well as innovation.

项目导读

壳体是机械结构中常见的一类零件，这类零件的特点是结构比较简单，零件整体外形成块状，零件上一般会有凹腔、壁薄，侧壁有拔模角等，加工时易变形。在编程与加工过程中要特别注意侧壁和底面的加工精度。

任务描述

学生以企业制造部门 MC 数控程序员的身份进入 NX CAM 功能模块，根据壳体的特征，制定合理的工艺路线，创建可变轴曲面轮廓铣加工操作、设置必要的加工参数，生成刀具轨迹，检验刀具轨迹是否正确合理，并对操作过程中存在的问题进行研讨，通过相应的后处理生成数控加工程序，并运用机床加工零件。

项目实施

按照零件加工要求，制定壳体加工工艺；编制壳体加工程序；完成壳体的仿真加工，通过相应的后处理生成数控加工程序，完成零件加工。

一、制定加工工艺

1. 壳体件结构分析

该壳体结构比较简单，主要有凹腔、壁薄，侧壁有拔模角等。

2. 毛坯选用

零件材料由 45# 钢板切割而成，尺寸为 145 mm × 100 mm × 30 mm。零件厚度方向为了保证零件的装夹，余量为 6 mm。

3. 加工工序卡制定

零件选用立式五轴联动机床加工（双摆台摇篮式），平口钳夹持。遵循先粗后精加工原则，粗加工均采用三轴联动加工，精加工采用五轴联动加工。制定的加工工序卡如表 5-1 所示。

Project Guidance

Shell is a kind of common parts in mechanical structure. The characteristics of this kind of parts are that the structure is relatively simple, the overall shape of the parts is block, the parts generally have cavities and thin walls, and there are draft angles in side walls, etc., which are easy to deform during processing. In the process of programming and machining, special attention should be paid to the machining accuracy of side wall and bottom surface.

Task Description

Students operate NX CAM functional modules as MC programmers in enterprise manufacturing department. According to the characteristics of model parts, establish a reasonable processing route, create variable axis surface contour milling operation and set necessary machining parameters, generate tool path and check the generated tool path. Besides, students should discuss the problems occurred in the operation. By selecting the corresponding post-processor to generate NC machining programs, students import them into machine tools to complete the parts processing.

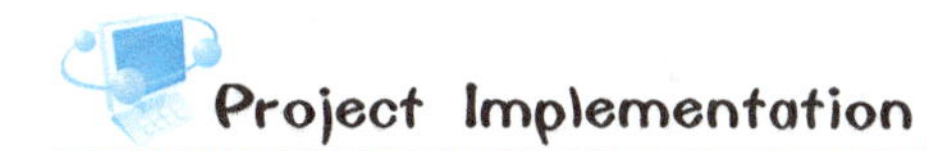

Project Implementation

Firstly, formulate processing technic under processing requirements. Secondly, program for machining model. Finally, complete simulation machining, and then import the NC machining program obtained by post processor to the machine tool to complete machining.

Ⅰ. Formulate Processing Technic

1. Structural analysis of shell parts

The shell structure is relatively simple, mainly including concave cavity and thin wall, and there are draft angles on the side walls.

2. Blank selection

The material of the part is 45# steel plate, and the size is 145 mm × 100 mm × 30 mm. In order to ensure the clamping of parts in the thickness direction of parts, the allowance is 6 mm.

3. Formulation of processing procedure card

The parts are processed by vertical 5-axis linkage machine tool (double swing table cradle type) and clamped by flat pliers. Follow the principle of rough machining before finish machining, 3-axis linkage machining is adopted for rough machining, and

表 5-1 加工工序卡

零件号：15012-3		工序名称：壳体铣削加工		工艺流程卡_工序单
材料：45#	页码：1		工序号：01	版本号：0
夹具：平口钳	工位：MC		数控程序号：	
刀具及参数设置				
加工内容	刀具号	刀具规格	主轴转速/rpm	进给速度/mmpm
型面粗加工	T01	D20R0.4	1800	1200
型面精加工-1	T02	D12R0	2600	1400
型面精加工-2	T02	D12R0	2600	1400
型面精加工-3	T02	D12R0	2600	1400
型面精加工-4	T03	B4	2600	1400
更改号	更改内容	批准	日期	
拟制：　日期：	审核：　日期：	批准：	日期：	

二、加工准备

（1）启动 NX，单击【打开】，打开【打开】对话框，选择【壳体.prt】文件，如图 5-1 所示，单击【OK】，打开零件模型。

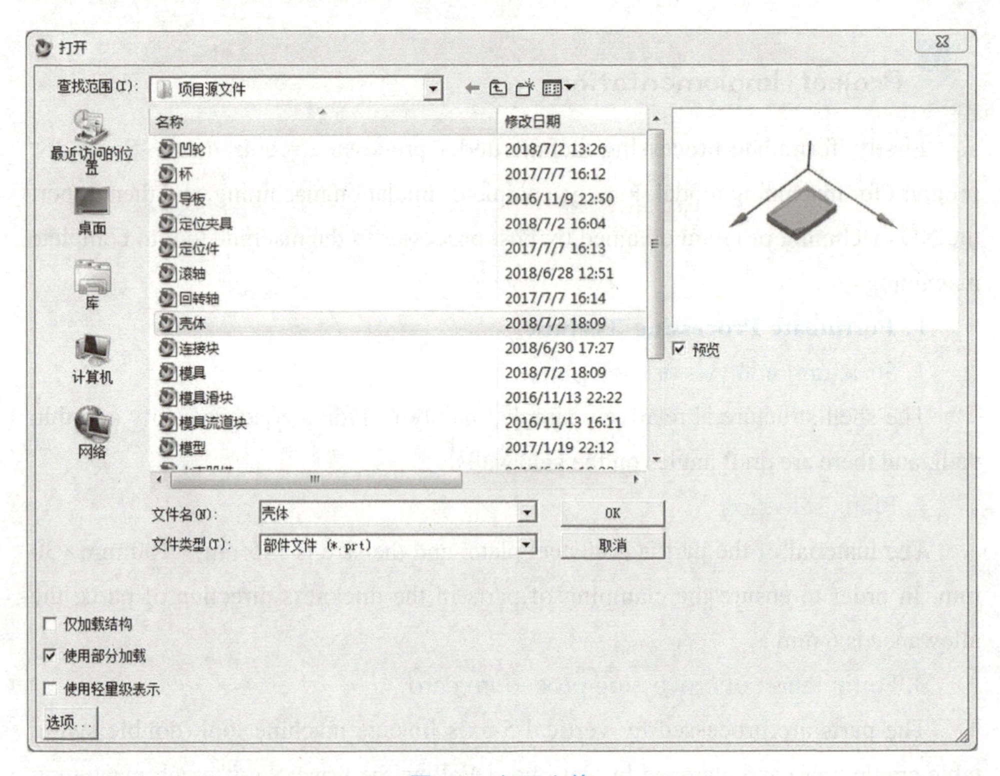

图 5-1 打开文件

5-axis linkage machining is adopted for finish machining. The processing procedure is shown in Table 5-1.

Table5-1　Processing Procedure Card

Part number: 15012-3			Name of process: Milling of shell			Process card - Process sheet	
Material: 45#			Page number: 1			Procedure number: 01	Version number: 0
Fixture: Parallel-jaw vice			Work station: MC			CNC program number:	
Tool and parameter setting							
Tool number	Tool specification		Processing content	Spindle speed /rpm	Feed speed /mmpm		
T01	D20R0.4		Rough machining	1800	1200		
T02	D12R0		Finish machining-1	2600	1400		
T02	D12R0		Finish machining-2	2600	1400		
T02	D12R0		Finish machining-3	2600	1400		
T03	B4		Finish machining-4	2600	1400		
02							
01							
Change number	Change content		Approve	Date			
Draws:	Date:	review:	Date:	Approve:	Date:		

Ⅱ. Preparation for Processing

(1) As shown in Fig. 5-1, start NX, click "OPEN", select model "Shell"(.prt file), and click "OK". Then, the part model can be seen in WCS.

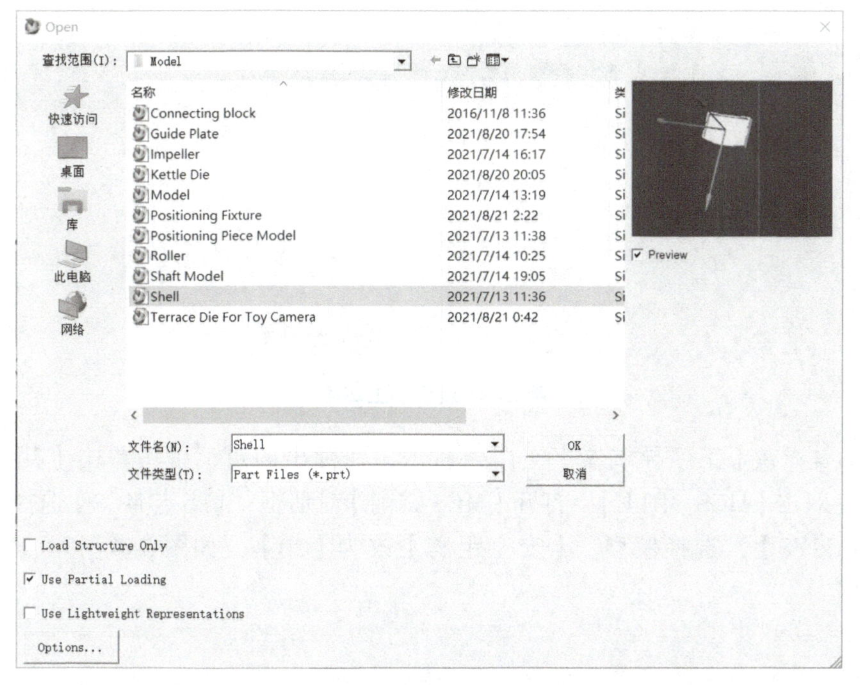

Fig. 5-1　Open the File

（2）选择【应用模块】选项卡，单击【加工】，进入加工环境，如图 5-2 所示。

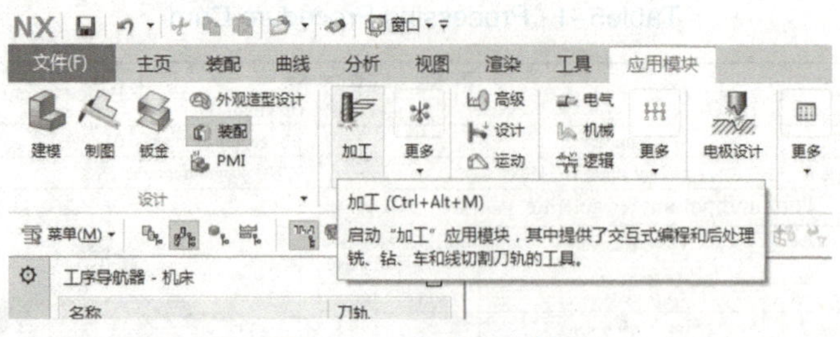

图 5-2 进入加工环境

（3）弹出【加工环境】对话框，在【CAM 会话配置】选项中选择【cam_general】，在【要创建的 CAM 设置】选项中选择【mill_planar】，如图 5-3 所示，单击【确定】。

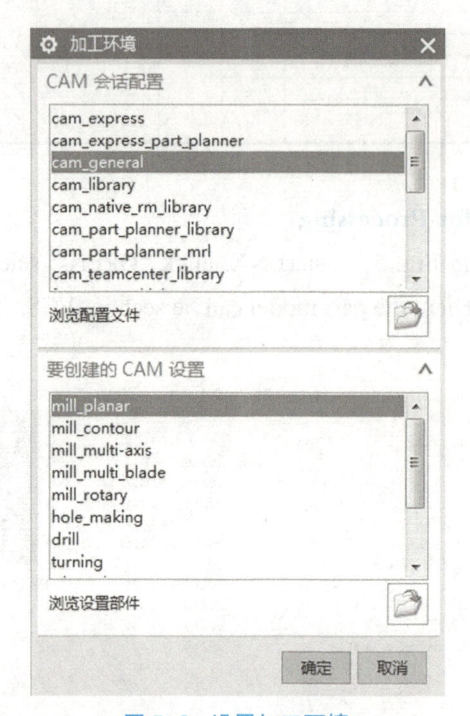

图 5-3 设置加工环境

（4）在【工序导航器】空白处右击，在弹出的对话框中单击【几何视图】，双击【MCS_MILL】，打开【MCS 铣削】对话框，【指定 MCS】选择【对象的 CSYS】，选择模型，【安全距离】改为【20】，如图 5-4 所示，单击【确定】。

（2）Select "Application", and click "Manufacturing". The manufacturing menu can be seen as shown in Fig. 5-2.

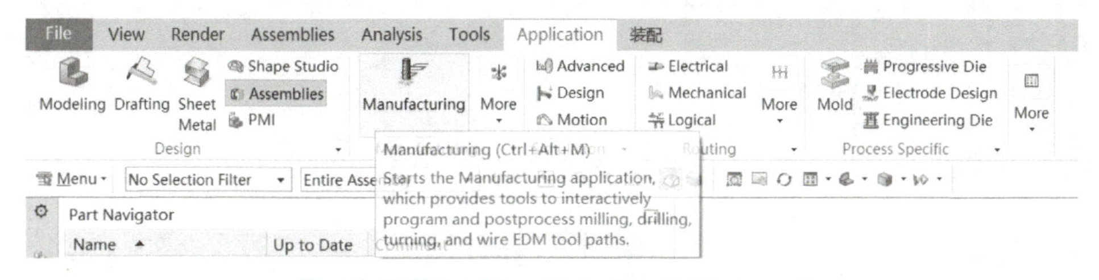

Fig. 5-2 Enter Manufacturing Environment

（3）As shown in Fig. 5-3, in "Machining Environment" window, select "cam_general" in "CAM Session Configuration" tab, and then, select "mill_planar" in "CAM Setup to Create" tab. Finally, click "OK".

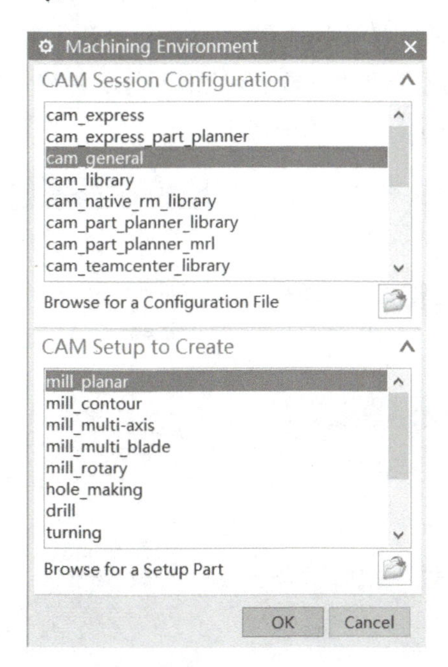

Fig. 5-3 Set Machining Environment

（4）Right click in the blank area of "Process Navigator", and click "Geometry View" in the window. Double click "MCS_MILL", in "MCS Mill" window, click "CSYS Dialog", select the "CSYS of Object" as "Specify MCS" as shown in Fig. 5-4. Modify "Safe Clearance Distance " as 20, and then click "OK" to close the window.

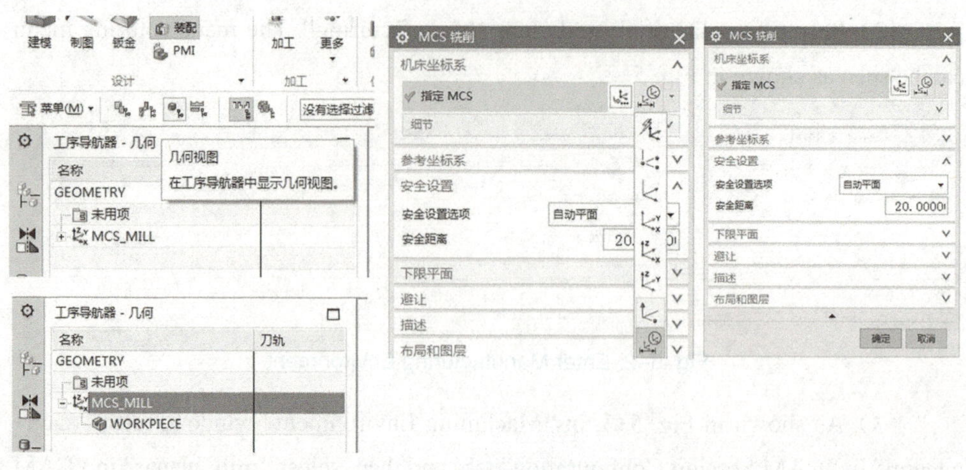

图 5-4　设定工件坐标系

（5）单击【工序导航器】中的【WORKPIECE】，在弹出【工件】对话框中单击【指定毛坯】中的【选择或编辑毛坯几何体】，弹出【毛坯几何体】对话框，选择毛坯，如图 5-5 所示，单击【确定】。

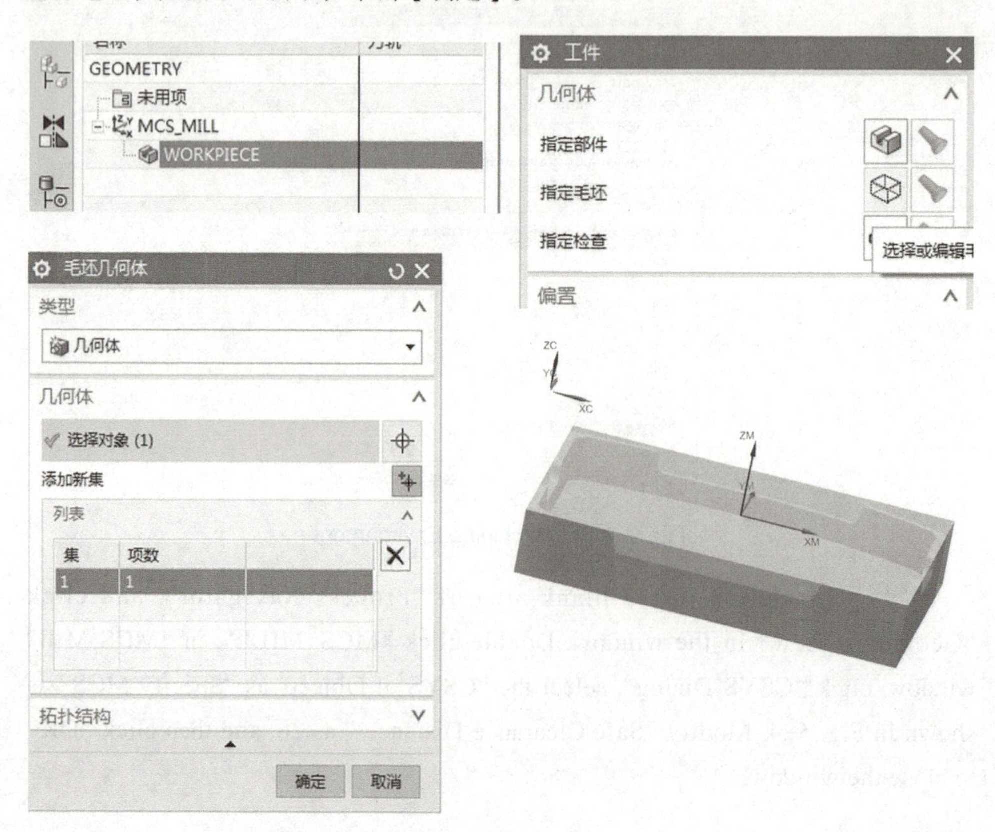

图 5-5　指定毛坯

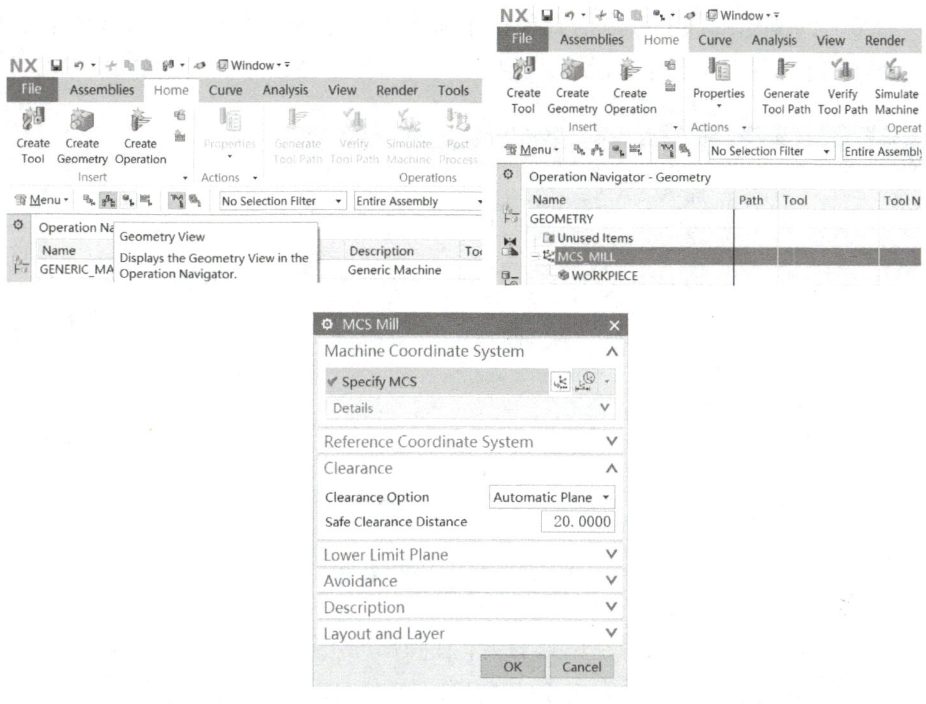

Fig. 5-4 Set MCS

(5) Click "WORKPIECE" and click "Specify Blank" in "Workpiece" window. In "Blank Geometry" window, select the blank as shown in Fig. 5-5. Click "OK".

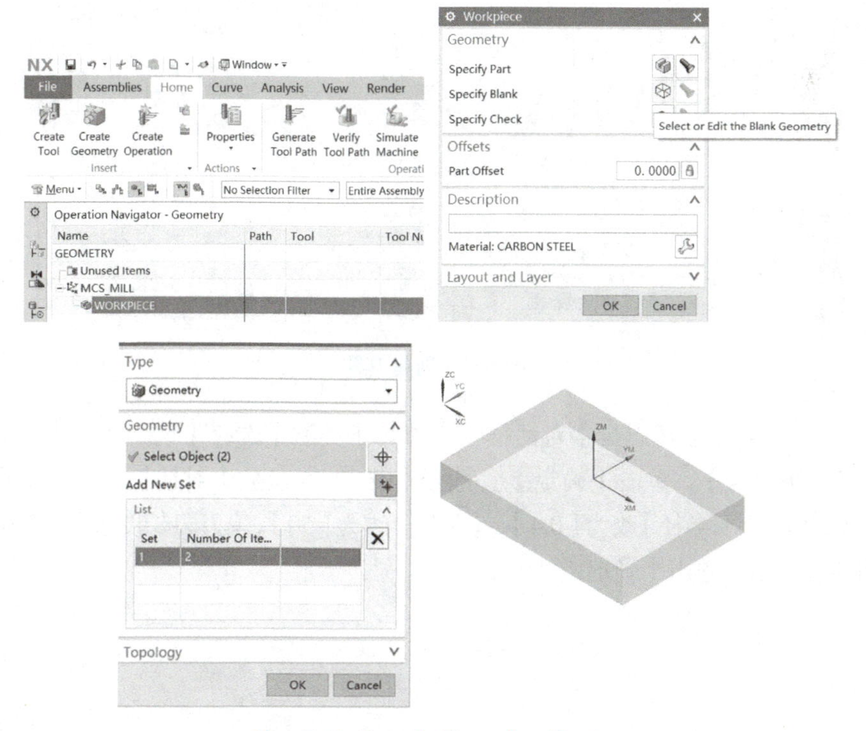

Fig. 5-5 Specify Bounding Block

（6）回到【工件】对话框，单击【指定部件】中的【选择或编辑部件几何体】，弹出【部件几何体】对话框，选择模型，如图 5–6 所示，依次单击【确定】。

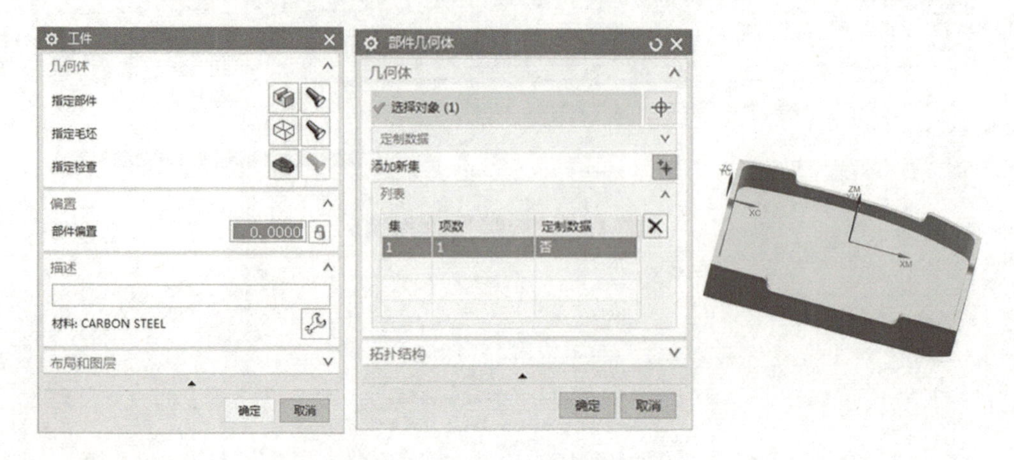

图 5–6　指定部件几何体

（7）在【工序导航器】空白处右击，在弹出的快捷菜单中单击【机床视图】，在【工序导航器 – 机床】中右击【GENERIC_MACHINE】，在弹出的快捷菜单中选择【插入】→【刀具】，如图 5–7 所示。

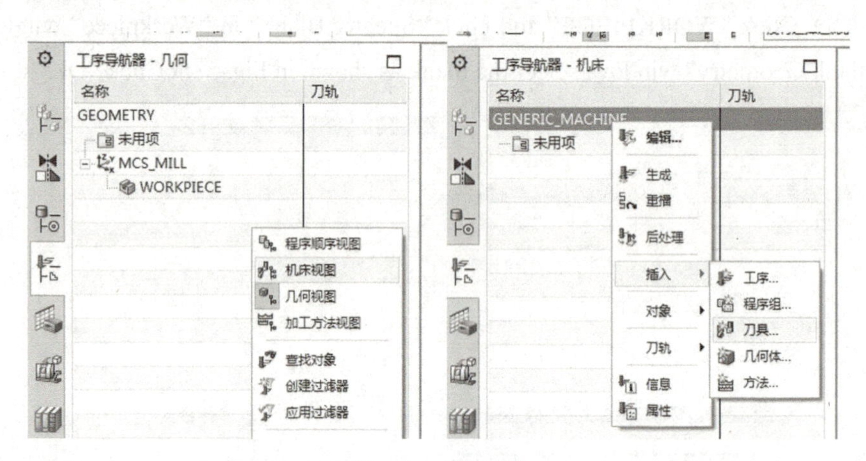

图 5–7　创建刀具

（8）打开【创建刀具】对话框，【刀具子类型】选择【铣刀】，【名称】改为【D20R0.4】，单击【确定】。打开【铣刀 –5 参数】对话框，将【直径】改为【20】，【下半径】改为【0.4】，【长度】改为【50】，【刀刃长度】改为【30】，如图 5–8 所示，单击【确定】。

(6) Back to the window of "Workpiece", click "Specify Part" in "Workpiece" window, and select the solid part as shown in Fig. 5-6. Click "OK" to close the window.

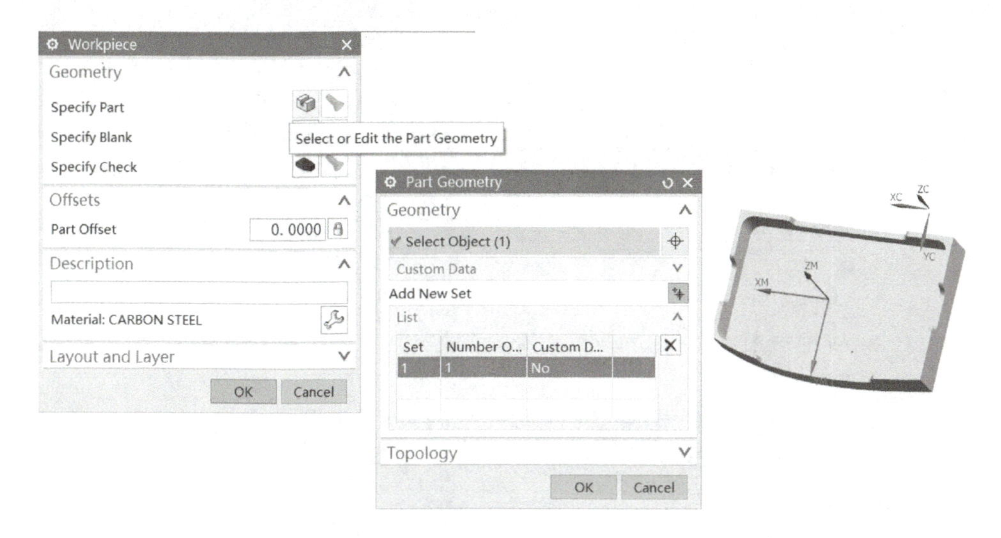

Fig. 5-6 Specify Part

(7) Right click in the blank area of "Process Navigator". Click "Machine Tool View" in the shortcut menu. Right click "GENERIC_MACHINE" in "Process navigator-Machine" and click "Insert"→"Tool" as shown in Fig. 5-7.

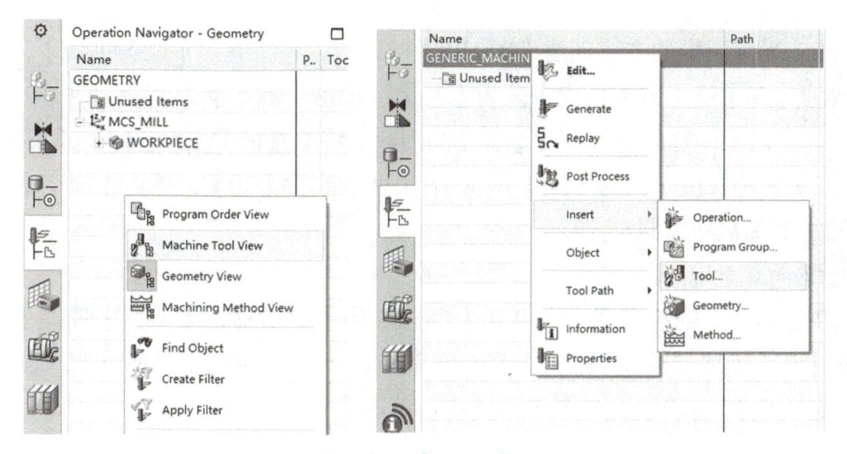

Fig. 5-7 Create Tool

(8) As shown in Fig. 5-8, in "Create Tool" window, select "mill" in "Tool Subtype" and modify "Name" as "D20R0.4". Then click "OK". In "Milling Tool-5 Parameters" window, "Diameter" value is 20, "Lower Radius" value is 0.4, "Length" is 50 and "Flute Length" is 30. Other parameters default, click "OK" to close the window.

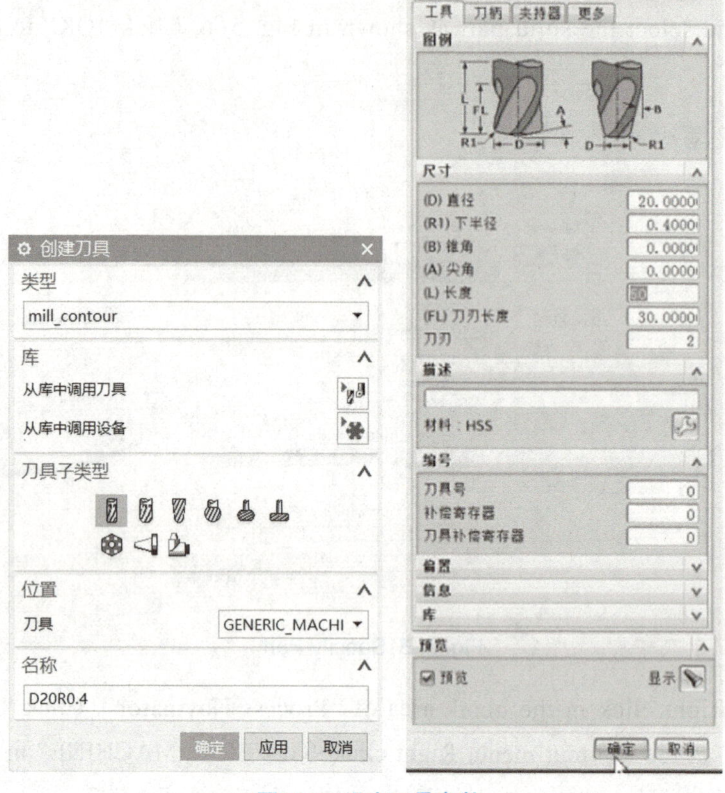

图 5-8 设定刀具参数

（9）采用同样的方法创建【D12R0】的刀具，将【直径】改为【12】，【长度】改为【50】，【刀刃长度】改为【30】，单击【确定】。

（10）采用同样的方法创建【B4】的刀具，将【直径】改为【4】，【下半径】改为【2】，【长度】改为【50】，【刀刃长度】改为【30】，单击【确定】。

三、加工程序编制

1. 型面粗加工

（1）单击【创建工序】，打开【创建工序】对话框，【类型】选择【mill_contour】，【工序子类型】选择【型腔铣】，【刀具】选择【D20R0.4】，【方法】选择【MILL_ROUGH】，如图 5-9 所示，单击【确定】。

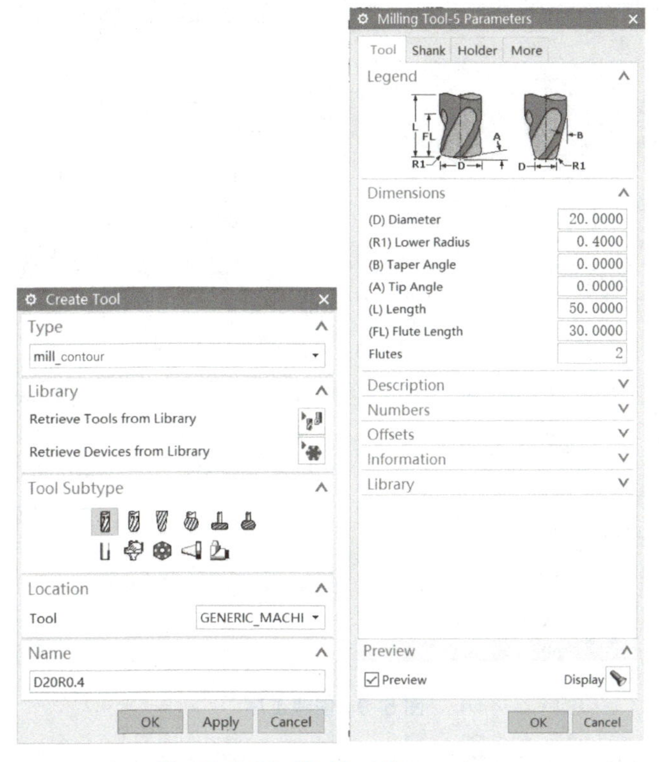

Fig. 5–8 Modify Tool Parameters

(9) The same method is used to create the tool "D12R0". "Diameter" value is 12, "Length" is 50 and "Flute Length" is 30. Other parameters default, click "OK" to close the window.

(10) Use the same method to create the tool "B4". Change "Diameter" to 4, "Lower Radius" to 2, "Length" to 50, "Blade Length" to 30. Click "OK" to close the window.

Ⅲ. Programming

1. Rough Machining

(1) As shown in Fig. 5-9, click "Create Operation", and then select "mill_contour" in pull-down box of "Type". Select operation subtype as "cavity mill", tool as "D20R0.4", method as "MILL_ROUGH". Click "OK".

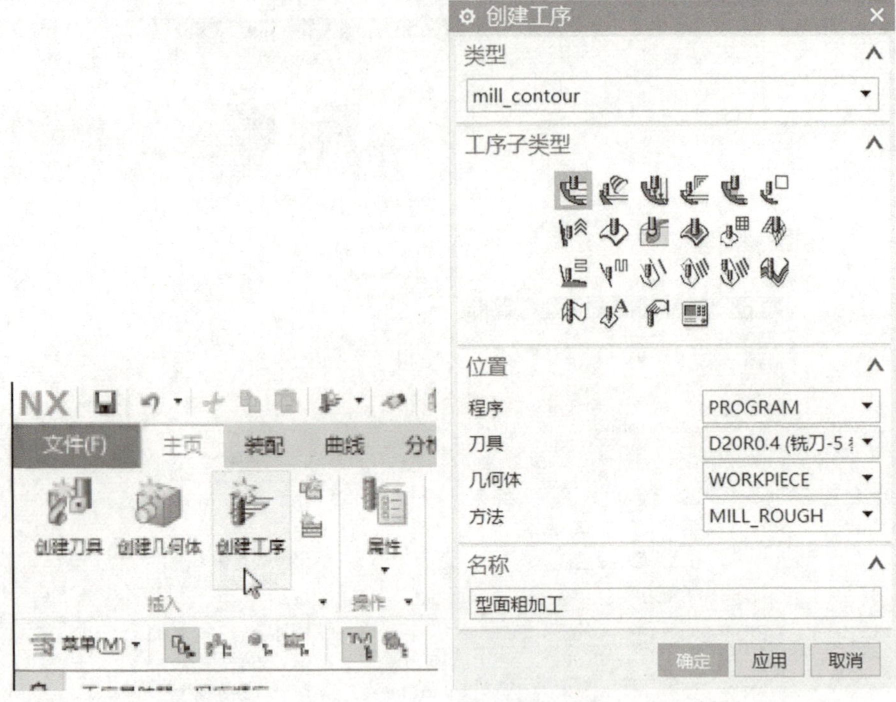

图 5-9 创建工序

（2）【切削模式】选择【跟随周边】，【最大距离】改为【1 mm】，如图 5-10 所示。

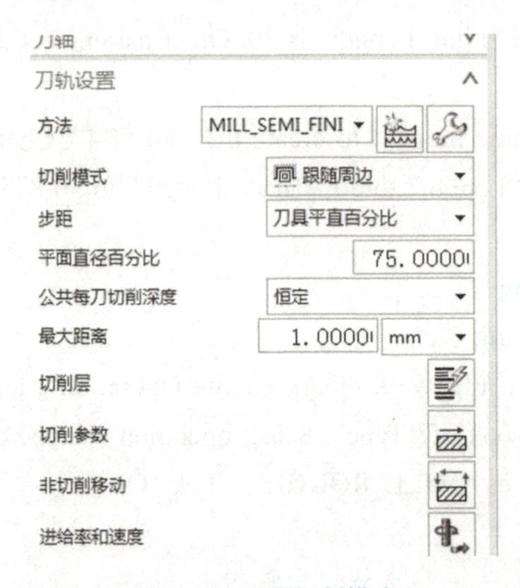

图 5-10 设置切削模式

（3）单击【切削参数】，打开【切削参数】对话框，单击【余量】选项卡，设置【部件底面余量】为【0.2】，如图 5-11 所示，单击【确定】。

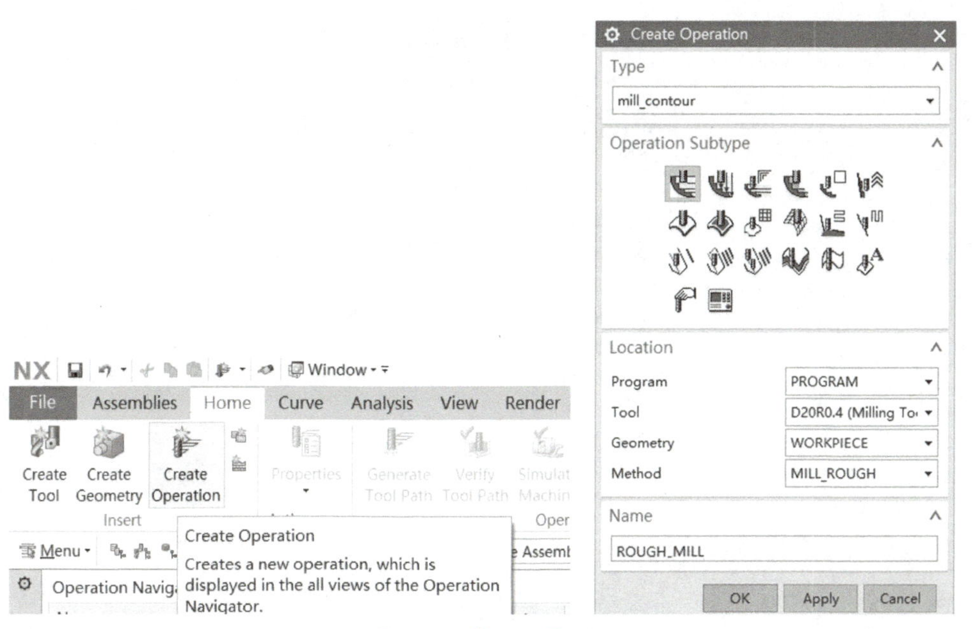

Fig. 5-9 Create Tool

(2) As shown in Fig. 5-10, modify "Cut Pattern" as "Follow Periphery" and "Maximum Distance" as 1 mm.

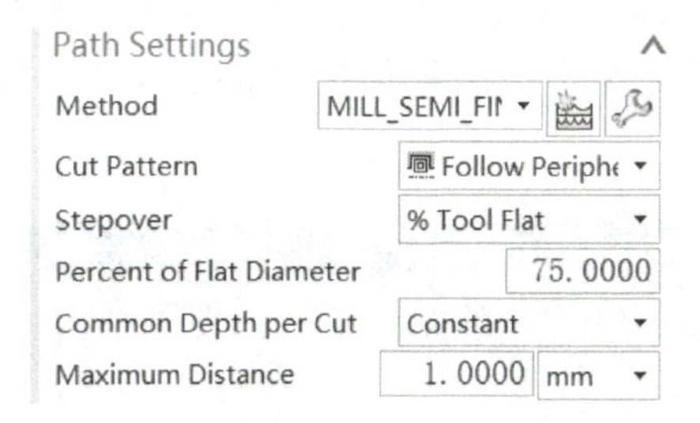

Fig. 5-10 Modify Path Settings

(3) As shown in Fig. 5-11, click "Cutting Parameters" button, and in "Stock" tab, modify "Part Floor Stock" as 0.2. Click "OK".

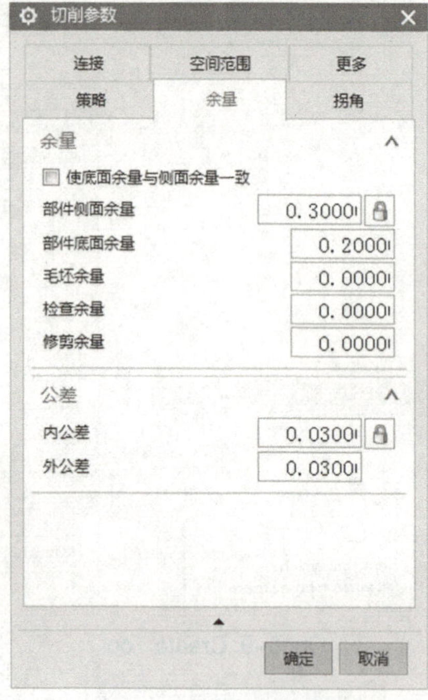

图 5-11 设置切削余量

（4）单击【进给率和速度】，打开【进给率和速度】对话框，设置【主轴速度】为【1800】，进给率【切削】为【1200 mmpm】，如图 5-12 所示，单击【确定】。

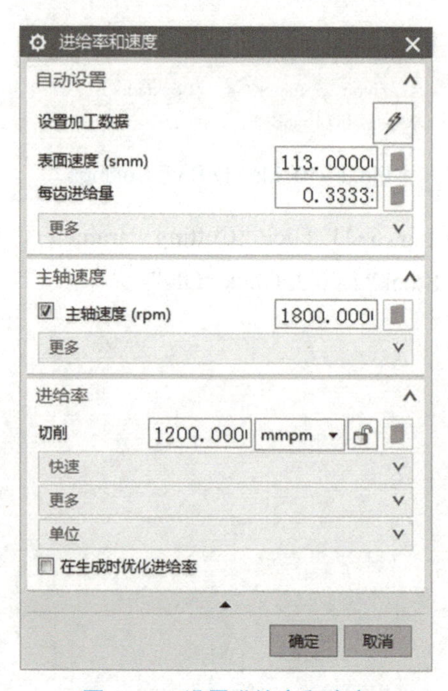

图 5-12 设置进给率和速度

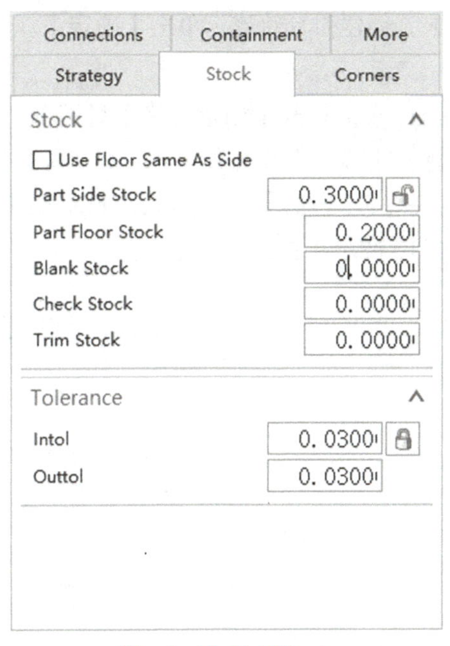

Fig. 5-11　Set Stock

(4) Click "Feeds and Speeds". Modify "Spindle Speed" as 1,800 and "Feed Rates" as 1,200 mmpm as shown in Fig. 5-12. Click "OK".

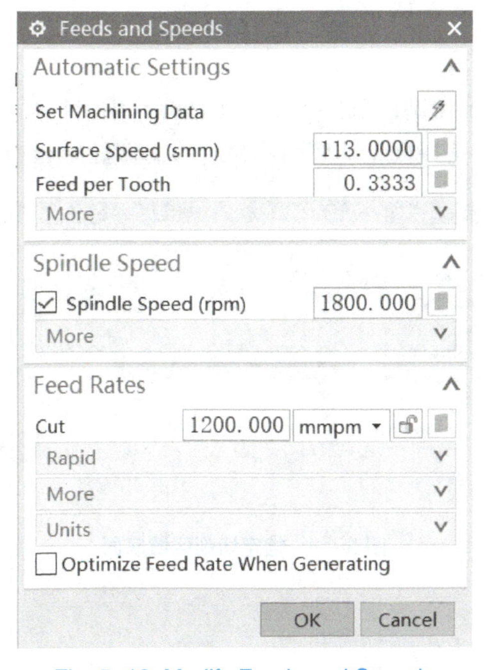

Fig. 5-12　Modify Feeds and Speeds

（5）单击【生成】，创建刀具轨迹，单击【确定】。

2. 型面精加工 -1

（1）单击【创建工序】，打开【创建工序】对话框，【类型】选择【mill_planar】，【工序子类型】选择【外形轮廓铣】，【刀具】选择【D12R0】，【几何体】选择【WORKPIECE】，【方法】选择【MILL_FINISH】，修改【名称】为【型面精加工 -1】，如图 5-13 所示，单击【确定】。

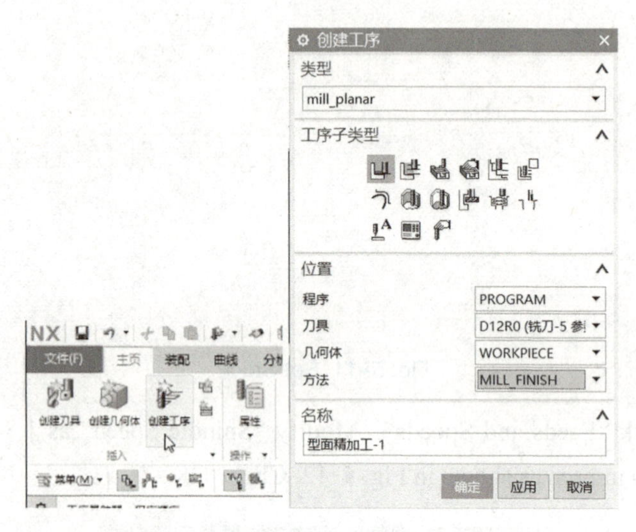

图 5-13 创建工序

（2）单击【指定切削区底面】中的【选择或编辑切削区域几何体】，弹出【切削区域】对话框，选择模型的底面和台阶面，如图 5-14 所示，单击【确定】。

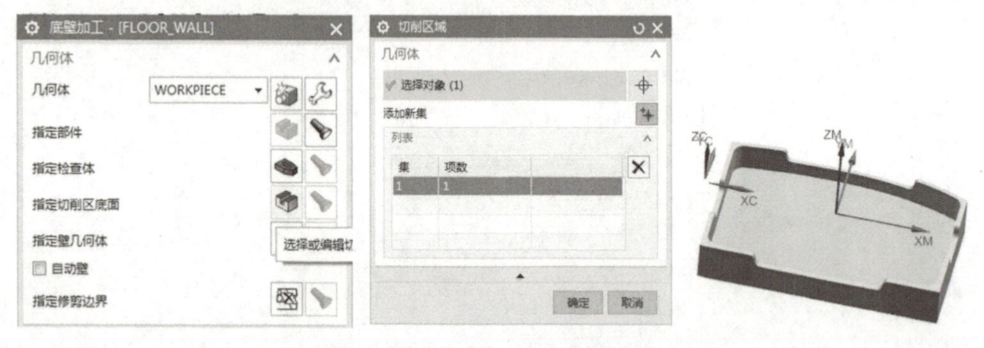

图 5-14 指定切削区域底面

（5）Click “Generate” to view the tool path and then click “OK”.

2. Finish Machining-1

（1）As shown in Fig. 5-13, click “Create Operation”, and then select “mill_planar” in pull-down box of “Type”. Select operation subtype as “Floor and Wall”, tool as “D12R0”, geometry as “ WORKPIECE” and method as “MILL_FINISH” and modify “Name” as “FINISH_MILL-1”. Click “OK”.

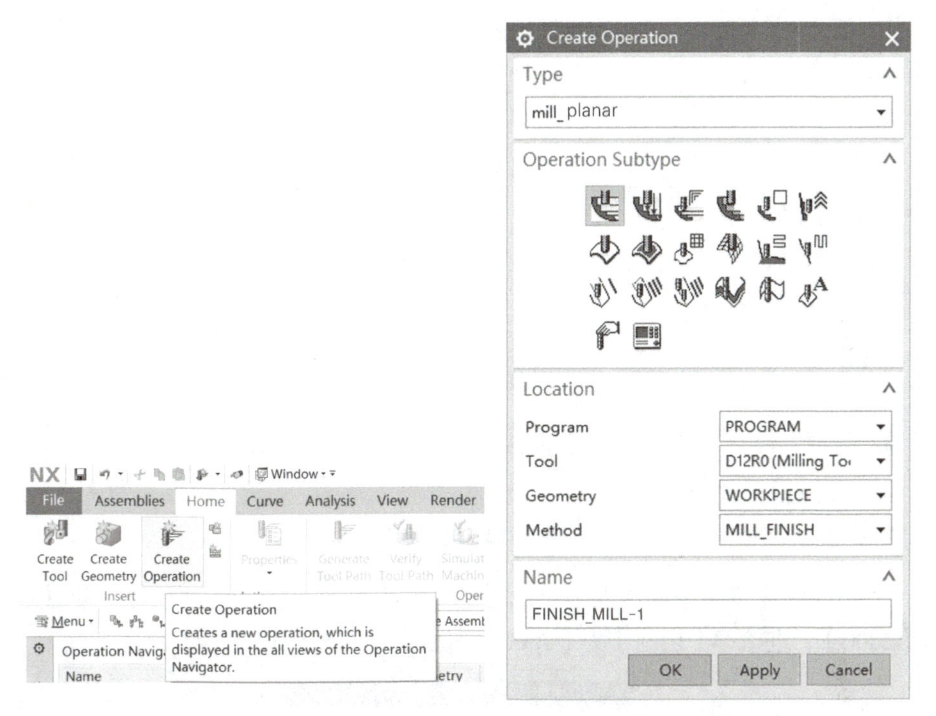

Fig. 5-13 Create Operation

（2）As shown in Fig. 5-14, click the button “Specify Floor” and select the surface. Click “OK”.

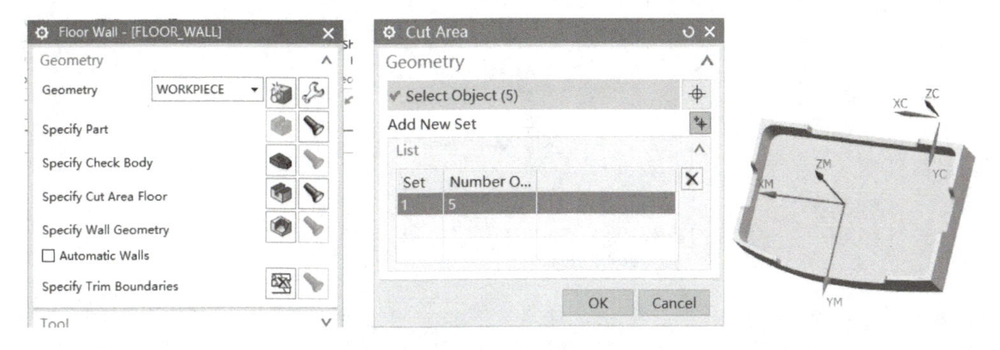

Fig. 5-14 Specify Cut Area

（3）在【刀轨设置】中，【切削模式】选为【跟随周边】，单击【切削参数】，打开【切削参数】对话框，在【策略】选项卡中，【切削方向】选择【顺铣】，【刀路方向】为【向外】，如图 5-15 所示，单击【确定】。

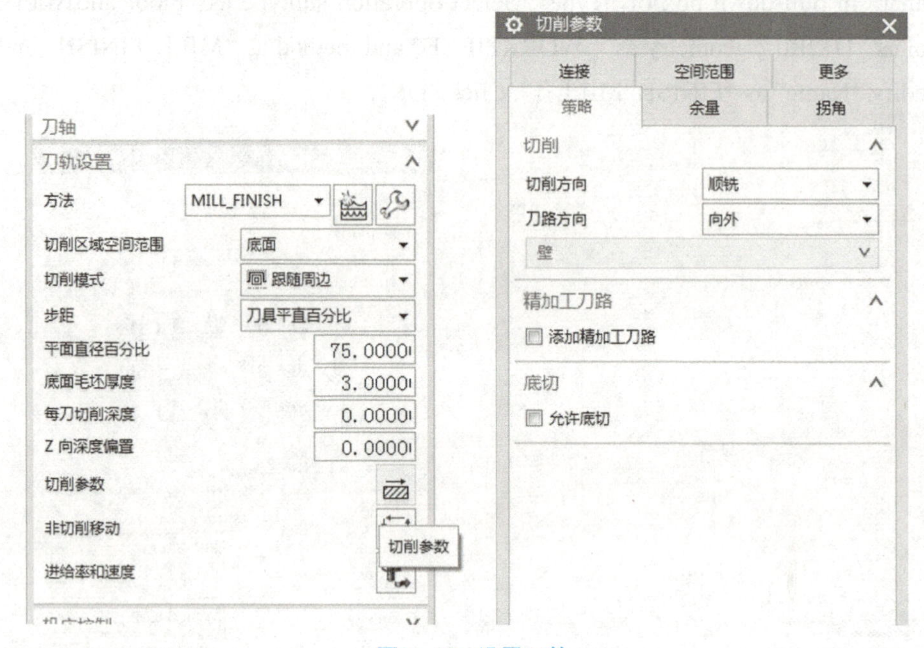

图 5-15 设置刀轨

（4）单击【进给率和速度】，打开【进给率和速度】对话框，设置【主轴速度】为【2800】，进给率【切削】为【1000 mmpm】，如图 5-16 所示，单击【确定】。

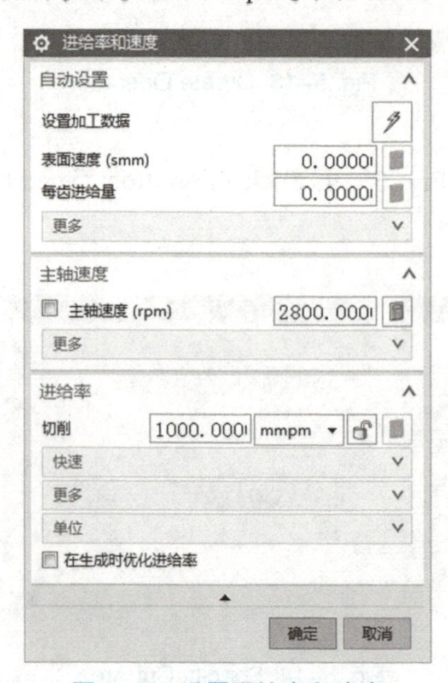

图 5-16 设置进给率和速度

(3) As shown in Fig. 5-15, modify "Cut Pattern" as "Follow Periphery". Click "Cutting Parameters" button, in "Strategy" tab, select "Climb Cut" as cut direction and "Outward" as pattern direction. Click "OK".

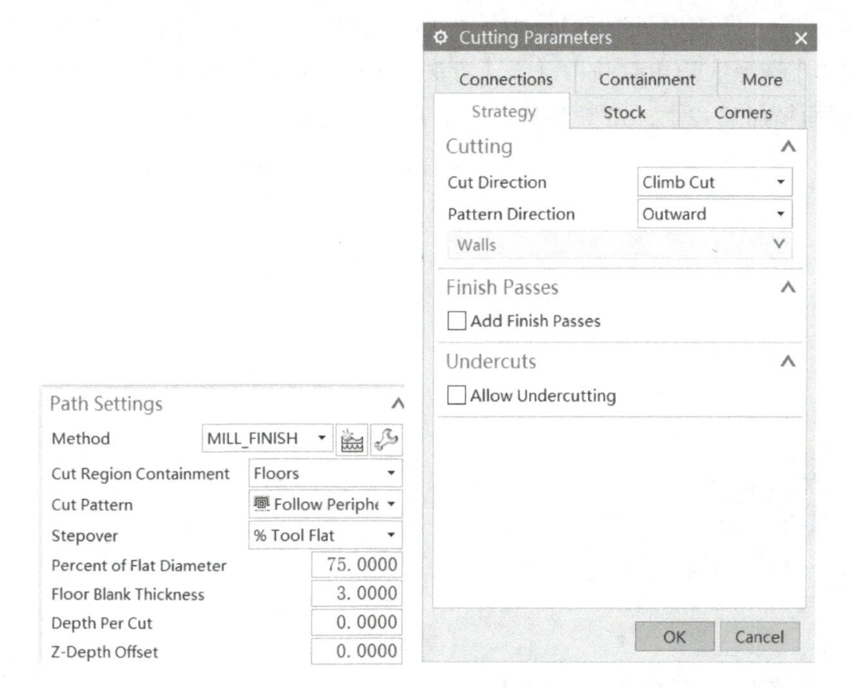

Fig. 5-15　Modify Path Settings

(4) Click "Feeds and Speeds", modify "Spindle Speed" as 2,800 and "Feed Rates" as 1,000 mmpm as shown in Fig. 5-16. Click "OK".

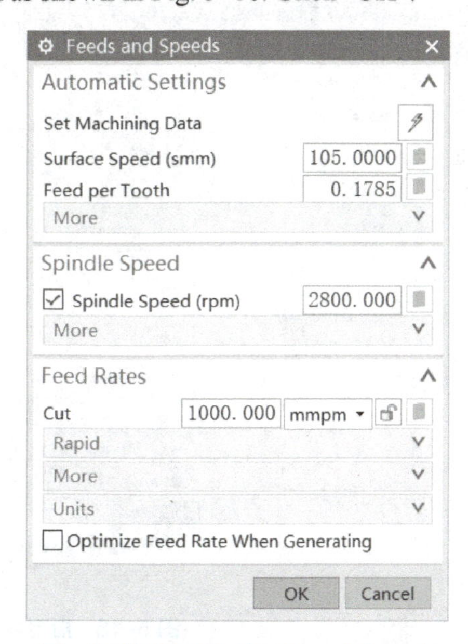

Fig. 5-16　Modify Feeds and Speeds

（5）单击【生成】，创建刀具轨迹，单击【确定】。

3. 型面精加工 -2

（1）单击【创建工序】，打开【创建工序】对话框，【类型】选择【mill_multi-axis】，【工序子类型】选择【外形轮廓铣】，【刀具】选择【D12R0】，【几何体】选择【WORKPIECE】，【方法】选择【MILL_FINISH】，修改【名称】为【型面精加工-2】，如图 5-17 所示，单击【确定】。

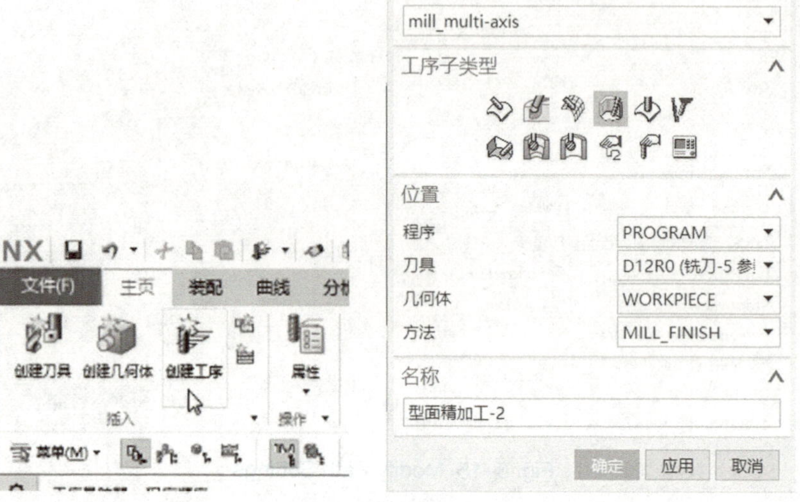

图 5-17　创建工序

（2）单击【指定底面】中的【选择或编辑底面几何体】，选择模型的底面作为底面几何体，如图 5-18 所示，单击【确定】。

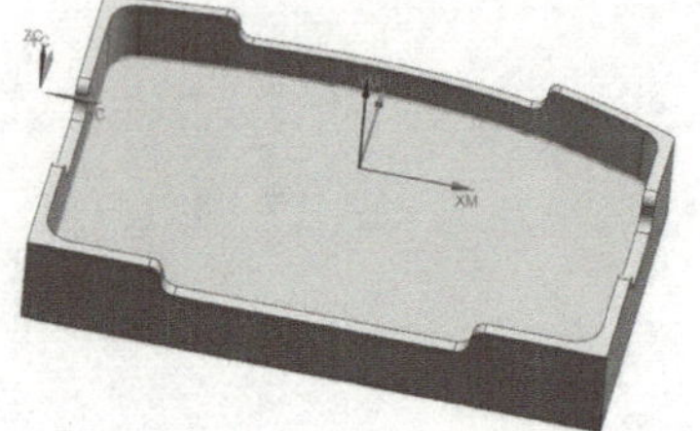

图 5-18　指定底面

（5）Click “Generate” to view the tool path and then click “OK”.

3. Finish Machining-2

（1）As shown in Fig. 5-17, click “Create Operation”, and then select “mill_multi-axis” in pull-down box of “Type”. Select operation subtype as “contour profile”, tool as “D12R0”, geometry as “WORKPIECE” and method as “MILL_FINISH”, modify “Name” as “FINISH_MILL-2” and then click “OK”.

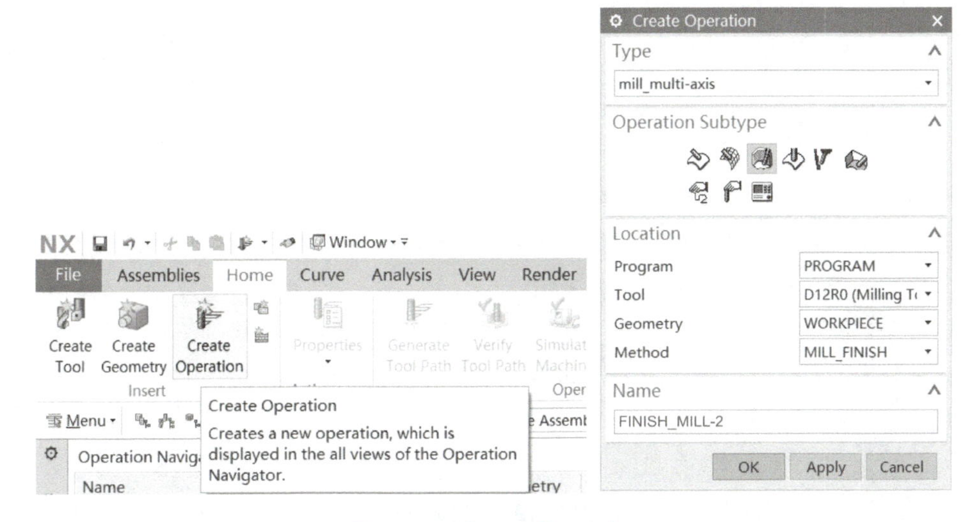

Fig. 5-17　Create Operation

（2）Click “Specify Floor” and select the bottom as shown in Fig. 5-18. Click “OK”.

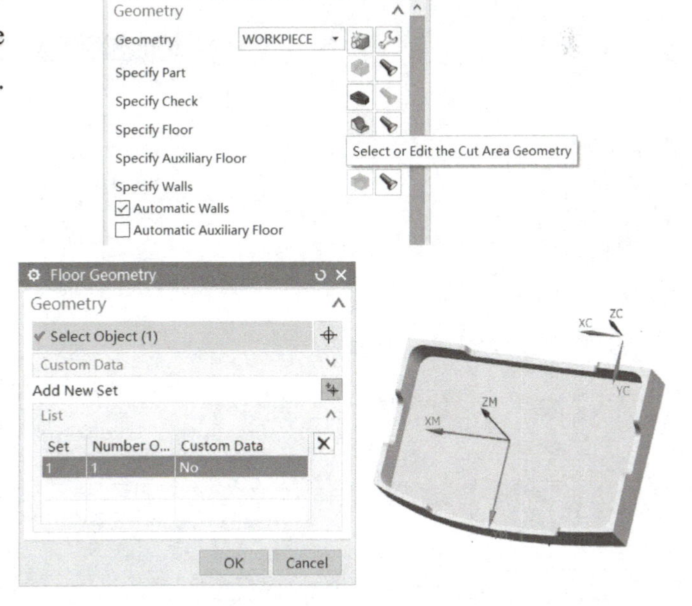

Fig. 5-18　Specify Floor

（3）在【刀轨设置】中，单击【非切削移动】，打开【非切削移动】对话框，在【进刀】选项卡中，【进刀类型】选择【圆弧 - 垂直于刀轴】，如图 5–19 所示，单击【确定】。

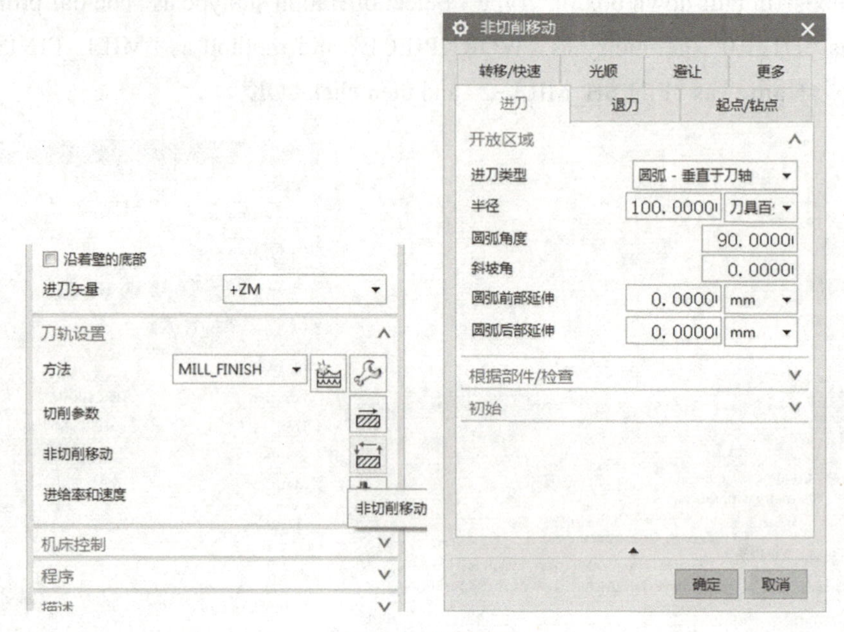

图 5–19　设置刀轨

（4）单击【进给率和速度】，打开【进给率和速度】对话框，设置【主轴速度】为【2200】，进给率【切削】为【1000 mmpm】，如图 5–20 所示，单击【确定】。

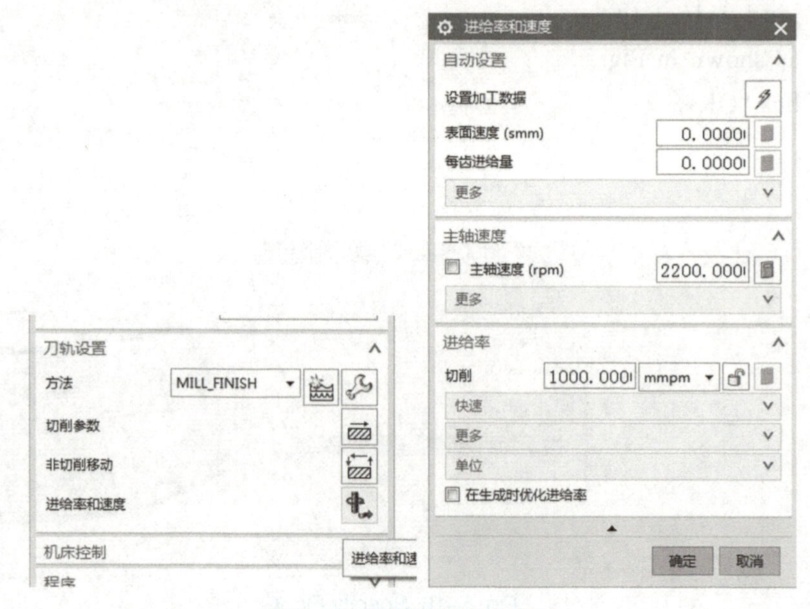

图 5–20　设置进给率和速度

（3）Click "Non Cutting Moves", in "Engage" tab, select "Arc-Normal to tool axis" as "Engage Type" as shown in Fig. 5-19. Click "OK".

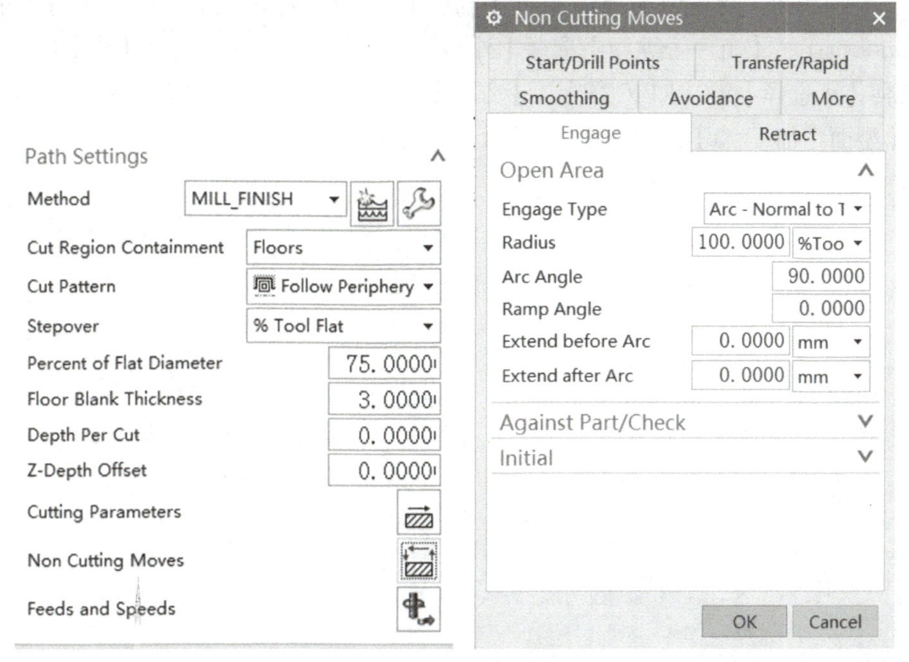

Fig. 5-19　Modify Path Settings

（4）Click "Feeds and Speeds", modify "Spindle Speed" as 2,200 and "Feed Rates" as 1,000 mmpm as shown in Fig. 5-20. Click "OK".

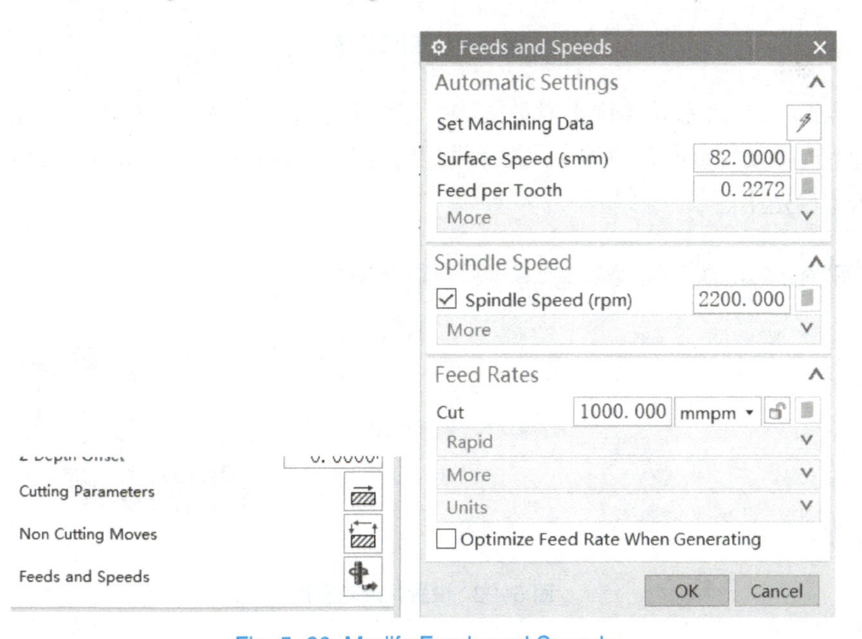

Fig. 5-20　Modify Feeds and Speeds

（5）单击【生成】，创建刀具轨迹，单击【确定】。

4. 型面精加工 -3

（1）单击【创建工序】，打开【创建工序】对话框，【类型】选择【mill-multi-axis】，【工序子类型】选择【外形轮廓铣】，【刀具】选择【D12R0】，【几何体】选择【WORKPIECE】，【方法】选择【MILL_FINISH】，修改【名称】为【型面精加工 -3】，如图 5-21 所示，单击【确定】。

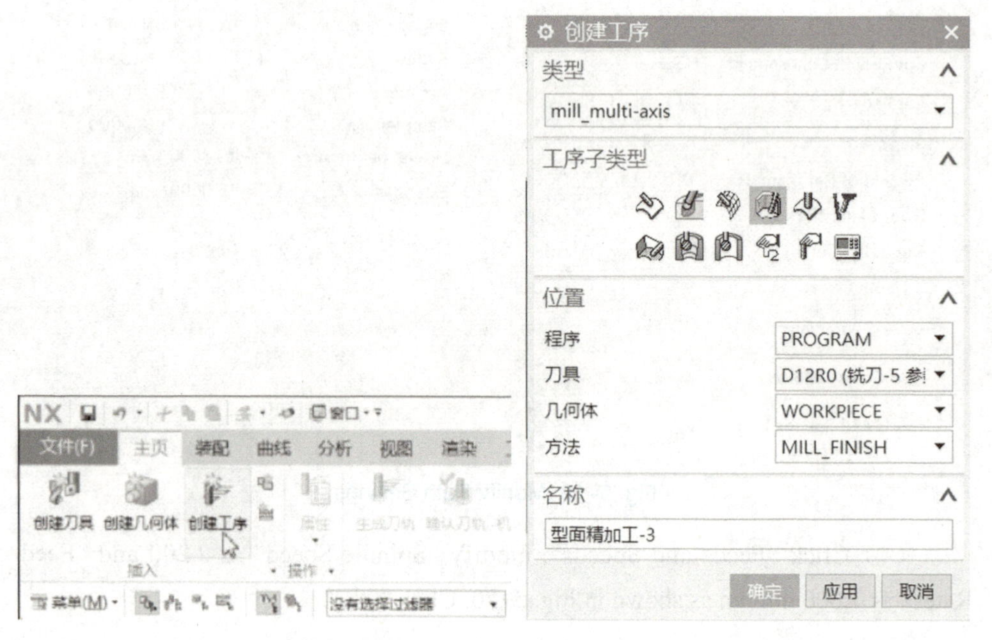

图 5-21　创建工序

（2）在【外形轮廓铣】对话框中，取消勾选【自动壁】，单击【指定壁】中的【选择或编辑壁几何体】，选择模型的外壁作为壁几何体，如图 5-22 所示，单击【确定】。

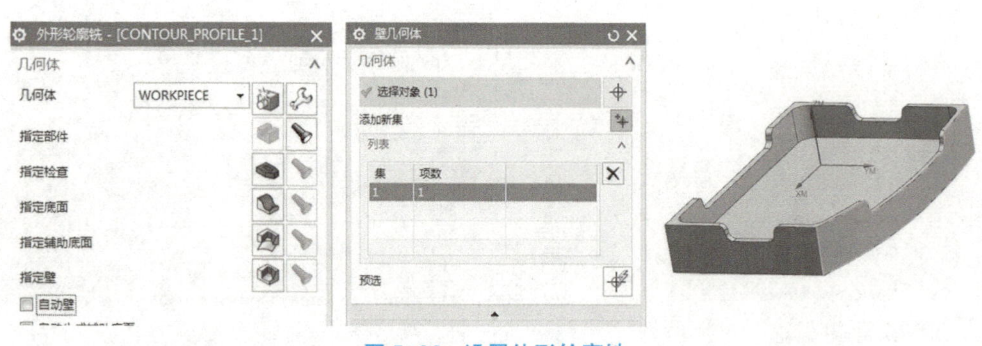

图 5-22　设置外形轮廓铣

(5) Click “Generate” to view the tool path and then click “OK”.

4. Finish Machining-3

(1) As shown in Fig. 5-21, click “Create Operation”, and then select “mill_multi-axis” in pull-down box of “Type”. Select operation subtype as “contour profile”, tool as “D12R0”, geometry as “WORKPIECE” and method as “MILL_FINISH”, and modify name as “FINISH_MILL-3” and then click “OK”.

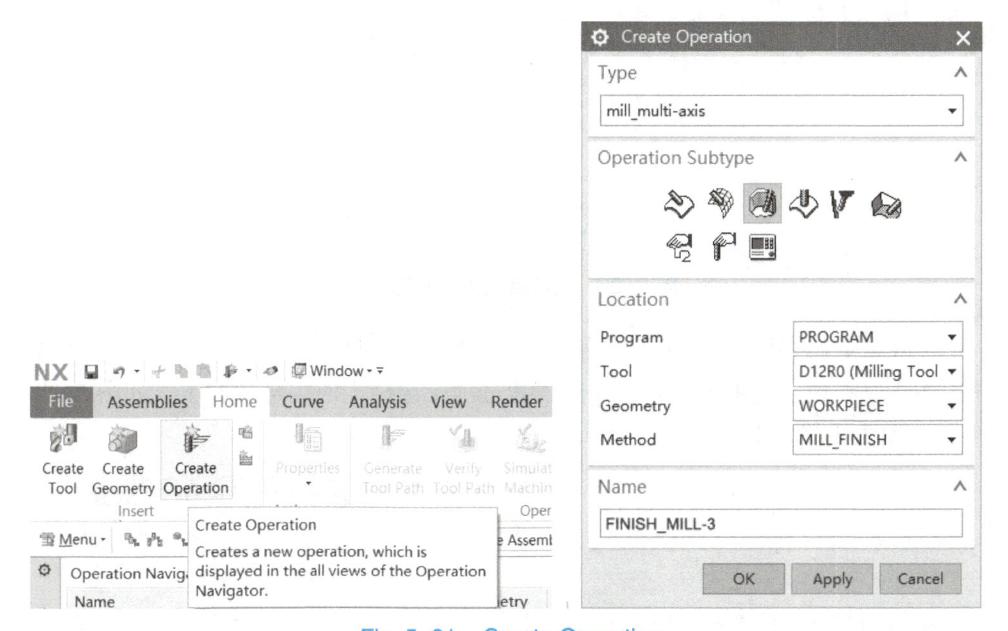

Fig. 5-21 Create Operation

(2) In the window of “Contour Profile”, uncheck the option “Automatic Walls”, click “Specify Walls” and then select side of the part as shown in Fig. 5-22. Click “OK”.

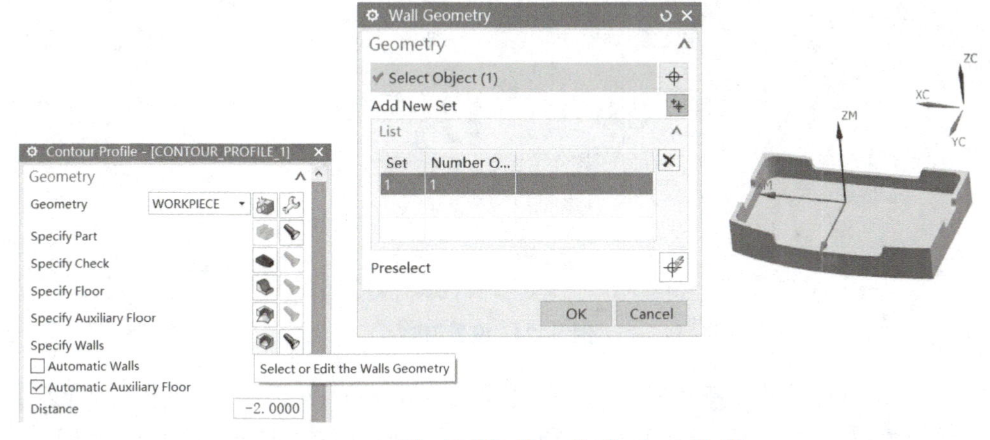

Fig. 5-22 Specify Contour Profile

（3）在【外形轮廓铣】对话框中，勾选【自动生成辅助底面】，【距离】改为【-2】，如图 5-23 所示。

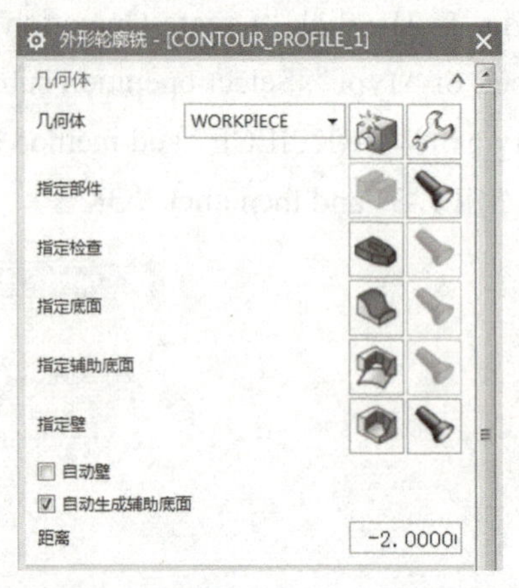

图 5-23 设置自动生成辅助底面

（4）在【刀轨设置】中，单击【非切削移动】，打开【非切削移动】对话框，在【转移 / 快速】选项卡中，【安全设置选项】选择【使用继承的】，如图 5-24 所示，单击【确定】。

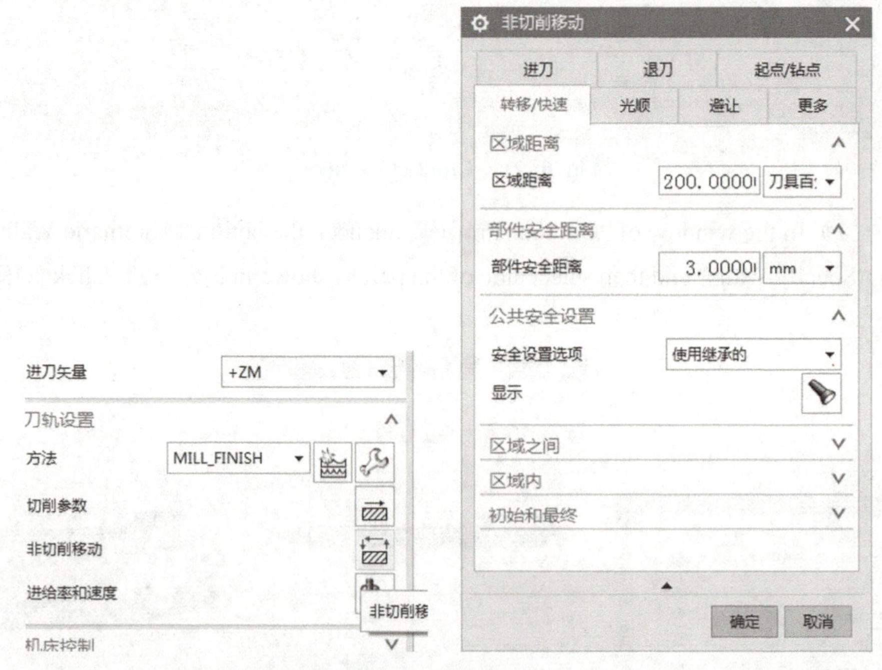

图 5-24 设置刀轨

（5）单击【进给率和速度】，打开【进给率和速度】对话框，设置【主轴速度】为【2200】，进给率【切削】为【1000 mmpm】，如图 5-25 所示，单击【确定】。

(3) Tick the option "Automatic Auxiliary Floor" and modify distance as – 2 as shown in Fig. 5-23.

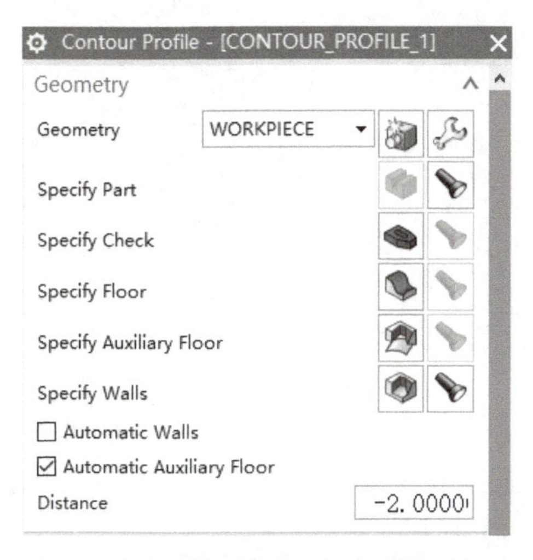

Fig. 5–23 Modify Automatic Auxiliary Floor

(4) As shown in Fig. 5-24, click "Non Cutting Moves" button, and in "Transfer/Rapid" tab, select "Use Inherited" as "Clearance Option". Click "OK".

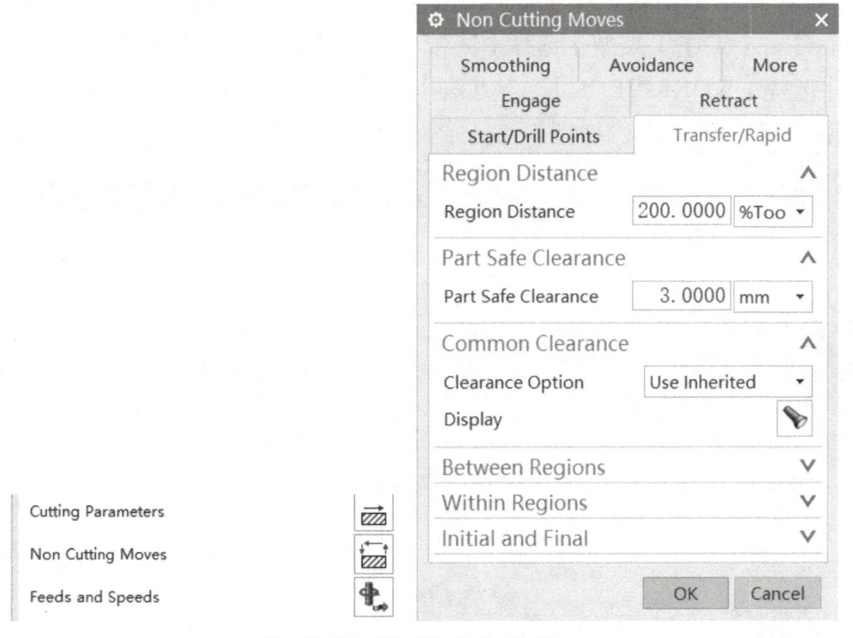

Fig. 5–24　Modify Path Settings

(5) Click "Feeds and Speeds", modify "Spindle Speed" as 2,200 and "Feed Rates" as 1,000 mmpm as shown in Fig. 5-25. Click "OK".

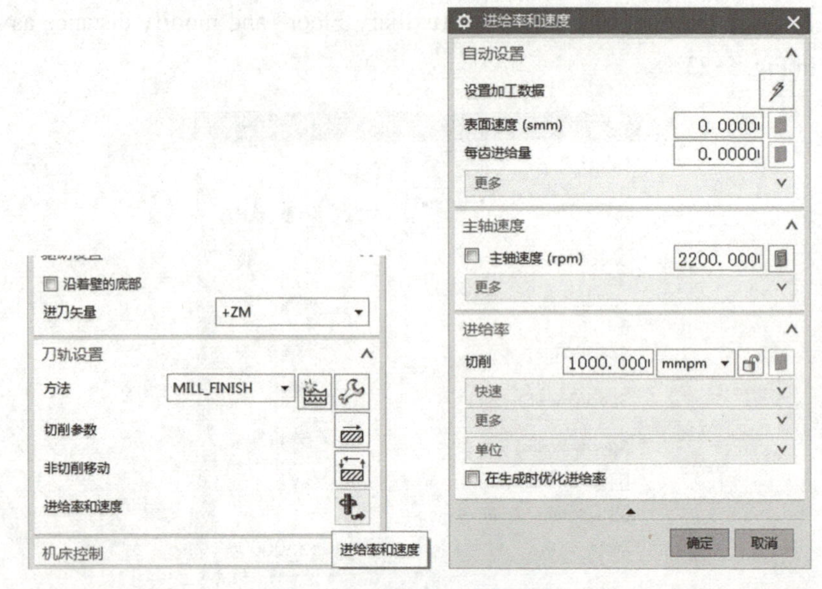

图 5-25 设置进给率和速度

（6）单击【生成】，创建刀具轨迹，单击【确定】。

5. 型面精加工 -4

（1）单击【创建工序】，打开【创建工序】对话框，类型选择【mill-contour】，【工序子类型】选择【区域轮廓铣】，【刀具】选择【B4】，【几何体】选择【WORKPIECE】，【方法】选择【MILL_FINISH】，修改【名称】为【型面精加工 -4】，如图 5-26 所示，单击【确定】。

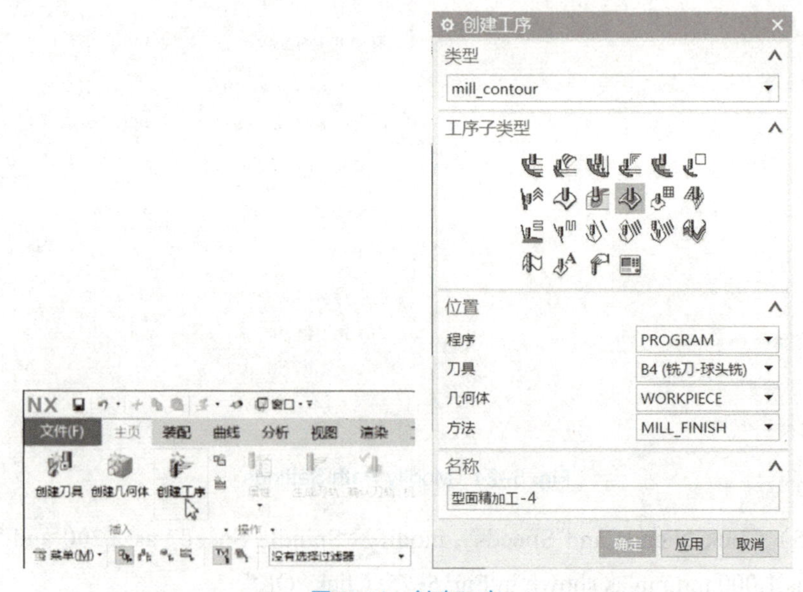

图 5-26 创建工序

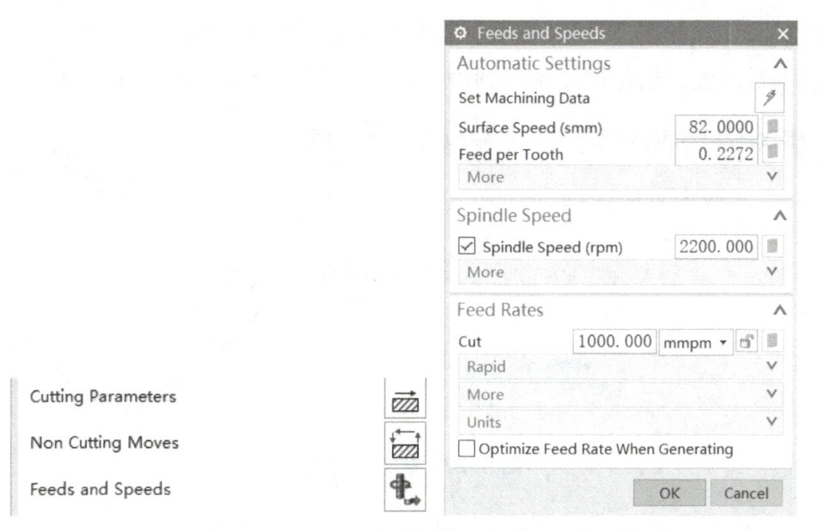

Fig. 5-25 Modify Feeds and Speeds

(6) Click "Generate" to view the tool path and then click "OK".

5. Finish Machining-4

(1) As shown in Fig. 5-26, click "Create Operation", and then select "mill_contour" in pull-down box of "Type". Select operation subtype as "contour area", tool as "BALL4", geometry as "WORKPIECE" and method as "MILL_FINISH", and modify "Name" as "FINISH_MILL-4". Click "OK".

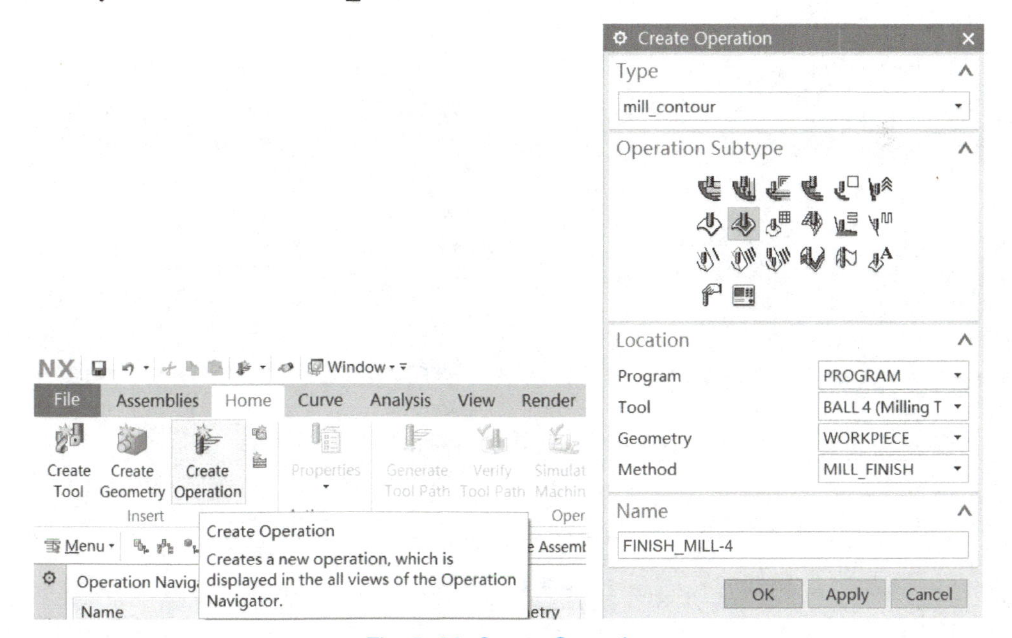

Fig. 5-26 Create Operation

（2）单击【指定切削区域】中的【选择或编辑切削区域几何体】，弹出【切削区域】对话框，选择模型中的圆角，如图 5-27 所示，单击【确定】。

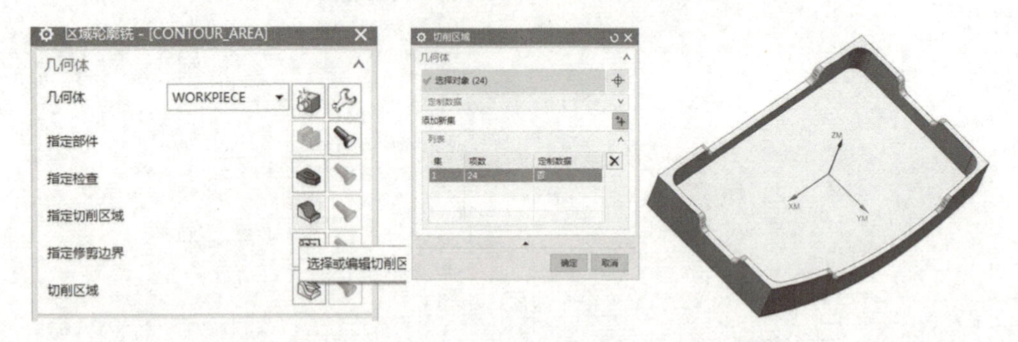

图 5-27 指定切削区域

（3）设置【驱动方法】为【区域铣削】，在【区域铣削驱动方法】对话框中，【步距】选为【残余高度】，【最大残余高度】改为【0.005】，如图 5-28 所示，单击【确定】。

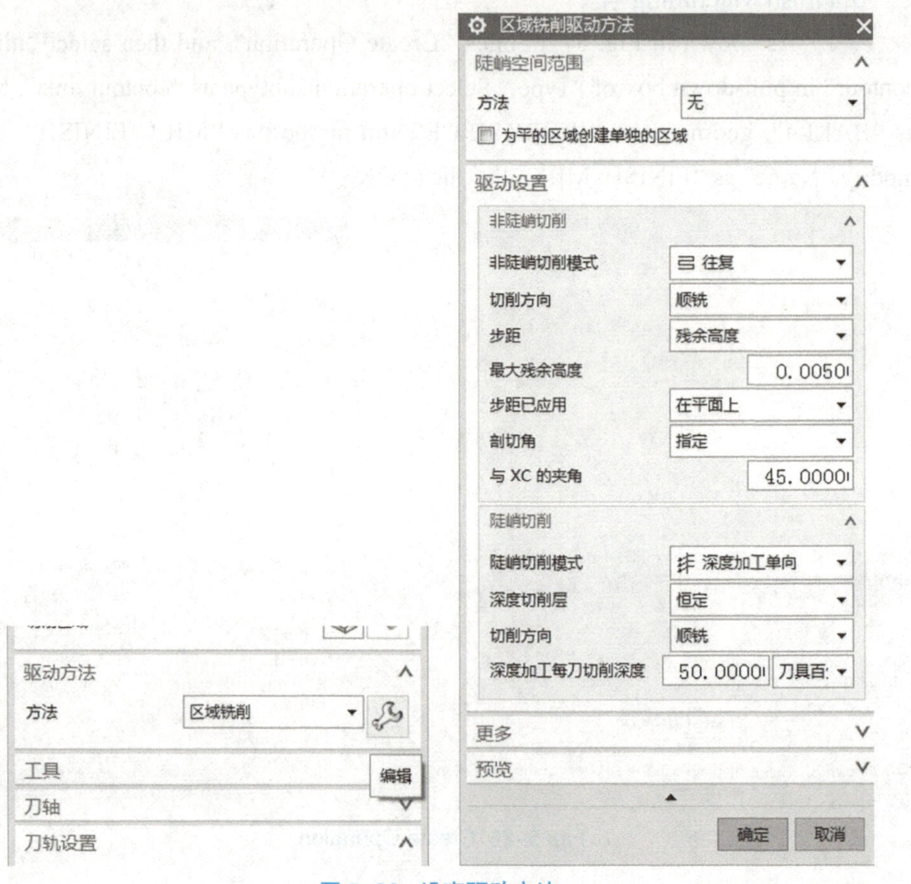

图 5-28 设定驱动方法

(2) Click "Specify Cut Area" and select fillets as shown in Fig 5-27. Click "OK".

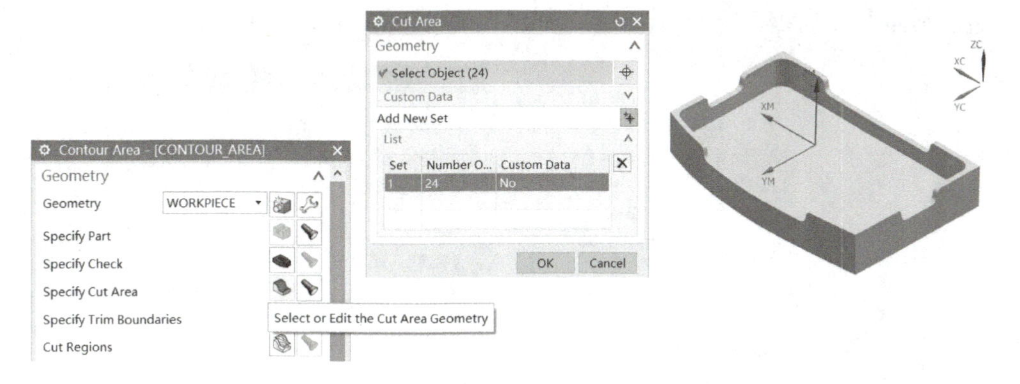

Fig. 5-27 Specify Cut Area

(3) Click "Edit" to set parameters of "Drive Method". Select "Scallop" as "Stepover", 0.005 as "Maximum Scallop Height" as shown in Fig. 5-28. Click "OK".

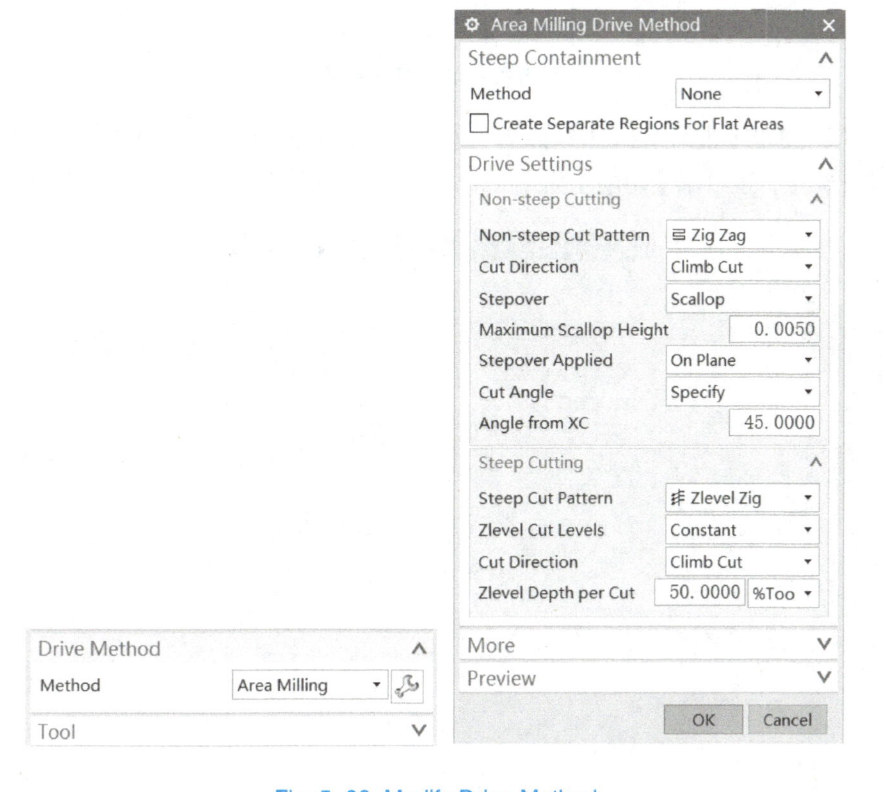

Fig. 5-28 Modify Drive Method

（4）单击【切削参数】，打开【切削参数】对话框，在【策略】选项卡中，【剖切角】选择【指定】，【与 XC 的夹角】改为【45】，【延伸路径】中勾选【在边上延伸】，【距离】改为【10】，如图 5-29 所示，单击【确定】。

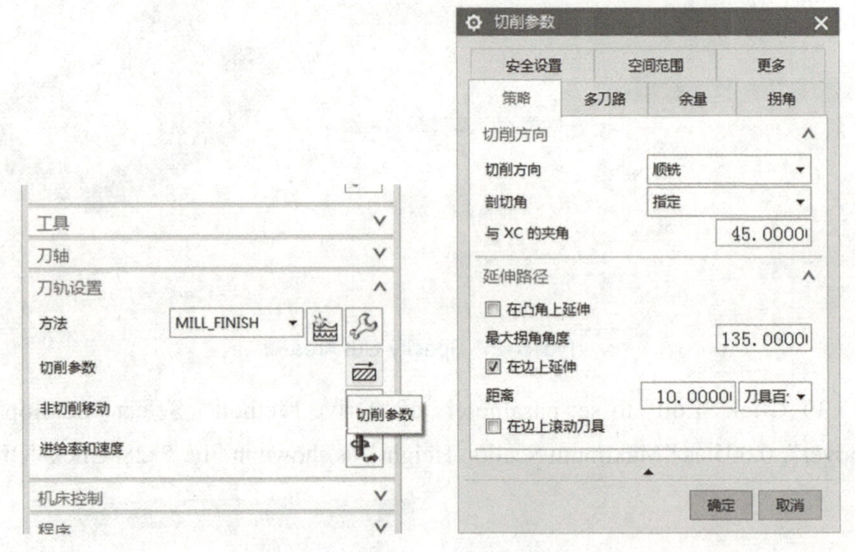

图 5-29　设定切削参数

（5）单击【进给率和速度】，打开【进给率和速度】对话框，设置【主轴速度】为【3600】，进给率【切削】为【800 mmpm】，如图 5-30 所示，单击【确定】。

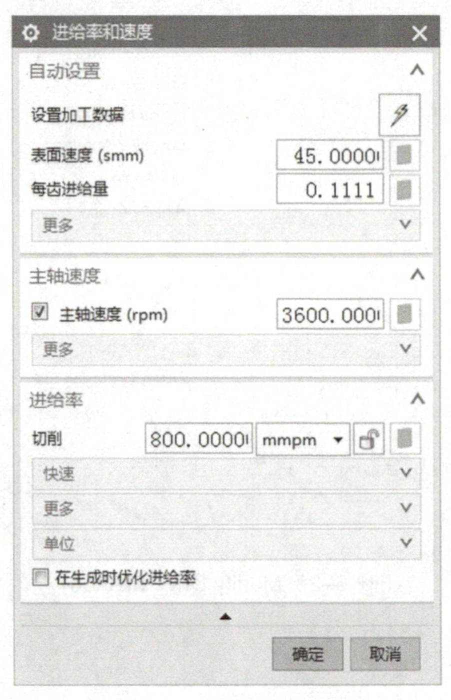

图 5-30　设置进给率和速度

(4) As shown in Fig. 5-29, click "Cutting Parameters" button, in "Strategy" tab, select "Specify" as cut angle and modify the angle from XC as 45°. Tick the option "Extend at Edges" and modify "Distance" as 10. Click "OK".

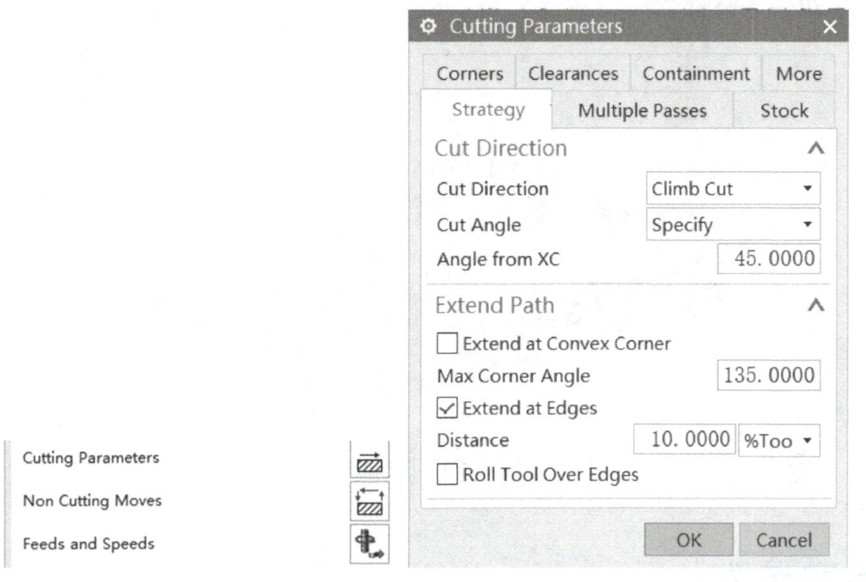

Fig. 5-29 Modify Cutting Parameters

(5) Click "Feeds and Speeds", modify "Spindle Speed" as 3,600 and "Feed Rates" as 800 mmpm as shown in Fig. 5-30. Click "OK".

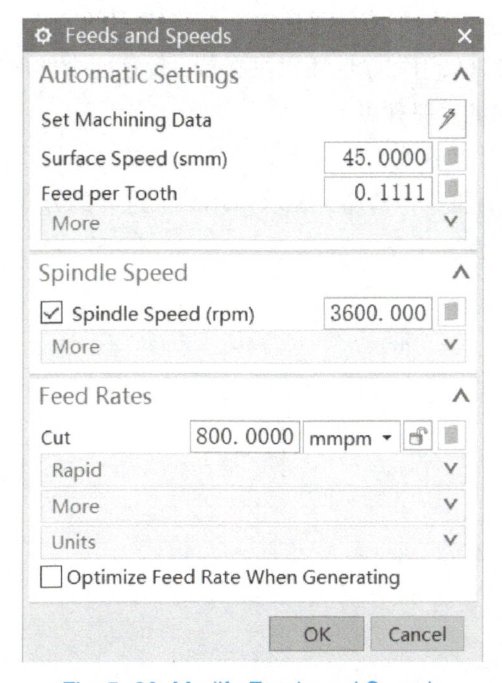

Fig. 5-30 Modify Feeds and Speeds

（6）单击【生成】，创建刀具轨迹，单击【确定】。

四、仿真加工

单击【NC_PROGRAM】，单击【确认刀轨】，选择【3D 动态】，如图 5-31 所示，单击【播放】。

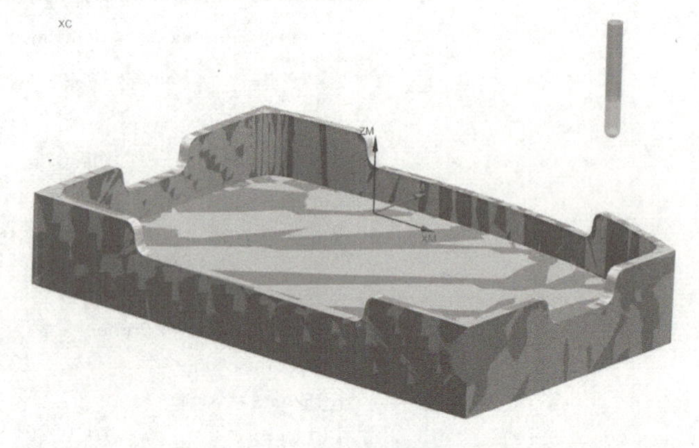

图 5-31　仿真结果

专家点拨

（1）在编制任何一个零件的加工程序前，必须仔细分析零件图样和零件模型，并编制合理的加工工艺。

（2）在 NX CAM 中编制零件程序时，要考虑零件的装夹。一般对于块类零件的小批量生产，可以采用加厚毛坯以供夹持，等正面全部加工完后，反身装夹，铣去夹持部分，并保证总厚即可。

（3）在粗加工时应尽可能提高效率，精加工时要保证质量。

（4）为保证表面质量，精加工要求采用圆弧进刀的方式。

(6) Click "Generate" to view the tool path and then click "OK".

Ⅳ. Simulation Machining

Click "NC_PROGRAM", and then "Verify Tool Path" button. Select "3D Dynamic" and click "Play" button to view the simulation result. The simulation result is shown as in Fig. 5-31.

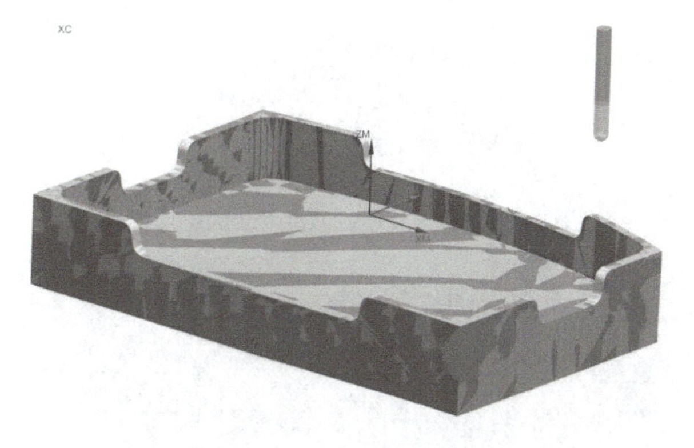

Fig. 5-31 Simulation Result

Expert Reviews

(1) Reasonable processing technic and deep analysis of part models as well as part drawings are prior to processing programming.

(2) When compiling part program in UG NX CAM, the clamping of parts should be considered. Generally, for small batch production of block parts, thickened blanks can be used for clamping. After all the front faces are processed, reverse clamping and mill off the clamping part, ensuring the total thickness.

(3) In rough machining, the efficiency should be improved as much as possible, and the quality should be ensured during finishing.

(4) To ensure the quality of surface, circular feed is required in finish machining.

课后训练

根据图 5-32 所示的多面体零件的特征，制定合理的工艺路线，设置必要的加工参数，生成刀具轨迹，通过相应的后处理生成数控加工程序，并运用机床加工零件。

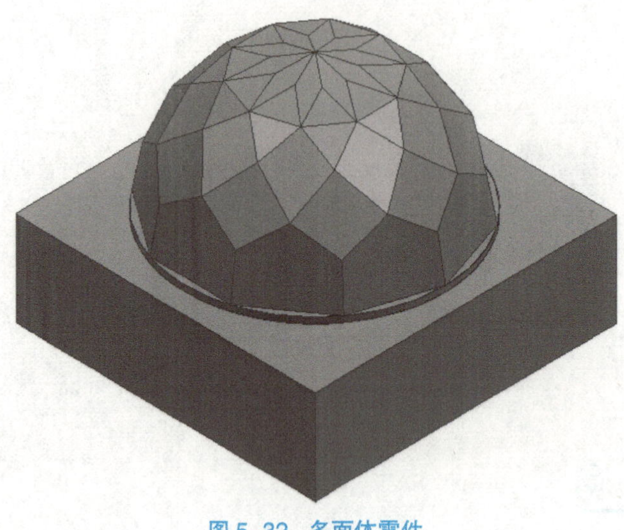

图 5-32　多面体零件

Practice

According to the characteristics of disc parts as shown in Fig. 5-32, make a reasonable processing technic, set necessary parameters, generate tool path, generate NC program through corresponding post-processor, and use machine tools to machine parts.

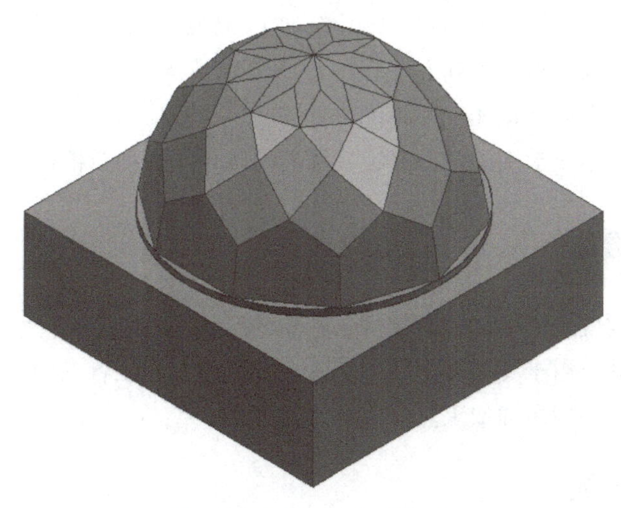

Fig. 5-32 Polyhedral Parts

项目 6 叶轮的数控编程与仿真加工

学习目标

能力目标：能运用 NX 软件完成叶轮的数控编程与仿真加工。

知识目标：掌握叶轮铣削几何体的设置方法；

掌握叶片粗加工、叶片精加工、轮毂精加工、圆角精加工方法；

掌握刀路阵列方法；

掌握叶轮加工刀路优化方法。

素质目标：激发学生的学习兴趣，培养团队合作和创新精神。

项目导读

叶轮是航空发动机中的核心部件，叶轮的形状比较复杂，叶片与叶片之间一般会有加工干涉，由于其零件形状的特殊性，采用车削或者三轴铣削都没法完成零件加工，只能采用多轴加工。

任务描述

学生以企业制造部门 MC 数控程序员的身份进入 NX CAM 功能模块，根据叶轮零件的特征，制定合理的工艺路线，创建型腔铣、可变轴轮廓铣等加工操作，设置必要的加工参数，生成刀具轨迹，检验刀具轨迹是否正确合理，并对操作过程中存在的问题进行研讨和交流，通过相应的后处理生成数控加工程序，并运用机床加工零件。

项目实施

按照零件加工要求，制定叶轮的加工工艺，编制叶轮加工程序，完成叶轮的仿真加工，后处理得到数控加工程序，完成零件加工。

Project 6 Numerical Control Programming and Simulation Machining of Impeller

Learning Objectives

Capacity Objective: Complete impeller programming and simulation machining with NX software.

Knowledge Objective: Master geometry settings of milling impeller;
Master the impeller rough machining, blade finishing, hub finishing and round corner finishing;
Master the array of the tool path;
Master the method of optimizing tool path in machining the impeller.

Quality Objective: Stimulate students' interest in learning and cultivate the spirit of teamwork as well as innovation.

Project Guidance

Impellers are the core parts of aeroplane engines. Due to their complicated shapes and the interference between blades, multi-axis machining is introduced instead of turning and 3-axis milling.

Task Description

Students operate NX CAM functional modules as MC programmers in enterprise manufacturing department. According to the characteristics of impellers, establish a reasonable process route, create machining operation of cavity mill and variable contour, set necessary machining parameters, generate tool path and check the generated tool path. Besides, students should discuss the problems occurred in the operation. By selecting the corresponding post processor to generate NC programs, students import them into machine tools to complete parts machining.

Project Implementation

Firstly, formulate processing technic of the impeller under processing requirements. Secondly, program for machining impellers. Thirdly, complete simulation machining, and then import the NC program obtained by post processor to the machine tool to complete parts machining.

一、制定加工工艺

1. 叶轮零件分析

该零件形状比较复杂，加工精度要求高，叶片属于薄壁零件，加工时容易产生变性，而且加工叶片时容易产生干涉。

2. 毛坯选用

零件材料为7075航空铝棒，尺寸为 ϕ120 mm × 50 mm。零件长度、直径尺寸已经精加工到位，无须再次加工。

3. 加工工序卡制定

零件选用立式五轴联动机床加工（双摆台摇篮式），三爪卡盘夹持，遵循先粗后精加工原则，粗加工采用“3+2”轴型腔铣方式，精加工采用五轴联动加工。制定的加工工序卡如表6-1所示。

表6-1 加工工序卡

零件号：1722536		工序名称：叶轮铣削加工		工艺流程卡_工序单
材料：AL7075	页码：1		工序号：01	版本号：0
夹具：三爪卡盘	工位：MC		数控程序号：	
刀具及参数设置				
加工内容	刀具号	刀具规格	主轴转速/rpm	进给速度/mmpm
叶片粗加工	T01	B8	1500	5000
叶片精加工	T02	B6	2500	8000
轮毂精加工	T02	B6	2500	8000
圆角精加工	T02	B6	4000	10000
更改号	更改内容	批准	日期	
拟制：　日期：	审核：　日期：	批准：	日期：	

二、加工准备

（1）启动NX，单击【打开】，在打开的【打开】对话框中，选择【叶轮.prt】文件，如图6-1所示，单击【OK】，打开叶轮模型。

Ⅰ. Formulate Processing Technic

1. Model part analysis

The shape of the impeller is complex and requires high machining precision. Because of the features of thin-walled parts, the shape of the impeller changes easily when it comes to machining. Besides, interference easily occurs during the machining.

2. Blank selection

Part material: 7075 aviation aluminum rods. Size: ϕ120 mm × 50 mm. Note that there is no need to finish the length and the diameter of the part.

3. Formulation of processing procedure card

Vertical 5-axis machine tool with double swing cradle is used to process the part. The fixture is three-jaw chuck. Before finish machining, rough machining should be carried out. Rough machining mode: cavity mill (i.e. "3+2"-axis side milling). Finish machining mode: 5-axis machining. The processing procedure card is shown in Table 6-1.

Table 6–1 Processing Procedure Card

Part number: 1722536		Name of process: Milling of impeller	Process card - Process sheet
Material: AL7075	Page number: 1	Procedure number: 01	Version number: 0
Fixture: Parallel-jaw vice	Work station: MC	CNC program number:	

Tool and parameter setting				
Tool number	Tool specification	Processing content	Spindle speed /rpm	Feed speed /mmpm
T01	B8	Blade rough machining	1500	5000
T02	B6	Blade finish machining	2500	8000
T02	B6	Hub finish machining	2500	8000
T02	B6	Blend finish	4000	10000

02					
01					
Change number	Change content		Approve		Date
Draws:	Date:	review:	Date:	Approve:	Date:

Ⅱ. Preparation for Processing

(1) As shown in Fig. 6-1, start NX, click "OPEN", select the "Impeller" (.prt file), and click"OK". Then, the impeller can be seen in WCS.

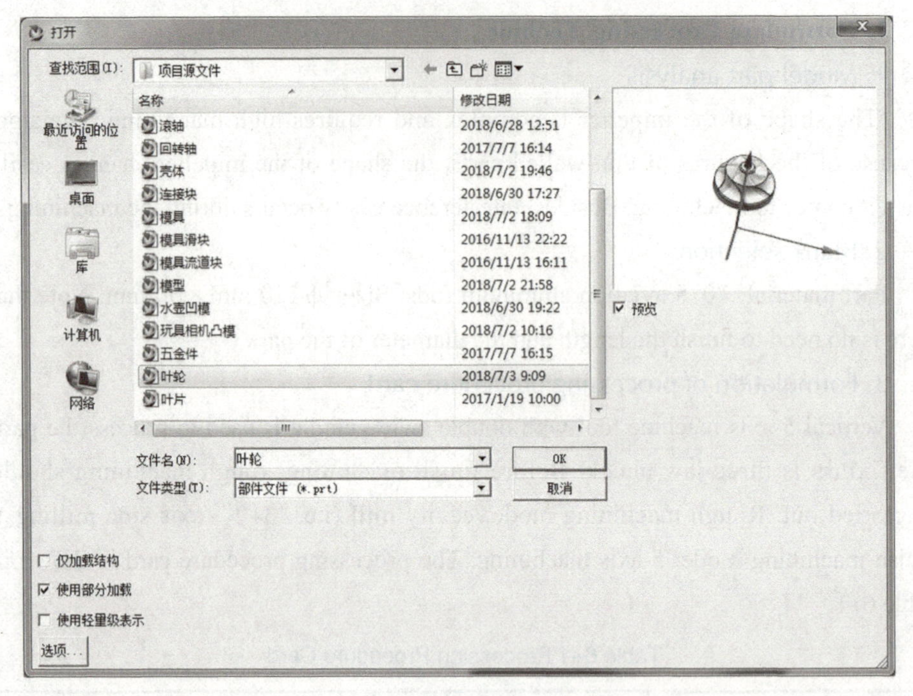

图 6–1　打开文件

（2）选择【应用模块】选项卡，单击【加工】，如图 6–2 所示，进入加工环境。

（3）弹出【加工环境】对话框，在【CAM 会话配置】选项中选择【cam_general】，【要创建的 CAM 设置】选项中选择【mill_multi_blade】，如图 6–3 所示，单击【确定】。

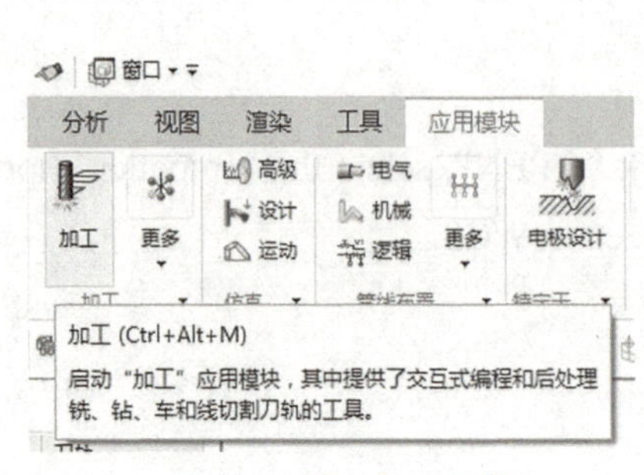

图 6–2　进入加工环境

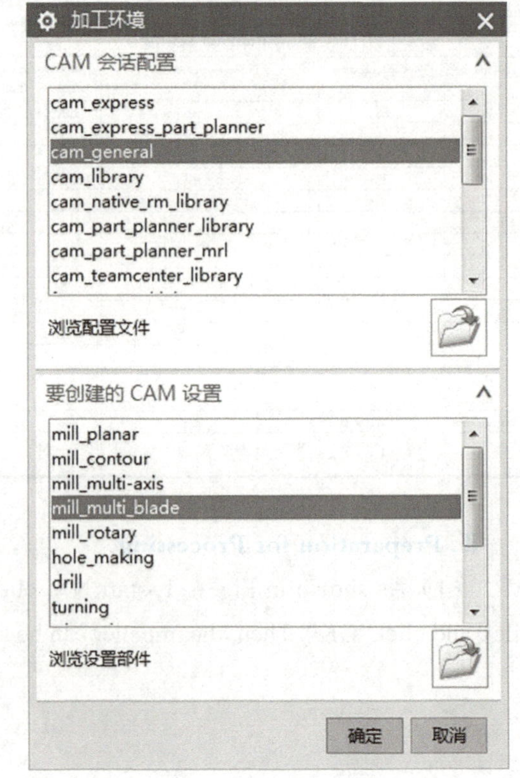

图 6–3　设计加工环境

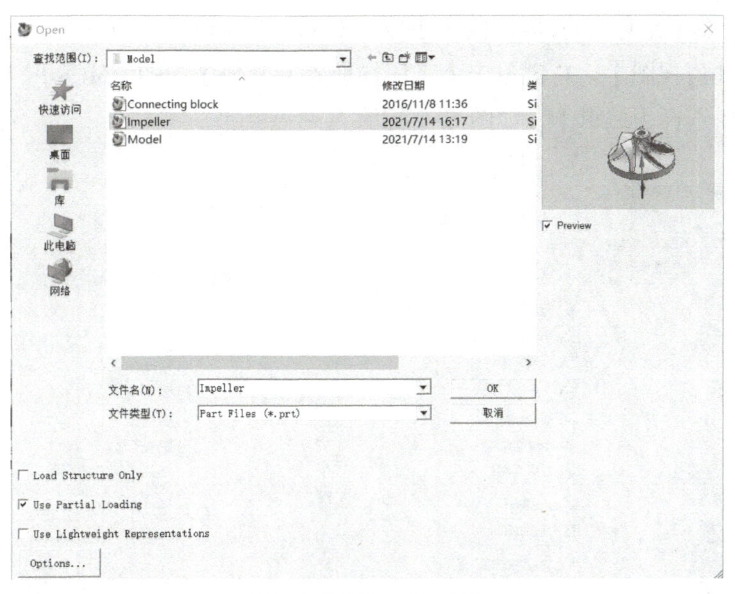

Fig. 6–1　Open the File

(2) Select "Application", and click "Manufacturing". The manufacturing menu can be seen as shown in Fig. 6-2.

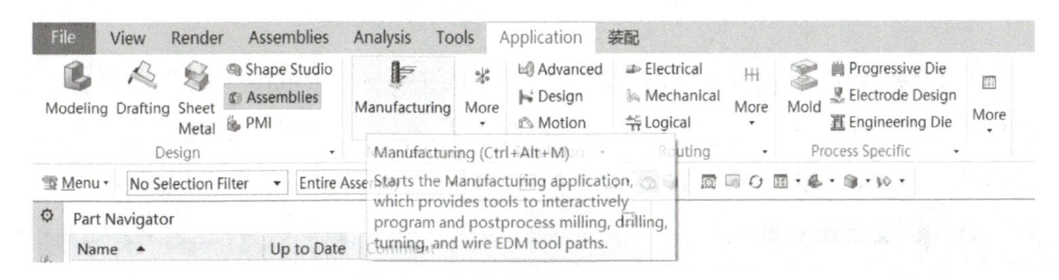

Fig. 6–2　Enter Manufacturing Environment

(3) As shown in Fig. 6-3, in "Machining Environment" window, select "cam_general" in "CAM Session Configuration", and then select "mill_multi_blade" in "CAM Setup to Create". Finally, click "OK".

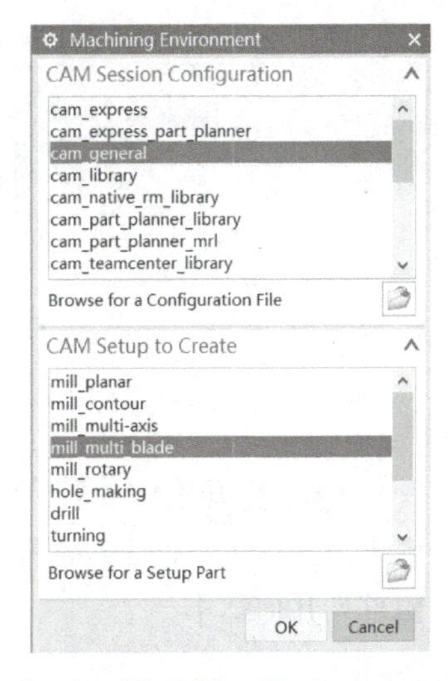

Fig. 6–3　Design Machining Environment

（4）在【工序导航器－机床】空白处单击鼠标右键，在弹出的快捷菜单中选择【几何视图】。右键单击【工序导航器－几何】中的【MCS】，在弹出的快捷菜单中选择【编辑】，如图 6-4 所示。

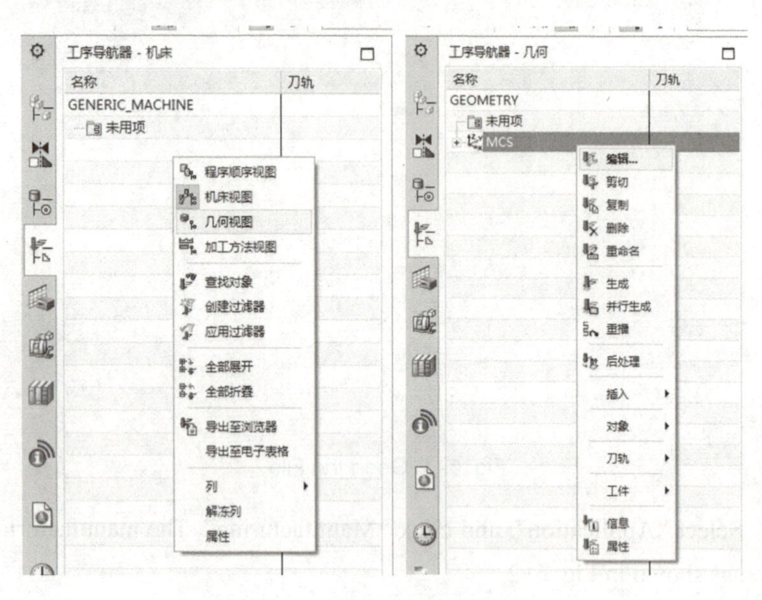

图 6-4　进入 MCS 坐标系

（5）弹出【MCS】对话框，单击工件上表面轮廓指定 MCS，【安全设置】选项中的【安全设置选项】选择【包容圆柱体】，【安全距离】设置为【10】，如图 6-5 所示，单击【确定】。

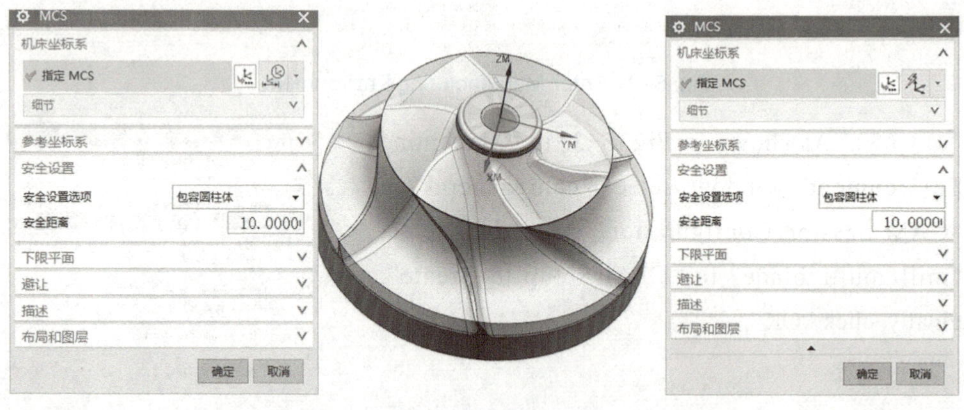

图 6-5　设置 MCS 坐标系

（6）打开【MCS】下的子选项，右键单击【WORKPIECE】选项，选择【编辑】。弹出【工件】对话框，如图 6-6 所示，单击【指定部件】中的【选择或编辑部件几何体】。

(4) As shown in Fig. 6-4, right click on the blank area of "Operation Navigator-Machine Tool", and select "Geometry View". In "Operation Navigator-Geometry" window, right click "MCS" and then click "Edit".

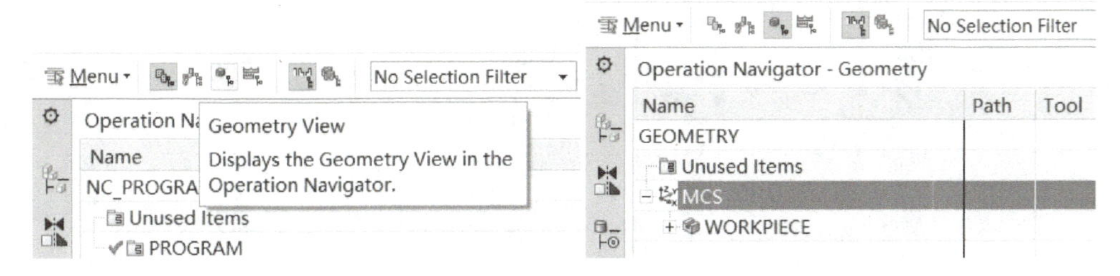

Fig. 6-4 Enter MCS

(5) In "MCS" window, specify the upper surface center of the part as MCS. In "Clearance" tab, select "Bounding Cylinder" in the pull-down box of "Clearance Option", and modify the value of "Safe Clearance Distance" as 10. Then, click "OK". The settings are shown in Fig. 6-5.

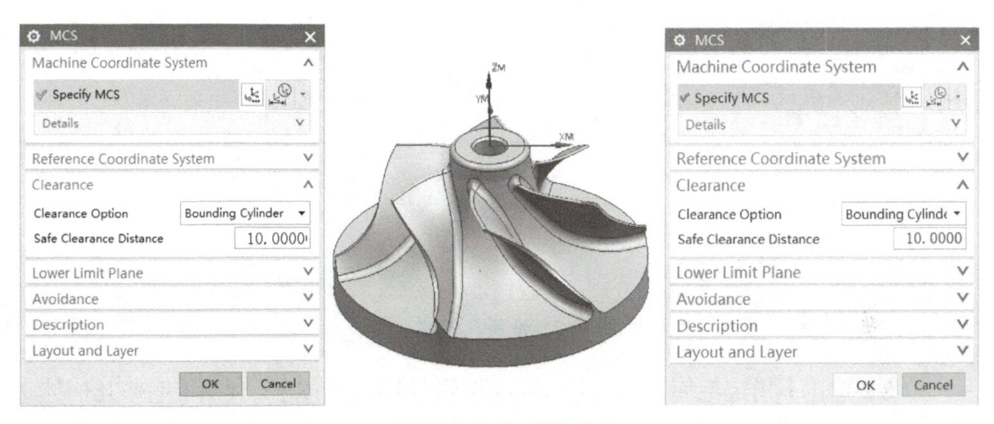

Fig. 6-5 Set MCS

(6) Open the child menu of "MCS" tree and right click on "WORKPIECE" to select "Edit". In "Workpiece" window, click "Specify Part". The steps are shown in Fig. 6-6.

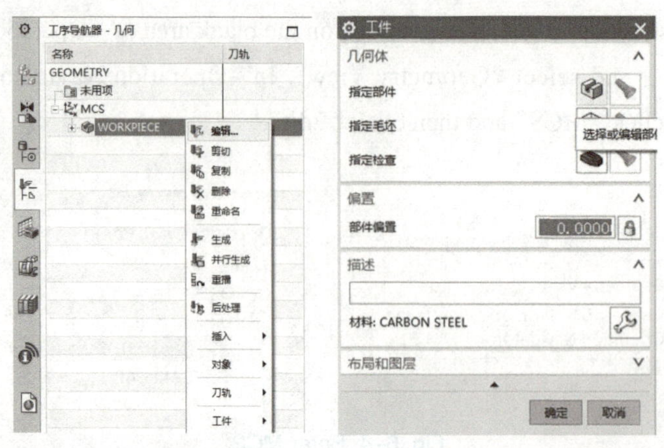

图 6-6 进入工件

（7）弹出【部件几何体】对话框，选择叶轮几何体，如图 6-7 所示，单击【确定】。

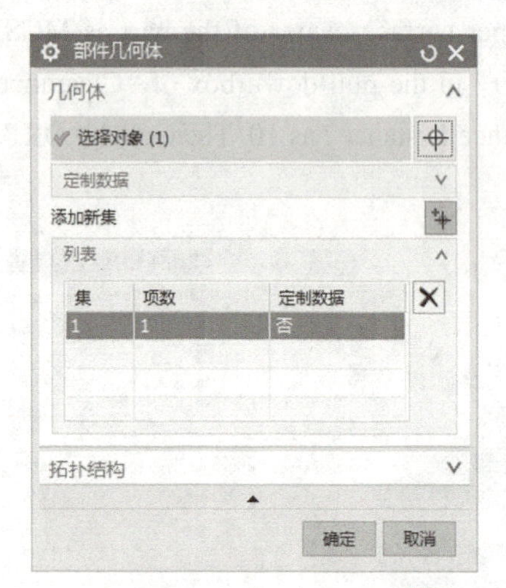

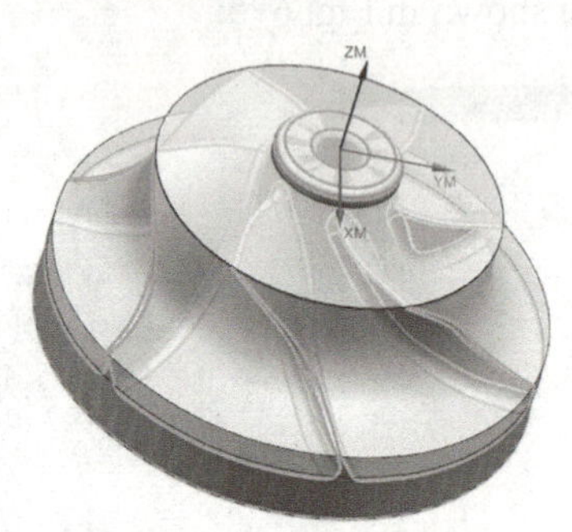

图 6-7 设置几何体

（8）单击【指定毛坯】中的【选择或编辑毛坯几何体】，弹出【毛坯几何体】对话框，选择毛坯几何体，如图 6-8 所示，依次单击【确定】。

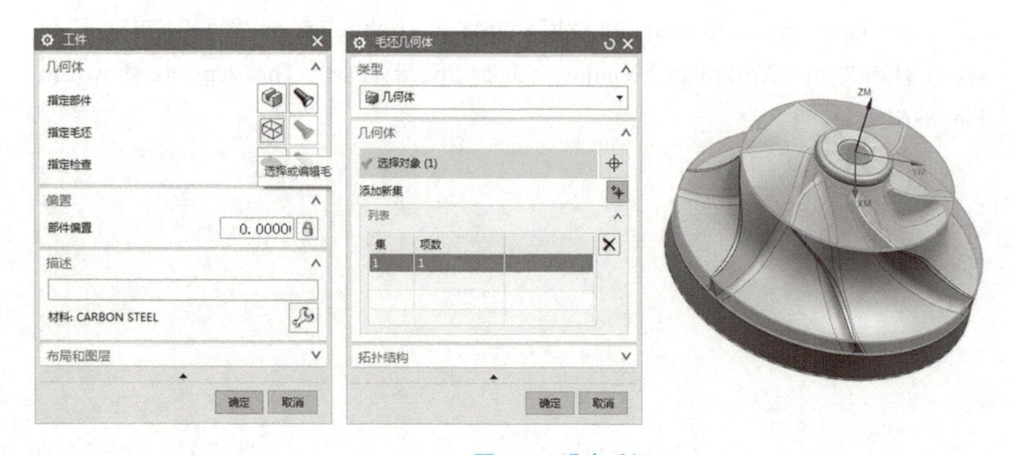

图 6-8 设定毛坯

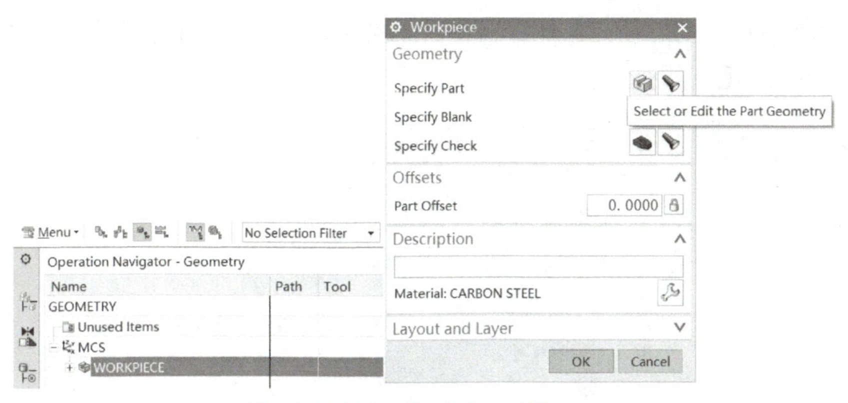

Fig. 6-6　Enter Workpiece Window

(7) In "Part Geometry" window, select the impeller as shown in Fig. 6-7 and click "OK".

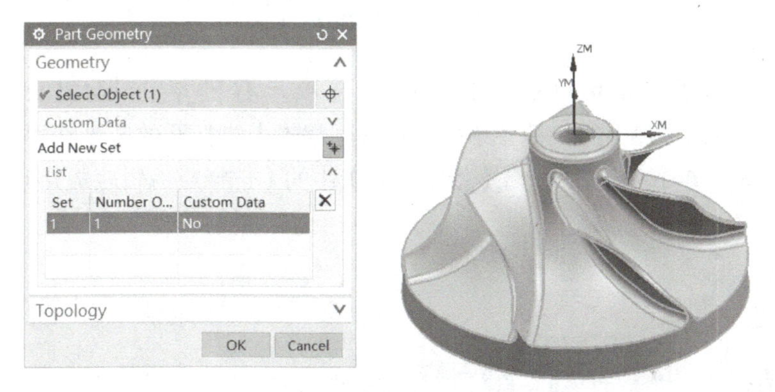

Fig. 6-7　Specify Part Geometry

(8) In "Workpiece" window, click "Specify Blank" to select the blank as shown in Fig. 6-8 and click "OK". Back to "Workpiece" window, click "OK" again.

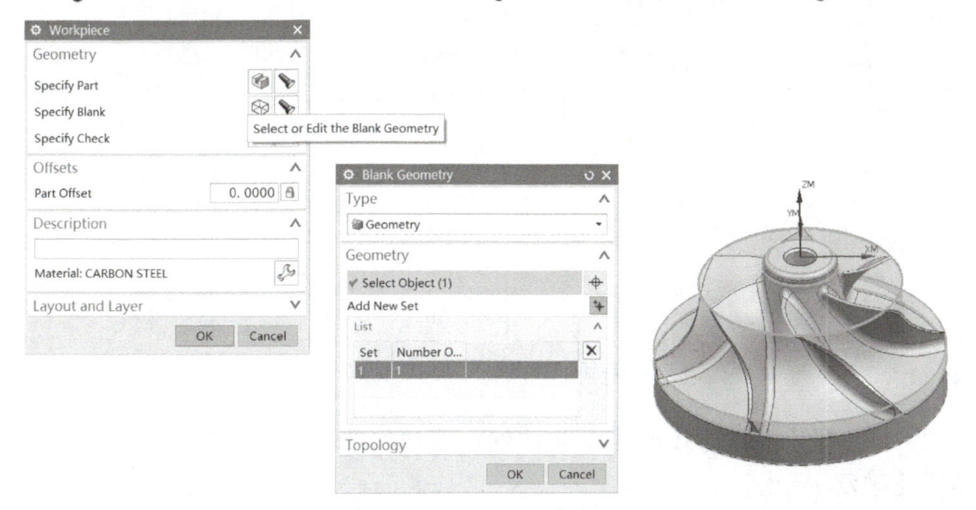

Fig. 6-8　Specify Blank

（9）右键单击毛坯几何体，在弹出的快捷菜单中选择【隐藏】，如图 6-9 所示，不显示毛坯几何体。

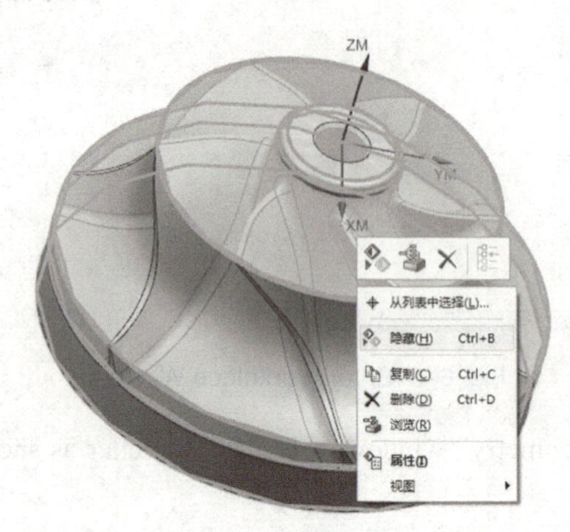

图 6-9 隐藏毛坯

（10）打开【WORKPIECE】下的子选项，右键单击【MULTI_BLADE_GEOM】选项，在弹出的快捷菜单中，选择【编辑】，打开的【多叶片几何体】对话框，如图 6-10 所示，单击【指定轮毂】中的【选择或编辑轮毂几何体】。

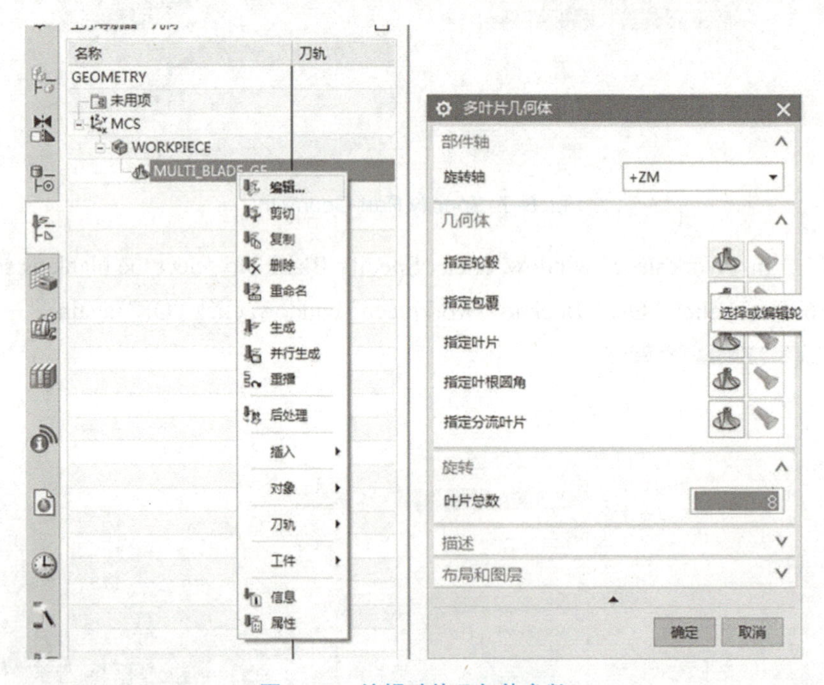

图 6-10 编辑叶片几何体参数

（11）打开【轮毂几何体】对话框，选择轮毂实体，如图 6-11 所示，单击【确定】。

(9) As shown in Fig. 6-9, right click on the blank and select "Hide" in pop-up menu to hide the blank.

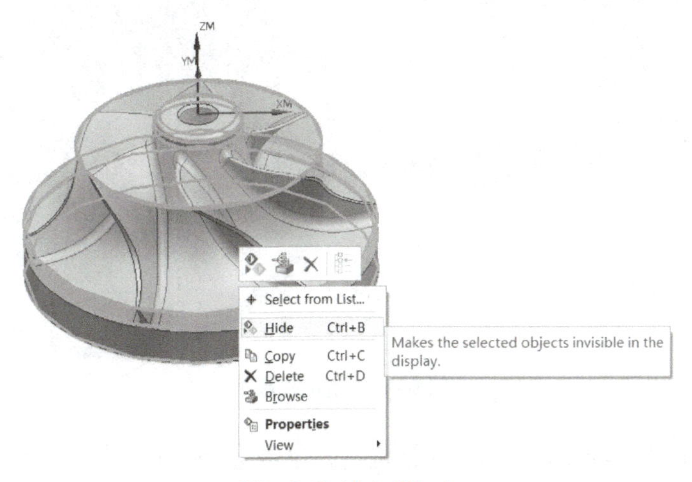

Fig. 6-9 Hide Blank

(10) Open the child menu of "WORKPIECE" and right click "MULTI_BLADE_GEOM" to select "Edit" in the pop-up menu. In "Multi Blade Geom" window, click "Specify Hub" button shown in Fig. 6-10.

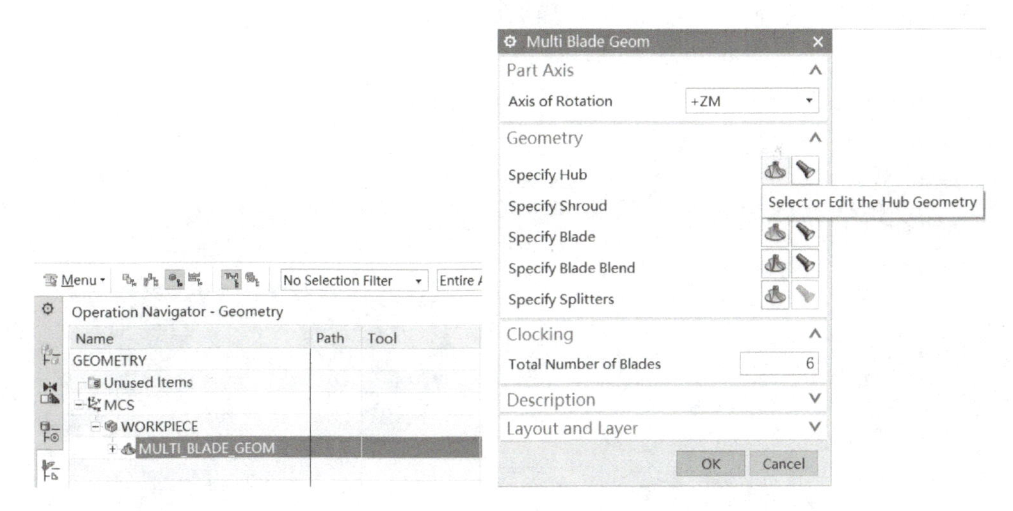

Fig. 6-10 Edit Blade Geometry

(11) In "Hub Geometry" window, select the hub as shown in Fig. 6-11, and click "OK".

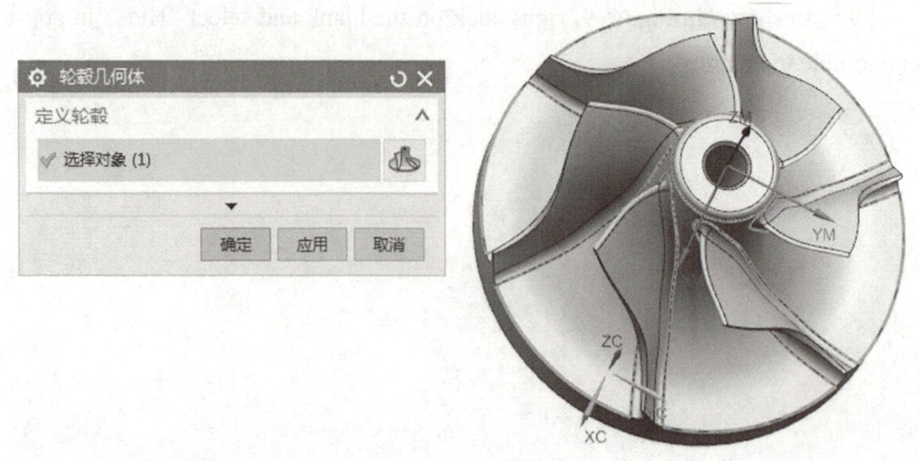

图 6-11　指定轮毂

（12）在【多叶片几何体】对话框中，单击【指定包覆】中的【选择或编辑包覆几何体】，弹出【包覆几何体】对话框，选择叶轮外包络面，如图 6-12 所示，单击【确定】。

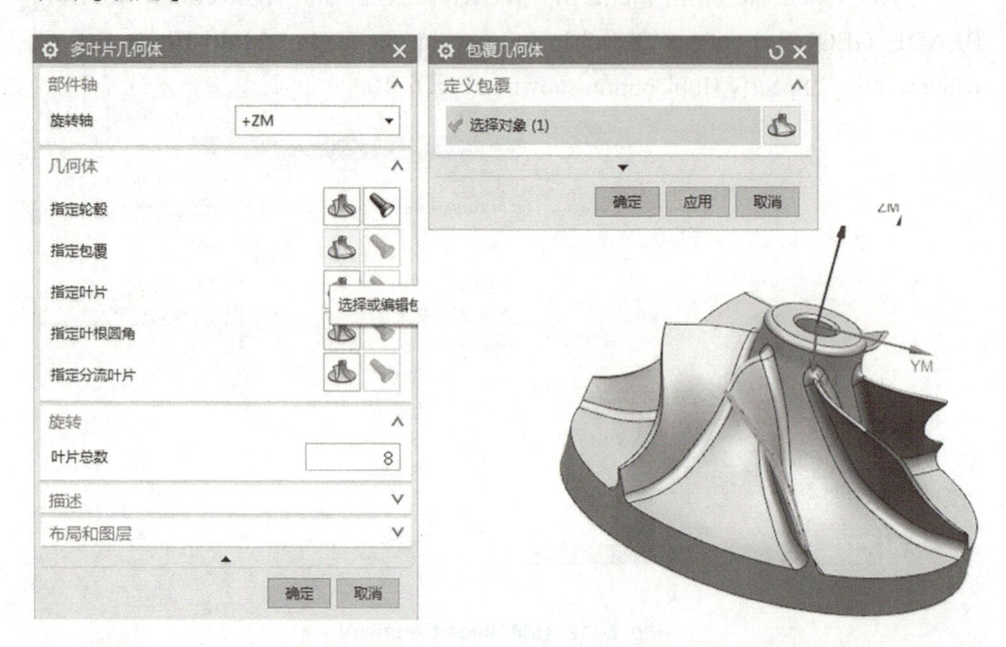

图 6-12　指定包覆

（13）在【多叶片几何体】对话框中，单击【指定叶片】中的【选择或编辑叶片几何体】，弹出【叶片几何体】对话框，选择叶片曲面（3 个曲面片），如图 6-13 所示，单击【确定】。

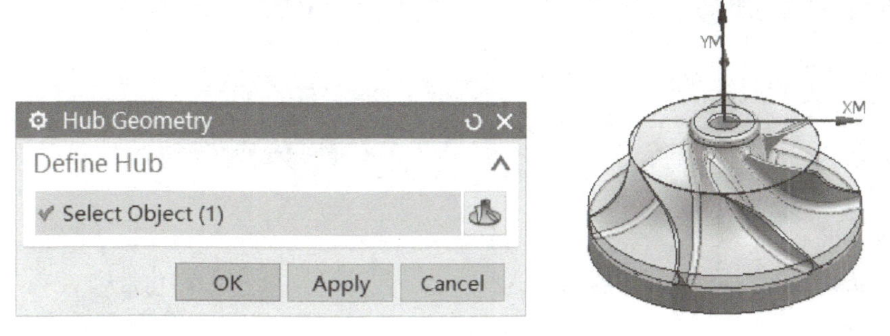

Fig. 6-11 Specify Hub

(12) In "Multi Blade Geom" window, click "Specify Shroud" button. In "Shroud Geometry" window, select outer surface of the impeller as shown in Fig. 6-12 and click "OK".

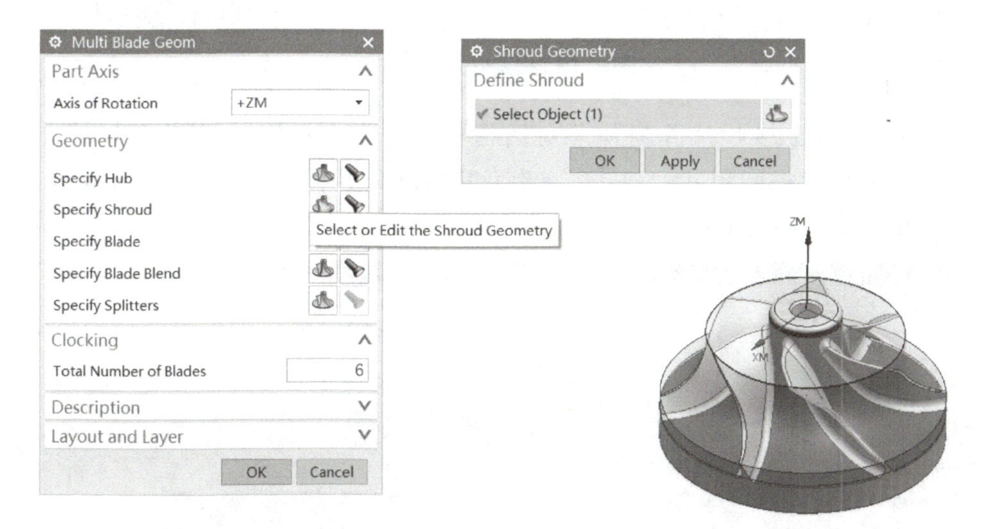

Fig. 6-12 Specify Shroud

(13) In "Multi Blade Geom" window, click "Specify Blade" button. In "Blade Geometry" window, select the blade (i.e. consist three surfaces) as shown in Fig. 6-13. Click "OK".

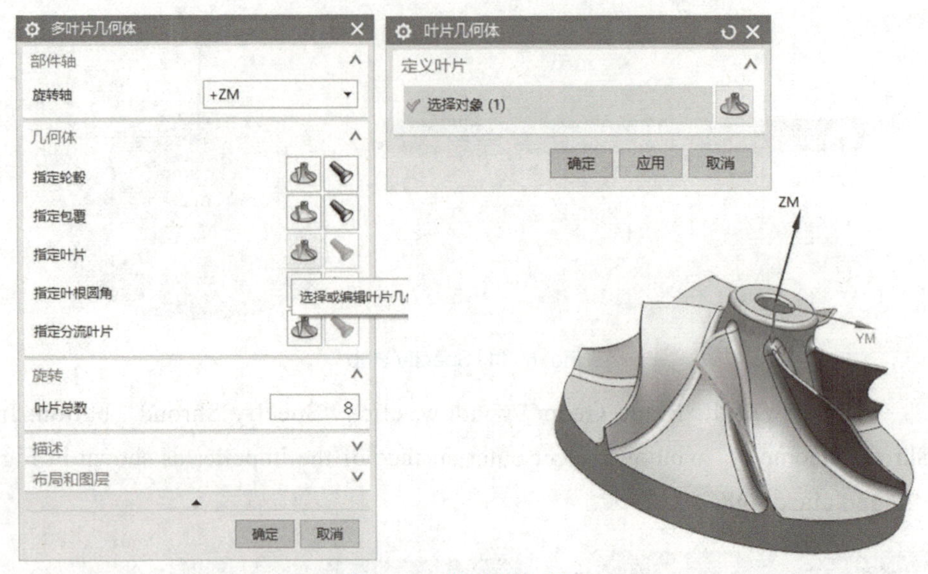

图 6-13 指定叶片曲面

（14）在【多叶片几何体】对话框中，单击【指定叶根圆角】中的【选择或编辑叶根圆角几何体】，弹出【叶根圆角几何体】，选择叶跟圆角（3 个曲面片），如图 6-14 所示，单击【确定】。

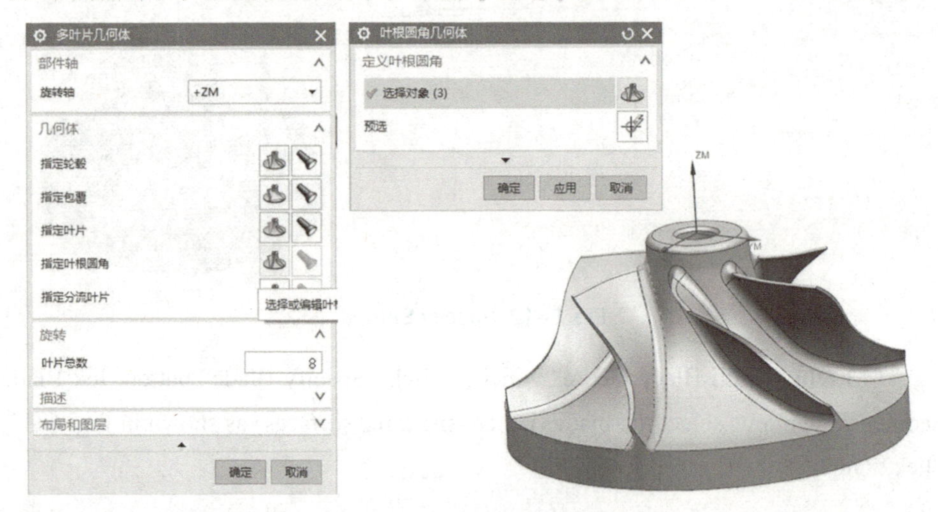

图 6-14 指定叶根圆角

（15）在【多叶片几何体】对话框中，修改【叶片总数】为【6】，如图 6-15 所示，单击【确定】。

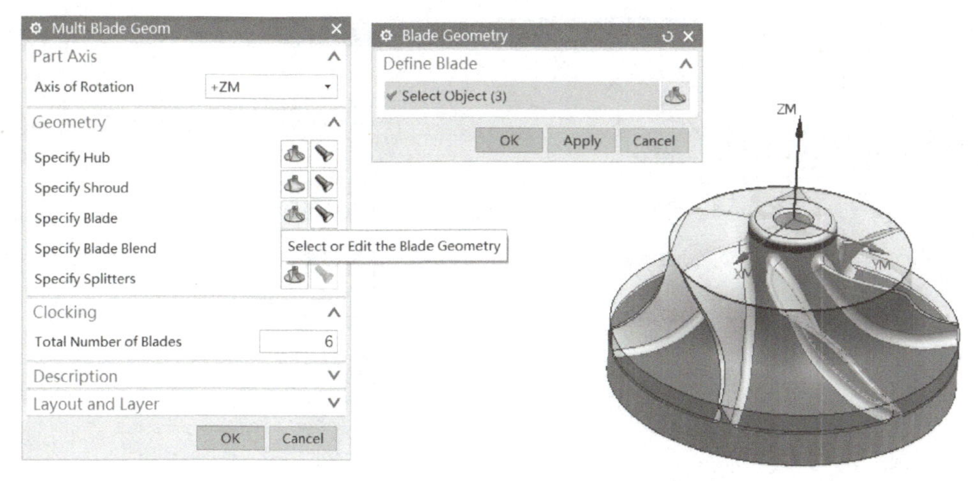

Fig. 6-13 Specify Blade

(14) In "Multi Blade Geom" window, click "Specify Blade Blend" button. In "Blade Blend Geometry" window, select the blade blend (i.e. consist three surfaces) as shown in Fig. 6-14. Click "OK".

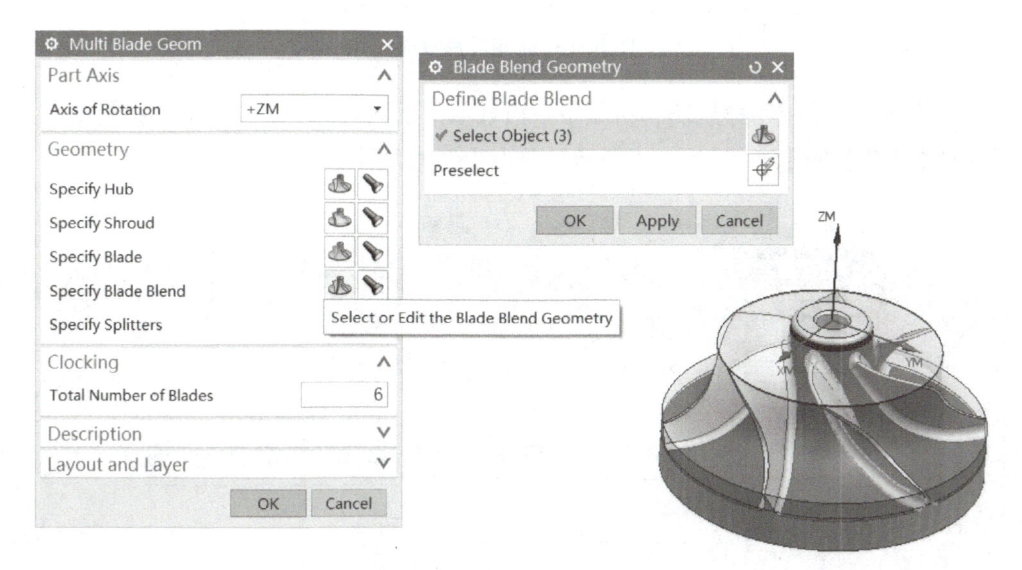

Fig. 6-14 Specify Blade Blend

(15) In "Multi Blade Geom" window, modify the value of "Total Number of Blades" as 6, as shown in Fig. 6-15 and click "OK".

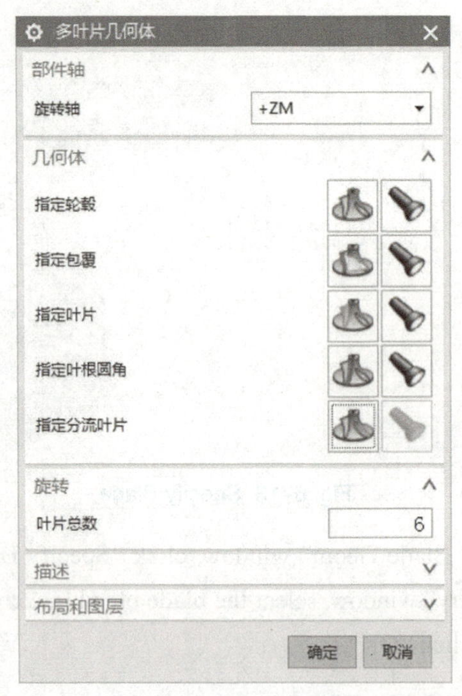

图 6-15 指定叶片数量

（16）在【工序导航器 - 机床】空白处单击鼠标右键，在弹出的快捷菜单中选择【机床视图】，如图 6-16 所示，单击【创建刀具】。

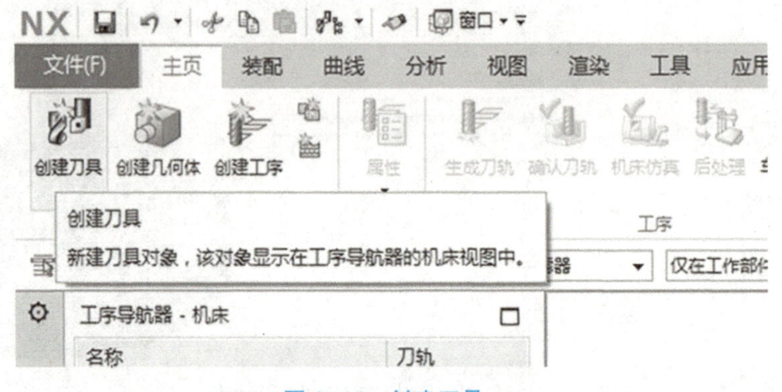

图 6-16 创建刀具

（17）在打开的【创建刀具】对话框中，【刀具子类型】选择【球头铣刀】，修改【刀具】名称【BALL_MILL_8】，单击【确定】。弹出【铣刀 - 球头铣】对话框，在【工具】选项卡中设置刀具参数：【球直径】为【8】，【长度】为【50】，【刀刃长度】为【30】，【刀具号】为【1】，【补偿寄存器】为【1】，【刀具补偿寄存器】为【1】，如图 6-17 所示。

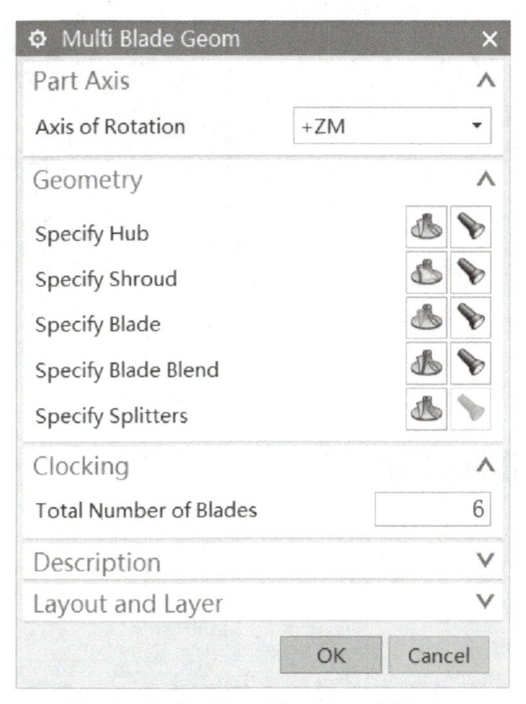

Fig. 6–15 Set Total Number of Blades

(16) Right click on the blank area of "Operation Navigator-Machine Tool", and select "Machine Tool View", then click "Create Tool" button, as shown in Fig. 6-16.

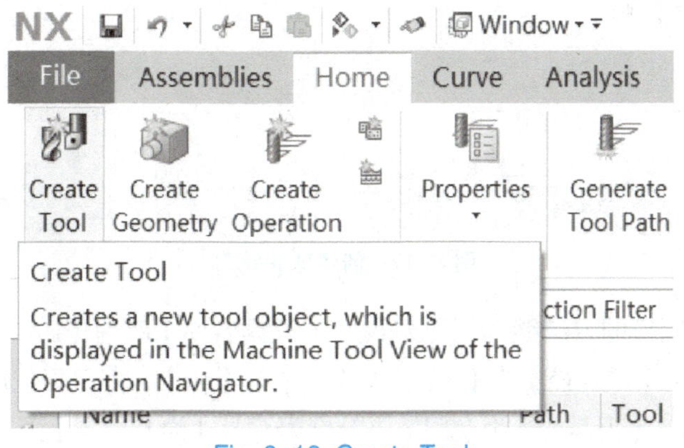

Fig. 6–16 Create Tool

(17) As shown as Fig. 6-17, in "Create Tool" window, select "ball mill" in "Tool Subtype" and modify "Name" as "BALL_MILL_8", then click "OK". In "Milling Tool-Ball Mill" window, "Ball Diameter" value is 8,"Length" value is 50, and "Flute Length" value is 30. In "Numbers" tab, "Tool Number" value is 1, "Adjust Register" value is 1 and "Cutcom Register" value is 1. Click "OK" to close the window.

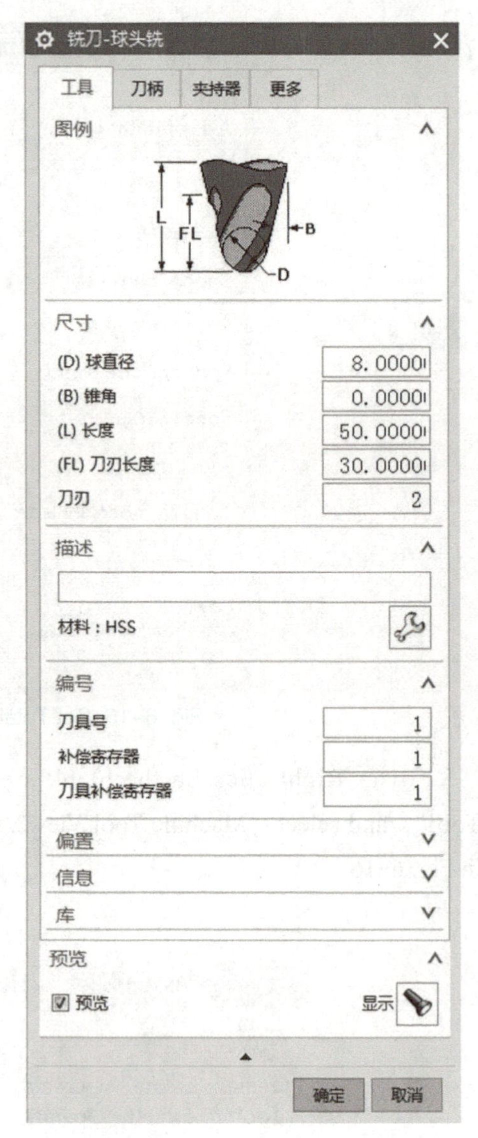

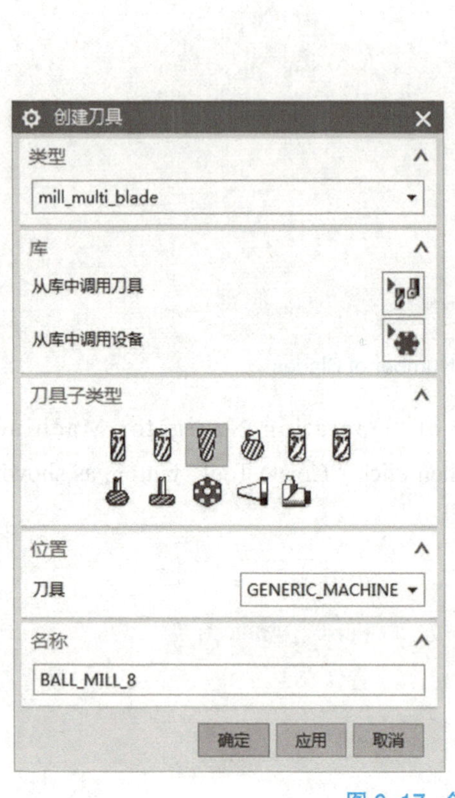

图 6–17 创建球头铣刀

（18）用同样的方法继续创建刀具，【刀具名称】改为【BALL_MILL_6】，设置【尺寸】中【球直径】为【6】，【长度】为【50】，【刀刃长度】为【20】，【刀刃】为【2】，【刀具号】为【2】，【补偿寄存器】为【2】，【刀具补偿寄存器】为【2】，如图 6–18 所示。单击【确定】，完成刀具设置。

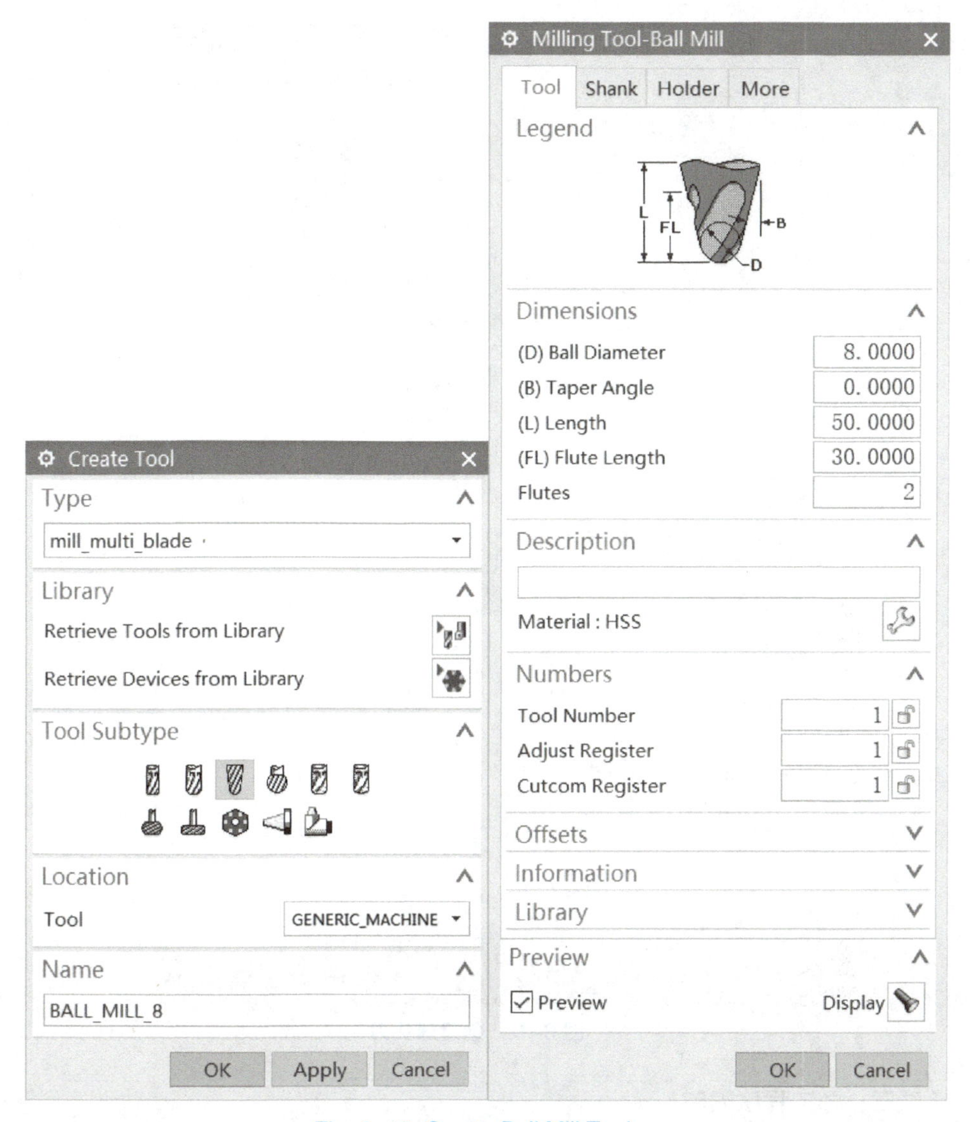

Fig. 6-17 Create Ball Mill Tool

(18) Similarly, create a new ball mill tool named as "BALL_MILL_6". In "Milling Tool-Ball Mill" window, "Ball Diameter" is 6, "Length" is 50, "Flute Length" is 20 and "Flute" is 2. In "Numbers" tab, "Tool Number" value is 2, "Adjust Register" is 2 and "Cutcom Register" is 2. The settings are shown in Fig. 6-18.

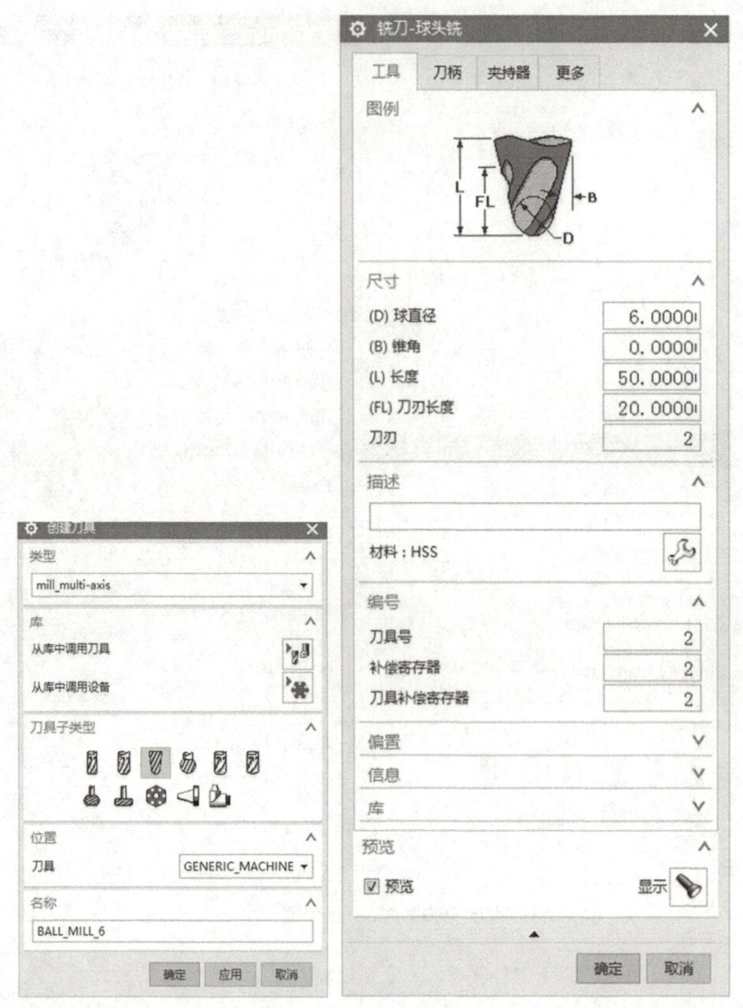

图 6-18　创建球头刀

三、加工程序编制

1. 叶片粗加工

（1）单击【创建工序】，弹出【创建工序】对话框，【工序子类型】选择【叶片粗加工】，【程序】选择【PROGRAM】，【刀具】选择【BALL_MILL_8】，【几何体】选择【MULTI_BLADE_GEOM】，【方法】选择【MILL_ROUGH】，如图 6-19 所示，单击【确定】。

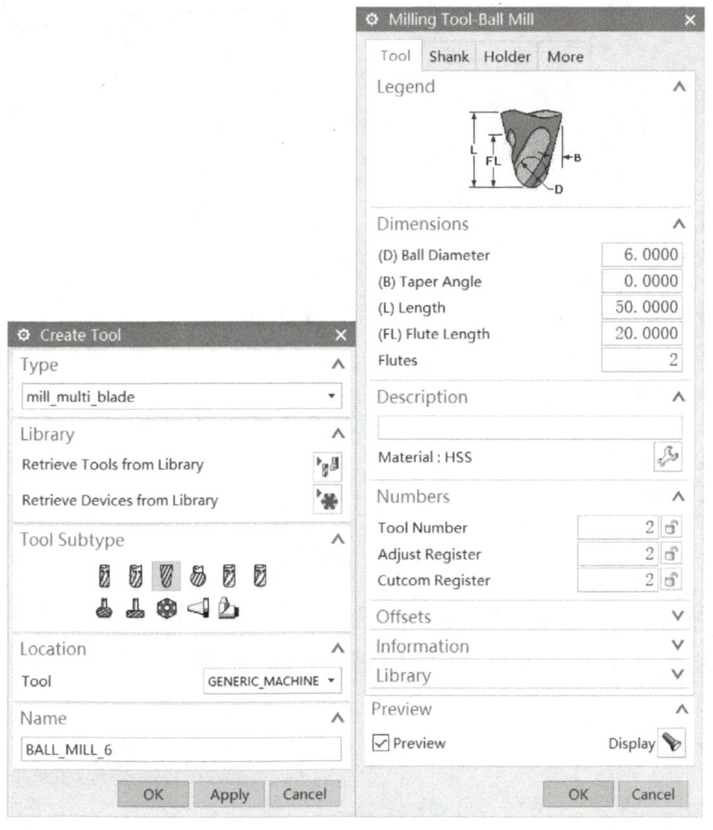

Fig. 6–18　Create Ball Mill Tool

Ⅲ. Programming

1. Blade Rough Machining

(1) As shown in Fig. 6-19, click “Create Operation”. Select “mill_multi_blade” in pull-down box of “Type”. Select operation subtype as “Multi Blade Rough”, program as “PROGRAM”, tool as “BALL_MILL_8”, geometry as “MULTI_BLADE_GEOM” and method as “MILL_ROUGH”. Then, click “OK”.

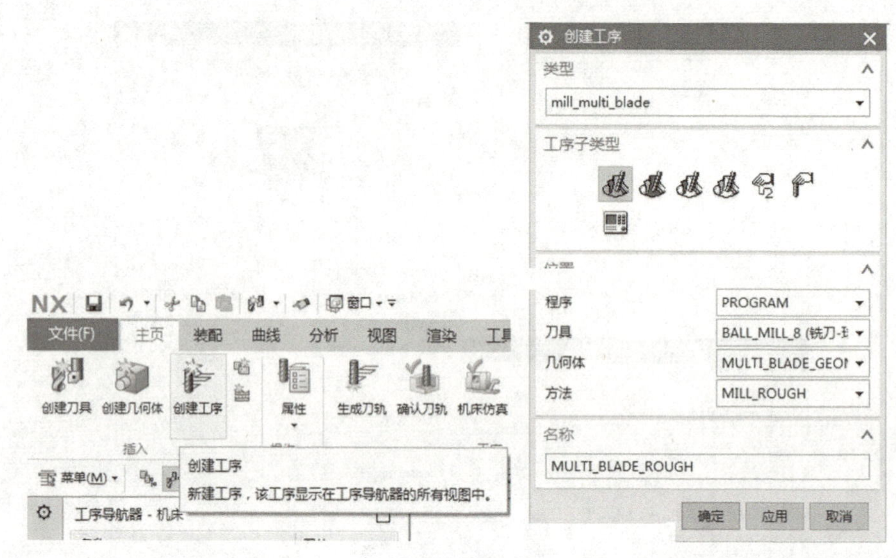

图 6-19 创建工序

（2）弹出【多叶片粗加工】对话框，【驱动方法】选项中【叶片粗加工】，弹出【叶片粗加工驱动方法】，默认窗口中的参数，如图 6-20 所示，单击【确定】。

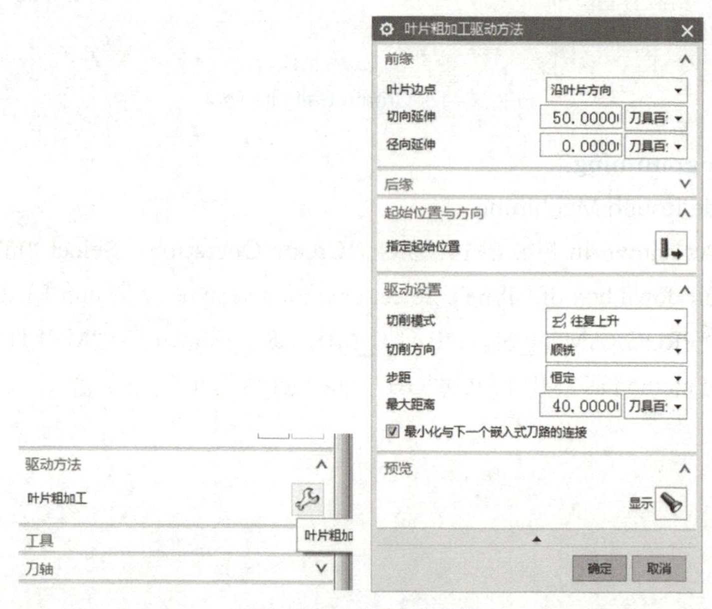

图 6-20 设置驱动方法

（3）【刀轴】选项中【轴】选择【自动】，单击【刀轨设置】中【切削层】。弹出【切削层】对话框，【深度选项】选项下的【深度模式】选择【从轮毂偏置】，【每刀切削深度】选择【恒定】，【距离】设置为【100】，如图 6-21 所示，单击【确定】。

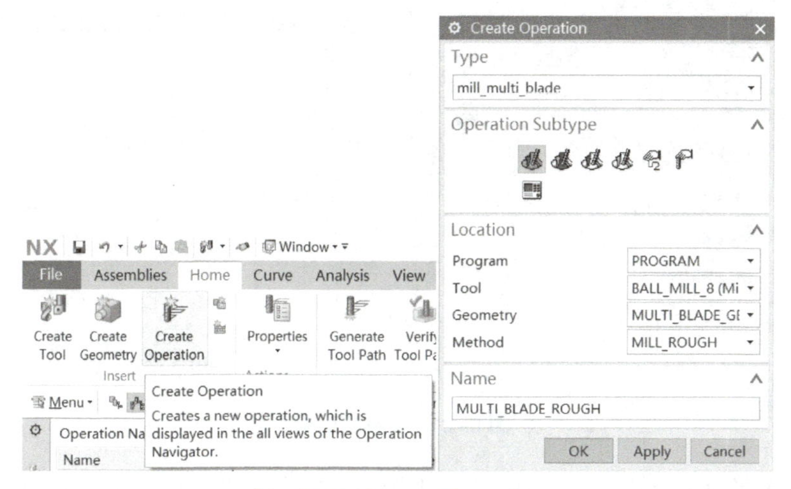

Fig. 6–19 Create Operation

(2) In “Multi Blade Rough” window, click “Blade Rough” button in “Drive Method” tab. In “Blade Rough Drive Method” window, the parameter settings are shown in Fig. 6-20. Then, click“OK”.

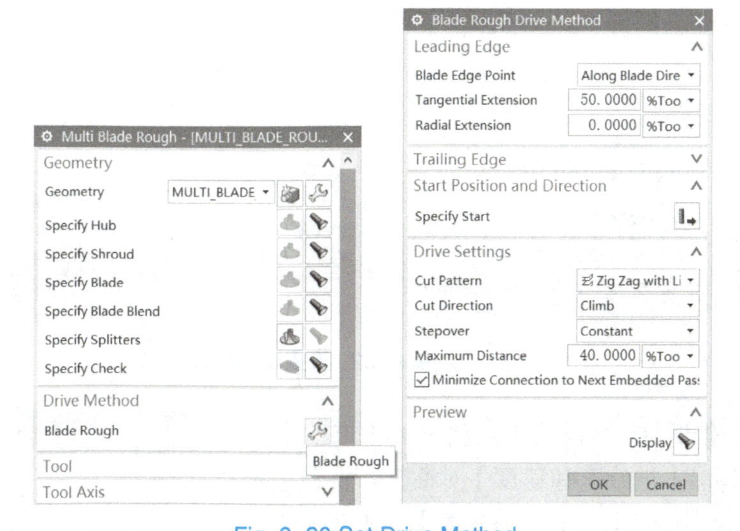

Fig. 6–20 Set Drive Method

(3) In “Tool Axis” tab, select “Automatic” in the pull-down box of “Axis”. Click “Cut Levels” button in “Path Settings” tab. In “Depth Options” tab of “Cut Levels” window, select “Offsets from Hub” in the pull-down box of “Depth Mode”. “Depth per Cut” is “Constant” and “Distance” value is 100. Click “OK”.The settings are shown in Fig. 6-21.

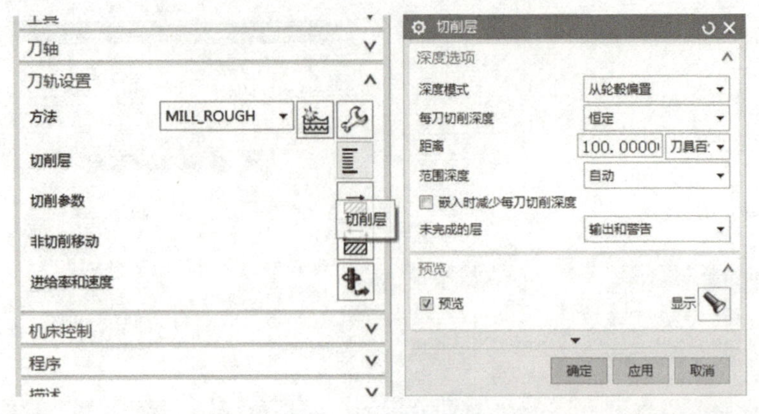

图 6-21　设置切削层

（4）单击【刀轨设置】选项中的【切削参数】，弹出【切削参数】对话框。单击【余量】选项卡，设置【叶片余量】为【0.3】，【轮毂余量】为【0.3】，【检查余量】为【0.3】，如图 6-22 所示，单击【确定】。

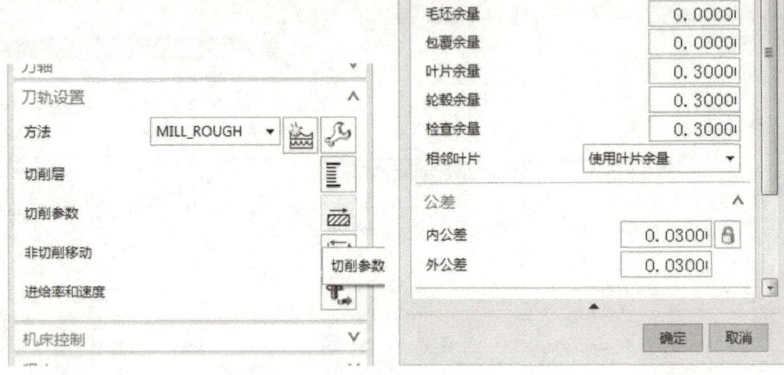

图 6-22　设置切削参数

（5）单击【刀轨设置】选项中的【进给率和速度】，弹出【进给率和速度】对话框。设置【表面速度】为【1500】，【主轴速度】为【5000】，如图 6-23 所示，单击【确定】。

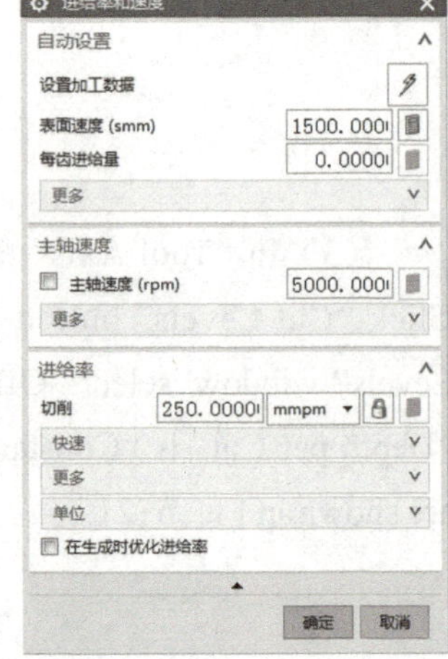

图 6-23　设置进给率和速度

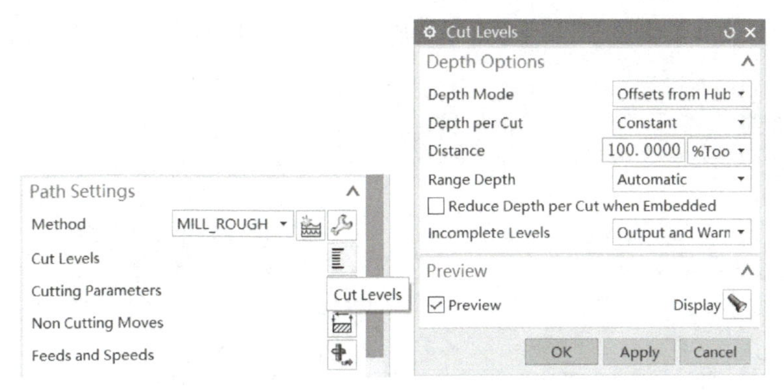

Fig. 6-21　Set Cut Levels

(4) Click "Cutting Parameters" button in "Path Settings" tab. In "Cutting Parameters" window, click "Stock" tab and set "Blade Stock" value as 0.3, "Hub Stock" value as 0.3 and "Check Stock" value as 0.3. Click "OK". The settings are shown in Fig. 6-22.

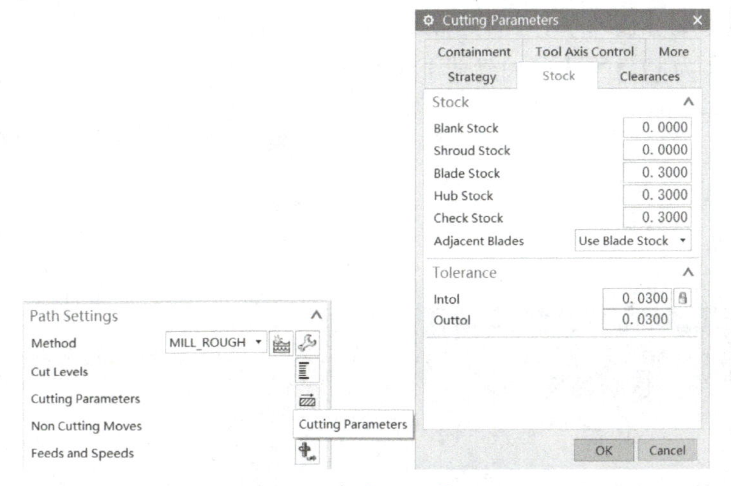

Fig. 6-22　Modify Cutting Parameters

(5) Click "Feeds and Speeds" button in "Path Settings" tab. In "Feeds and Speeds" window, "Surface Speed" value is 1,500 and "Spindle Speed" value is 5,000. Click "OK". Feeds and speeds settings are shown in Fig. 6-23.

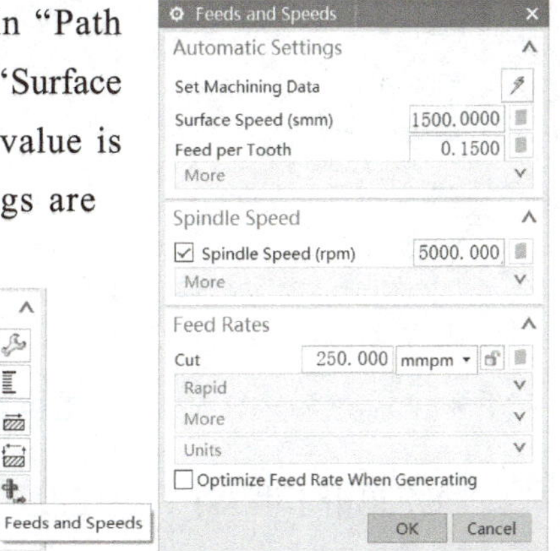

Fig. 6-23　Modify Feeds and Speeds

（6）单击【生成】，创建刀具轨迹，如图 6-24 所示，检查刀具轨迹。

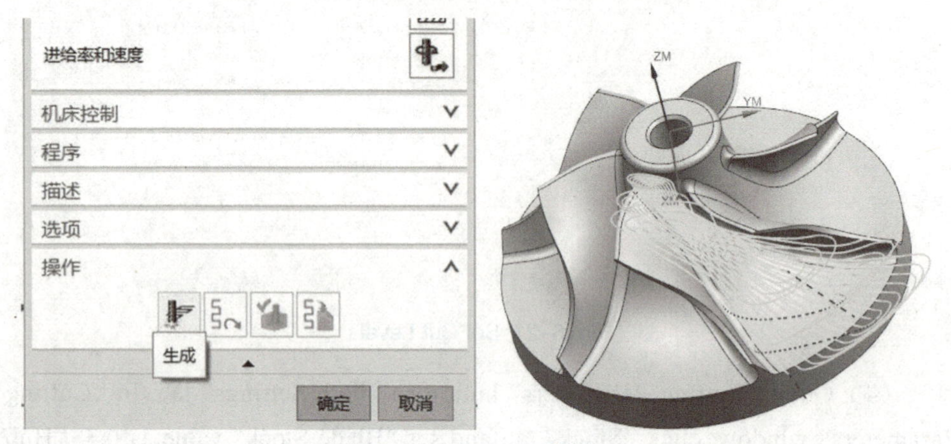

图 6-24　生成刀具轨迹

（7）单击【驱动方法】选项中【叶片粗加工】，弹出【叶片粗加工驱动方法】对话框。【前缘】选项中【叶片边点】选择【沿部件轴】，【距离】设置为【15】，【切向延伸】为【0】，【径向延伸】为【150】，如图 6-25 所示，单击【确定】。

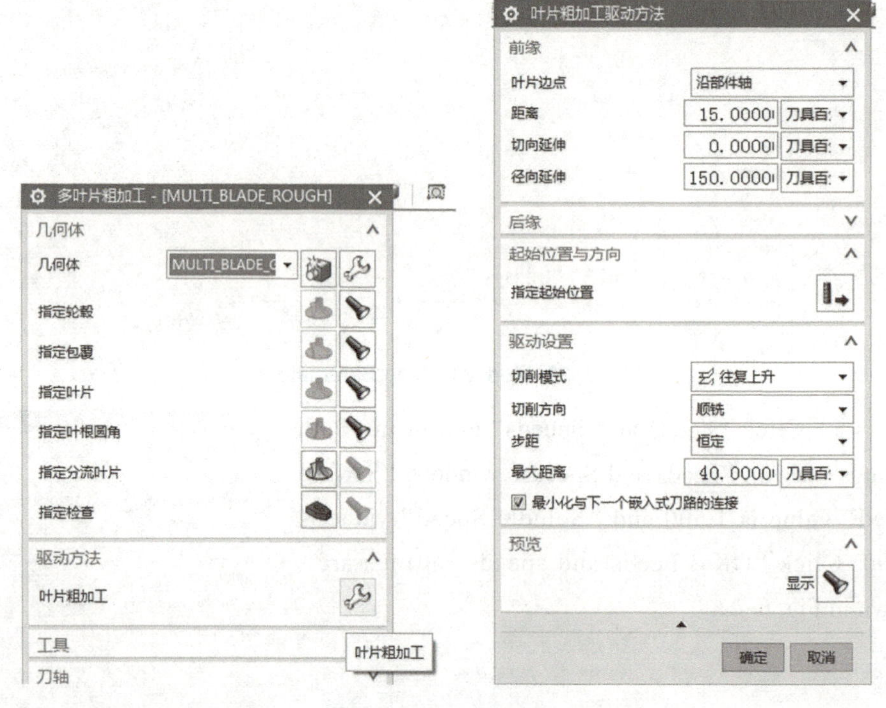

图 6-25　设置驱动方法

（8）单击【生成】，创建刀具轨迹，如图 6-26 所示，检查刀具轨迹。

(6) Click "Generate" to view the tool path as shown in Fig. 6-24.

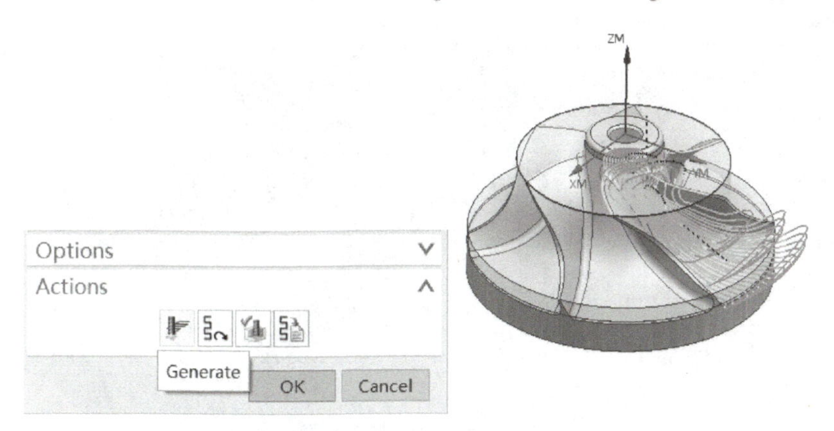

Fig. 6-24 Generate Tool Path

(7) Click "Blade Rough" button in "Drive Method" tab. In "Leading Edge" tab of "Blade Rough Drive Method" window, select "Along Part Axis" in the pull-down box of "Blade Edge Point", and "Distance" value is 15, "Tangential Extension" value is 0 and "Radial Extension" value is 150. Click "OK". Drive method settings are shown in Fig. 6-25.

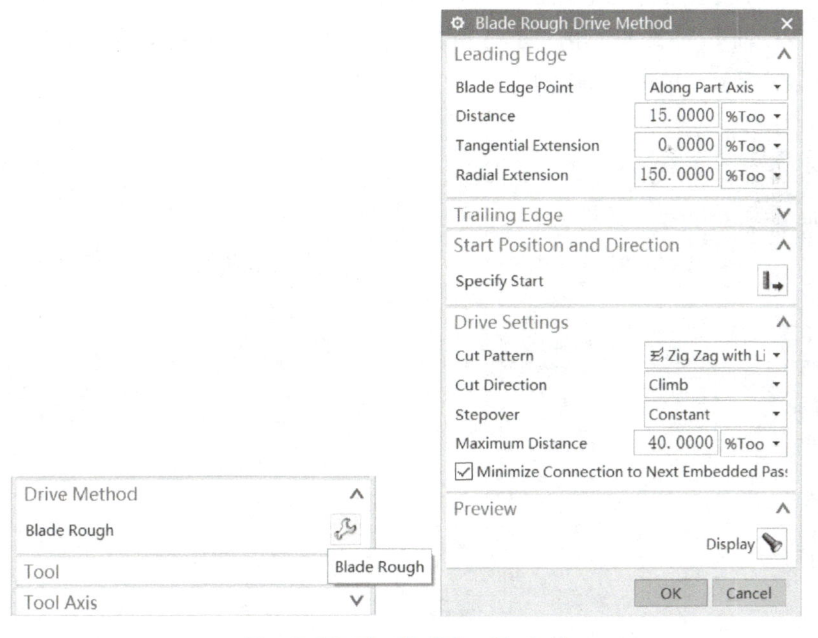

Fig. 6-25 Modify Drive Method

(8) Click "Generate" button to view the tool path as shown in Fig. 6-26.

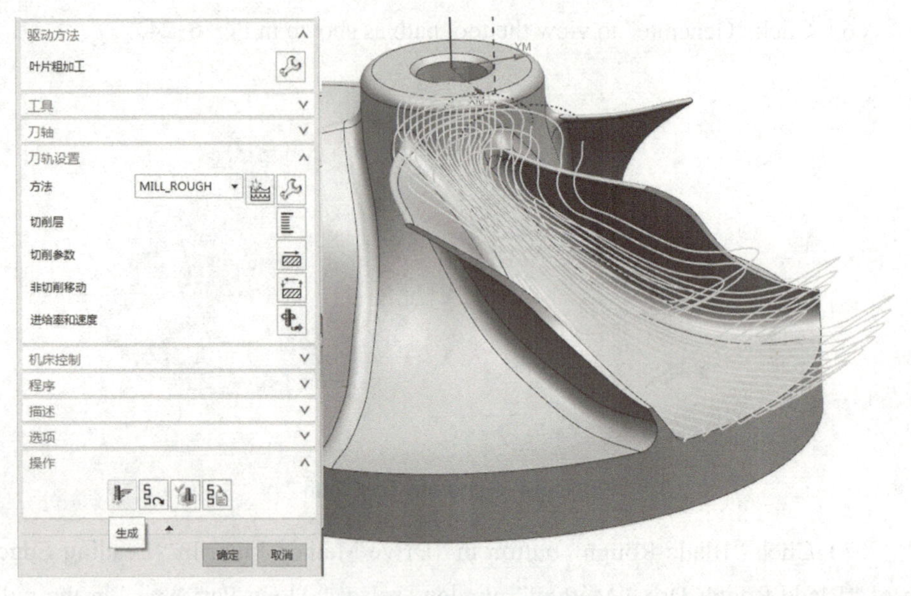

图 6–26 生成刀具轨迹

（9）右键单击【工序导航器 - 程序顺序】中的【MULTI_BLADE_GEOM】，在弹出的快捷菜单中选择【对象】→【变换】。在弹出的【变换】对话框中，设置【类型】为【绕直线旋转】，【指定点】选择坐标原点，【指定矢量】选择【+ZC 轴】，【距离 / 角度分割】设置为【6】，【非关联副本数】设置为【5】，如图 6–27 所示，单击【确定】。

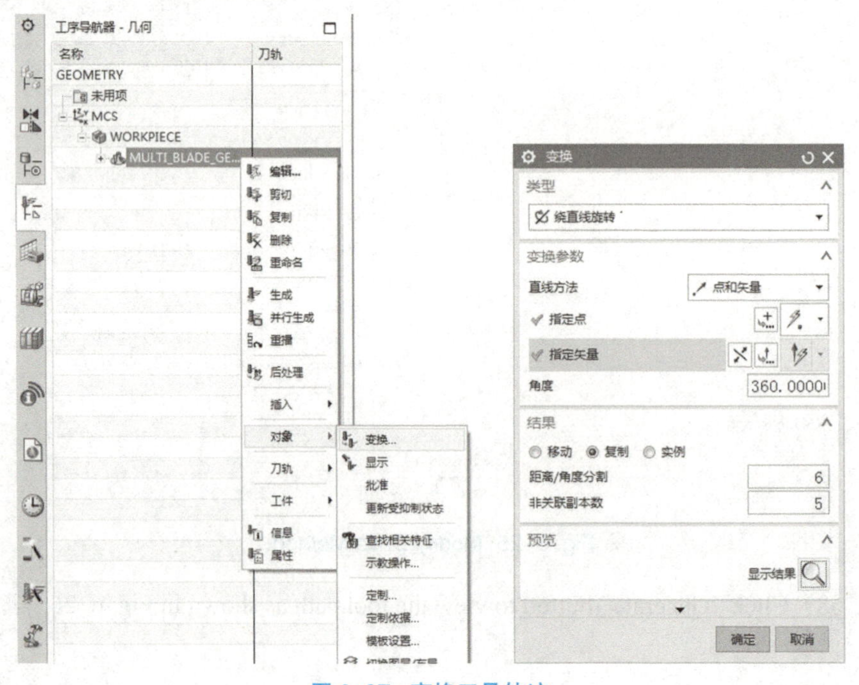

图 6–27 变换刀具轨迹

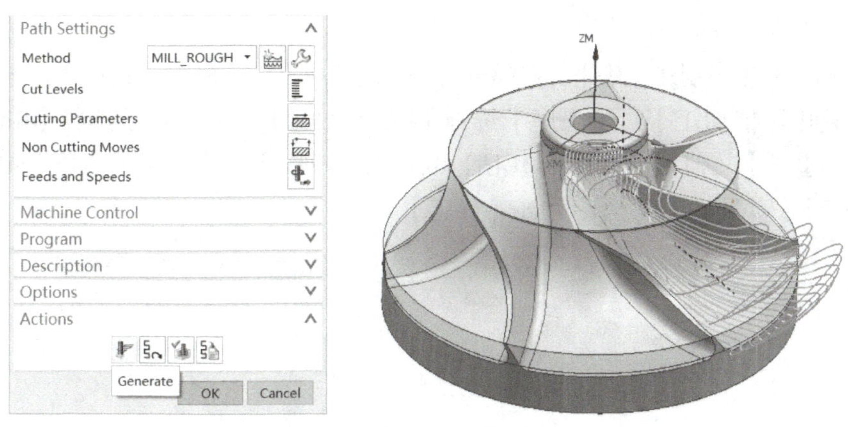

Fig. 6-26 Generate Tool Path

(9) Right click "MULTI_BLADE_GEOM" in "Operation Navigator-Geometry" and move the cursor to the "Object" to select "Transform". In "Transformations" window, select "Rotate About a Line" in the pull-down box of "Type". In "Transformations Parameters" tab, "Line Method" is "Point and Vector". Then, specify the origin of coordinate as the specified point and "+ZC Axis" as the specified vector. The "Distance/Angle Divisions" value is 6 and "Number of Non-associative Copies" value is 5. Click "OK". The settings are shown in Fig. 6-27.

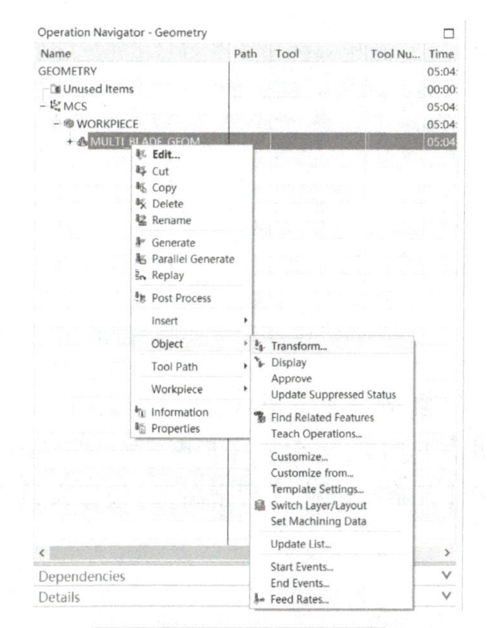

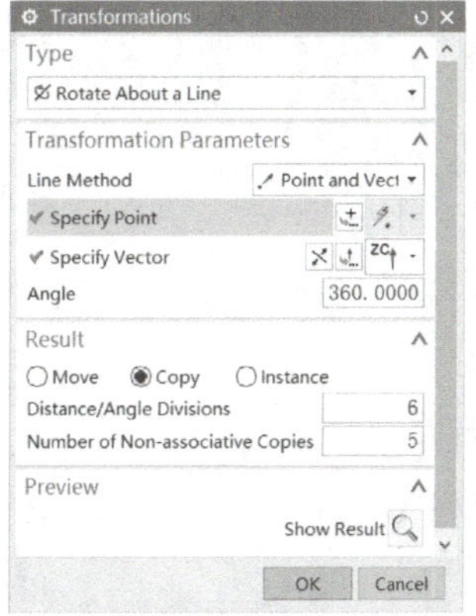

Fig 6-27 Change Tool Path

2. 叶片精加工

（1）单击【创建工序】，在弹出的【创建工序】对话框中，【工序子类型】选择【叶片精加工】，【程序】选择【PROGRAM】，【刀具】选择【BALL_MILL_6】，【几何体】选择【MULTI_BLADE_GEOM】，【方法】选择【MILL_FINISH】，如图 6-28 所示，单击【确定】。

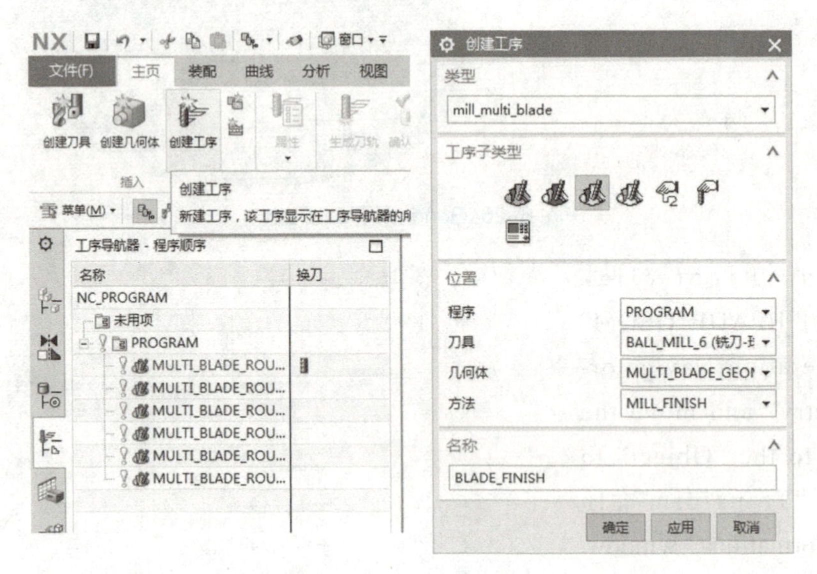

图 6-28 创建工序

（2）弹出【叶片精加工】对话框，单击【刀轨设置】选项中的【切削层】，弹出【切削层】对话框，【深度模式】选择【从包覆插补至轮毂】，【每刀切削深度】选择【残余高度】，【残余高度】设置为【0.05】，如图 6-29 所示，单击【确定】。

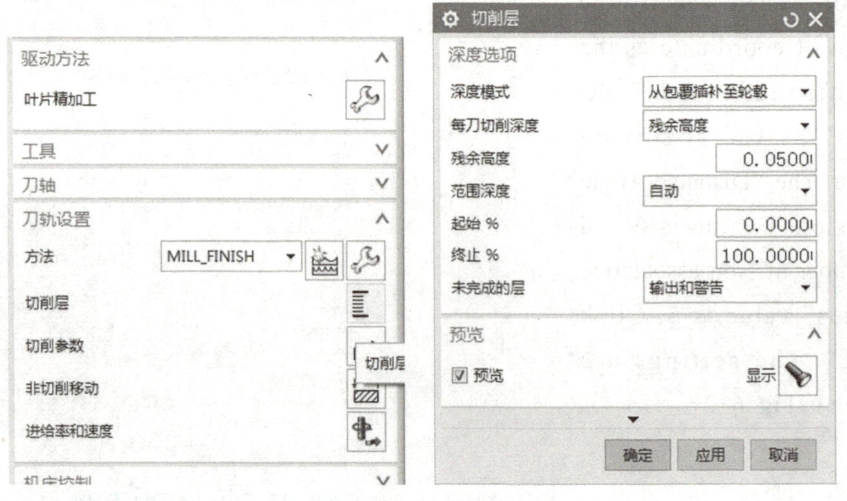

图 6-29 设置切削层

2. Blade Finish Machining

(1) As shown in Fig. 6-28, click "Create Operation", and then select "mill_multi_blade" in pull-down box of "Type". Select operation subtype as "blade finish", program as "PROGRAM", tool as "BALL_MILL_6", geometry as "MULTI_BLADE_GEOM" and method as " MILL_FINISH". Then, click "OK".

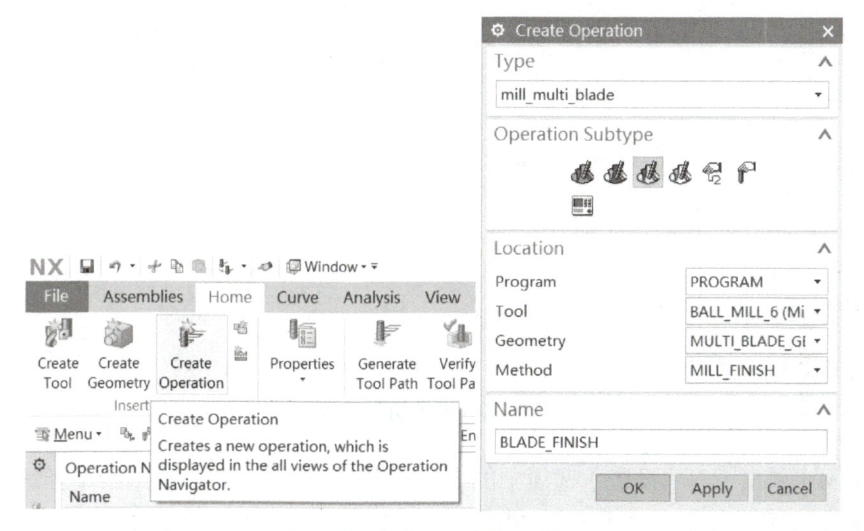

Fig. 6-28　Create Operation

(2) In "Blade Finish" window, click "Cut Levels" button in "Path Settings" tab. In "Depth Opinions" tab of "Cut Levels" window, select "Interpolate from Shroud to Hub" in the pull-down box of "Depth Mode". "Depth per Cut" is "Scallop" and "Scallop Height" value is 0.05. Click "OK". The settings are shown in Fig. 6-29.

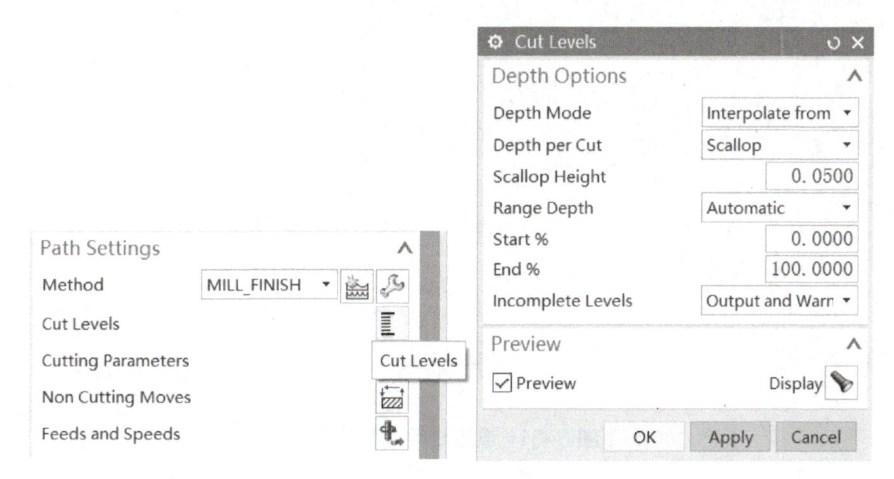

Fig. 6-29　Modify Cut Levels

（3）单击【刀轨设置】选项中的【切削参数】，弹出【切削参数】对话框，默认设置，如图 6-30 所示，单击【确定】。

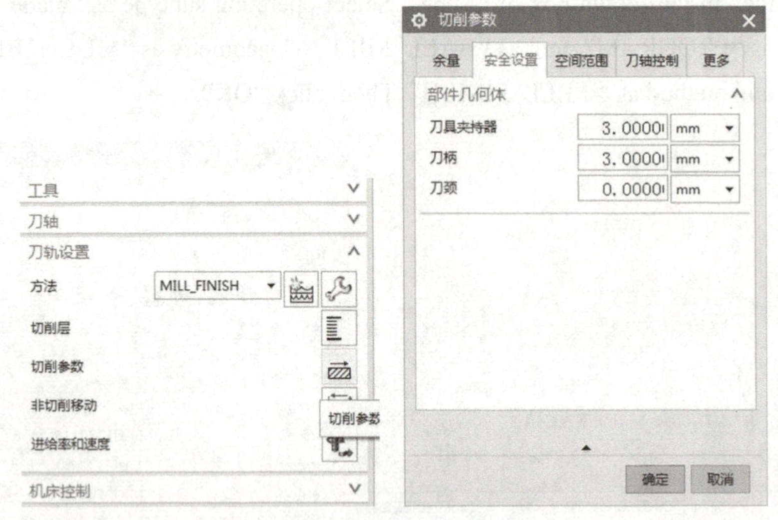

图 6-30 设置切削参数

（4）单击【刀轨设置】选项中的【进给率和速度】，弹出【进给率和速度】对话框，设置【表面速度】为【2500】，【主轴速度】为【8000】，单击【确定】，如图 6-31 所示。

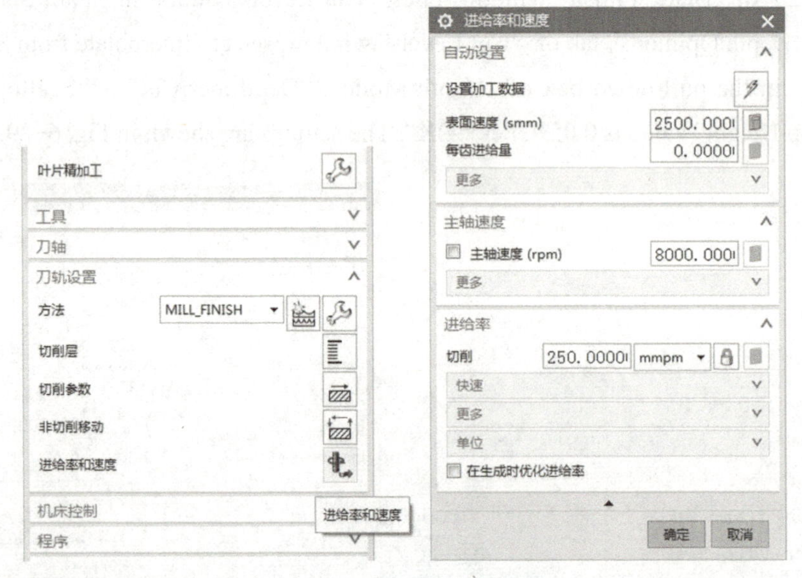

图 6-31 设置进给率和速度

（5）单击【生成】，生成叶片精加工刀具轨迹，检查生成的刀具轨迹，如图 6-32 所示。

(3) Click "Cutting Parameters" button in "Path Settings" tab. Parameter settings are shown in Fig. 6-30. Click "OK" to close the window.

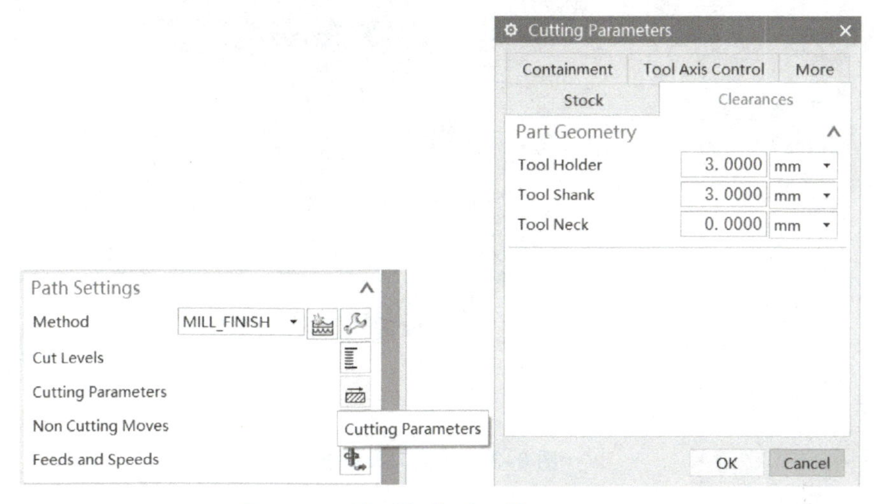

Fig. 6–30 Modify Cutting Parameters

(4) Click "Feeds and Speeds" button in "Path Settings" tab. In "Feeds and Speeds" window, "Surface Speed" value is 2,500 and "Spindle Speed" value is 8,000. Click "OK". Feeds and speeds settings are shown in Fig. 6-31.

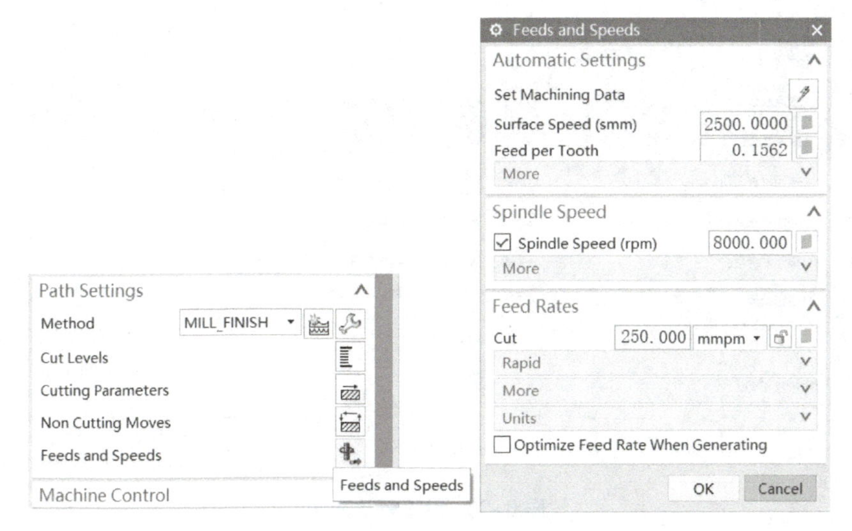

Fig. 6–31 Modify Feeds and Speeds

(5) Click "Generate" button to view the tool path as shown in Fig. 6-32.

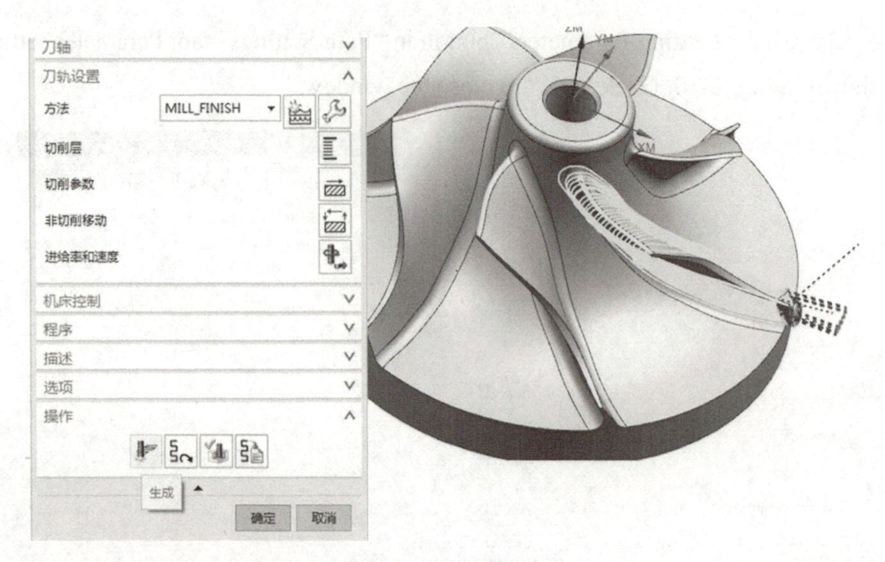

图 6-32 生成刀具轨迹

（6）单击【驱动方法】选项中的【叶片精加工】，弹出【叶片精加工驱动方法】对话框，设置【后缘】选项中【切向延伸】为【75】，单位为【刀具百分比】，如图 6-33 所示，单击【确定】。

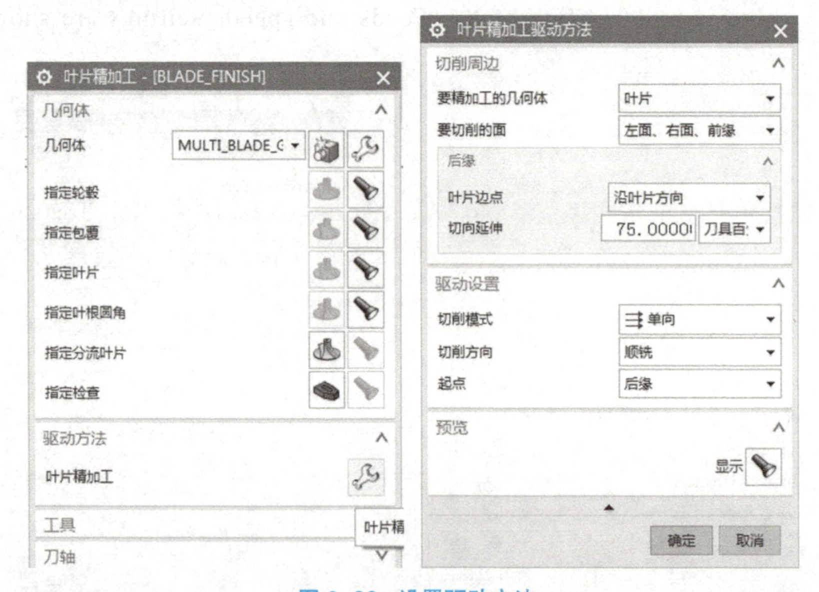

图 6-33 设置驱动方法

（7）单击【刀轨设置】选项中【非切削移动】，弹出【非切削移动】对话框，在【进刀】选项卡中【开放区域】下【进刀类型】选择【无】，在【转移 / 快速】选项卡中【区域之间】下【逼近方法】选择【无】，【离开方法】选择【无】，【移刀类型】选择【光顺】，如图 6-34 所示，单击【确定】。

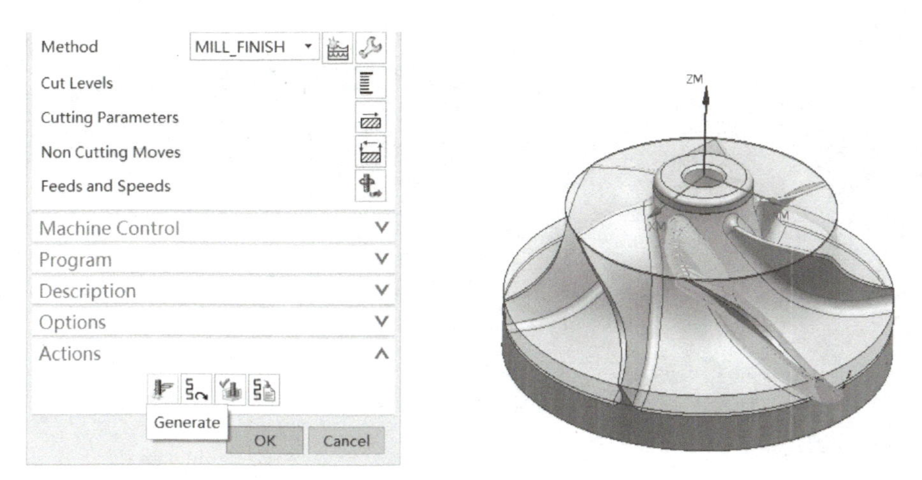

Fig. 6-32 Generate Tool Path

(6) Click "Blade Finish" button in "Drive Method" tab. In "Blade Finish Drive Method" window, click "Trailing Edge" tab to modify "Tangential Extension" value as 75 as shown in Fig. 6-33.

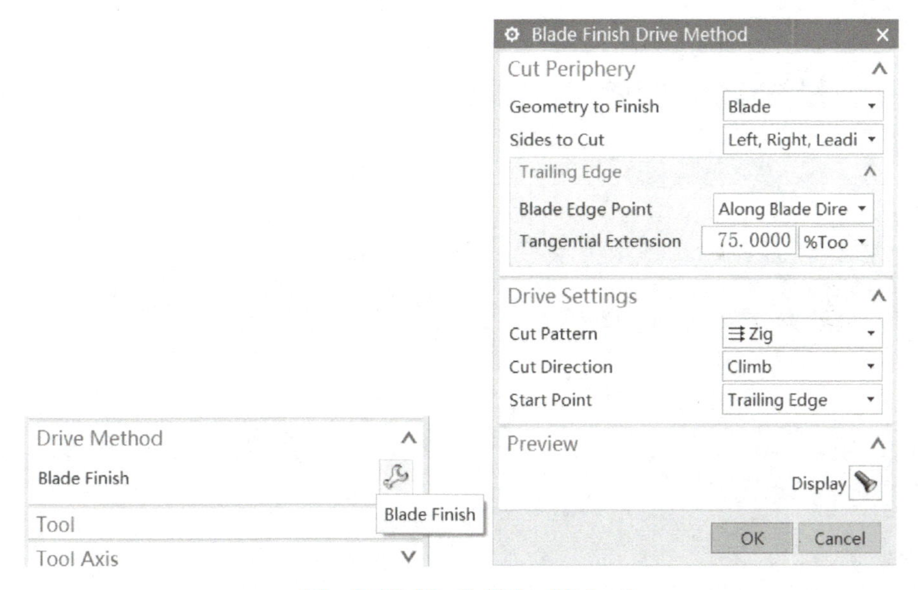

Fig. 6-33 Modify Drive Method

(7) Click "Non Cutting Moves" button in "Path Settings" tab. In "Engage" tab of "Non Cutting Moves" window, select "None" in the pull-down box of "Engage Type". Further, click "Transfer/Rapid", and in "Between Regions" tab, select "None" in the pull-down box of "Approach Method" and "Departure Method". Similarly, select "Smooth" in the pull-down box of "Traverse Type". Click "OK". The settings are shown in Fig. 6-34.

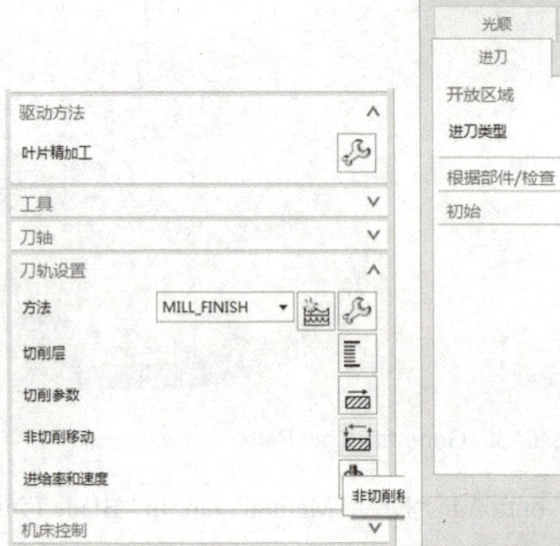

图 6-34 设置非切削移动

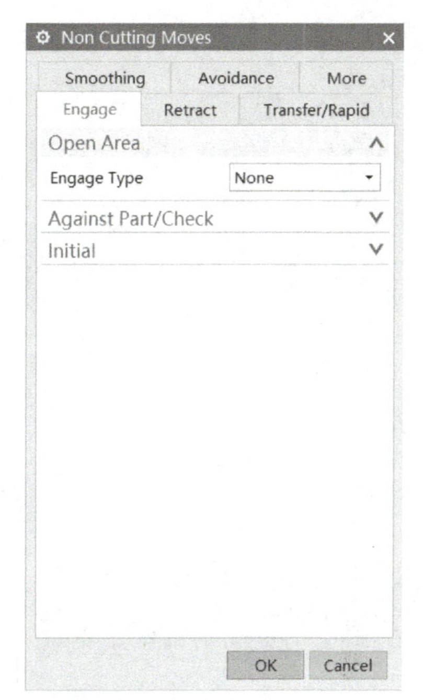

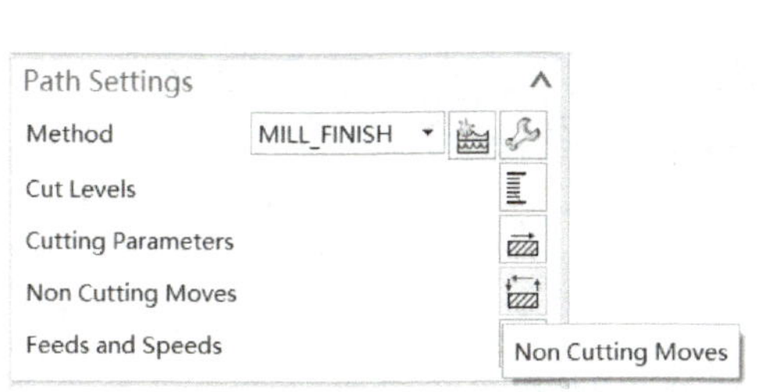

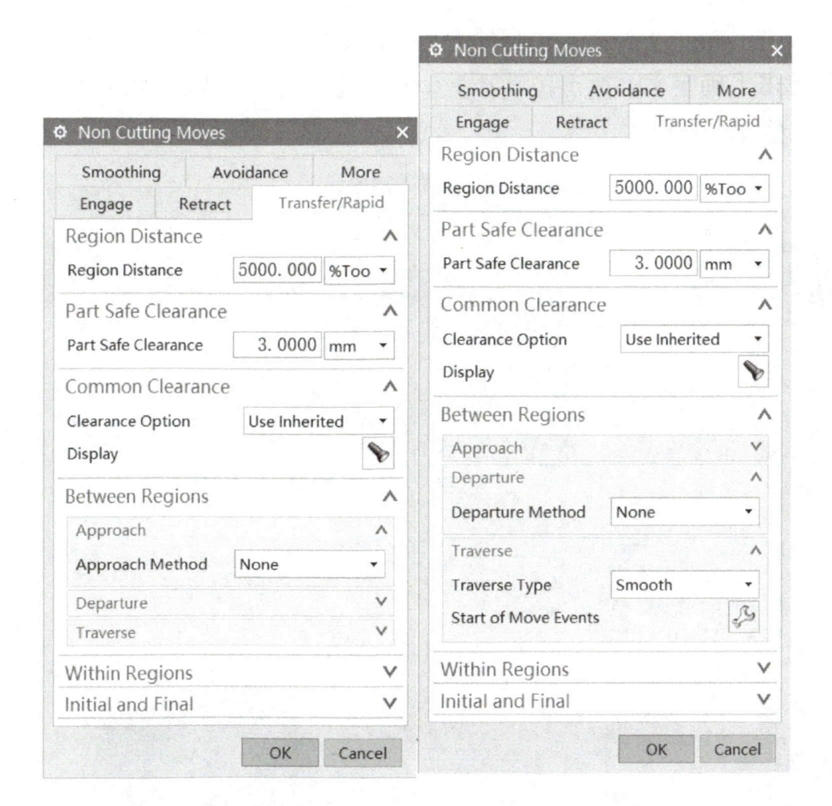

Fig. 6–34 Modify Non Cutting Moves

（8）单击【生成】，创建刀具轨迹，如图6-35所示，检验重新生成的刀具轨迹。

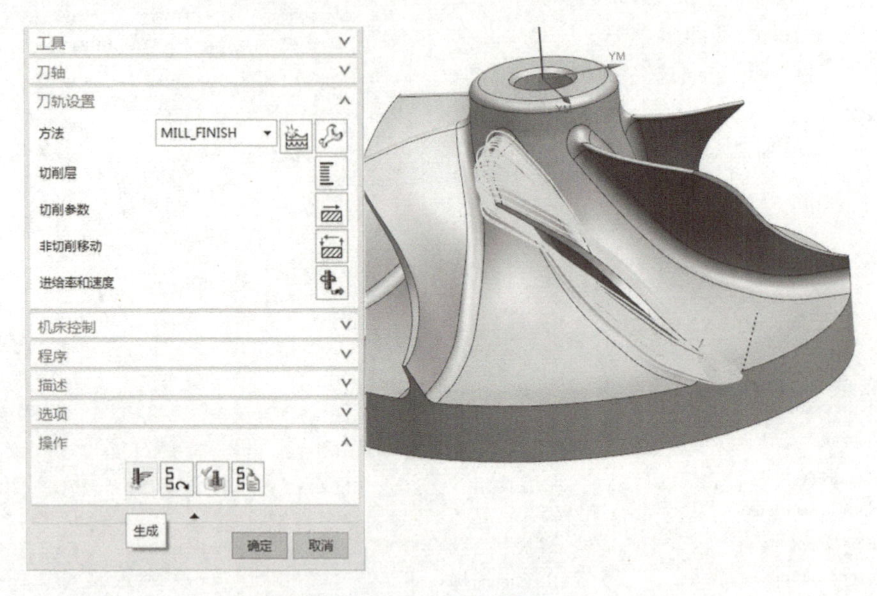

图6-35 生成刀具轨迹

（9）单击【确认】，进入仿真环境，单击【播放】，开始加工仿真，仿真结束后，单击【确定】，退出仿真环境，单击【确定】，如图6-36所示，完成单个叶片精加工编程。

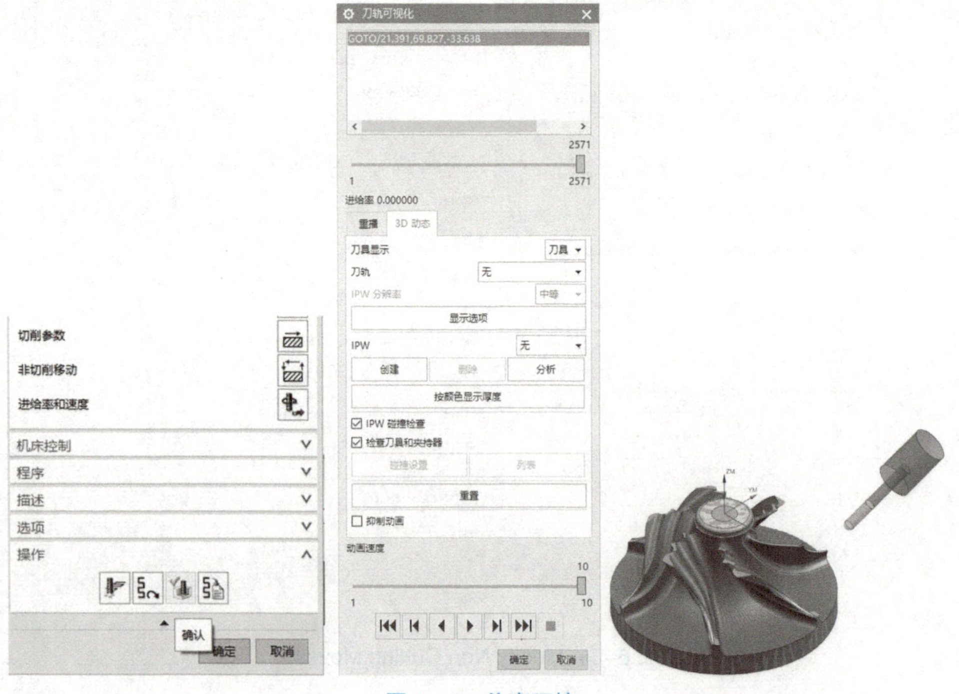

图6-36 仿真环境

(8) Click "Generate" button to view the new tool path as shown in Fig. 6-35.

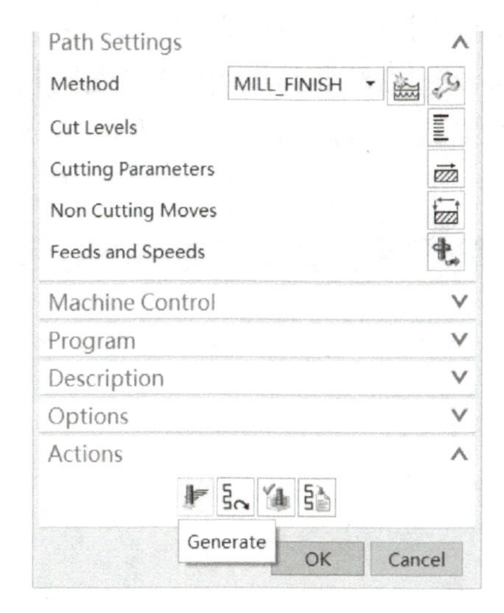

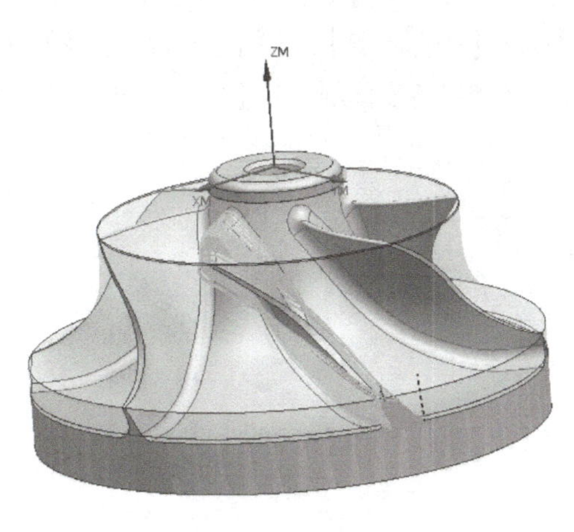

Fig. 6-35　Generate Tool Path

(9) As shown in Fig. 6-36, click "Verify" button to enter simulation environment, and then click "Play" to start simulation. Click "OK" to close the simulation window. Then, click "OK".

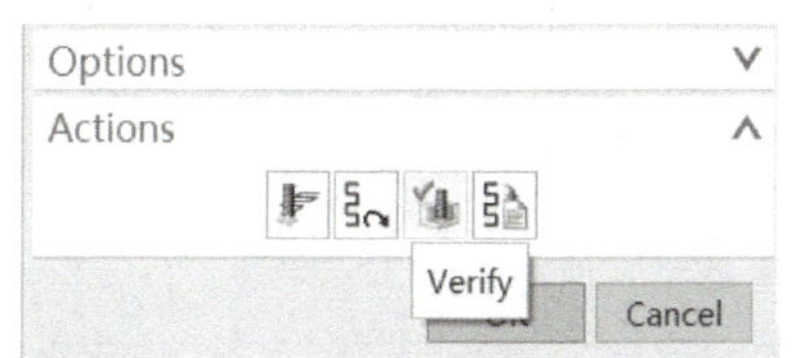

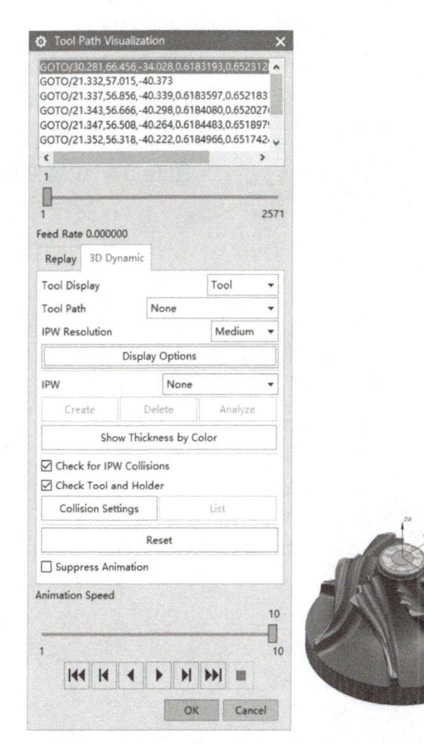

Fig. 6-36　Simulation Environment

（10）右键单击【BLADE_FINISH】程序，在弹出的快捷菜单中依次选择【对象】→【变换】，弹出【变换】对话框，【类型】选择【绕直线旋转】，【指定点】选择【坐标原点】，【指定矢量】选择【+ZC 轴】，【结果】选项中选择【复制】，设置【距离 / 角度分割】为【6】，【非关联副本数】为【5】，如图 6-37 所示，单击【确定】。

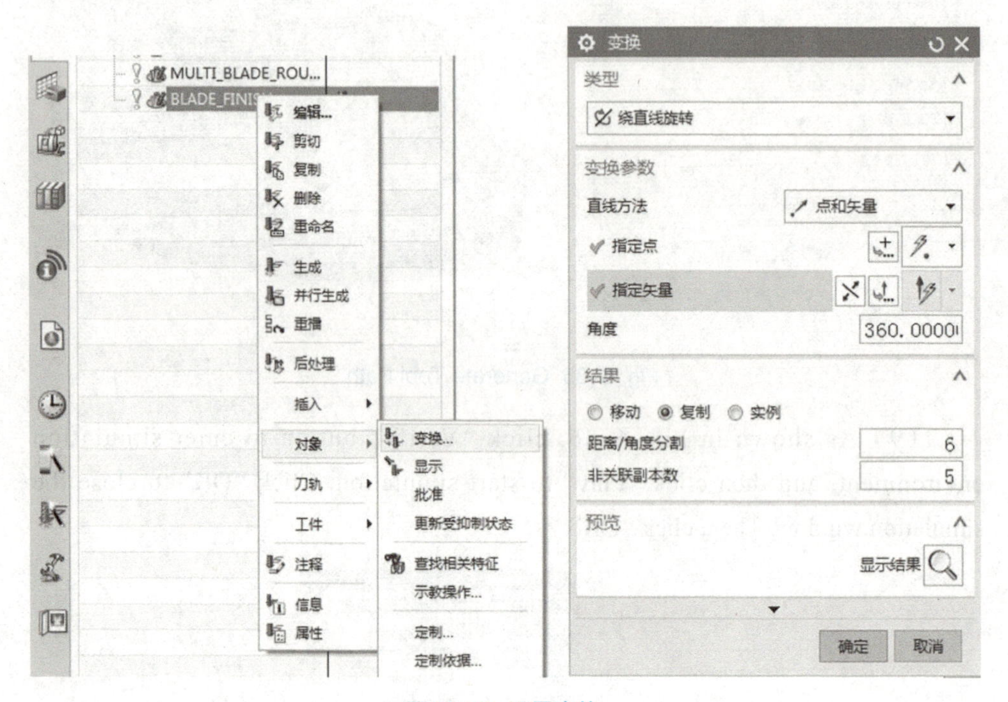

图 6-37 设置变换

(10) Right click on "BLADE_FINISH" program and move the cursor to the "Object" to select "Transform". In "Transformations" window, select "Rotate About a Line" in the pull-down box of "Type". In "Transformations Parameters" tab, "Line Method" is "Point and Vector". Then, specify the origin of coordinate as the specified point and "+ZC Axis" as the specified vector. Click "Copy" radio in "Result" tab. The "Distance/Angle Divisions" value is 6 and "Number of Non-associative Copies" value is 5. Click "OK". The settings are shown in Fig. 6-37.

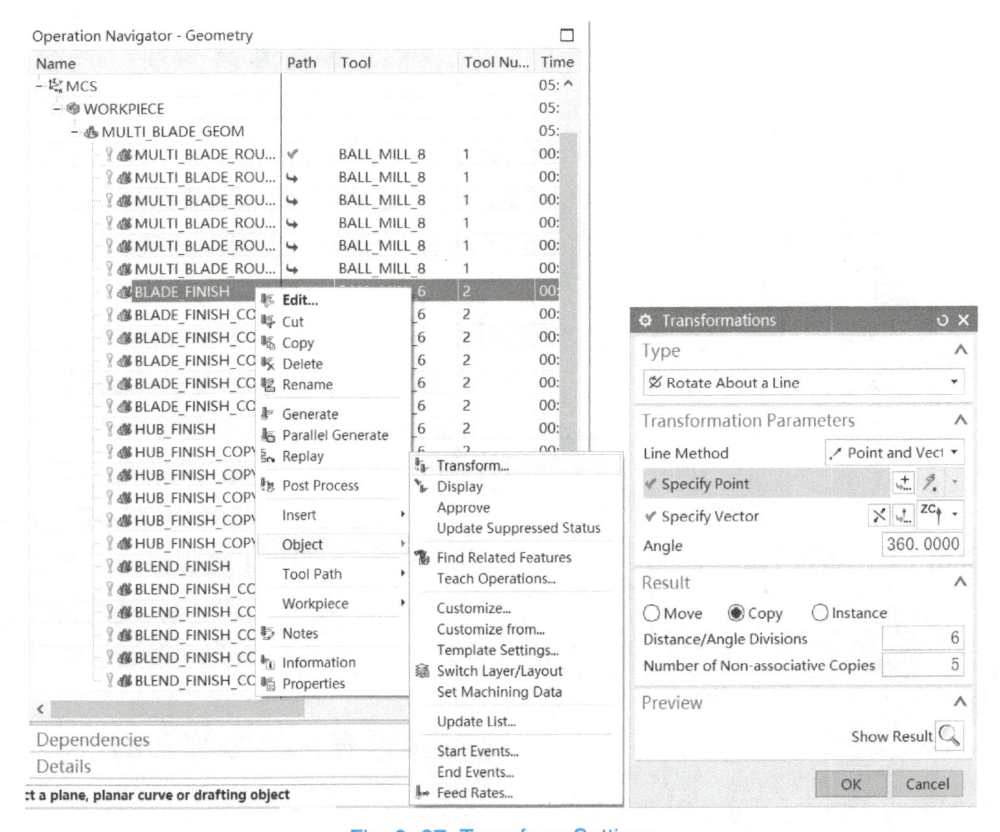

Fig. 6-37 Transform Settings

3. 轮毂精加工

（1）单击【创建工序】，弹出【创建工序】对话框，【工序子类型】选择【轮毂精加工】，【程序】选择【PROGRAM】，【刀具】选择【BALL_MILL_6】，【方法】选择【MILL_FINISH】，单击【确定】，如图 6-38 所示。

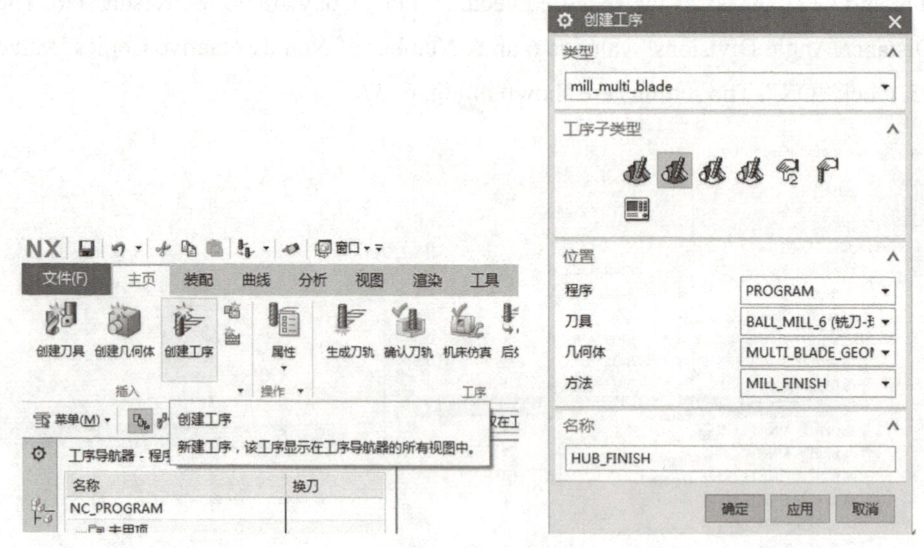

图 6-38 创建工序

（2）弹出【轮毂精加工】对话框，单击【刀轨设置】选项中的【切削参数】，弹出【切削参数】对话框，默认设置，如图 6-39 所示，单击【确定】。

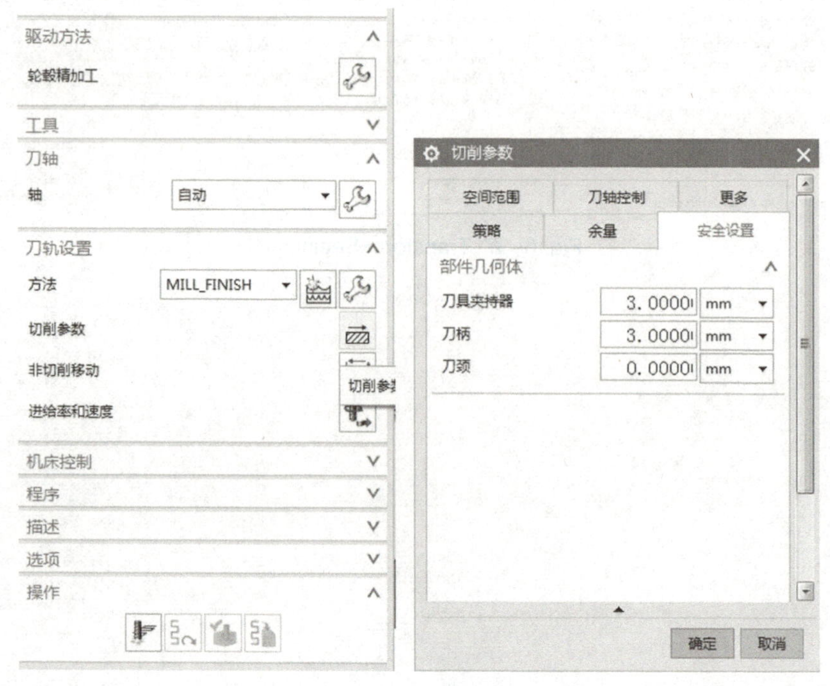

图 6-39 设置刀轨

3. Hub Finish Machining

(1) As shown in Fig. 6-38, click "Create Operation". Then, select "mill_multi_blade" in pull-down box of "Type". Select operation subtype as "hub finish", program as "PROGRAM", tool as "BALL_MILL_6", and method as "MILL_FINISH". Then, click "OK".

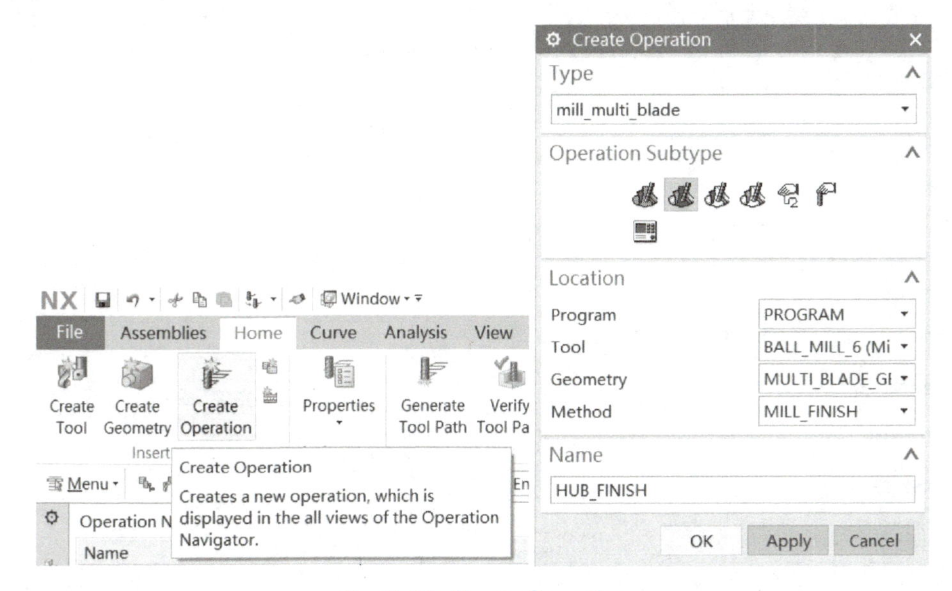

Fig. 6-38 Create Operation

(2) In "Hub Finish" window, click "Cutting Parameters" button and the parameter settings as shown in Fig. 6-39 are all default.

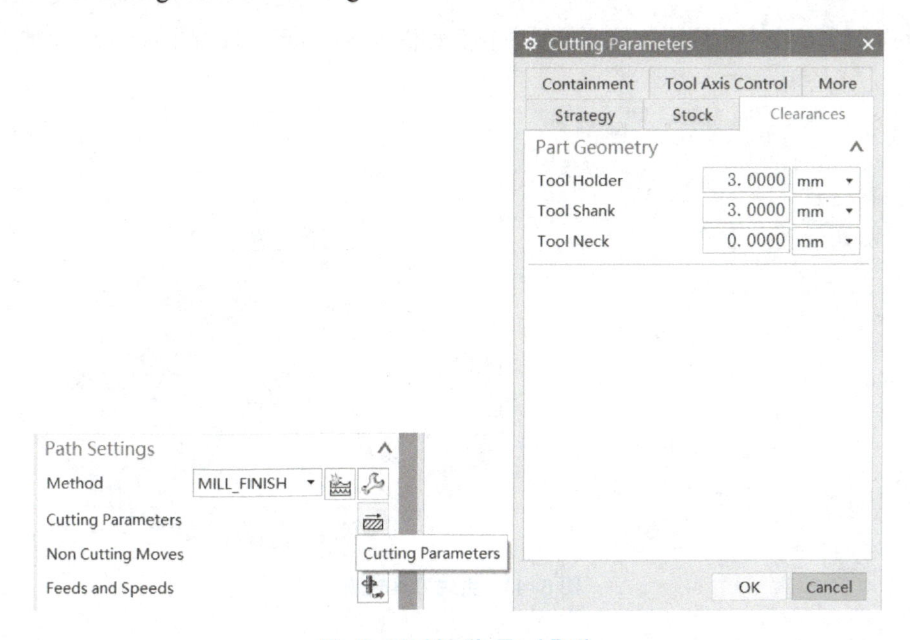

Fig.6-39 Modify Tool Path

（3）单击【刀轨设置】选项中的【进给率和速度】，弹出【进给率和速度】对话框，设置【表面速度】为【2500】，【主轴速度】为【8000】，如图 6-40 所示，单击【确定】。

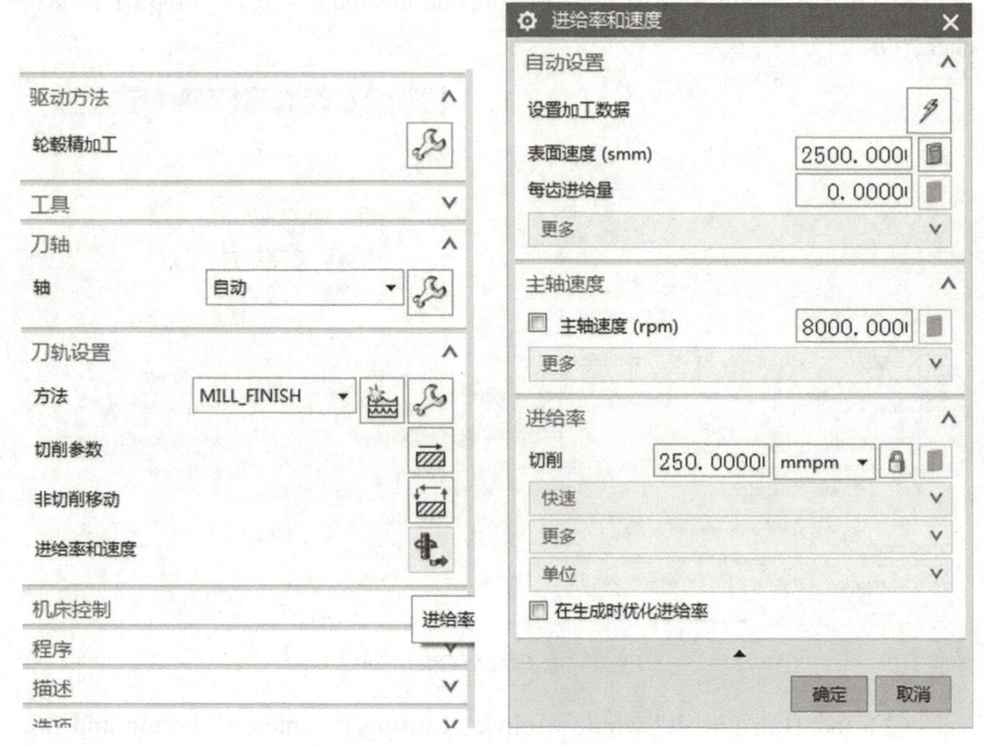

图 6-40　设置进给率和速度

（4）单击【生成】，创建刀具轨迹，如图 6-41 所示，检查生成的刀具轨迹。

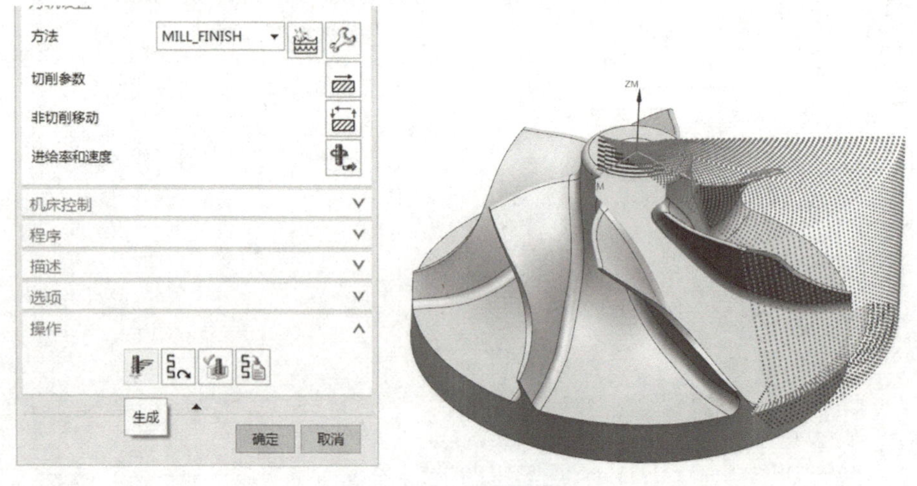

图 6-41　生成刀具轨迹

(3) Click "Feeds and Speeds" button in "Path Settings" tab. In "Feeds and Speeds" window, "Surface Speed" value is 2,500 and "Spindle Speed" value is 8,000. Click "OK". Feeds and speeds settings are shown in Fig. 6-40.

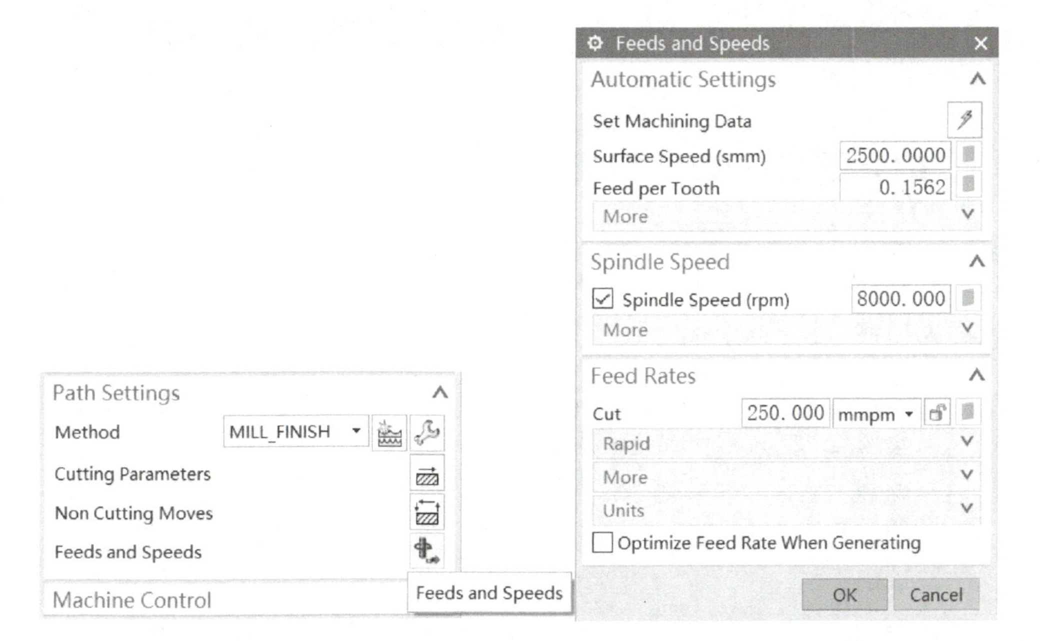

Fig. 6-40　Modify Feeds and Speeds

(4) Click "Generate" button to view the tool path as shown in Fig. 6-41.

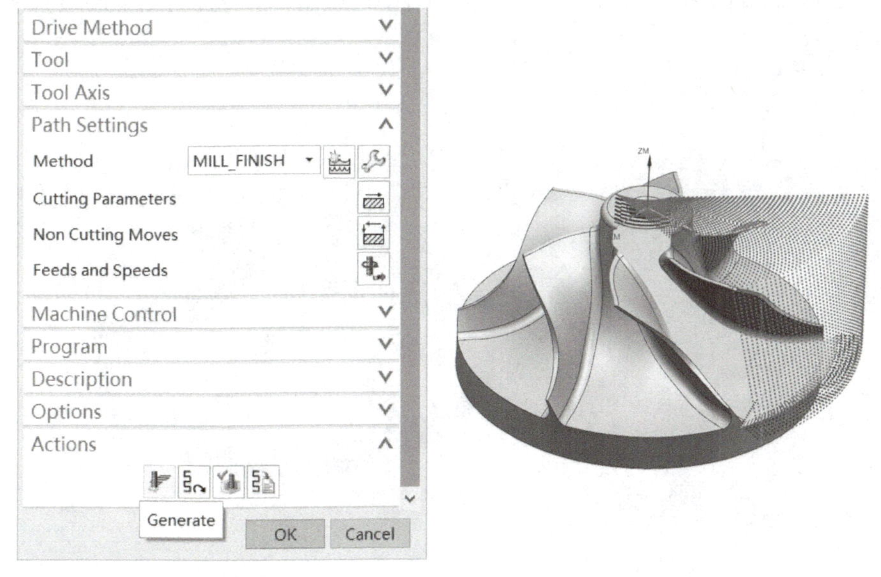

Fig. 6-41　Generate Tool Path

（5）单击【驱动方法】选项中的【轮毂精加工】，弹出【轮毂精加工】对话框，【驱动设置】中的【切削模式】选择【往复上升】，如图 6-42 所示，单击【确定】。

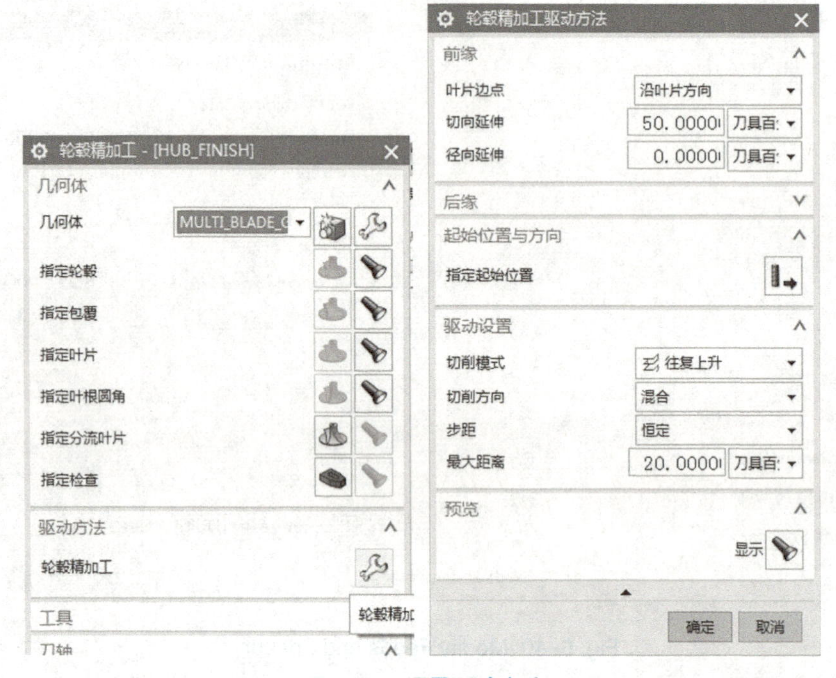

图 6-42　设置驱动方法

（6）单击【生成】，创建刀具轨迹，如图 6-43 所示，检查新生成的刀具轨迹。

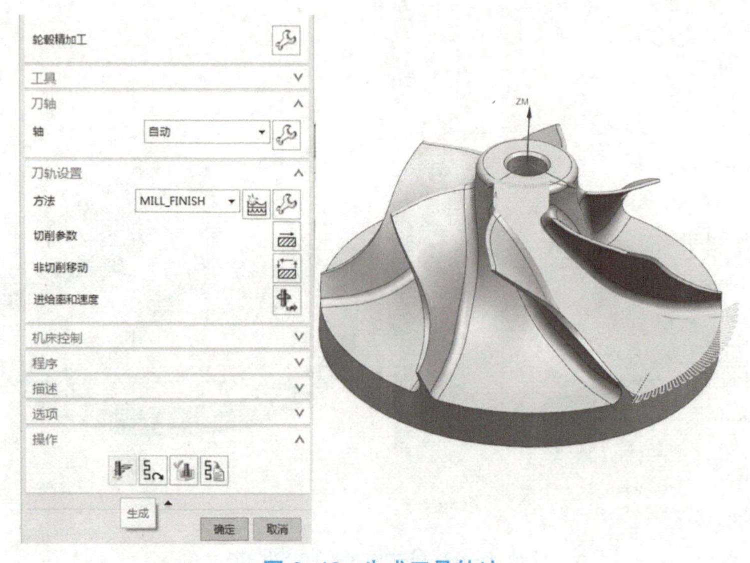

图 6-43　生成刀具轨迹

(5) Click "Hub Finish" button in "Drive Method" tab. In "Drive Settings" tab of "Hub Finish Drive Method" window, select "Zig Zag with Lifts" in the pull-down box of "Cut Pattern". Click "OK". The drive method settings are shown in Fig. 6-42.

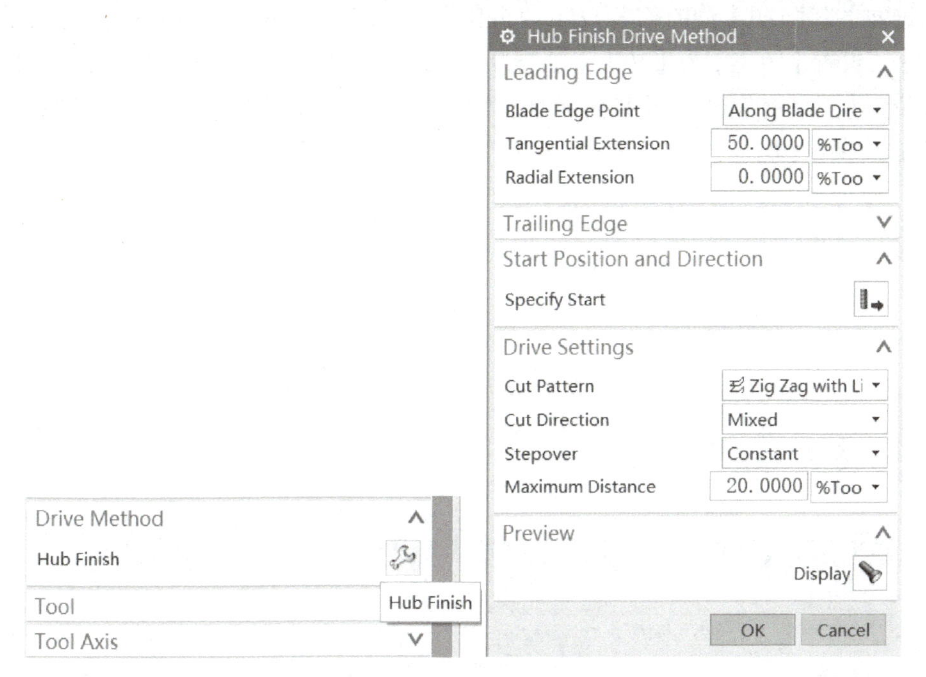

Fig. 6-42 Modify Drive Method

(6) Click "Generate" button to view the new tool path as shown in Fig. 6-43.

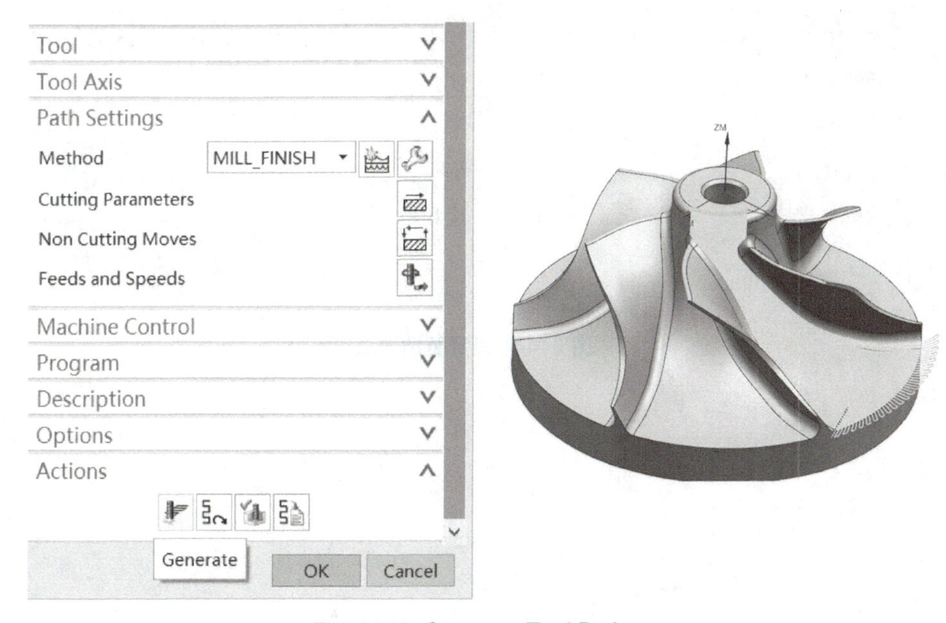

Fig. 6-43 Generate Tool Path

（7）单击【驱动方法】选项中的【轮毂精加工】，弹出【轮毂精加工驱动方法】对话框，在【前缘】中的【叶片边点】选择【沿部件轴】，设置【距离】为【10】，单位为【刀具百分比】，【切向延伸】为【0】，单位为【刀具百分比】，【径向延伸】为【120】，单位为【刀具百分比】，在【后缘】中的【边定义】选择【指定】，设置【距离】为【10】，单位为【刀具百分比】，【切向延伸】为【25】，单位为【刀具百分比】，【径向延伸】为【0】，单位为【刀具百分比】，如图 6-44 所示，单击【确定】。

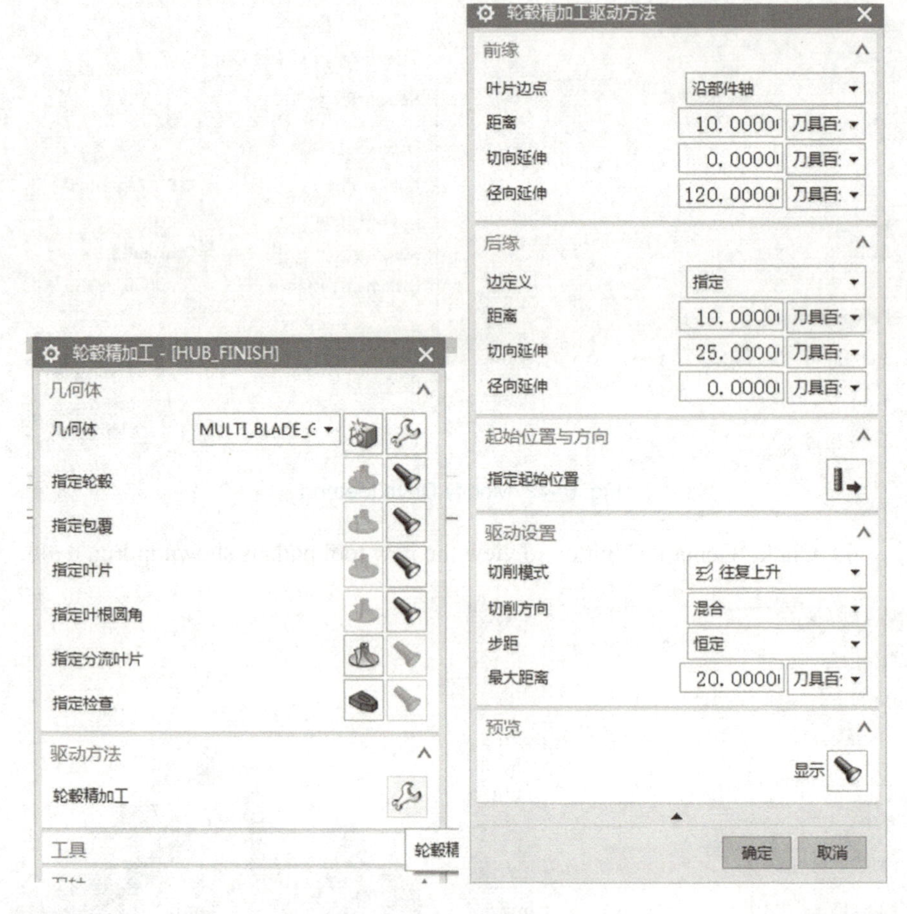

图 6-44　设置驱动方法

（7）Click "Hub Finish" button in "Drive Method" tab. In "Leading Edge" tab of "Blade Rough Drive Method" window, select "Along Part Axis" in the pull-down box of "Blade Edge Point", and "Distance" value is 10, "Tangential Extension" value is 0 and "Radial Extension" value is 120. In "Trailing Edge" tab, select "Specify" in the pull-down box of "Edge Definition", and "Distance" value is 10, "Tangential Extension" value is 25 and "Radial Extension" value is 0. Click "OK". Drive method settings are shown in Fig. 6-44.

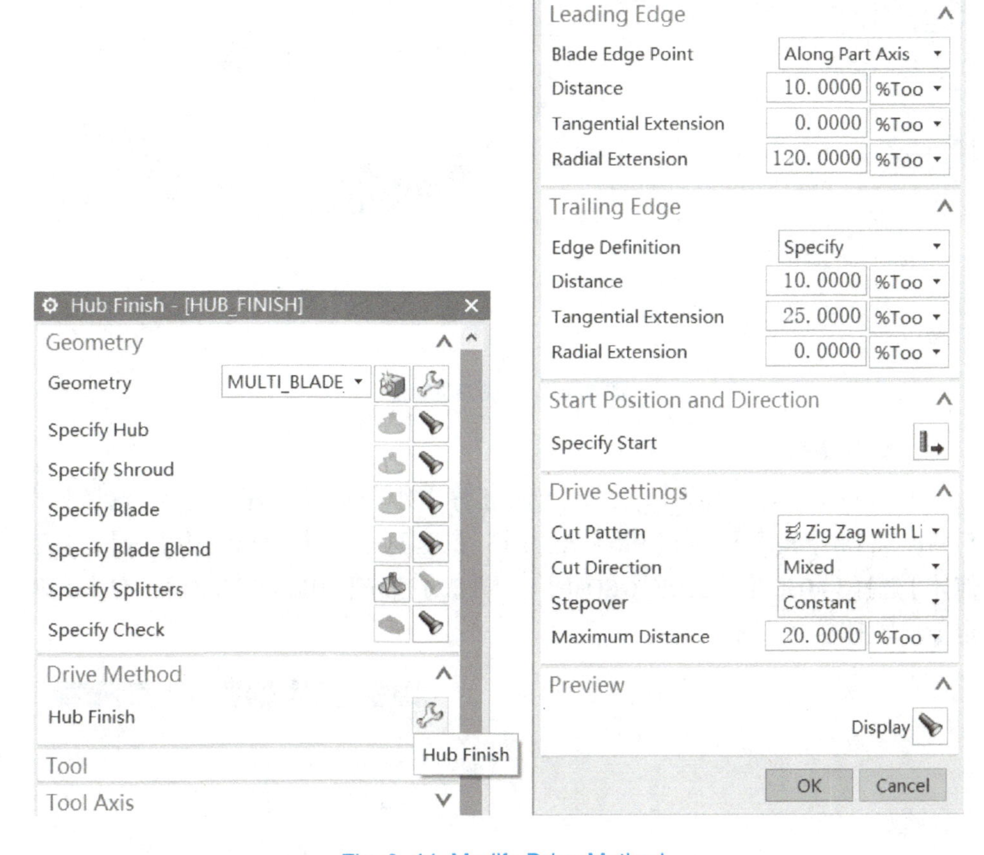

Fig. 6-44　Modify Drive Method

（8）单击【生成】，创建刀具轨迹，如图 6–45 所示，检查生成的刀具轨迹。

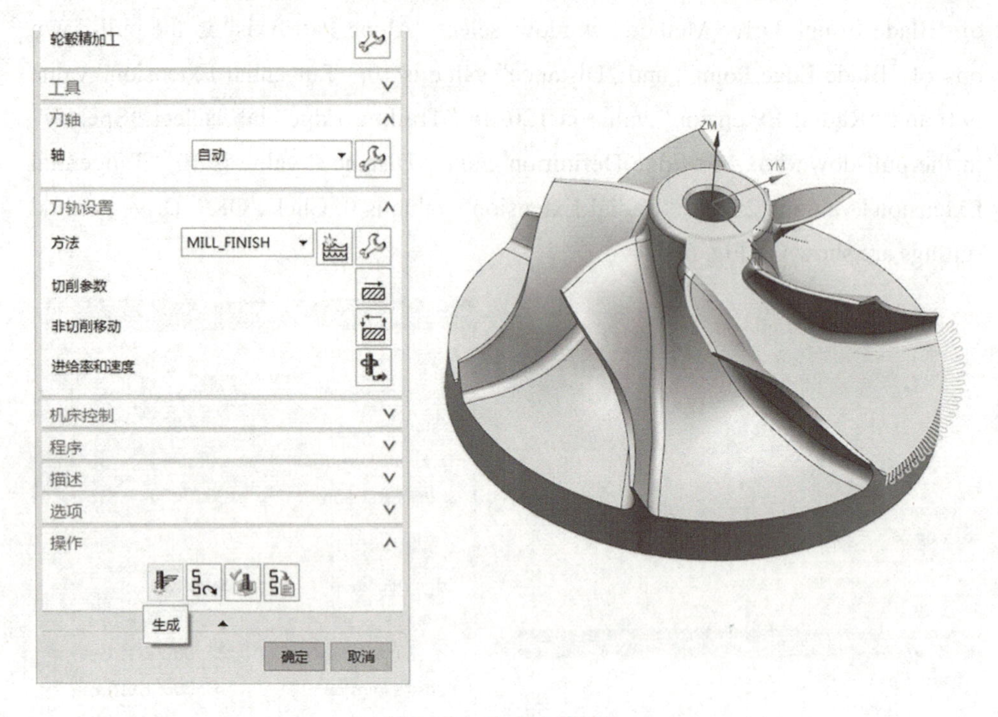

图 6–45　生成刀具轨迹

4. 圆角精加工

（1）单击【创建工序】，打开【创建工序】对话框，【工序子类型】选择【圆角精加工】，【程序】选择【PROGRAM】，【刀具】选择【BALL_MILL_6】，【几何体】选择【MULTI_BLADE_GEOM】，【方法】选择【MILL_FINISH】，单击【确定】，如图 6–46 所示。

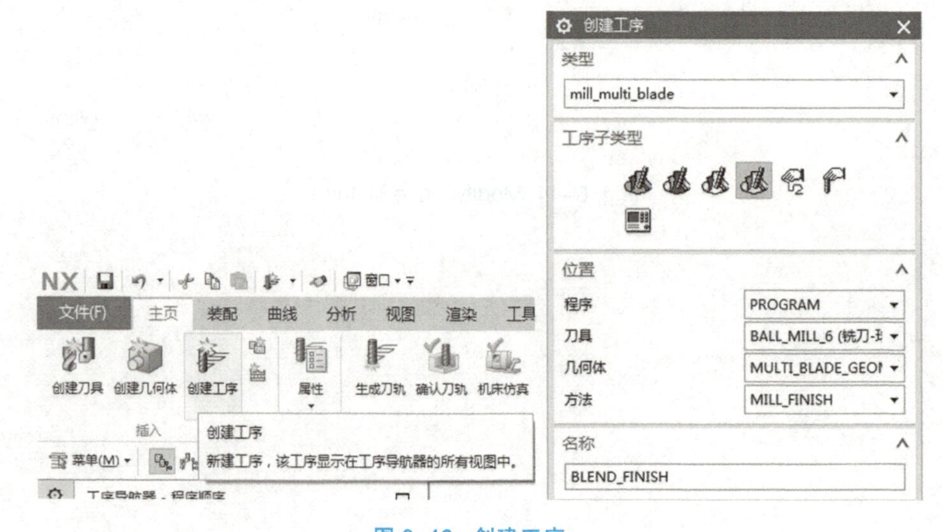

图 6–46　创建工序

(8) Click "Generate" button to view the tool path as shown in Fig. 6-45.

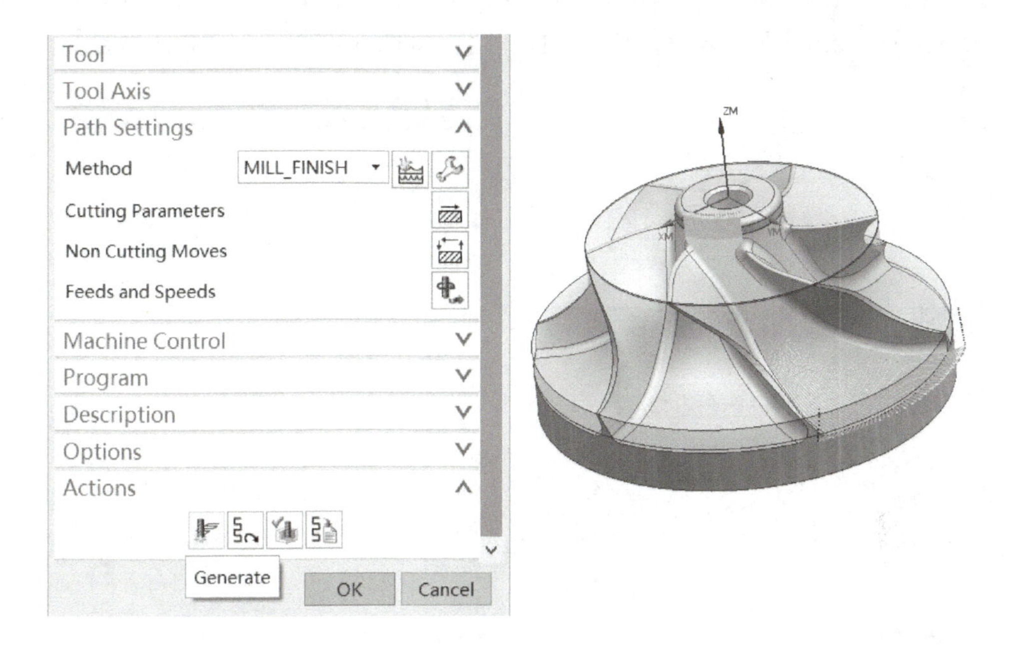

Fig. 6–45 Generate Tool Path

4. Blend Finish

(1) As shown in Fig. 6-46, click "Create Operation", then select "mill_multi_blade" in pull-down box of "Type". Select operation subtype as "blend finish", program as "PROGRAM", tool as "BALL_MILL_6", geometry as "MULTI_BLADE_GEOM" and method as "MILL_FINISH". Then click "OK".

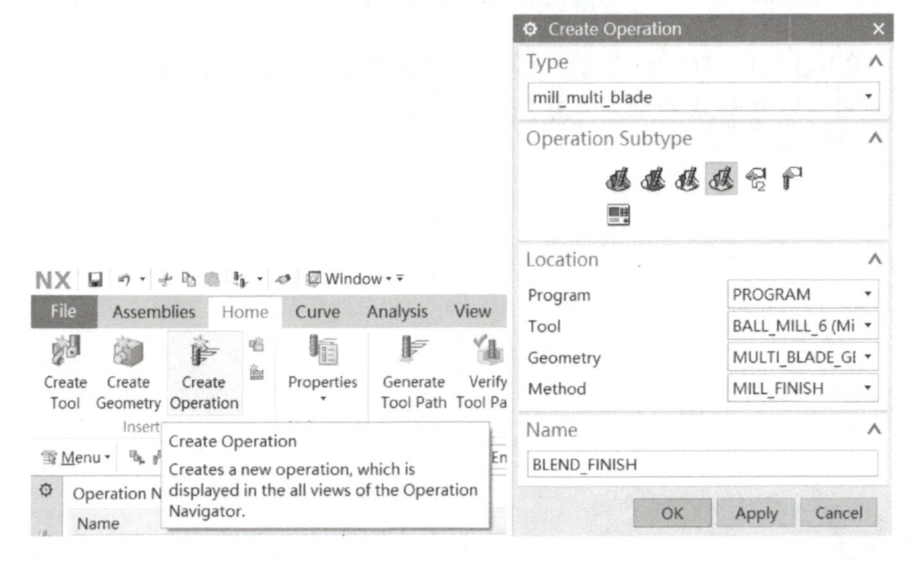

Fig. 6–46 Create Operation

（2）弹出【圆角精加工】对话框，单击【驱动方法】选项下的【圆角精加工】，弹出【圆角精加工驱动方法】对话框，修改【驱动设置】选项下的【步距】为【残余高度】，设置【最大残余高度】为【0.01】，如图 6-47 所示，单击【确定】。

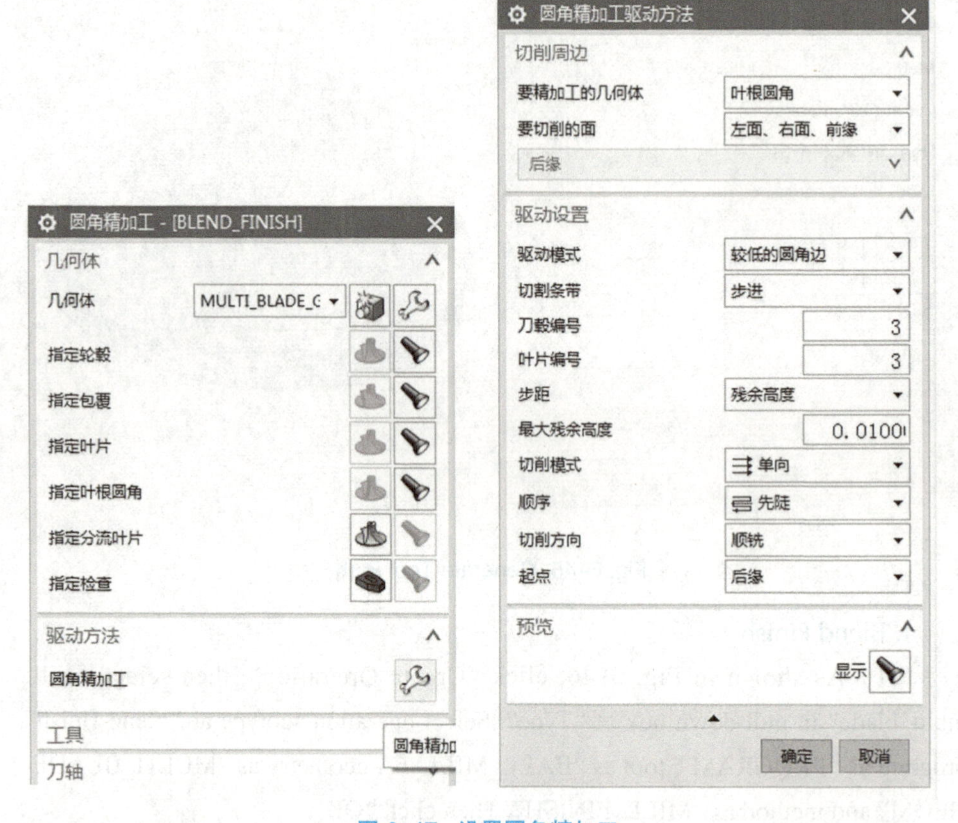

图 6-47 设置圆角精加工

（3）单击【刀轨设置】选项下的【切削参数】，弹出【切削参数】对话框，默认设置，如图 6-48 所示，单击【确定】。

（2）Click “Blend Finish” button in “Drive Method” tab. In “Drive Settings” tab of “Blend Finish Drive Method” window, select “Scallop” in the pull-down box of “Stepover”, and “Maximum Scallop Height” value is 0.01. Then click “OK”. The drive method settings are shown in Fig. 6-47.

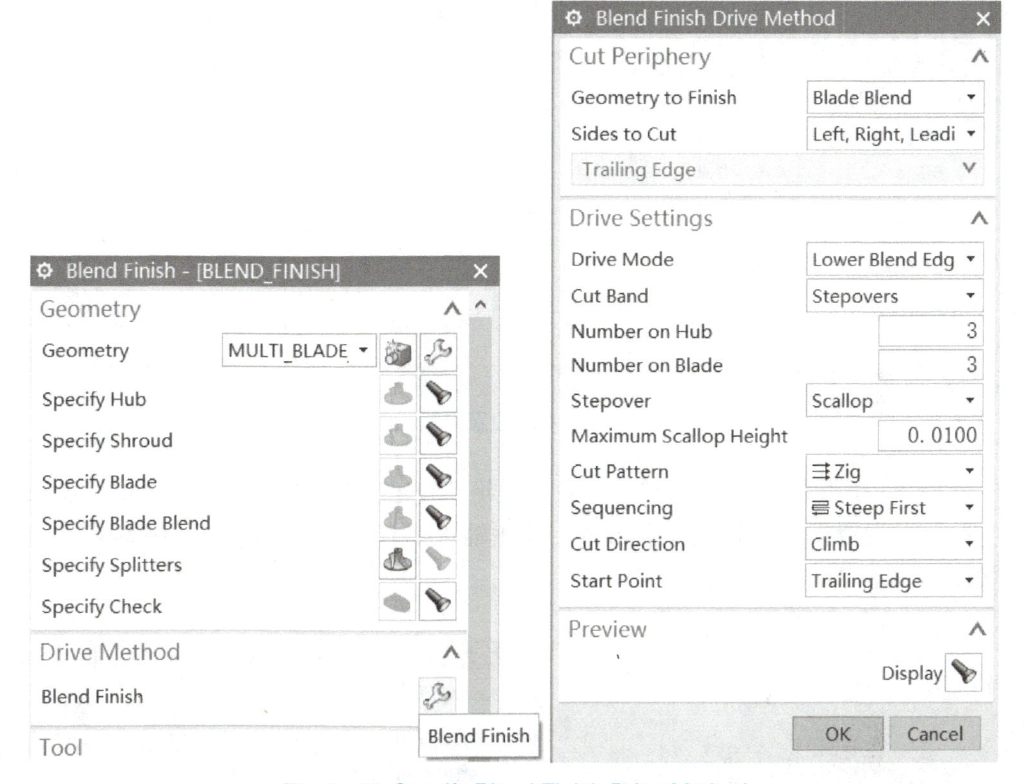

Fig. 6-47　Specify Blend Finish Drive Method

（3）In “Blend Finish” window, click “Cutting Parameters” button and the parameter settings as shown in Fig. 6-48 are all default . Click “OK”.

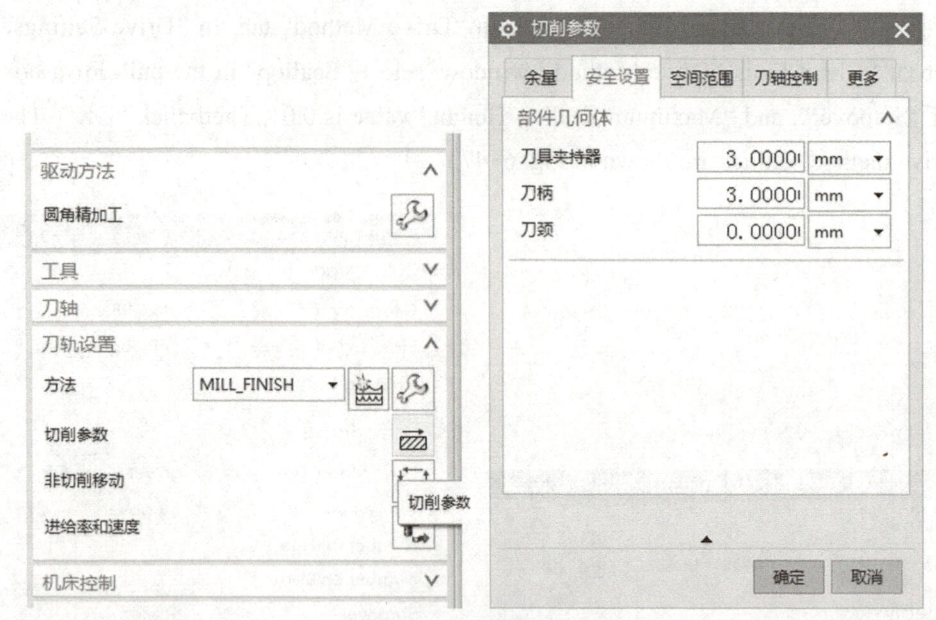

图 6-48　设置刀轨

（4）单击【刀轨设置】选项下的【进给率和速度】，弹出【进给率和速度】对话框，设置【表面速度】为【4000】，【主轴速度】为【10000】，如图 6-49 所示，单击【确定】。

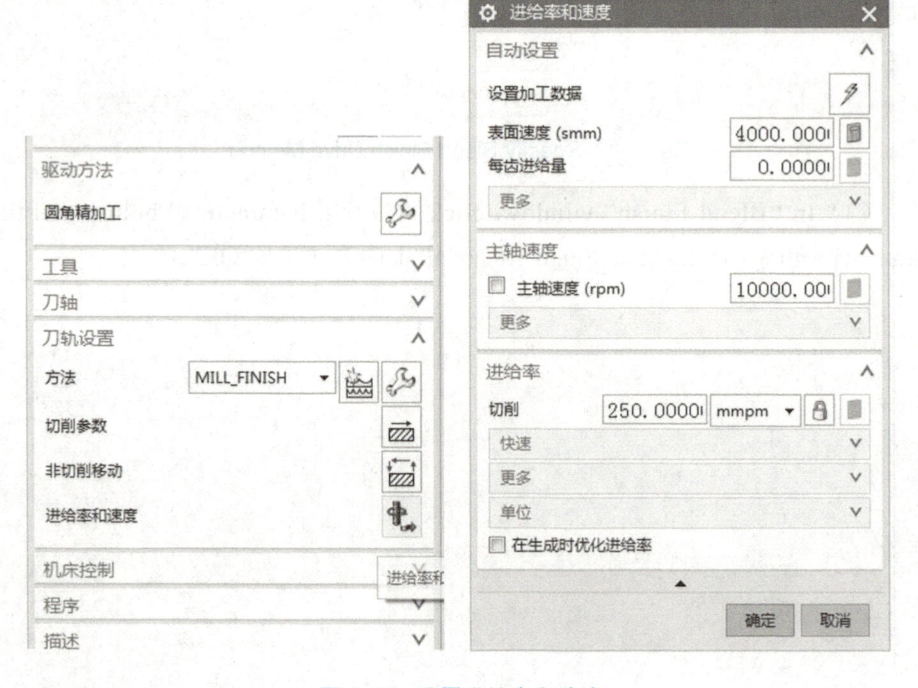

图 6-49　设置进给率和速度

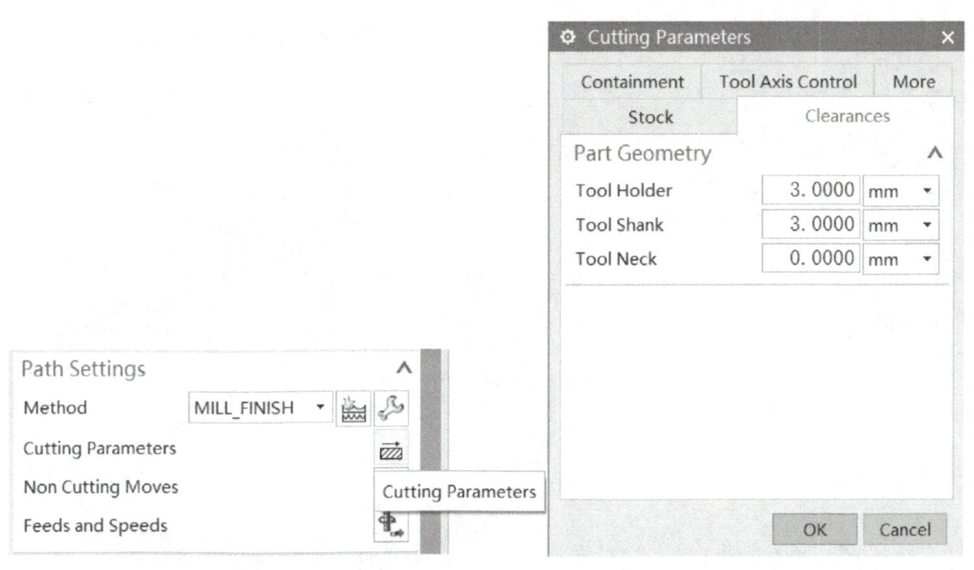

Fig. 6–48　Modify Path Settings

（4）Click “Feeds and Speeds” button in “Path Settings” tab. In “Feeds and Speeds” window, “Surface Speed” value is 4,000 and “Spindle Speed” value is 10,000. Click “OK”. Feeds and speeds settings are shown in Fig. 6-49.

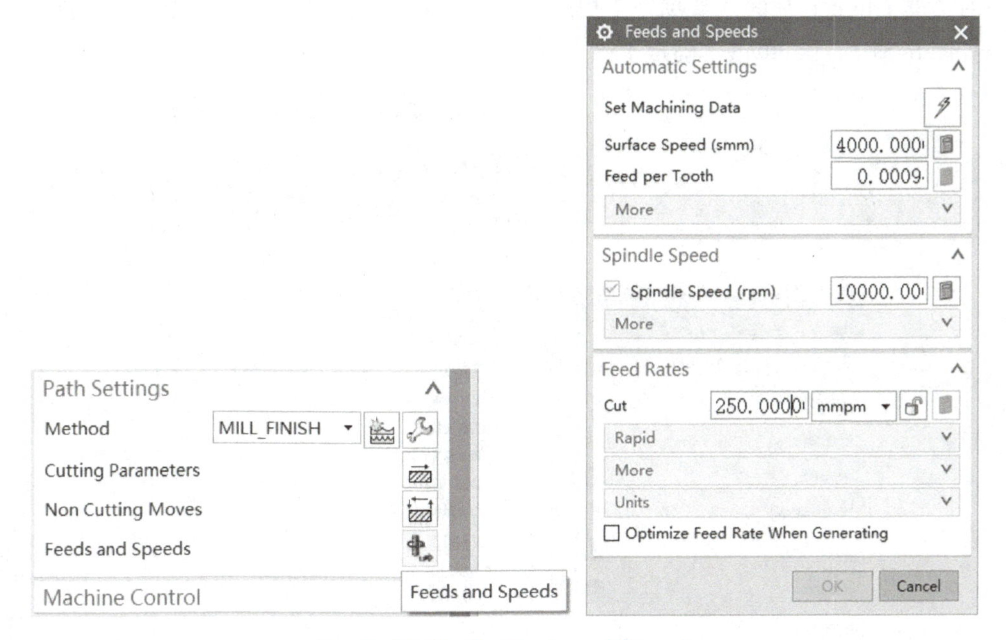

Fig. 6–49　Modify Feeds and Speeds

（5）单击【生成】，创建刀具轨迹，如图 6–50 所示，检查刀具轨迹。

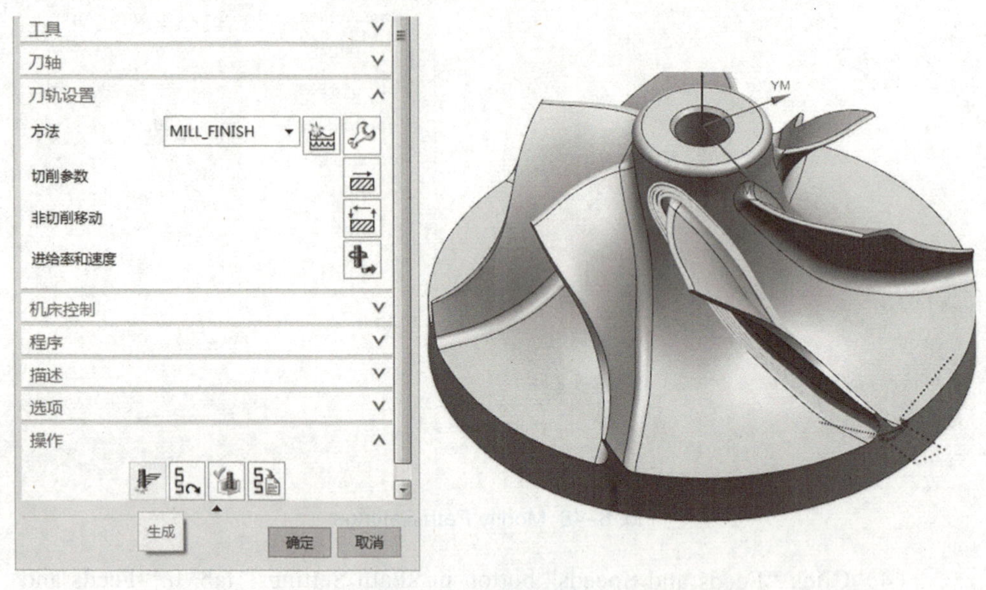

图 6–50　生成刀具轨迹

（6）单击【驱动方法】选项中的【圆角精加工】，弹出【圆角精加工驱动方法】对话框，设置【后缘】选项中【切向延伸】为【75】，单位为【刀具百分比】，如图 6–51 所示，单击【确定】。

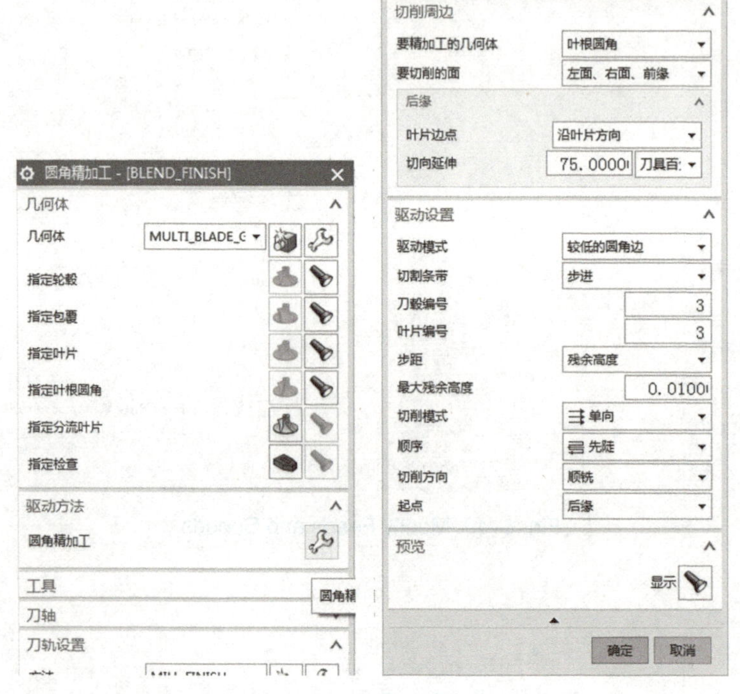

图 6–51　设置驱动方法

(5) Click "Generate" button to view the tool path as shown in Fig. 6-50.

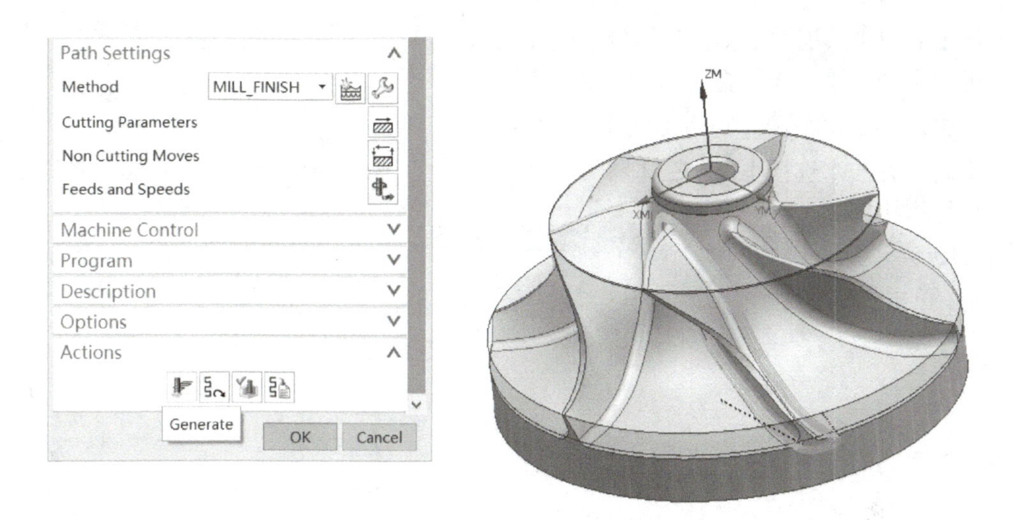

Fig. 6-50 Generate Tool Path

(6) Click "Blend Finish" button in "Drive Method" tab. In "Trailing Edge" tab, "Tangential Extension" value is 75. Click "OK". Drive Method Settings are shown in Fig. 6-51.

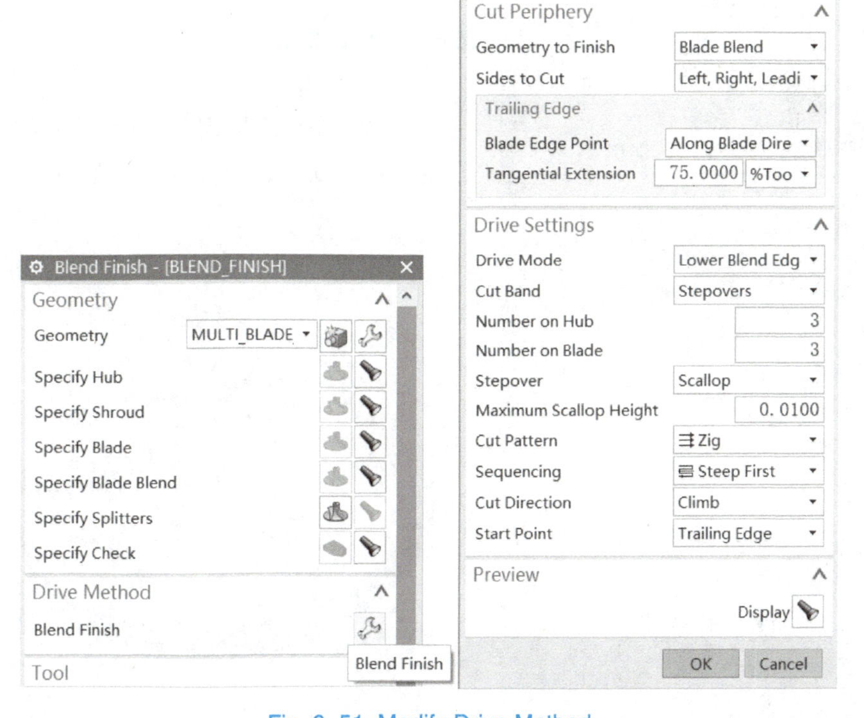

Fig. 6-51 Modify Drive Method

（7）单击【刀轨设置】选项中的【非切削移动】，弹出【非切削移动】对话框，在【进刀】选项卡中【开放区域】下【进刀类型】选择【无】，在【转移/快速】选项卡中【区域之间】下【逼近方法】选择【无】，【离开方法】选择【无】，【移刀类型】选择【光顺】，如图 6-52 所示，单击【确定】。

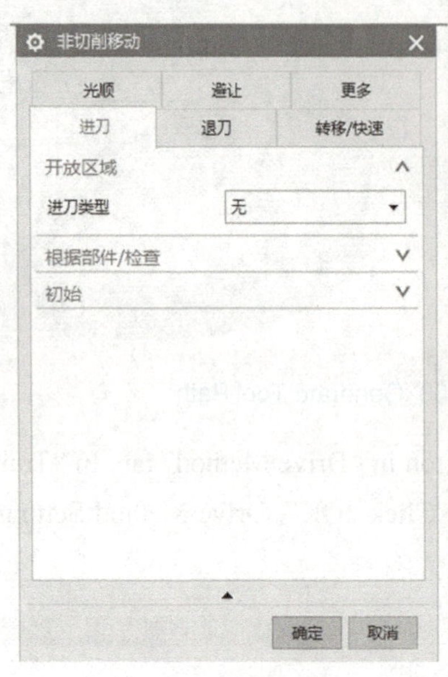

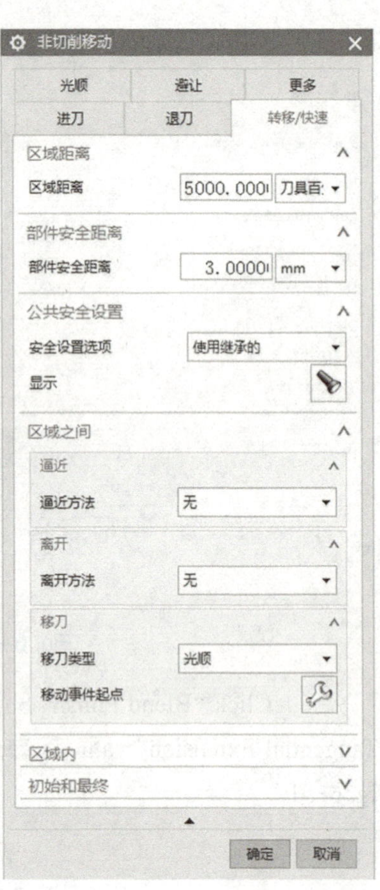

图 6-52 设置刀轨

驱动方法
圆角精加工
工具
刀轴
刀轨设置
方法 MILL_FINISH
切削参数
非切削移动
进给率和速度
机床控制
程序
描述
选项
操作
生成
确定 取消

（8）单击【生成】，创建刀具轨迹，如图 6-53 所示，检查刀具轨迹。

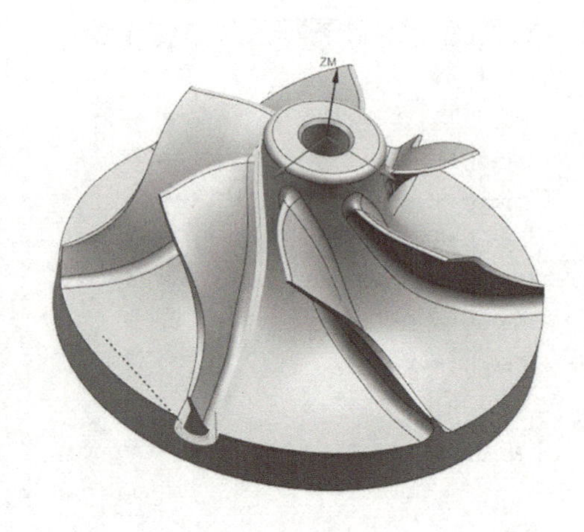

图 6-53 生成刀具轨迹

（7）Click “Non Cutting Moves” button in “Path Settings” tab. In “Engage” tab of “Non Cutting Moves” window, select “None” in the pull-down box of “Engage Type”. Further, click “Transfer/Rapid”, and in “Between Regions” tab , select “None” in the pull-down box of “Approach Method” and “Departure Method”. Similarly, select “Smooth” in the pull-down box of “Traverse Type”. Click “OK”. The settings are shown in Fig. 6-52.

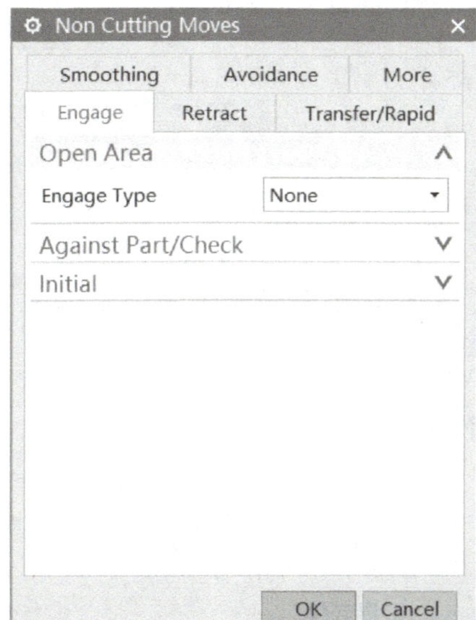

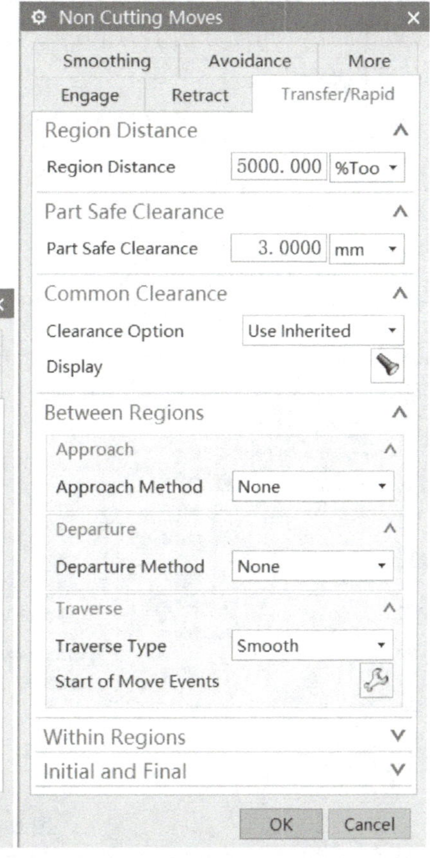

Fig. 6-52 Specify Path Settings

（8） Click “Generate” button to view the tool path as shown in Fig. 6-53.

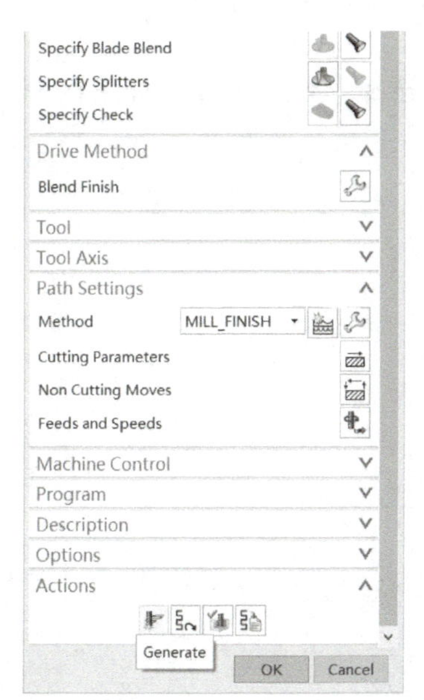

Fig. 6-53 Generate Tool Path

（9）右键单击【BLEND_FINISH】程序，在弹出的快捷菜单中选择【对象】→【变换】，弹出【变换】对话框，【类型】选择【绕直线旋转】，【指定点】选择【坐标原点】，【指定矢量】选择【+ZC 轴】，【结果】选项中选择【复制】，设置【距离 / 角度分割】为【6】，【非关联副本数】为【5】，如图 6-54 所示，单击【确定】。

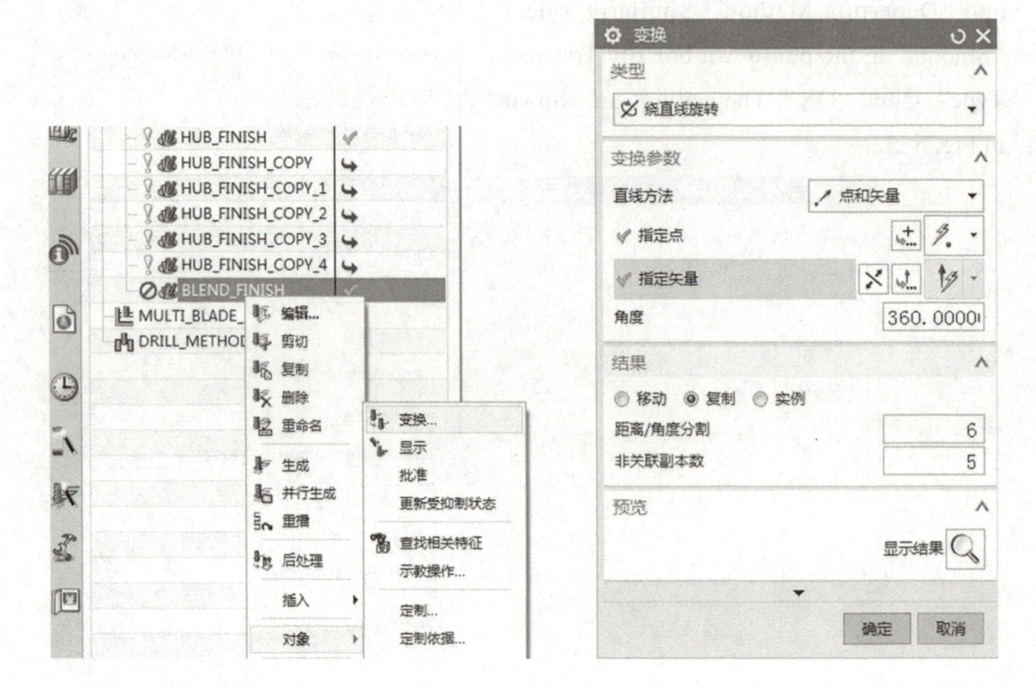

图 6-54　设置变换

四、仿真加工

单击【NC_PROGRAM】，单击【确认刀轨】，选择【3D 动态】，如图 6-55 所示，单击【播放】。

(9) Right click on "BLEND_FINISH" in "Operation Navigator-Geometry" and move the cursor to the "Object" to select "Transform". In "Transformations" window, select "Rotate About a Line" in the pull-down box of "Type". In "Transformations Parameters" tab, "Line Method" is "Point and Vector". Then, specify the origin of coordinate as the specified point and "+ZC Axis" as the specified vector. Click "Copy" radio in "Result" tab. The "Distance/Angle Divisions" value is 6 and "Number of Non-associative Copies" value is 5. Click "OK". The settings are shown in Fig. 6-54.

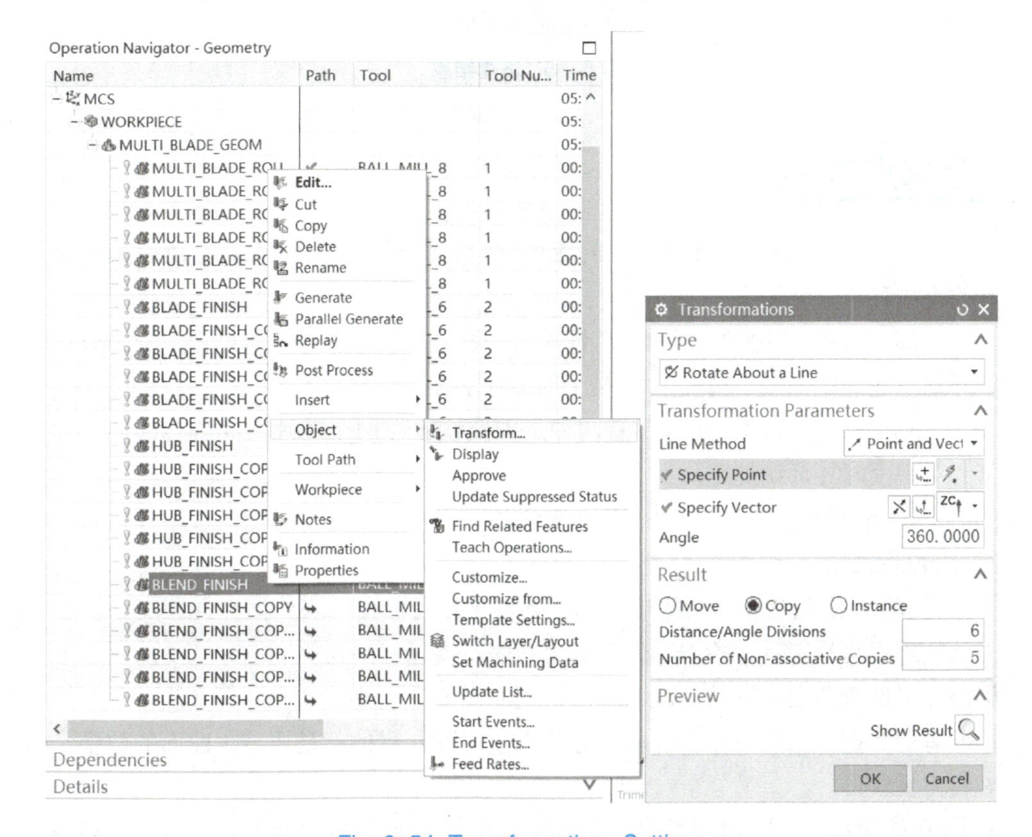

Fig. 6-54 Transformations Settings

Ⅳ. Simulation Machining

Click "NC_PROGRAM", and then click "Verify" button. Select "3D Dynamic" and click "Play" button to view the simulation result. The simulation result is shown in Fig. 6-55.

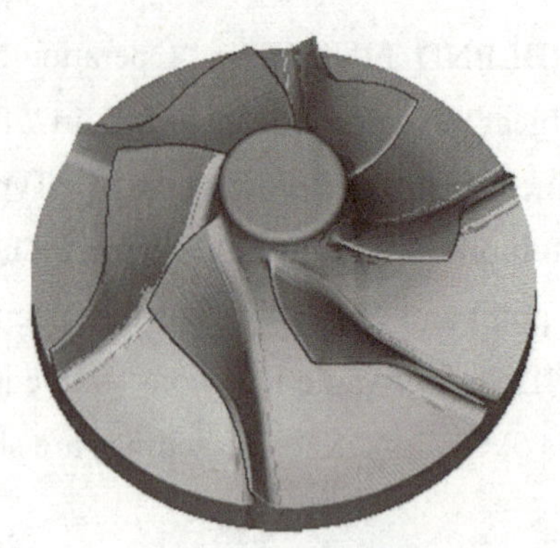

图 6-55 仿真结果

专家点拨

（1）加工叶轮，当叶片之间的流道很窄、材料又不好加工时，可以采用插铣的方法。

（2）在加工叶轮时，开槽和扩槽未必一次到底。根据情况可以分步骤完成。即开到一定深度先做半精加工，然后再继续开槽。

（3）叶轮叶片的半精加工有两种走刀路线：在两个叶片之间进行区域式加工和对一个叶片环绕加工。

（4）清根是叶片、叶轮加工的难点之一。经常出现的问题是过切和抬刀。解决的方法如下：

① 光顺根部曲面。叶根与轮毂相交的部分通常是几张曲面在此相交，所以刀具轨迹很容易凌乱，抬刀、下刀都会增多。要想解决这类问题，就要从根本上解决叶根部分曲面光顺的问题，可以采用曲面缝合、拼接、光顺等方法，甚至根据原有的数据重新生成整个叶根部分圆弧过渡面，目的就是要减少曲面的数量，规整刀具轨迹。

② 优化程序。有时刀具沿某一方向切削就会有许多次抬刀和下刀，而沿另一垂直方向就会少得多。另外连续环绕加工就比单向或往复加工的质量好，因为后者总会留下接刀痕。加工时的切削用量，即主轴转速、进给量和切削深度也会影响加工质量。如果选择不好，有可能刀具在叶根部分产生颤振，当刀具长径比较长时，这种颤振会引起过切。因此合理的走刀路线和切削用量对于叶根部分的加工是非常重要的。

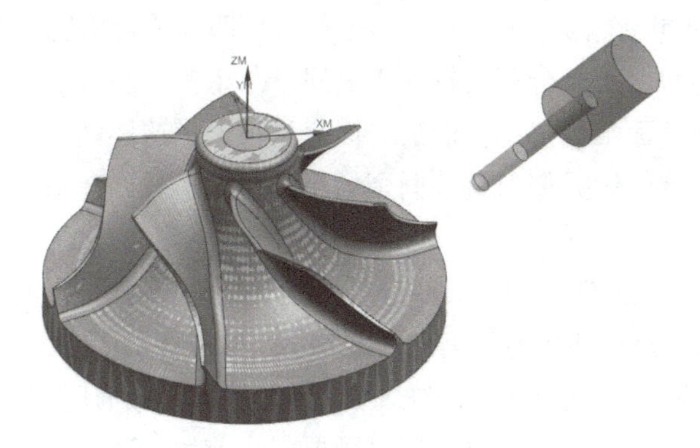

Fig. 6–55 Simulation Result

Expert Reviews

(1) Plunge milling is recommended when the material is hard to machine or the passage between impeller and blade is narrow.

(2) During the processing of impeller, grooving can be done step by step, which means, when it grooves to a certain depth, semi-finishing can replace grooving. After semi-finishing, continue grooving.

(3) There are two paths in semi-finishing. One is regional processing between two blades. The other is surround processing of one blade.

(4) Flowcut is one of the core challenges in blade or impeller processing. Over cutting and frequently tool lifting often occur in flowcut. The solutions are as follows:

① Smooth root surface: Due to the intersection of blend surface and hub surface, there will be an obvious increase on non-cutting moves and a messy tool path. To solve this problem, smoothing on blend should be taken into consideration. With the goal to reduce the number of surfaces and regularize tool paths, surface stitching and surface smoothing are brilliant ways to smooth the blend. Note that regenerating the arc transition surface of the blend with original data is also considered as an effective solution.

② Optimize Program: Machining in a certain direction will cause non-cutting moves many times. On the contrary, non-cutting moves will decrease in the other vertical direction. Besides, continuous surround processing is better than zig or zig-zag processing because tool marks frequently occurs in zig or zig-zag processing. Cutting parameters, which consist of spindle speed, feed and cutting depth also have impact on processing quality. If the cutting parameters is carelessly set, chatter will occur on the blend. When the major diameter of the tool is large, chatter will cause over cutting. Thus, it is of great importance to select the reasonable tool path and cutting parameters when it comes to blend processing.

③ 合理安排粗精加工工序。叶根部分与轮毂部分紧密相连，但又不是一张面，不能一刀加工出来。所以合理安排加工顺序十分重要。叶根和轮毂的加工余量要均衡。不要造成一面余量很小，而另一面余量很大，这样就会造成在切削余量大的一面时，由于反作用力使刀具偏向余量小的一面，从而在余量小的一面产生过切现象。

④ 合理选择切削刀具。叶根部分的圆角过渡一般都不会很大。有时球头铣刀半径稍微偏大，就会使抬刀下刀的现象增多。如果在不影响刀具刚性的情况下，适当更换稍小直径的球头铣刀，这种现象就会大大减少。有时为了加工叶根部分，还可以采用锥形的球头铣刀，目的就是既要减小前面的球头半径，又要使刀具后面的部分粗一些，保证刀具有足够的刚性。

课后训练

根据图 6-56 所示的头像零件的特征，制定合理的工艺路线，设置必要的加工参数，生成刀具轨迹，通过相应的后处理生成数控加工程序，并运用机床加工零件。

图 6-56 头像零件

③ Reasonable Arrangement：Although blend is closely connected to hub, it is impossible to machine both of them at the same time, which means reasonable processing sequence is necessary. The stock of blend and hub need a balance. Unbalanced stock is not allowed because over cutting caused by reactive force will occur. Note that over cutting can be observed on either blend or hub which has less stock.

④ Reliable Tools：Generally, the rounded corners of blend is small. Sometimes, non-cutting moves will increase due to the larger radius of ball-end mill. Without affecting tool rigidity, replacing the ball-end mill with a smaller diameter one can greatly avoid this phenomenon. In order to reduce the radius of the ball head and ensure the rigidity of tool, taper ball-end cutter can be used in blend processing.

Practice

According to the characteristics of disc parts as shown in Fig. 6-56, make a reasonable processing technic, set necessary processing parameters, generate tool path, generate NC processing program through corresponding post-processor, and use machine tools to process parts.

Fig. 6-56　Head Sculpture Part

微课 二维码索引

项目 1

1-1 加工准备　1-2 外轮廓粗加工　1-3 开口腔粗加工　1-4 导轨槽粗加工

1-5 顶面和腔底面精加工　1-6 外轮廓精加工　1-7 导轨槽精加工　1-8 钻中心孔

1-9 钻 D10 通孔　1-10 铣 D24 通孔

项目 2

2-1 加工准备　2-2 型面粗加工 -1　2-3 型面粗加工 -2　2-4 型面粗加工 -3

2-5 型面半精加工　2-6 精铣平面　2-7 型面精加工 -1　2-8 型面精加工 -2

QR Code Index

Project 1

1-1 Preparation for Processing

1-2 Contour Rough Machining

1-3 Part Opening Rough Machining

1-4 Rough Machining of Rail Groove

1-5 Finish Machining of Part Top Surface and Cavity Bottom Surface

1-6 Part Contour Finishing

1-7 Finishing of Rail Groove

1-8 Drill Center Hole

1-9 Drill D10 Thru Hole

1-10 Mill D24 Thru Hole

Project 2

2-1 Preparation for Processing

2-2 Rough Machining-1

2-3 Rough Machining-2

2-4 Rough Machining-3

2-5 Semi-finishing

2-6 Finishing Face Milling

2-7 Finishing-1

2-8 Finishing-2

2-9 清根

项目 3

3-1 加工准备

3-2 型面粗加工

3-3 型面侧壁精加工

3-4 型面底面精加工

项目 4

4-1 加工准备

4-2 型面粗加工

4-3 型面精加工 -1

4-4 型面精加工 -2

4-5 型面精加工 -3

4-6 中心孔

4-7 钻孔

4-8 铰孔

2-9 Flowcut

Project 3

3-1 Preparation for Processing

3-2 Rough Machining

3-3 Side Finishing

3-4 Bottom Finishing

Project 4

4-1 Preparation for Processing

4-2 Rough Machining

4-3 Finish Machining-1

4-4 Finish Machining-2

4-5 Finish Machining-3

4-6 Drill Center Hole

4-7 Drill

4-8 Reaming

项目 5

5-1 加工准备

5-2 型面粗加工

5-3 型面精加工-1

5-4 型面精加工-2

5-5 型面精加工-3

5-6 型面精加工-4

项目 6

6-1 加工准备

6-2 叶片粗加工

6-3 叶片精加工

6-4 轮毂精加工

6-5 圆角精加工

Project 5

5-1 Preparation for Processing

5-2 Rough Machining

5-3 Finish Machining-1

5-4 Finish Machining-2

5-5 Finish Machining-3

5-6 Finish Machining-4

Project 6

6-1 Preparation for Processing

6-2 Blade Rough Machining

6-3 Blade Finish Machining

6-4 Hub Finish Machining

6-5 Blend Finish